Y0-FZX-095
NORTH DAKOTA
STATE UNIVERSITY

AUG 1 7 1993

SERIALS DEPT.
LIBRARY

WITHDRAWN

White Noise

Mathematics and Its Applications

Managing Editor:

M. HAZEWINKEL
Centre for Mathematics and Computer Science, Amsterdam, The Netherlands

Volume 253

White Noise

An Infinite Dimensional Calculus

by

Takeyuki Hida

Department of Mathematics,
Meijo University,
Nagoya, Japan

Hui-Hsiung Kuo

Department of Mathematics,
Louisiana State University,
Baton Rouge, U.S.A.

Jürgen Potthoff

Lehrstuhl für Mathematik V,
Universität Mannheim,
Mannheim, Germany

and

Ludwig Streit

BiBos, Universität Bielefeld,
Bielefeld, Germany, and
Universidade da Madeira, Funchal, Madeira

KLUWER ACADEMIC PUBLISHERS

DORDRECHT / BOSTON / LONDON

A C.I.P. Catalogue record for this book is available from the Library of Congress.

ISBN 0-7923-2233-9

Published by Kluwer Academic Publishers,
P.O. Box 17, 3300 AA Dordrecht, The Netherlands.

Kluwer Academic Publishers incorporates
the publishing programmes of
D. Reidel, Martinus Nijhoff, Dr W. Junk and MTP Press.

Sold and distributed in the U.S.A. and Canada
by Kluwer Academic Publishers,
101 Philip Drive, Norwell, MA 02061, U.S.A.

In all other countries, sold and distributed
by Kluwer Academic Publishers Group,
P.O. Box 322, 3300 AH Dordrecht, The Netherlands.

Printed on acid-free paper

All Rights Reserved
© 1993 Kluwer Academic Publishers
No part of the material protected by this copyright notice may be reproduced or
utilized in any form or by any means, electronic or mechanical,
including photocopying, recording or by any information storage and
retrieval system, without written permission from the copyright owner.

Printed in the Netherlands

QA
274.4
.H54
1993

To

Minami, Fukuko, Tomiko, Clara and Jan

Contents

1. Gaussian Spaces 1

2. $\mathcal{T}$ and $\mathcal{S}$ Transformation and the Decomposition Theorem 10

3. Generalized Functionals 35

3.A The Spaces $(L^2)^{\pm}$ 36

3.B Triples of Functional Spaces 52

3.C Constructions with Second Quantized Operators 61

4. The Spaces $(\mathcal{S})$ and $(\mathcal{S})^*$ 74

4.A Basic Properties and Examples 74

4.B Positive Generalized Functionals 109

4.C Characterization Theorems and Applications 119

4.D Analytic Version of $(\mathcal{S})$ and Growth Estimates 136

5. Calculus of Differential Operators 146

5.A Pointwise Calculus of Differential Operators 147

5.B Differential Operators on $(\mathcal{S})$ 154

5.C Differential Operators and Their Adjoints on (L^2) and $(\mathcal{S})^*$ 163

5.D Table of Rules 179

6. Laplacian Operators 184

6.A White Noise Kernel Operators 184

6.B The Number Operator 196

6.C The Gross and Volterra Laplacians 199

6.D The Lévy Laplacian 211

6.E The Commutation Relations 228

7. The Spaces $\mathfrak{D}$ and $\mathfrak{D}^*$ 232
7.A The Ornstein–Uhlenbeck Semigroup 233
7.B Littlewood–Paley–Stein Inequalities 249
7.C Meyer's Inequalities and Their Applications 263

8. Stochastic Integration 277
8.A Introduction 277
8.B Integrals of Itô Type 283
8.C Stochastic Integrals with Anticipating Integrands 289
8.D Properties of Anticipating Stochastic Integrals 300

9. Fourier and Fourier-Mehler Transforms 317
9.A A Heuristic Motivation 318
9.B Definition and Examples of the Fourier Transform 319
9.C Elementary Properties 324
9.D The Fourier Transform as an Adjoint Operator 332
9.E A Characterization of the Fourier Transform 336
9.F Relations with Laplacians, Wick Products, and Convolutions 339
9.G Relation with the Finite Dimensional Fourier Transform 348
9.H Fourier–Mehler Transforms 356

10. Dirichlet Forms 366
10.A Construction of Energy Forms over $\mathcal{S}'(\mathbb{R})$ 367
10.B Closability, and the Associated Diffusion Processes 382

11. Applications to Quantum Field Theory 399
11.A Introduction 399
11.B Free Fields 405
11.C Interacting Fields and the Role of φ – Bounds 418

12. Feynman Integrals 435
12.A Construction of the Feynman Integrand: Free Particles 436
12.B Construction of the Feynman Integrand: Interacting Particles 439
12.C Manipulating the Feynman Integrand: Schwinger's Variational Principle... 448

Appendices 451
A.1 Hermite Polynomials 451
A.2 Fock Space 456
A.3 Reproducing Kernel Hilbert Spaces 468
A.4 Fréchet and Gâteaux Derivatives on $\mathcal{S}(\mathbb{R}^d)$ 472
A.5 Topologies on $\mathcal{S}(\mathbb{R}^d)$ and $\mathcal{S}'(\mathbb{R})$ 477

Bibliography 486

Notations and Conventions 507

Index 513

Preface

Many areas of applied mathematics call for an efficient calculus in infinite dimensions. This is most apparent in quantum physics and in all disciplines of science which describe natural phenomena by equations involving stochasticity. With this monograph we intend to provide a framework for analysis in infinite dimensions which is flexible enough to be applicable in many areas, and which on the other hand is intuitive and efficient. Whether or not we achieved our aim must be left to the judgment of the reader.

This book treats the theory and applications of analysis and functional analysis in infinite dimensions based on *white noise*. By white noise we mean the generalized Gaussian process which is (informally) given by the time derivative of the Wiener process, i.e., by the velocity of Brownian motion. Therefore, in essence we present analysis on a Gaussian space, and applications to various areas of science.

Calculus, analysis, and functional analysis in infinite dimensions (or dimension–free formulations of these parts of classical mathematics) have a long history. Early examples can be found in the works of Dirichlet, Euler, Hamilton, Lagrange, and Riemann on variational problems. At the beginning of this century, Fréchet, Gâteaux and Volterra made essential contributions to the calculus of functions over infinite dimensional spaces. The important and inspiring work of Wiener and Lévy followed during the first half of this century. Moreover, the articles and books of Wiener and Lévy had a view towards probability theory.

Almost from its very beginning quantum physics has been using infinite dimensional calculus and it became an important (though sometimes informal) tool in modern theoretical and mathematical physics through the pioneering work of Berezin, Cook, Feynman, Fock, Schwinger, and Segal. During the sixties and seventies, in the context of Euclidian and constructive quantum field theory, many structures and deep results in infinite dimensional analysis have been worked out by Albeverio and Høegh–Krohn, Glimm and Jaffe, Gross, Nelson, Simon, and

Symanzik (to name just a few). In parallel, first approaches to a systematic (functional) analysis over Gaussian spaces have been developed. Among others, there was the influential work of Berezanskii and Kondratiev, Gross, Krée, and Malliavin, and their respective schools. The approach which is presented in this book has its roots in a paper by Hida in 1975, and its essential difference to other formulations is the specific choice of the white noise space as the underlying Gaussian space.

This choice is motivated by the following facts. On one hand white noise is a very simple process: it is stationary centered Gaussian, and it is independent at different moments of time. The combination of these properties singles it out as a natural object to start from: in a sense, white noise is a "universal" source of randomness, and as such it appears in many fields of applications. In particular, the independence at different moments of time suggests an interpretation of white noise as a (by the time parameter) continuously indexed coordinate system in infinite dimensions.

The theoretical core of this book consists of the first five chapters. In the first two chapters we present some general facts concerning Gaussian spaces and integrable random variables. We shall review there the Wiener–Itô chaos decomposition theorem which will play an essential role as we develop the analysis of functions of white noise. In the third chapter we introduce various (classes of) spaces of smooth and generalized functionals over a Gaussian space. The fourth chapter is devoted to the detailed study of one specific dual pair of spaces of smooth and generalized random variables over the white noise space. Chapter five deals with the theory of differential operators for functions of white noise.

In the remainder of this book, the framework which is formulated in the first five chapters is developed in the directions of harmonic and stochastic analysis. Furthermore, there are chapters with applications in potential theory and mathematical physics. In a number of appendices we collected some (more or less well–known) facts which the reader might find useful.

Large parts of the first three chapters are written for general Gaussian spaces which are given by the dual of a nuclear space equipped with a Gaussian measure, because this generality comes here at no extra costs. However, from the fourth chapter on, we work almost exclusively with the choice of the white noise probability space. On one hand, many of the structures which are developed need more than just a general Gaussian space as the underlying probability space. (For example, it is sometimes necessary that the sample space consists of (generalized) functions.) On the other hand: although often structures can be formulated and results hold for more general set–ups, we felt that it is more transparent and economical to present them in the white noise case.

We would like to take the occasion to express our deep gratitude to the following friends and colleagues for their encouragement, their help and for the many fruitful discussions we had with them: S. Albeverio, F. Asadian, J. Asch, P.L. Chow, G. Cochran, D. de Falco, M. de Faria, M. Fukushima, L. Gross, M. Hazewinkel, D.C. Khandekar, J. Kondratiev, I. Kubo, J.–S. Lee, K.–S. Lee, Y.–J. Lee, P. Leukert, P.A. Meyer, N. Obata, B. Øksendal, M. Röckner, A. Sengupta, P. Sundar, N. Takeda, S. Takenaka, M. Timpel, H. Watanabe, W. Westerkamp, and J.–A. Yan. Moreover, we are very grateful to the following institutions for various forms of support: Bhabha Atomic Research Centre, Bielefeld–Bochum–Stochastics, Center of Interdisciplinary Research Bielefeld, Council on Research of Louisiana State University, Deutsche Forschungsgemeinschaft, Japan Society for the Promotion of Science, Louisiana Education Quality Support Fund, and National Science Foundation. Last but not least we want to thank the staff at Kluwer and the series editor Michiel Hazewinkel for the fast and efficient handling of the manuscript.

T. Hida
H.–H. Kuo
J. Potthoff
L. Streit

January 1993

1. Gaussian Spaces

As is well-known, one can not have a Lebesgue measure on infinite dimensional vector spaces. On the other hand, one can not set up a concrete functional analysis without reference to a measure. Gaussian measures offer a good choice as a reference measure because of several reasons. Firstly, they *do exist* on infinite dimensional spaces. Moreover, they have simple properties, and appear in many applications. Therefore, we shall use Gaussian measures as reference measures throughout this book. For our purposes, the measure of white noise (Example 1.2) is the most important example. Since the "sample paths" of white noise are not given by functions but rather by generalized functions, this measure is constructed on the Schwartz space $\mathcal{S}'(\mathbb{R})$ of tempered distributions. $\mathcal{S}'(\mathbb{R})$ is the dual of the Schwartz space of test functions $\mathcal{S}(\mathbb{R})$, and the topological structure of $\mathcal{S}(\mathbb{R})$ is the one of a nuclear countably Hilbert space (see, e.g., Gel'fand and Vilenkin (1968), and below). Thus we introduce in this chapter Gaussian spaces as spaces which are the topological dual of a countably Hilbert space, and which are equipped with a Gaussian measure. Such spaces are the basis of the whole book.

Now we recall Minlos' theorem and refer the interested reader to Gel'fand and Vilenkin (1964), Hida (1970b, 1980a), Minlos (1962), and Simon (1979) for detailed proofs. First we quickly sketch the notion of a countably Hilbert space (cf. Gel'fand and Vilenkin (1968)).

Let $\mathfrak{D}$ be a real linear space which is equipped with a sequence $\{(\cdot\,,\cdot)_n, n\in\mathbb{N}\}$ of scalar products which induce norms $|\cdot|_n$, $n\in\mathbb{N}$, on $\mathfrak{D}$. Assume that the system of scalar products is *compatible*, in the sense that if $\{\xi_n, n\in\mathbb{N}\}$ is a sequence in $\mathfrak{D}$ which converges to zero with respect to a norm $|\cdot|_n$ and is Cauchy with respect to $|\cdot|_m$, $n,m\in\mathbb{N}$, then it converges also to zero with respect to $|\cdot|_m$. (Compatibility is discussed in some more detail in Chapter 3.B.) Suppose furthermore that the norms obey $|\cdot|_n \leq |\cdot|_m$ for $n\leq m$, so that we have $\mathcal{N}_m\subset\mathcal{N}_n$, where $\mathcal{N}_n$ denotes the completion of $\mathfrak{D}$ under $|\cdot|_n$, $n\in\mathbb{N}$. Let $\mathcal{N}$ denote the intersection of the Hilbert spaces $\mathcal{N}_n$:

$$\mathcal{N} = \bigcap_n \mathcal{N}_n. \tag{1.1}$$

Let $\mathcal{N}$ be equipped with the *projective limit topology* τ_p. An open neighborhood base at zero of this topology is given by the choice of $\epsilon>0$ and $n\in\mathbb{N}$, and the set $U_{\epsilon,n} = \{\xi\in\mathcal{N}, |\xi|_n<\epsilon\}$. A sequence $\{\xi_n, n\in\mathbb{N}\}$ in $\mathcal{N}$ converges to $\xi\in\mathcal{N}$ with respect to τ_p, if and only if it converges to ξ in every Hilbert space $\mathcal{N}_n$. $\mathcal{N}$ endowed with τ_p is called a *countably Hilbert space.* It turns out (cf., e.g., Gel'fand and Vilenkin (1968)) that $(\mathcal{N}, \tau_p)$ is metrizable and complete, and hence it is a Fréchet space.

From now on we shall assume in addition that $\mathcal{N}$ is *nuclear.* This means that for every $n\in\mathbb{N}$ there exists $m\in\mathbb{N}$, so that the injection ι_n^m of $\mathcal{N}_m$ into $\mathcal{N}_n$ is a Hilbert–Schmidt operator.

Let $\mathcal{N}_n^* \equiv \mathcal{N}_{-n}$ (with norm $|\cdot|_{-n}$) denote the dual of $\mathcal{N}_n$ and $\mathcal{N}^*$ the dual of $\mathcal{N}$. The dual pairing between $\mathcal{N}^*$ and $\mathcal{N}$ is denoted by $<\cdot\,,\,\cdot>$. We have (e.g., Gel'fand and Vilenkin (1968))

$$\mathcal{N}^* = \bigcup_n \mathcal{N}_n^*. \tag{1.2}$$

Unless stated otherwise, we shall consider $\mathcal{N}^*$ as equipped with the weak topology σ. By $\mathfrak{B}$ we shall denote the Borel σ-algebra on $\mathcal{N}^*$, i.e. the σ–algebra generated by the weak topology (or the 'cylinder sets') of $\mathcal{N}^*$.

Now we are ready to state Minlos' theorem.

Theorem 1.1. *Let* C *be a function on* $\mathcal{N}$ *with the following properties:*

(i) C *is continuous on* $\mathcal{N}$,

(ii) C *is positive definite,*

(iii) $C(0) = 1$.

I.e., C is a characteristic function on $\mathcal{N}$. *Then there exists a unique probability measure* μ_C on $(\mathcal{N}^*,\mathfrak{B})$, *so that for all* $\xi\in\mathcal{N}$

$$\int_{\mathcal{N}^*} \exp(i\langle x, \xi\rangle)\, d\mu_C(x) = C(\xi) \tag{1.3}$$

Moreover, if C *is continuous with respect to* $|\cdot|_p$, $p\in\mathbb{N}$, *and if* $n\ (>p)$ *is such that the injection* $\iota_p^n : \mathcal{N}_n \to \mathcal{N}_p$ *is of Hilbert–Schmidt type, then* $\mu_C(\mathcal{N}_{-n}) = 1$.

As mentioned above, in this book we shall most of the time choose $\mathcal{N}$ to be the Schwartz space $\mathcal{S}(\mathbb{R})$ of test functions. Throughout, we shall assume in addition that $\mathcal{N}$ is a separable pre–Hilbert space with a scalar product $(\cdot\,,\cdot)$, which is compatible with the topology of $\mathcal{N}$ (i.e., compatible with the scalar products $(\cdot\,,\cdot)_n$, $n\in\mathbb{N}$, in the above sense). The norm induced by this scalar product is denoted by $|\cdot|$ and the completion of $\mathcal{N}$ with respect to this norm is denoted by $\mathcal{H}$. In particular, $\mathcal{H}$ can be a distinguished choice of one of the spaces $\mathcal{N}_n$, $n\in\mathbb{N}$. Consider now the following function C on $\mathcal{N}$:

$$C(\xi) := \exp(-\tfrac{1}{2}\,|\,\xi\,|^2) \tag{1.4}$$

It is not hard to see that C is a characteristic function on $\mathcal{N}$. By Theorem 1.1 we are given a probability space $(\mathcal{N}^*,\mathcal{B},\mu)$, where μ is the Gaussian measure defined by (1.3) and (1.4). We call $(\mathcal{N}^*,\mathcal{B},\mu)$ the *Gaussian space associated with* $(\mathcal{N},|\cdot|)$.

Here are some examples.

Example 1.2. Choose $\mathcal{N} = \mathcal{S}(\mathbb{R}^d)$, the Schwartz space of rapidly decreasing C^∞-functions on $\mathbb{R}^d$, $d\in\mathbb{N}$, endowed with its usual topology. It is well-known that $\mathcal{S}(\mathbb{R}^d)$ is countably Hilbert (cf. Appendix A.5), and that it is nuclear (e.g. Hida (1980)). Furthermore let $\mathcal{H} = L^2(\mathbb{R}^d)$ (real, with Lebesgue measure). Then $(\mathcal{S}'(\mathbb{R}^d),\mathcal{B},\mu)$ is called *white noise with* d-*dimensional time parameter.* If d=1, we call it simply *white noise.*

Informally speaking, in this case the covariance of μ is given by

$$\int_{\mathcal{S}'(\mathbb{R}^d)} x(t)x(s)\, d\mu(x) = \delta(t-s), \tag{1.5}$$

with t, $s \in \mathbb{R}^d$, so that in physicists' terminology, all its Fourier components contribute equally: this is the reason for the adjective 'white' in the name. Also the associated coordinate process X over $\mathcal{N}$, given by $X_\xi(x) := \langle x, \xi \rangle$, will be called white noise and sometimes we shall treat it informally as in (1.5), i.e. as if it were a regular process over $\mathbb{R}^d$.

Example 1.3. Let again $\mathcal{N} = \mathcal{S}(\mathbb{R}^d)$, but take now $\mathcal{H} = \mathcal{H}_{-1}(\mathbb{R}^d, dt)$, the Sobolev space of order -1 over $\mathbb{R}^d$. The resulting centered Gaussian measure μ on $\mathcal{B}$ has covariance

$$\int_{\mathcal{S}'(\mathbb{R}^d)} \langle x, \xi \rangle \langle x, \eta \rangle\, d\mu(x) = (\xi, (1-\Delta)^{-1} \eta)_{L^2(\mathbb{R}^d)} \tag{1.6}$$

where ξ, $\eta \in \mathcal{S}(\mathbb{R}^d)$, and Δ is the Laplacian in d dimensions. In this case the probability space $(\mathcal{S}'(\mathbb{R}^d), \mathcal{B}, d\mu)$ (as well as the associated coordinate process) is called the *free Euclidian Bose field of mass* 1 *in* d *dimensions.* This name is due to the fact that the resolvent kernel $(1-\Delta)^{-1}(u,v)$, $u,v \in \mathbb{R}^d$, describes the propagation of non-interacting ('free') quantum fields in d dimensional Euclidian space-time (cf., e.g., Glimm and Jaffe (1981), or Simon (1974)).

Example 1.4. Choose $\mathcal{N} = \mathcal{S}(\mathbb{R}^{d-1})$, $d \in \mathbb{N}$, and $\mathcal{H} = \mathcal{H}_{-1/2}(\mathbb{R}^{d-1})$. Let us denote the measure defined by this choice by μ_0. $(\mathcal{S}'(\mathbb{R}^{d-1}), \mathcal{B}, \mu_0)$ is called the *free time zero Bose field of mass* 1 *in* d-1 *space dimensions.* The name is explained as follows. Denote a point in $\mathbb{R}^d$ by (t,u), $t \in \mathbb{R}$, $u \in \mathbb{R}^{d-1}$, t being interpreted as time. If $f \in \mathcal{S}(\mathbb{R}^{d-1})$, then it makes sense to consider the coordinate process X of example 1.3 for 'fixed time t', i.e. one may consider $X_{\delta_t \otimes f} \equiv X_f(t)$, cf. Nelson (1972, 1973). Then the measure μ_0 is the measure associated with the process $f \mapsto X_f(0)$ over

$\mathcal{S}(\mathbb{R}^{d-1})$. For more details we refer the reader to Nelson (1973) or Simon (1974).

Example 1.5. $\mathcal{N} = \mathcal{D}([0,1])$, the space of C^∞-functions supported in [0,1] with its usual Fréchet topology. As Hilbert space $\mathcal{H}$ we choose

$$\mathcal{H} = L^2([0,1], dt)/\{\text{constant functions}\}, \tag{1.7}$$

with scalar product $(f,g) = (f_C,g_C)_{L^2}$, where

$$f_C(t) := f(t) - \int_0^1 f(s)\, ds. \tag{1.8}$$

The resulting probability space $(\mathcal{D}^*([0,1]),\mathcal{B},\mu)$ is a model for the noise (i.e. the informal time derivative) of the Brownian bridge $W(t)$, $0 \leq t \leq 1$, $W(0) = W(1) = 0$.

If $p \in [1,+\infty]$, we shall denote the (complex) space $L^p(\mathcal{N}^*,\mathcal{B},\mu)$ by (L^p) with norm $\|\cdot\|_p$, unless there is danger of confusion. Let us give some examples of $\mathcal{N}^*$-functionals in (L^p). The scalar product of (L^2) will be denoted by $(\cdot, \cdot)_2$, or simply by $(\cdot, \cdot)$ if there is no danger of confusion.

Given a complex-valued polynomial P on $\mathbb{R}^n$, $n \in \mathbb{N}$, and $\xi_1, \xi_2, \dots, \xi_n \in \mathcal{N}$ we may form the $\mathcal{N}^*$-functional P_ξ given by

$$P_\xi(x) = P(\langle x, \xi_1 \rangle, \dots, \langle x, \xi_n \rangle)\,,\ x \in \mathcal{N}^*. \tag{1.9}$$

Proposition 1.6. $P_\xi \in (L^p)$ *for all* $p \in [1,+\infty)$.

Proof. Using the triangular and Hölder's inequality it is sufficient that we show that for all $m \in \mathbb{N}$ and all $\xi \in \mathcal{N}$,

$$\int \langle x, \xi \rangle^{2m}\, d\mu(x)$$

is finite. But $<\cdot, \xi>$ is normally distributed with variance $|\xi|^2$. Thus the last integral is equal to[1] $(2m-1)!! \, |\xi|^{2m}$. □

As in Example (1.2), for $\xi \in \mathcal{N}$ we shall denote the random variable $x \mapsto <x, \xi>$ by X_ξ. The mapping $\xi \mapsto X_\xi$ is called *(canonical) coordinate process over* $\mathcal{N}$.

Consider exponential functionals on $\mathcal{N}^*$ of the form $\exp(\alpha X_\xi)$ with $\xi \in \mathcal{N}$ and $\alpha \in \mathbb{C}$.

Proposition 1.7. *For every* $\alpha \in \mathbb{C}$, $\xi \in \mathcal{N}$, $p \in [1,+\infty)$, $\exp(\alpha X_\xi) \in (L^p)$. *If* $\mathrm{Re}\alpha = 0$, *then also* $\exp(\alpha X_\xi) \in (L^\infty)$.

Proof. The second statement is obvious. For the first statement, we use again our knowledge about the distribution of $X_\xi = <\cdot, \xi>$ to compute

$$\int_{\mathcal{N}^*} |\exp(\alpha <x, \xi>|^p \, d\mu(x) = \int_{\mathcal{N}^*} \exp(p \, \mathrm{Re}\alpha <x, \xi>) \, d\mu(x)$$

$$= \exp(\tfrac{1}{2}(p \, \mathrm{Re}\alpha)^2 \, |\xi|^2) < \infty.$$ □

Definition 1.8. $\mathcal{P}$ *denotes the algebra of polynomials, i.e., generated by* X_ξ, $\xi \in \mathcal{N}$. $\mathcal{E}$ *is the algebra of trigonometric exponentials, i.e., generated by* $\exp(iX_\xi)$, $\xi \in \mathcal{N}$, $\mathcal{E}_r$ *is the algebra of real exponentials, i.e., generated by* $exp(X_\xi)$, $\xi \in \mathcal{N}$. $\mathcal{A}$ *is the algebra generated by polynomials and exponentials, it is generated by* X_ξ *and* $\exp(\alpha X_\xi)$, $\xi \in \mathcal{N}$, $\alpha \in \mathbb{C}$.

Remark. Above (and throughout), we consider vector spaces (and therefore algebras) of random variables as over the field $\mathbb{C}$, unless stated otherwise. However note that the vector spaces $\mathcal{H}$, $\mathcal{N}$, and $\mathcal{N}^*$ are real.

---

[1] For $n \in \mathbb{N}$, we denote $(2n-1)!! = \frac{(2n)!}{n! 2^n}$.

The following result is fairly standard. However, since we shall make frequently use of it, we shall provide it together with the sketch of a proof for the convenience of the reader. For more details we refer to Badrikian (1970), Schwartz (1973). (For the corresponding result in the case p=2, cf. also Hida (1980, Theorem 4.1).)

Theorem 1.9. *The algebras* $\mathcal{P}$, $\mathcal{E}$, $\mathcal{E}_r$, $\mathcal{A}$ *are dense in* (L^p), $p\in[1,+\infty)$. *Furthermore,* $\mathcal{E}$ *is dense in* (L^∞).

Sketch of the Proof. We consider $\mathcal{N}^*$ as being equipped with the inductive limit topology τ_{ind} of the chain $\{\mathcal{N}_{-n}, n\in\mathbb{N}_0\}$ of Hilbert spaces $\mathcal{N}_{-n}$. Thus $\mathcal{N}^*$ is a Lusin space and consequently a Suslin space (cf. e.g. Badrikian (1970), Exposé no. 8). Let $\hat{\mathcal{B}}$ be the σ-algebra generated by τ_{ind}. Then it is also proved there, that actually $\hat{\mathcal{B}} = \mathcal{B}$. By a theorem of Meyer (cf., e.g., Schwartz (1973), p. 122), it follows that μ is a Radon measure on $(\mathcal{N}^*, \hat{\mathcal{B}}) = (\mathcal{N}^*, \mathcal{B})$, because μ is finite.

Due to the regularity properties of Radon measures, the set of functions on $\mathcal{N}^*$ which are bounded, and continuous on every compact of $\mathcal{N}^*$, are dense in every (L^p), $1\leq p\leq+\infty$. It also follows from the inner regularity (with respect to compacts) of the Radon measure μ that it is sufficient to prove that every bounded function, which is continuous on every compact, can be approximated uniformly on every compact by functions in any of the above algebras. But this follows now from the Stone–Weierstrass theorem, because all the algebras in the statement consist of continuous functions, they separate points, contain 1, and they are stable under complex conjugation. □

Remark. Assume that $e = \{e_k, k\in\mathbb{N}\}$ is a complete orthonormal system of $\mathcal{H}$ which belongs to $\mathcal{N}$. Denote by $\mathcal{P}_e$ the algebra generated by the random variables $\{X_{e_k}, k\in\mathbb{N}\}$. Then by the same arguments one finds that $\mathcal{P}_e$ is dense in all (L^p), $p\geq1$.

Before we close this section by explaining the connection of white noise (cf.

Example 1.2) and Brownian motion, we have to prove a lemma about the coordinate process X: $\xi \mapsto \langle\cdot, \xi\rangle$ on $\mathcal{N}$. Because of Proposition 1.6 it is clear that

$$X : \mathcal{N} \to \bigcap_{1\leq p<\infty} (L^p).$$

Lemma 1.10. *The coordinate process on* $\mathcal{N}$ *has an extension to* $\mathcal{H}$, *denoted also by* X: $f \mapsto X_f(\cdot) = \langle\cdot, f\rangle$, $f\in\mathcal{H}$, *such that this mapping is linear from* $\mathcal{H}$ *into* $\bigcap_{p\geq 1}(L^p)$ *and continuous in every* $\|\cdot\|_p$, $p\geq 1$. *Moreover this process is centered Gaussian with covariance*

$$\int_{\mathcal{N}^*} \langle x, f\rangle\langle x, g\rangle \, d\mu(x) = (f, g). \tag{1.10}$$

Proof. Consider $f\in\mathcal{H}$ and let $(f_k, k\in\mathbb{N})$ be a sequence in $\mathcal{N}$ converging to f in $\mathcal{H}$. By construction we have for $p\geq 1$, η, $\xi\in\mathcal{N}$

$$\int_{\mathcal{N}^*} |\langle x, \xi\rangle - \langle x, \eta\rangle|^p \, d\mu(x) = \pi^{-\frac{1}{2}}\, 2^{\frac{p}{2}}\, \Gamma(\tfrac{p+1}{2})\, |\xi-\eta|^p. \tag{1.11}$$

Thus we see that $(X_{f_k}, k\in\mathbb{N})$ is a Cauchy sequence in every (L^p), $p\geq 1$. Moreover, it follows from (1.11) and Hölder's inequality that for any p, q, $1\leq p\leq q<\infty$, there exists a constant $c_{p,q}$ so that

$$\|\langle\cdot, \xi\rangle\|_p \leq \|\langle\cdot, \xi\rangle\|_q \leq c_{p,q}\, \|\langle\cdot, \xi\rangle\|_p. \tag{1.12}$$

This implies that the limit functional of this Cauchy sequence in (L^p) is μ–a.e. equal to its limit in (L^q) for any q, $p\geq 1$. Denote this limit by $X_f = \langle\cdot, f\rangle$. It is clear that $f \mapsto \langle\cdot, f\rangle$ is continuous from $\mathcal{H}$ into every (L^p). Hence it is linear, because its restriction to $\mathcal{N}$ is linear. Similarly, the fact that $f \mapsto X_f$ is centered Gaussian with covariance given by the scalar product in $\mathcal{H}$ follows by the continuity of the extension. □

Now let us come back to the white noise space $(\mathcal{S}'(\mathbb{R}),\mathcal{B},\mu)$ of Example 1.2 (with d=1). From the preceding lemma we know that we may form the random variables $\{ X_{1_{[0,t)}} = <\cdot, 1_{[0,t)}>, t\in\mathbb{R}_+\}$, 1_A being the indicator of $A\in\mathcal{B}(\mathbb{R})$, and that this family is centered Gaussian with covariance

$$\int_{\mathcal{S}'(\mathbb{R})} <x, 1_{[0,t)}><x, 1_{[0,s)}> d\mu(x) = (1_{[0,t)}, 1_{[0,s)}) = s \wedge t. \tag{1.13}$$

Hence this family is a one-dimensional Brownian motion starting at the origin at time zero. Denote this Brownian motion by $B(\cdot)$. Informally we may write

$$B(t,x) = \int_0^t x(s)\, ds. \tag{1.14}$$

Therefore we may consider $x(t)$ as the time derivative of Brownian motion at time t. Of course, it does not exist at any time t in a pointwise sense. However the time derivative of B is a distribution, which fails to be a function in $L^2(\mathbb{R})$ only slightly: it follows from Theorem 1.1, that the μ-measure of the weighted Sobolev space $\mathcal{S}_{-\alpha}(\mathbb{R})$ (cf. Appendix A.5) is equal to 1 for all $\alpha>1/2$. A similar consideration for the noise of the Brownian bridge (Example 1.3) is left as an exercise to the interested reader.

2. $\mathcal{T}$ and $\mathcal{S}$ Transformation and the Decomposition Theorem

In this section we analyze further the structure of $(L^2) = L^2(\mathcal{N}^*, \mathcal{B}, \mu)$. We shall establish the well-known Wiener–Itô decomposition theorem which states that (L^2) has a direct sum decomposition into homogeneous chaos' (see below). This theorem is due to Wiener (1938) and Itô (1951) in the case of the Wiener space (or the white noise space). In its general form this result has been proved first by Segal (1956). We refer the reader also to Hida (1980a) and Simon (1974). We shall prove the theorem here with the help of two important transformations, denoted by $\mathcal{T}$ and $\mathcal{S}$, which will be introduced first. At the end of this chapter we shall specialize to the white noise space. In this case we shall provide the very powerful representation of elements in (L^2) in terms of Wick powers of distributions. The ideas are closely related to "folklore wisdom" in quantum physics (e.g., Glimm and Jaffe (1981), Simon (1974)). The presentation of some parts of this chapter owes much to the article by Nelson (1973c).

First we introduce the following transformation $\mathcal{T}$ on (L^2). For $\xi \in \mathcal{N}$ and $\varphi \in (L^2)$ define

$$(\mathcal{T}\varphi)(\xi) := \int \exp(i\langle x, \xi\rangle)\, \varphi(x)\, d\mu(x). \tag{2.1}$$

Thus $\mathcal{T}$ transforms a functional on $\mathcal{N}^*$ into an $\mathcal{N}$-functional. We denote the image of (L^2) under $\mathcal{T}$ by $\mathcal{R}$. Since $\mathcal{T}$ is linear, it is clear that $\mathcal{R}$ is a vector space of complex valued functions on $\mathcal{N}$.

Proposition 2.1. *$\mathcal{T}$ is an isomorphism (of vector spaces) from (L^2) onto $\mathcal{R}$.*

Proof. By definition, $\mathcal{T}$ is linear and by definition of $\mathcal{R}$, $\mathcal{T}$ is surjective. It remains to check that $\mathcal{T}$ is injective. In Theorem 1.8 we have shown that the algebra $\mathcal{E}$ generated by the functions $\exp(iX_\xi)$, $\xi \in \mathcal{N}$, is dense in (L^2). Thus if $\varphi \in (L^2)$ is such that $\mathcal{T}\varphi(\xi) = (\overline{\varphi}, \exp(iX_\xi)) = 0$ for all $\xi \in \mathcal{N}$, then $\varphi = 0$, i.e., $\ker\mathcal{T} = \{0\}$. □

Next let us compute the $\mathcal{T}$ transform of elements φ, ψ in the algebra $\mathcal{E}$ of trigonometric exponentials. Let

$$\varphi(x) = \sum_{j=1}^{n} a_j \exp(-i\langle x, \eta_j\rangle),$$

and

$$\psi(x) = \exp(-i\langle x, \zeta\rangle),$$

with $a_j \in \mathbb{C}$, η_j, $\zeta \in \mathcal{N}$, j=1,2, … ,n. We easily find for $\xi \in \mathcal{N}$

$$(\mathcal{T}\varphi)(\xi) = \sum_{j=1}^{n} a_j\, C(\eta_j - \xi),$$

and in particular

$$(\mathcal{T}\psi)(\xi) = C(\zeta - \xi),$$

where C is the characteristic functional on $\mathcal{N}$ given by (1.4). On the other hand we have

$$(\psi, \varphi) = \sum_{j=1}^{n} a_j\, C(\eta_j - \zeta).$$

Consider the reproducing kernel Hilbert space $\mathcal{R}(\mathcal{N},C)$, defined in Appendix A.3, which contains $\mathcal{T}\varphi$ and $\mathcal{T}\psi$. The scalar product of these two elements in $\mathcal{R}(\mathcal{N},C)$ is given by

$$(\mathcal{T}\psi, \mathcal{T}\varphi)_{\mathcal{R}(\mathcal{N},C)} = \sum_{j=1}^{n} a_j\, C(\eta_j - \zeta).$$

Thus we equip $\mathcal{R}$ with this scalar product and identify $\mathcal{R}$ and $\mathcal{R}(\mathcal{N},C)$. Since $\mathcal{E}$ is dense in (L^2), we have shown that $\mathcal{T}$ is a unitary map from (L^2) onto $\mathcal{R}(\mathcal{N},C)$.

For the next step we need the following result.

Proposition 2.2. *Let $\varphi \in (L^2)$ and $\xi \in \mathcal{N}$. Then the mapping*

$$\lambda \mapsto \mathcal{T}\varphi(\lambda\xi) \equiv \int \exp(i\lambda<x, \xi>) \, \varphi(x) \, d\mu(x), \quad \lambda \in \mathbb{R},$$

has an entire analytic extension.

Proof. Let $\lambda \in \mathbb{R}$, $\varphi \in (L^2)$, and $\xi \in \mathcal{N}$. Expand the exponential in

$$\mathcal{T}\varphi(\lambda\xi) = \int \exp(i\lambda<x, \xi>) \, \varphi(x) \, d\mu(x),$$

and interchange the resulting sum with the integral. This is justified by the following argument. By Schwarz' inequality we have

$$\int |<x, \xi>|^n \, |\varphi(x)| \, d\mu(x) \leq \| <\cdot, \xi>^n \|_2 \, \|\varphi\|_2$$

$$= \sqrt{(2n-1)!!} \, | \, \xi \, |^n \, \|\varphi\|_2.$$

From this estimate and Stirling's theorem we conclude that the sum

$$\sum_{n=0}^{\infty} \frac{(i\lambda)^n}{n!} \int <x, \xi>^n \, \varphi(x) \, d\mu(x), \tag{2.2}$$

converges absolutely for every $\lambda \in C$ and uniformly on each compact subset of C. Thus for real λ, $\mathcal{T}\varphi(\lambda\xi)$ is equal to (2.2), and hence (2.2) is for $\lambda \in C$ the entire analytic extension of $\mathcal{T}\varphi(\lambda\xi)$. □

We shall continue to denote the entire extension of the mapping $\lambda \mapsto \mathcal{T}\varphi(\lambda\xi)$ by the same symbol.

We use Proposition 2.2 to define another transformation on (L^2). This transformation was also used in Gross (1967a), Krée (1977), and Kubo and Takenaka (1980a). For $\varphi\in(L^2)$ and $\xi\in\mathcal{N}$ set

$$\mathcal{S}\varphi(\xi) := C(\xi)\ \mathcal{T}\varphi(-i\xi). \tag{2.3}$$

Thus we have

$$\begin{aligned}\mathcal{S}\varphi(\xi) &= \int \varphi(x) : e^{<x,\ \xi>} : d\mu(x)\\ &= (\overline{\varphi}, : e^{<\cdot,\ \xi>} :)_2,\end{aligned} \tag{2.4}$$

where we have set

$$: e^{<x,\ \xi>} : \equiv e^{<x,\xi>-\frac{1}{2}|\ \xi\ |^2}, \quad x\in\mathcal{N}^*,\ \xi\in\mathcal{N}. \tag{2.5}$$

Proposition 2.3. *Let* $\varphi\in(L^2)$ *and* $\xi\in\mathcal{N}$. *Then*

$$\mathcal{S}\varphi(\xi) = \int \varphi(x+\xi)\ d\mu(x). \tag{2.6}$$

Furthermore,

$$\mathcal{S} : e^{<\cdot,\ \eta>} : (\xi) = e^{(\eta,\ \xi)}. \tag{2.7}$$

Proof. By definition

$$\begin{aligned}(\mathcal{S}\varphi)(\xi) &= \int \exp(-\tfrac{1}{2}|\ \xi\ |^2+<x,\xi>)\ \varphi(x)\ d\mu(x)\\ &= \int \varphi(x+\xi)\ d\mu(x),\end{aligned}$$

due to the translation formula for the Gaussian measure, cf., e.g., Hida (1980a,

Chapt. 5.7) or Kuo (1975). Formula (2.7) follows by a simple computation. □

Note that the mapping $\lambda \mapsto C(\lambda\xi)=\exp(-\frac{\lambda^2}{2}\,|\,\xi\,|^2)$, $\lambda\in\mathbb{R}$, has an entire extension, too. Therefore we have the following result.

Proposition 2.4. *For all* $\varphi\in(L^2)$, $\xi\in\mathcal{N}$, *the mapping* $\lambda \mapsto \mathcal{S}\varphi(\lambda\xi)$ *has an entire extension. Moreover, the mapping* $\varphi \mapsto \mathcal{S}\varphi$ *is injective.*

Proof. This follows directly from Propositions 2.1 and 2.2 and equation (2.3). □

It is clear, that we have for all $\lambda\in\mathbb{C}$, η, $\xi\in\mathcal{N}$,

$$\mathcal{S}\!: e^{\lambda<\cdot,\,\eta>} : (\xi) = e^{\lambda(\eta,\,\xi)}. \tag{2.8}$$

As the next step towards the proof of the Wiener–Itô decomposition theorem, we construct a special basis of (L^2).

Let $f\in\mathcal{H}$. We know from Lemma 1.10 that we may form the random variables

$$: <\cdot, f>^n : \equiv : <\cdot, f>^n :_{|f|^2},$$

with $n\in\mathbb{N}$, and $: u^n :_{\sigma^2}$ is the Hermite polynomial in u of order n and parameter σ, cf. Appendix A.1.

Warning. Let $\xi\in\mathcal{N}$, $n\in\mathbb{N}$, and consider the random variable $x \mapsto : <x, \xi>^n :$, $x\in\mathcal{N}^*$. If $\lambda\in\mathbb{R}$, one might be tempted to think that $: <x, \lambda\xi>^n :$ and $: <\lambda x, \xi>^n :$ are the same. But they are *not.* This is so, because in this notation, the parameter of the Hermite polynomial in the variable $<x, \lambda\xi> = <\lambda x, \xi>$ is determined by the $L^2(\mathbb{R})$–norm of the element in $\mathcal{N}$, which is the second quantity in the pairing. For example, if n = 2, then

$$:<x, \lambda\xi>^2: = \lambda^2 (<x, \xi>^2 - |\xi|^2),$$

while

$$:<\lambda x, \xi>^2: = \lambda^2 <x, \xi>^2 - |\xi|^2.$$

Also note that formulae (A.1.16) and (A.1.17) can be used to establish a formula relating $:<x, \lambda\xi>^n:$ and $:<\lambda x, \xi>^n:$. We continue to use this notation in spite of this danger of confusion, because it has become quite common and because it allows for an efficient calculus.

To construct an orthogonal basis of (L^2), we use the notation of Appendix A.2: I^n is the set of naturally ordered n-tuples α in $\mathbb{N}^n$, $n_k(\alpha)$ is the number of entries in α equal to $k \in \mathbb{N}$ and

$$n(\alpha)! = \prod_{k=1}^{\infty} n_k(\alpha)!$$

with the convention that an empty product equals 1.

Let $(e_k, k \in \mathbb{N})$ be an orthonormal basis of $\mathcal{H}$ and let $n \in \mathbb{N}$, $\alpha \in I^n$. Set

$$H_\alpha(x) := \prod_{k=1}^{\infty} :<x, e_k>^{n_k(\alpha)}: . \tag{2.9}$$

Lemma 2.5. *The collection* $\{ H_\alpha, \alpha \in I^n, n \in \mathbb{N}_0 \}$ *forms an orthogonal basis of* (L^2).

Proof. First we show orthogonality. Note that the orthogonality of e_k and e_l, $k \neq l$, in $\mathcal{H}$ implies by (1.3) and (1.4) the independence of the random variables $<\cdot, e_k>$ and $<\cdot, e_l>$ and measurable functions thereof. Hence

$$(H_\alpha, H_\beta) = \int \prod_{k,l} :\langle x, e_k\rangle^{n_k(\alpha)}: :\langle x, e_l\rangle^{n_l(\beta)}: d\mu(x)$$

$$= \prod_k \int :\langle x, e_k\rangle^{n_k(\alpha)}: :\langle x, e_k\rangle^{n_k(\beta)}: d\mu(x)$$

$$= \prod_k \int_{\mathbb{R}} :u^{n_k(\alpha)}:_1 :u^{n_k(\beta)}:_1 \frac{\exp(-\frac{1}{2}u^2)}{\sqrt{2\pi}} du$$

$$= \prod_k \delta_{n_k(\alpha),n_k(\beta)}\, n_k(\alpha)!$$

$$= \delta_{\alpha,\beta}\, n(\alpha)!,$$

where we used the orthogonality relation (A.1.23) for the Hermite polynomials.

The proof is concluded by the following remark. We know by Theorem 1.9 that the polynomials are dense in (L^2) and by the Remark after Theorem 1.9, this is so for the algebra of polynomials in the variables $\langle\cdot, e_k\rangle$, $k\in\mathbb{N}$. Rewriting polynomials as finite linear combinations of Hermite polynomials, (cf. (A.1.17)), we see that the system $\{H_\alpha, \alpha\in I^n, n\in\mathbb{N}_0\}$ has a dense linear span in (L^2). □

Note that we have derived

$$\| H_\alpha \|_2^2 = n(\alpha)!. \tag{2.10}$$

Therefore we know that $\varphi\in(L^2)$ has the representation

$$\varphi(x) = \sum_{n=0}^{\infty} \varphi^{(n)}(x) \tag{2.11a}$$

$$\varphi^{(n)}(x) = \sum_{\alpha \in I^n} a_\alpha \, H_\alpha(x) \tag{2.11b}$$

$$\| \varphi \|_2^2 = \sum_{n=0}^{\infty} \sum_{\alpha \in I^n} n(\alpha)! \, | a_\alpha |^2 \tag{2.11c}$$

with coefficients $a_\alpha \in \mathbb{C}$.

Let $n \in \mathbb{N}_0$. The closed linear span of $\{H_\alpha, \alpha \in I^n\}$ will be denoted by $\mathcal{H}^{(n)}$. Thus we have that $\varphi^{(n)} \in \mathcal{H}^{(n)}$, and that (L^2) is the direct sum of the Hilbert subspaces $\mathcal{H}^{(n)}$, $n \in \mathbb{N}_0$. $\mathcal{H}^{(n)}$ is often called n*–th homogeneous chaos,* and the decomposition (2.11a) of φ will be called its *chaos decomposition.*

Now we shall prove that $\mathcal{S}$ implements the isomorphism between (L^2) and the symmetric Fock space $\Gamma(\mathcal{H}_c)$ over the standard complexification $\mathcal{H}_c = \mathcal{H} \oplus i\mathcal{H}$ of $\mathcal{H}$. As a first step we compute the $\mathcal{S}$–transform of the basis elements H_α. Let $\xi \in \mathcal{N}$, then

$$(\mathcal{S}H_\alpha)(\xi) = \int e^{\langle x, \xi \rangle - \frac{1}{2} | \xi |^2} \prod_k : \langle x, e_k \rangle^{n_k(\alpha)} : d\mu(x).$$

Decompose ξ in $\mathcal{H}$ as

$$\xi = \sum_k (e_k, \xi) \, e_k + \xi',$$

where the sum is over those k for which $n_k(\alpha)$ is non-zero and ξ' is orthogonal to all the corresponding e_k's. Then

$$(\mathcal{S}H_\alpha)(\xi) = e^{-\frac{1}{2}(| \xi |^2 - | \xi' |^2)} \prod_k \int e^{(e_k, \xi) \langle x, e_k \rangle} : \langle x, e_k \rangle^{n_k(\alpha)} : d\mu(x).$$

The value of the one-dimensional integral is given by (cf. (A.1.24))

$$(e_k, \xi)^{n_k(\alpha)} \exp\left(\tfrac{1}{2}(e_k, \xi)^2\right),$$

and therefore

$$(\mathcal{S}H_\alpha)(\xi) = \prod_k (e_k, \xi)^{n_k(\alpha)}. \tag{2.12}$$

We rewrite (2.12) as follows

$$\begin{aligned}(\mathcal{S}H_\alpha)(\xi) &= \left(\hat{\otimes}_{k=1}^{\infty} e_k^{\hat{\otimes} n_k(\alpha)}\right)(\xi, \dots, \xi) \\ &= \left(\hat{\otimes}_{k=1}^{n} e_{\alpha_k}\right)(\xi, \dots, \xi) \\ &\equiv e_\alpha(\xi, \dots, \xi),\end{aligned} \tag{2.13}$$

using the notation of Appendix A.2 for the basis elements e_α, $\alpha \in I^n$, of the symmetric Fock space $\Gamma(\mathcal{H}_c)$. By taking limits of (complex) linear combinations, we find that for $\varphi \in (L^2)$, represented as in (2.11), the following formulae hold

$$\mathcal{S}\varphi(\xi) = \sum_{n=0}^{\infty} \mathcal{S}\varphi^{(n)}(\xi), \tag{2.14a}$$

$$\mathcal{S}\varphi^{(n)}(\xi) = \Big(\sum_{\alpha \in I_n} a_\alpha\, e_\alpha\Big)(\xi, \dots, \xi). \tag{2.14b}$$

Let us define for $n \in \mathbb{N}_0$,

$$F^{(n)} = \sum_{\alpha \in I_n} a_\alpha\, e_\alpha \in \mathcal{H}_c^{\hat{\otimes} n}, \tag{2.15}$$

and note that by (A.2.15)

$$\| F^{(n)} \|^2_{\Gamma^{(n)}(\mathcal{H}_c)} = \sum_{\alpha \in I^n} n(\alpha)! \, | a_\alpha |^2. \tag{2.16}$$

(Cf. Appendix A.2 for the definition of the norm on the left hand side of (2.16).) Since both, $\varphi^{(n)} \in \mathcal{H}^{(n)}$ and $F^{(n)} \in \Gamma^{(n)}(\mathcal{H}_c) \equiv \mathcal{H}_c^{\hat{\otimes} n}$, are uniquely determined by the sequence $\{a_\alpha, \alpha \in I^n\}$, it is clear that they are in one-to-one correspondence. Moreover, by comparison of (2.16) and (2.11) we have

$$\| \varphi^{(n)} \|_2 = \| F^{(n)} \|_{\Gamma^{(n)}(\mathcal{H}_c)}.$$

It is often more convenient to consider the Hilbert–Schmidt norm of the n–multilinear form $F^{(n)}$ on the dual $\mathcal{H}_c^*$ of $\mathcal{H}_c$. Here, we shall identify these two spaces. The Hilbert–Schmidt norm of $F^{(n)}$ is given by (cf. also Appendix A.2)

$$\| F^{(n)} \|^2_{H.S.} = \sum_{k_1, \dots, k_n = 1}^{\infty} | F^{(n)}(e_{k_1}, \dots, e_{k_n}) |^2,$$

where $\{e_k, k \in \mathbb{N}\}$ is any complete orthonormal system of $\mathcal{H}$. It is easily checked that

$$\| F^{(n)} \|_{\Gamma^{(n)}(\mathcal{H}_c)} = \sqrt{n!} \, \| F^{(n)} \|_{H.S.}.$$

We have proved the following result.

Proposition 2.6. *Every* $\varphi \in (L^2)$ *is in one–to–one correspondence with a sequence* $\{F^{(n)}, n \in \mathbb{N}_0\}$ *in* $\Gamma(\mathcal{H}_c)$, *and*

$$\| \varphi \|_2^2 = \sum_{n=0}^{\infty} n! \, \| F^{(n)} \|^2_{H.S.}. \tag{2.17}$$

Moreover, for every $\xi \in \mathcal{N}$,

$$\mathcal{S}\varphi(\xi) = \sum_{n=0}^{\infty} F^{(n)}(\xi, \dots, \xi). \tag{2.18}$$

Remark. Observe that $F^{(n)}$ above can actually be constructed from $\mathcal{S}\varphi^{(n)}$ as follows. By multilinearity of $F^{(n)}$ it is sufficient to determine it on the real subspace $\mathcal{H}$ of $\mathcal{H}_c$. Since $F^{(n)}$ is multilinear and symmetric, the polarization identity gives

$$F^{(n)}(\xi_1, \dots, \xi_n) = \frac{1}{n!}\sum_{k=1}^{n} (-1)^{n-k} \sum_{i_1< \cdots <i_k} \mathcal{S}\varphi^{(n)}(\xi_{i_1} + \cdots + \xi_{i_k}),$$

with $\xi_1, \dots, \xi_n \in \mathcal{N}$. This specifies the symmetric n–multilinear form $F^{(n)}$ on the dense subspace $\mathcal{N}$ of $\mathcal{H}$, and therefore by continuity on the space $\mathcal{H}$ itself. Thus we may identify $\mathcal{S}\varphi$ with the sequence $\{F^{(n)}, n\in\mathbb{N}_0\}$, and will often do so without further explanation. In particular, we shall consider the one–to–one correspondence stated in Proposition 2.6 as being given by the $\mathcal{S}$–transform.

We summarize our preceding discussion in the following theorem which is essentially due to Itô (1951), Segal (1956) and Wiener (1938).

Theorem 2.7.

(i)

$$\begin{array}{ccc} (L^2) & \xrightarrow{\ \mathcal{T}\ } & \mathcal{R}(\mathcal{N},C) \\ \mathcal{S}\downarrow & \nearrow\ \mathcal{T}\circ\mathcal{S}^{-1} & \\ \Gamma(\mathcal{H}_c) & & \end{array} \tag{2.19}$$

where the arrows indicate unitary mappings.

(ii) (L^2) *admits a direct sum decomposition*

$$(L^2) = \bigoplus_{n=0}^{\infty} \mathcal{H}^{(n)}, \tag{2.20}$$

$$\mathcal{H}^{(n)} = \mathcal{S}^{-1}\, \Gamma^{(n)}(\mathcal{H}_c) = \mathcal{T}^{-1}\, \mathcal{R}^{(n)}(\mathcal{N},C), \tag{2.21}$$

such that every $\varphi\in(L^2)$ *is in one–to–one correspondence with a sequence* $(F^{(n)},$

$n\in\mathbb{N}_0)\in\Gamma(\mathcal{H}_c)$, *and furthermore*

$$\| \varphi \|_2^2 = \sum_{n=0}^{\infty} n! \, \| F^{(n)} \|^2_{H.S.}. \tag{2.22}$$

Remarks. 1. Also in cases where one deals with non–Gaussian measures one can have decomposition theorems of the above type. For example, in the article of Hida and Ikeda (1967) the theory of reproducing kernel Hilbert spaces has been applied to the space of the noise of the Poisson process, and a similar decomposition theorem was proven.

2. In the sequel it will be convenient to suppress the subscript "c" of $\mathcal{H}_c$ which indicates the complexification of $\mathcal{H}$. Thus the symbol $\mathcal{H}$ will be used for the real Hilbert space which determines the Gaussian measure μ on $(\mathcal{N}^*,\mathcal{B})$, and for the complex Hilbert space which is isomorphic to $\mathcal{H}^{(1)}$ according to Theorem 2.7. Since it will be clear from the context which of these two roles $\mathcal{H}$ is playing, there will be no danger of confusion. However, let us stress that we shall always consider $\mathcal{N}$ and $\mathcal{N}^*$, their concrete examples, and their typical elements $\xi,\eta, \ldots \in\mathcal{N}$, $x,y, \ldots \in\mathcal{N}^*$ as real, unless mentioned otherwise.

We shall now consider the chaos decomposition of products of the coordinate process X_f, $f\in\mathcal{H}$, in more detail. For complex $f\in\mathcal{H}$, we define $X_f = X_{Re(f)} + iX_{Im(f)}$, and we shall continue to use also the notation $X_f(x) = \langle x, f\rangle = \langle x, Re(f)\rangle + i\langle x, Im(f)\rangle$, $f\in\mathcal{H}$, $x\in\mathcal{N}^*$.

Let us denote the orthogonal projection onto $\mathcal{H}^{(n)}$ by P_n, $n\in\mathbb{N}_0$. Assume that $f\in\mathcal{H}$, $f \neq 0$, is real. $P_n\langle\cdot\, , f\rangle^n$, $n\in\mathbb{N}$, is given by

$$\sum_{\alpha\in I^n} \frac{1}{n(\alpha)!} H_\alpha \int \langle x, f\rangle^n H_\alpha(x) \, d\mu(x).$$

Choose a CONS $\{e_k, k\in\mathbb{N}\}$ of $\mathcal{H}$ with $e_1 = |f|^{-1} f \equiv \hat{f}$. Then it follows by a similar computation as in the proof of Lemma 2.5, that the only non–zero term in the above sum is the one where $\alpha = (1,1, \dots ,1)$. Moreover, for this α, $H_\alpha = \, :<\cdot , \hat{f}>^n:$, and therefore we have

$$P_n<\cdot , f>^n = \frac{1}{n!} :<\cdot , \hat{f}>^n: \int <x, f>^n : <x , \hat{f}>^n : d\mu(x)$$

$$= \frac{1}{n!} | f |^n :<\cdot , \hat{f}>^n: \int <x, \hat{f}>^n : <x , \hat{f}>^n : d\mu(x)$$

$$= \frac{1}{n!} | f |^n :<\cdot , \hat{f}>^n: \int_{\mathbb{R}} u^n : u^n :_1 e^{-\frac{1}{2}u^2} \frac{du}{\sqrt{2\pi}}.$$

Recall that $: u^n :_{\sigma^2}$ is the n–Hermite polynomial in u with parameter σ^2, cf. Appendix A.1. The last integral can be calculated easily, e.g., by using relations (A.1.17) and (A.1.23), and its value is given by n!. Thus we obtain the following result:

$$P_n <\cdot , f>^n = | f |^n :<\cdot , \hat{f}>^n:.$$

Now we remark the following formula for Hermite polynomials, which can be shown with the help of (A.1.16), or of the defining relation (A.1.13):

$$\sigma^n : u^n :_1 \; = \; : (\sigma u)^n :_{\sigma^2}. \tag{2.23}$$

Thus we have for any real $g \in \mathcal{H}$, $\alpha > 0$, $n \in \mathbb{N}$, $\alpha^n : <\cdot , g>^n : \, = \, : <\cdot , \alpha g>^n :$, and in particular we get

$$P_n <\cdot, f>^n = \, :<\cdot , f>^n:. \tag{2.24}$$

This relation justifies the following notation. For general (complex) $f_1, f_2, \dots , f_n \in \mathcal{H}$, we denote

$$P_n <\cdot , f_1> <\cdot , f_2> \cdots <\cdot , f_n> \equiv \, : <\cdot , f_1> <\cdot , f_2> \cdots <\cdot , f_n> :. \tag{2.25}$$

Lemma 2.8. *The $\mathcal{S}$–transform of* $:\langle\cdot, f_1\rangle \cdots \langle\cdot, f_n\rangle:$ *at* $\xi \in \mathcal{N}$ *is equal to*

$$(\overline{f_1}, \xi) \cdots (\overline{f_n}, \xi) = \left(\hat{\otimes}_{k=1}^n f_k\right)(\xi, \ldots, \xi). \tag{2.26}$$

Proof. First let $f_1 = f_2 = \ldots = f_n = f \in \mathcal{H}$, and assume that f is real. We use (cf. Appendix A.1)

$$:\langle x, f\rangle^n: = \frac{d^n}{d\lambda^n} e^{\lambda\langle x, f\rangle - \frac{1}{2}\lambda^2 |f|^2}\Big|_{\lambda=0}$$

to compute as follows. For $\xi \in \mathcal{N}$, we have

$$\begin{aligned} \mathcal{S} :\langle\cdot, f\rangle^n: (\xi) &= e^{-\frac{1}{2}|\xi|^2} \int :\langle x, f\rangle^n: e^{\langle x,\xi\rangle}\, d\mu(x) \\ &= e^{-\frac{1}{2}|\xi|^2} \frac{d^n}{d\lambda^n} e^{-\frac{\lambda^2}{2}|f|^2} \int e^{\langle x, \xi+\lambda f\rangle}\, d\mu(x)\Big|_{\lambda=0} \\ &= (f, \xi)^n, \end{aligned}$$

where we have interchanged the evaluation of the λ–derivatives and the integral, as can readily be justified. Therefore, in this special case the statement of the lemma holds true.

Now consider the case of real $f_1, f_2, \ldots, f_n$ in $\mathcal{H}$. Then we can compute the $\mathcal{S}$–transform of (2.25) with the help of the polarization identity

$$\langle x, f_1\rangle \cdots \langle x, f_n\rangle = \frac{1}{n!}\sum_{k=1}^n (-1)^{n-k} \sum_{i_1<\cdots<i_k} \langle x, f_{i_1} + \cdots + f_{i_k}\rangle^n .$$

In fact, since the projection $:\cdot:$ and $\mathcal{S}$ are linear, we obtain

$$\mathcal{S}(:\langle\cdot, f_1\rangle \cdots \langle\cdot, f_n\rangle:)(\xi) =$$

$$= \frac{1}{n!}\sum_{k=1}^n (-1)^{n-k} \sum_{i_1<\cdots<i_k} \mathcal{S} :\langle\cdot, f_{i_1} + \cdots + f_{i_k}\rangle^n: (\xi)$$

$$= \frac{1}{n!}\sum_{k=1}^{n} (-1)^{n-k} \sum_{i_1<\cdots<i_k} (f_{i_1}+ \cdots +f_{i_k}, \xi)^n$$

$$= (f_1, \xi) \cdots (f_n, \xi),$$

by using the result of the first step, and the polarization identity in reverse. Thus we find that the lemma is true in this case, too.

Finally, in the case that $f_1, f_2, \dots, f_n \in \mathcal{H}$ are complex, we obtain (2.26) from the previous result by decomposing the f's into real and imaginary parts, and by using again linearity of $\mathcal{S}$ and $: \cdot :$. (Note that

$$(f_1, f_2, \dots, f_n) \mapsto \mathcal{S} : <\cdot , f_1> <\cdot , f_2> \cdots <\cdot , f_n> : (\xi)$$

has to be n-linear, so that we obtain $(\overline{f_1}, \xi)\, (\overline{f_2}, \xi) \cdots (\overline{f_n}, \xi)$ as a result.) □

Next we derive a useful formula for the projection (2.25). To this end we define for $f \in \mathcal{H}$,

$$: e^{<x, f>} : = e^{<x, f> - \frac{1}{2}(\overline{f}, f)}, \quad x \in \mathcal{N}^*, \tag{2.27}$$

which extends (2.5). Note that for all $f \in \mathcal{H}$, we have

$$\int : e^{<x, f>} : d\mu(x) = 1, \tag{2.28}$$

as a straightforward computation shows.

Lemma 2.9. *For all* $f_1, f_2, \dots, f_n \in \mathcal{H}$, *the following holds*

$$: <x, f_1> \cdots <x, f_n> : = \prod_{j=1}^{n} \frac{\partial}{\partial \lambda_j} : e^{<x, \sum_k \lambda_k f_k>} : \Big|_{\lambda_1= \dots =\lambda_n=0} \tag{2.29}$$

Proof. Since $\mathcal{S}$ is an isomorphism between (L^2) and $\Gamma(\mathcal{H})$ it is sufficient to show that the right hand side of (2.29) has an $\mathcal{S}$-transform equal to (2.26). We have

$$\mathcal{S}\prod_{j=1}^{n}\frac{\partial}{\partial\lambda_j} :e^{<\cdot\,,\,\sum_k \lambda_k f_k>}: \Bigg|_{\lambda_1=\ldots\lambda_n=0}(\xi) =$$

$$=\prod_{j=1}^{n}\frac{\partial}{\partial\lambda_j}\int :e^{<x,\,\sum_k \lambda_k f_k>}::e^{<x,\,\xi>}:d\mu(x)\Bigg|_{\lambda_1=\ldots=\lambda_n=0}$$

$$=e^{-\frac{1}{2}|\,\xi\,|^2}\prod_{j=1}^{n}\frac{\partial}{\partial\lambda_j}\,e^{-\frac{1}{2}\sum_{k,l}\lambda_k\lambda_l\,(\overline{f_k},\,f_l)}\times$$

$$\times\int e^{<x,\,\sum_k \lambda_k f_k+\xi>}\,d\mu(x)\Bigg|_{\lambda_1=\ldots=\lambda_n=0},$$

where we have interchanged the evaluation of the λ-derivatives and the integral. For example, this can be justified by an application of the dominated convergence theorem. We use (2.27) and (2.28) to evaluate the last integral and the result is

$$\mathcal{S}\prod_{j=1}^{n}\frac{\partial}{\partial\lambda_j} :e^{<\cdot\,,\,\sum_k \lambda_k f_k>}: \Bigg|_{\lambda_1=\ldots\lambda_n=0}(\xi) =$$

$$=\prod_{j=1}^{n}\frac{\partial}{\partial\lambda_j}\,e^{\sum_k \lambda_k(\overline{f_k},\,\xi)}\Bigg|_{\lambda_1=\ldots=\lambda_n=0}.$$

Now it is obvious that the right hand side of (2.29) has $\mathcal{S}$-transform equal to (2.26). □

Another convenient formula for (2.25) is the following recursion relation.

Lemma 2.10. *For all* $f_1, f_2, \ldots, f_n \in \mathcal{H}$,

$$:\langle x, f_1\rangle \cdots \langle x, f_n\rangle : \; = \; : \langle x, f_1\rangle \cdots \langle x, f_{n-1}\rangle : \langle x, f_n\rangle - \\ - \sum_{j=1}^{n-1} (\overline{f_n}, f_j) : \langle x, f_1\rangle \cdots \langle \not{x}, f_j\rangle \cdots \langle x, f_{n-1}\rangle : \tag{2.30}$$

Proof. This follows directly from (2.27), (2.29) and explicit evaluation of the derivatives $\frac{\partial}{\partial \lambda_j}$. □

For the remainder of this section we shall work with the important special choice $\mathcal{N} = \mathcal{S}(\mathbb{R})$, $\mathcal{H} = L^2(\mathbb{R})$, i.e., with white noise. In this case, we can work out the representation of (L^2) described above in more detail. Most of the following is readily generalized to the situation of white noise with d–dimensional time parameter or to other triples of function spaces, such as the free Euclidian quantum field (Example 1.3).

Theorem 2.7, Proposition 2.6 and the Remark following it, tell us that in the general situation an element $\varphi^{(n)} \in \mathcal{H}^{(n)}$ is uniquely characterized by its $\mathcal{S}$–transform, which is given by

$$(\mathcal{S}\varphi^{(n)})(\xi) = F^{(n)}(\xi, \ldots, \xi), \tag{2.31}$$

where $F^{(n)} \in \mathcal{H}^{\hat{\otimes} n}$, i.e., $F^{(n)}$ is a complex-valued symmetric n–multilinear form on $\mathcal{H}$ with finite Hilbert–Schmidt norm. In case of the choice $\mathcal{H} = L^2(\mathbb{R})$ we have

$$\mathcal{H}^{\hat{\otimes} n} \simeq L^2(\widehat{\mathbb{R}^n}, dt), \tag{2.32}$$

where the right hand side of (2.32) is the space of (complex-valued) a.e. symmetric L^2–functions on $\mathbb{R}^n$. (2.32) is an isomorphism of vector spaces, but it is also an isometry if we equip $L^2(\mathbb{R})^{\hat{\otimes} n}$ with the Hilbert–Schmidt norm. Therefore $\varphi^{(n)} \in \mathcal{H}^{(n)}$ is in one–to–one correspondence with $f^{(n)} \in L^2(\widehat{\mathbb{R}^n})$ and $\varphi^{(n)}$ has the *integral representation*

$$\mathcal{S}\varphi^{(n)}(\xi) = \int_{\mathbb{R}^n} f^{(n)}(t_1, \ldots, t_n)\, \xi(t_1) \cdots \xi(t_n)\, d^n t. \tag{2.33}$$

Moreover, we have (cf. (2.17))

$$\| \varphi^{(n)} \|_2^2 = n! \| f^{(n)} \|_{L^2(\mathbb{R}^n)}^2. \tag{2.34}$$

We shall call $f^{(n)}$ the *integral kernel of* $\varphi^{(n)}$.

It is important to note the close relationship of $\mathcal{H}^{(n)}$ and its elements $\varphi^{(n)}$ as given above to the *n-fold Wiener integral* or *n-th homogeneous chaos*.

Let $f \in L^2(\mathbb{R}, dt)$ with support in $\mathbb{R}_+$. It was shown in Hida (1980a) that $X_f = \langle \cdot, f \rangle$ is a version of the elementary stochastic or Wiener–integral of f:

$$\langle \cdot, f \rangle \overset{d}{=} \int f(t)\, dB(t), \tag{2.35}$$

where B is a one dimensional Brownian motion starting at the origin at time zero[1]. By comparison with the article of Itô (1951) we conclude that

$$: \langle x, f_1 \rangle \cdots \langle x, f_n \rangle : \overset{d}{=} \int (\hat{\otimes}_{i=1}^n f_i)(t_1, \ldots, t_n)\, dB(t_1) \cdots dB(t_n), \tag{2.36}$$

for functions $f_1, \ldots, f_n \in L^2(\mathbb{R}_+, dt)$. Here the right hand side of (2.36) is the multiple Wiener integral of degree n as defined by Itô in the above quoted article. Note that by Lemma 2.5 the integrand on the right hand side of (2.36) is precisely the integral kernel of the $\mathcal{S}$-transform of the left hand side. Taking linear combinations and the (mean square) limit of both sides of equation (2.36) we obtain

$$\varphi^{(n)} \overset{d}{=} \int f^{(n)}(t_1, \ldots, t_n)\, dB(t_1) \cdots dB(t_n) \tag{2.37a}$$

---

1 . $\overset{d}{=}$ · means equality in distribution

$$\stackrel{d}{=} \quad n! \int\limits_{0\le t_1\le \cdots \le t_n} f^{(n)}(t_1, \dots ,t_n)\, dB(t_1) \cdots dB(t_n), \tag{2.37b}$$

for general $\varphi^{(n)} \in \mathcal{H}^{(n)}$ with integral kernel $f^{(n)}$ supported in $\{(t_1, \dots ,t_n)\in \mathbb{R}^n, t_i \ge 0\}$. In particular, if we choose the Brownian motion B in (2.37) as being given by its version in Chapter 1 on the white noise probability space, we obtain (2.37) as a μ—a.e. equality.

Now we introduce another form of the chaos representation of $\varphi \in (L^2)$. According to Theorem 2.7 and (2.33), φ is in one-to-one correspondence with a sequence $(f^{(n)}, n\in \mathbb{N}_0)$, $f^{(n)} \in L^2(\hat{\mathbb{R}}^n)$, $n\in\mathbb{N}$, $f_0\in\mathbb{C}$. In view of the representation (2.11) of φ, we find that $f^{(n)}$ is the symmetrization of the following function on $\mathbb{R}^n$:

$$u = (u_1, \dots ,u_n) \mapsto \sum_{\alpha\in I^n} a_\alpha\, e_{\alpha_1}(u_1) \cdots e_{\alpha_n}(u_n). \tag{2.38}$$

Our aim is to represent φ as a function of $x\in \mathcal{S}'(\mathbb{R})$ directly in terms of the $f^{(n)}$'s. Let us first consider the case n=2, and try to find such a representation for $\varphi^{(2)}$. By (2.11b) and (2.9), we have

$$\varphi^{(2)}(x) = \sum_{\substack{k_1,k_2\\ k_1\le k_2}} a_{k_1,k_2} :<x,e_{k_1}> <x,e_{k_2}>: .$$

Let us assume for the moment, that we have chosen a CONS $(e_k, k\in\mathbb{N})$ which belongs to $\mathcal{S}(\mathbb{R})$, and that the last sum has only a finite number of non-zero terms. Since

$$f^{(2)} = \sum_{\substack{k_1,k_2\\ k_1\le k_2}} a_{k_1,k_2}\, e_{k_1} \hat{\otimes} e_{k_2},$$

we have then $f^{(2)} \in \mathcal{S}_c(\hat{\mathbb{R}^2})$, where the subscript "c" indicates complexification. Extending the action of $x \in \mathcal{S}'(\mathbb{R})$ to $f \in \mathcal{S}_c(\mathbb{R})$ as $<x, f> = <x, Re(f)> + i<x, Im(f)>$, the following computation using (2.29) or (2.30) makes sense for every $x \in \mathcal{S}'(\mathbb{R})$.

$$\varphi^{(2)}(x) = \sum_{\substack{k_1,k_2 \\ k_1 \leq k_2}} a_{k_1,k_2} (<x, e_{k_1}> <x, e_{k_2}> - (e_{k_1}, e_{k_2}))$$

$$= \sum_{\substack{k_1,k_2 \\ k_1 \leq k_2}} a_{k_1,k_2}(< x^{\otimes 2}, e_{k_1} \hat{\otimes} e_{k_2}> - \int_{\mathbb{R}} <\delta_t^{\otimes 2}, e_{k_1} \hat{\otimes} e_{k_2}> dt)$$

$$= < x^{\otimes 2} - \int_{\mathbb{R}} dt\ \delta_t^{\otimes 2}, \Big(\sum_{\substack{k_1,k_2 \\ k_1 \leq k_2}} a_{k_1,k_2} e_{k_1} \hat{\otimes} e_{k_2} \Big)>$$

$$= < x^{\otimes 2} - \int_{\mathbb{R}} dt\ \delta_t^{\otimes 2}, f^{(2)} >$$

$$\equiv < : x^{\otimes 2} :, f^{(2)} >,$$

where we have denoted the distribution $(x^{\otimes 2} - \int dt\ \delta_t^{\otimes 2}) \in \mathcal{S}'(\mathbb{R}^2)$ by $: x^{\otimes 2} :$ (and again we extended it in the obvious way to complex test functions). Informally, we may write

$$: x^{\otimes 2} : (u_1,u_2) = : x(u_1)\ x(u_2) :$$

$$= x(u_1)\ x(u_2) - \delta(u_1 - u_2).$$

For general $n \in \mathbb{N}$, one can use the recursion (2.30), in order to find a definition for $: x^{\otimes n} :$ such that

$$\varphi^{(n)}(x) = < : x^{\otimes n} :, f^{(n)} >.$$

$$= \langle : x^{\otimes n} :, \mathrm{Re}(f^{(n)}) \rangle + i \langle : x^{\otimes n} :, \mathrm{Im}(f^{(n)}) \rangle. \tag{2.39}$$

Namely, we define

$$: x^{\otimes 0} : = 1 \tag{2.40a}$$

$$: x^{\otimes 1} : = x \tag{2.40b}$$

$$: x^{\otimes n} : = : x^{\otimes n-1} : \hat{\otimes}\, x - (n-1) : x^{\otimes n-2} : \hat{\otimes}\, \mathrm{Tr}, \tag{2.40c}$$

where $\mathrm{Tr} \in \mathcal{S}'(\hat{\mathbb{R}}^2)$ is the distribution given by

$$\mathrm{Tr} := \int \delta_t^{\otimes 2}\, dt, \tag{2.41}$$

i.e., if $f \in \mathcal{S}(\mathbb{R}^2)$, then $\langle \mathrm{Tr}, f \rangle = \int f(t,t)\, dt$. In physics, $: x^{\otimes n}:$ is known as the n-*th Wick power* of the distribution $x^{\otimes n}$ (with respect to μ, cf., e.g., Simon (1974)). It follows directly from the above recursion relation that $: x^{\otimes n}:$ belongs indeed to $\mathcal{S}'(\hat{\mathbb{R}}^n)$. Informally, the recursive definition (2.40) reads

$$: 1 : = 1 \tag{2.42a}$$

$$: x(u_1) : = x(u_1) \tag{2.42b}$$

$$\begin{aligned} : x(u_1) \cdots x(u_n) : &= : x(u_1) \cdots x(u_{n-1}) : x(u_n) - \\ &- \sum_{i=1}^{n-1} \delta(u_n - u_i) : x(u_1) \cdots \not{x}(u_i) \cdots x(u_{n-1}) :. \end{aligned} \tag{2.42c}$$

It is easy to repeat the computations which proved the Lemmas 2.8 and 2.9 in an informal way for $: x(u_1) \cdots x(u_n) :$ to find that

$$: x(u_1) \cdots x(u_n) : = \prod_{i=1}^{n} \frac{\delta}{\delta\eta(u_i)} \exp\Big(\int x(u)\eta(u)du - \int \eta(u)^2 du\Big)\Big|_{\eta=0}, \tag{2.43}$$

where $\frac{\delta}{\delta\eta(u)}$ is the Frêchet functional derivative.

Now let us drop the assumption that $f^{(n)} = \sum_\alpha a_\alpha e_\alpha \in L^2(\hat{\mathbb{R}^n})$ has only a finite number of non-zero terms: while the sum converges to $f^{(n)}$ in $L^2(\mathbb{R}^n)$, $<: x^{\otimes n}:, \sum_\alpha a_\alpha e_\alpha >$ converges to $\varphi^{(n)}$ in $\mathcal{H}^{(n)}$, and we shall use the same notation for the limit element. Thus our representation for $\varphi \in (L^2)$, which is in one–to–one correspondence with the sequence $(f^{(n)}, n \in N_0) \in \Gamma(L^2(\mathbb{R}))$ of kernel functions, reads

$$\varphi(x) = \sum_{n=0}^{\infty} < : x^{\otimes n} : , f^{(n)} >, \quad \mu\text{–a.e. } x \in \mathcal{S}'(\mathbb{R}). \tag{2.44}$$

Moreover, we have

$$\| \varphi \|_2^2 = \sum_{n=0}^{\infty} n! \, \| f^{(n)} \|^2_{L^2(\mathbb{R}^n)}, \tag{2.45}$$

$$\mathcal{S}\varphi(\xi) = \sum_{n=0}^{\infty} \int_{\mathbb{R}^n} f^{(n)}(u) \, \xi^{\otimes n}(u) \, du. \tag{2.46}$$

Because of the last relation, we also sometimes write informally for (2.44)

$$\varphi(x) = \sum_{n=0}^{\infty} \int_{\mathbb{R}^n} f^{(n)}(u) : x^{\otimes n} : (u) \, du$$

$$= \sum_{n=0}^{\infty} \int_{\mathbb{R}^n} f^{(n)}(u_1, \ldots, u_n) : x(u_1) \cdots x(u_n) : du.$$

In general, for $\varphi \in (L^2)$ we shall also refer to (2.44), or to the sequence $(f^{(n)}, n \in N_0)$

itself, as the *chaos expansion of* φ.

It will often be useful to have the following alternative description of the Wick powers $:x^{\otimes n}:\in\mathcal{S}'(\hat{\mathbb{R}^n})$ of $x\in\mathcal{S}'(\mathbb{R})$ (cf. also Potthoff and Yan (1992)). Since $:x^{\otimes n}:$ has to be defined as a continuous linear form on $\mathcal{S}(\hat{\mathbb{R}^n})$, it is sufficient to define this distribution on the dense subspace of (complex) finite linear combinations of elements of the form $\xi_1\hat{\otimes}\cdots\hat{\otimes}\xi_n$, $\xi_k\in\mathcal{S}(\mathbb{R})$, $k=1,2,\ldots,n$. (Here, as before, we identify $\xi_1\hat{\otimes}\cdots\hat{\otimes}\xi_n$ with the function on $\mathbb{R}^n$ which is the symmetrization of $(u_1,\ldots,u_n)\mapsto\xi_1(u_1)\cdots\xi_n(u_n)$.) By linearity and the polarization identity, it is sufficient to give $:x^{\otimes n}:$ as acting on $\xi^{\otimes n}$, and we define

$$<:x^{\otimes n}:, \xi^{\otimes n}> = :<x,\xi>^n:.$$

As usual, the right hand side of the last equality is the n–th Hermite polynomial in $<x,\xi>$ with parameter $|\xi|_2^2$, $|\cdot|_2$ denoting the norm of $L^2(\mathbb{R},dt)$. I.e.,

$$<:x^{\otimes n}:, \xi^{\otimes n}> = \frac{d^n}{d\lambda^n}\, e^{\lambda<x,\xi>-\frac{\lambda^2}{2}|\xi|_2^2}\Big|_{\lambda=0}. \tag{2.47}$$

Using (2.47) one verifies readily that $:x^{\otimes n}:$ as defined here satisfies the recursion relation (2.40), and therefore both definitions coincide. Working out the derivatives in (2.47), one gets the following formula.

Lemma 2.11. *Let* $x\in\mathcal{S}'(\mathbb{R})$ *and* $n\in\mathbb{N}$. *Then* $:x^{\otimes n}:\in\mathcal{S}'(\hat{\mathbb{R}^n})$ *is given by*

$$:x^{\otimes n}: = \sum_{k=0}^{[\frac{n}{2}]}\binom{n}{2k}(2k-1)!!\,(-1)^k\, x^{\otimes(n-2k)}\hat{\otimes}\mathrm{Tr}^{\otimes k}. \tag{2.48}$$

We conclude this chapter by preparing some technical results on the Wick powers (2.40), (2.42) and the decomposition (2.44) which will be of use later. It will be convenient to work for a moment with in a slightly informal manner. First we have to introduce some notation.

For natural numbers m,n, ... we let **m**,**n**, ... denote the sets $\{1, \dots ,m\}$, $\{1, \dots ,n\}$, ... Naturally ordered subsets of k elements of these sets will be denoted by $\mathbf{r}(k)\subset\mathbf{m}$, $\mathbf{s}(k)\subset\mathbf{n}$, etc. $\mathfrak{S}(k)$ is the permutation group of k letters and its elements are denoted by π, σ, ... $\pi\mathbf{s}(k)$ is the subset $\mathbf{s}(k)$ of **m**,**n**, ... with the ordering induced by $\pi\in\mathfrak{S}(k)$. If $m,n,k\in\mathbb{N}$, $k\leq m\wedge n$, then $P(m,n;k)$ denotes the set of all pairings of k elements in **m** with k elements in **n**. $p\in P(m,n;k)$ is uniquely determined by two subsets $\mathbf{r}(k)\subset\mathbf{m}$, $\mathbf{s}(k)\subset\mathbf{n}$ and a permutation $\pi\in\mathfrak{S}(k)$. Note that $P(m,n;k)$ has $k!\binom{m}{k}\binom{n}{k}$ elements. Finally for $p\in P(m,n;k)$ we set

$$\delta(u_{\mathbf{r}(k)}-v_{\mathbf{s}(k)}) := \prod_{(i,j)\in p} \delta(u_i-v_j), \tag{2.49}$$

where p corresponds to $\mathbf{r}(k)$, $\mathbf{s}(k)$, π and u_i, v_j are real variables.

The next lemma is proved by induction over the relations (2.42) and we leave the proof to the interested reader.

Lemma 2.12. (*The Re–Ordering Formula*). *Let* $m,n\in\mathbb{N}$, $u = (u_1, \dots ,u_m)\in\mathbb{R}^m$ *and* $v = (v_1, \dots ,v_n)\in\mathbb{R}^n$. *Then*

$$:x^{\otimes m}:(u):x^{\otimes n}:(v) = \sum_{k=0}^{m\wedge n} \sum_{\mathbf{r},\mathbf{s},\pi} \delta(u_{\mathbf{r}(k)}-v_{\mathbf{v}(k)}) \times$$
$$\times : \prod_{i\in\mathbf{m}\setminus\mathbf{r}(k)} x(u_i) \prod_{i\in\mathbf{n}\setminus\mathbf{s}(k)} x(v_i) :, \tag{2.50}$$

where the second sum in (2.50) *ranges over all* $\mathbf{r}(k)\in\mathbf{m}$, $\mathbf{s}(k)\in\mathbf{n}$ *and all* $\pi\in\mathfrak{S}(k)$.

Corollary 2.13. *Let* $\varphi\in\mathcal{H}^{(m)}$ *and* $\psi\in\mathcal{H}^{(n)}$, $m,n\in\mathbb{N}$, *having integral kernels* $f^{(m)}$, $g^{(n)}$ *resp. Then the pointwise product (which belongs to* (L^2)*) of* φ *and* ψ *is given by*

$(\varphi\cdot\psi)(x) =$

$$= \sum_{k=0}^{m\wedge n} k!\binom{m}{k}\binom{n}{k} \langle :x^{\otimes(m+n-2k)}:, f^{(m)}\hat{\otimes}_k g^{(n)}\rangle, \quad \mu\text{-a.e. } x\in\mathcal{S}'(\mathbb{R}), \tag{2.51a}$$

where $f^{(m)}\hat{\otimes}_k g^{(n)}$ *is the symmetrization of*

$$\int_{\mathbb{R}^k} f^{(m)}(s_1, \dots, s_k, \cdot)\, g^{(n)}(s_1, \dots, s_k, \cdot)\, ds_1 \dots ds_k. \tag{2.51b}$$

3. Generalized Functionals

In this chapter we construct spaces of generalized functionals on $\mathcal{N}^*$. Generalized functions over infinite dimensional spaces have applications in various domains. They appear in hydrodynamics and turbulence theory (e.g., Chow (1989), Lindstrøm et al. (1991)), quantum mechanics (Hida and Streit (1983), de Faria et al. (1990)), quantum field theory (Albeverio, Hida et al. (1990a), Potthoff and Streit (1990)), potential theory in infinite dimensions (Albeverio, Hida et al. (1990b), Hida, Potthoff, and Streit (1988), Potthoff and Röckner (1990), Potthoff and Streit (1990)), the theory of anticipating stochastic differential equations (Kuo and Potthoff (1990)), non-linear filtering (Hida (1975b)), in Malliavin calculus (e.g., Watanabe (1983), Sugita (1985), Potthoff (1988)), in the theory of stochastic partial differential equations (Chow (1989b), Nualart and Zakai (1989)).

The methods we use here to construct such spaces are quite similar to those in finite dimensional analysis: we exploit the concepts of duality and Gel'fand triples.

Our constructions are mainly motivated by the development and applications of the "causal" calculus of differential operators acting on functions of white noise, cf., e.g., Hida (1975a), Kubo and Takenaka (1980, 1981, 1982), Kuo (1983a,d), Potthoff and Yan (1992), and Chapter 5. However, it should be noted that recently many other authors have constructed and applied certain spaces of smooth and generalized functionals on infinite dimensional spaces. Without aiming at a complete list of references, we mention the works of Berezansky and Kondratiev (1988), Hida (1975a, 1980a), Hida and Streit (1981, 1983), Itô (1990), Kondratiev (1980), Kondratiev and Samoilenko (1976), Korezlioglu and Ustunel (1990), M. Krée (1977), M. Krée and P. Krée (1983), Kubo and Takenaka (1980), Kuo et al. (1991), Meyer (1983), Meyer and Yan (1987, 1989, 1990), Nualart and Zakai (1989), Potthoff (1987, 1988), Potthoff and Streit (1991a,b), Potthoff and Yan (1992), Shigekawa (1980), Sugita (1985a, 1985b, 1988), Watanabe (1983, 1984), Yan (1990a,b,c), Yokoi (1990) and references quoted there.

In Section 3.A we shall construct 'by hand' a class of spaces of test and generalized $\mathcal{N}^*$-functionals in the case of white noise which has appeared in the formulation of the causal calculus and its applications (Hida (1975a)). In Section 3.B we shall present a systematic construction scheme using 'second quantization' and Fock space techniques: the ideas here are closely related to those of Berezansky and Kondratiev (1988), Kondratiev (1980), Kubo and Takenaka (1980). In section 3.C we shall specialize to spaces constructed by using 'second quantized operators'. This will lead to two triples of spaces of functionals over $\mathcal{S}'(\mathbb{R})$ which play central roles in infinite dimensional analysis. One of these triples will be analyzed in detail in the next chapter, the other is discussed in Chapter 7.

3.A The Spaces $(L^2)^{\pm}$

Throughout this section we shall work with white noise, Example 1.2, but most of the ideas are readily generalized to white noise with d-dimensional time parameter, or to other choices of function spaces for $\mathcal{N}$ and $\mathcal{H}$. We shall follow Hida (1975a) quite closely (cf. also Kuo (1983d)).

The key observation is the following: consider the integral representation (2.33) of $\varphi^{(n)} \in \mathcal{H}^{(n)}$. Think of the right hand side of (2.33) as the realization of the dual pairing of $\xi^{\otimes n} \in \mathcal{S}(\hat{\mathbb{R}}^n)$ with $f^{(n)} \in L^2(\hat{\mathbb{R}}^n)$ viewed as a regular distribution in $\mathcal{S}'(\hat{\mathbb{R}}^n)$. Then clearly the right hand side makes sense, if we extend it to elements $f^{(n)}$ in $\mathcal{S}'(\hat{\mathbb{R}}^n)$ in the sense of dual pairing. This defines corresponding generalized white noise functionals, if we extend the inverse of the $\mathcal{S}$-transform appropriately. This idea will now be carried out in a slightly restricted way.

Let $\mathcal{H}_{a_n}(\mathbb{R}^n)$ be the standard (complex) Sobolev space of order $a_n \geq 0$ over $\mathbb{R}^n$ (e.g., Adams (1975), Reed and Simon (1975)), and denote by $\mathcal{H}_{a_n}(\hat{\mathbb{R}}^n)$ its symmetric subspace

$$\mathcal{H}_{a_n}(\hat{\mathbb{R}}^n) = \mathcal{H}_{a_n}(\mathbb{R}^n) \cap L^2(\hat{\mathbb{R}}^n, dt).$$

Identifying $L^2(\hat{\mathbb{R}}^n)$ with its dual, we have then the Gel'fand triple

$$\mathcal{H}_{-a_n}(\hat{\mathbb{R}}^n) \supset L^2(\hat{\mathbb{R}}^n) \supset \mathcal{H}_{a_n}(\hat{\mathbb{R}}^n), \tag{3.1}$$

where $\mathcal{H}_{-a_n}(\hat{\mathbb{R}}^n)$ is the dual of $\mathcal{H}_{a_n}(\hat{\mathbb{R}}^n)$. According to (2.39) and Theorem 2.7, an element $f^{(n)}$ in $\mathcal{H}_{a_n}(\hat{\mathbb{R}}^n)$ defines a white noise functional $\varphi^{(n)}$ in the complete, dense subspace $\mathcal{H}_{a_n}^{(n)} \equiv \mathcal{I}^{-1}\mathcal{H}_{a_n}(\hat{\mathbb{R}}^n)$:

$$\varphi^{(n)}(x) = <: x^{\otimes n} :, f^{(n)}>,$$

in the notation and conventions of Chapter 2. For $g^{(n)} \in \mathcal{H}_{-a_n}(\hat{\mathbb{R}}^n)$ we define a generalized white noise functional $\Phi^{(n)} \equiv \mathcal{I}^{-1}g^{(n)}$ by setting

$$<\Phi^{(n)}, \varphi^{(n)}> = n! <g^{(n)}, f^{(n)}>, \tag{3.2}$$

where the pairing on the right hand side is the one between $\mathcal{H}_{-a_n}(\mathbb{R}^n)$ and $\mathcal{H}_{a_n}(\mathbb{R}^n)$. We denote the corresponding Hilbert space $\mathcal{I}^{-1}\mathcal{H}_{-a_n}(\hat{\mathbb{R}}^n)$ of generalized white noise functionals of degree n by $\mathcal{H}_{-a_n}^{(n)}$:

$$\begin{array}{ccccc} \mathcal{H}_{-a_n}^{(n)} & \supset & \mathcal{H}^{(n)} & \supset & \mathcal{H}_{a_n}^{(n)} \\ \downarrow & & \downarrow \mathcal{I} & & \downarrow \\ \mathcal{H}_{-a_n}(\hat{\mathbb{R}}^n) & \supset & L^2(\hat{\mathbb{R}}^n) & \supset & \mathcal{H}_{a_n}(\hat{\mathbb{R}}^n). \end{array} \tag{3.3}$$

The $\mathcal{I}$–transform is extended to $\mathcal{H}_{-a_n}^{(n)}$ in such a way that the exterior arrows in diagram (3.3) become unitary maps, too.

The next step is clear: we take (weighted) direct sums of the spaces $\mathcal{H}_{\pm a_n}^{(n)}$. Let $\mathbf{a} = (a_n, n \in \mathbb{N}_0)$, $\mathbf{c} = (c_n, n \in \mathbb{N}_0)$ be two sequences with $a_n \geq 0$, and $c_n \geq 1$, for all $n \in \mathbb{N}_0$. Then we define the space $(L^2)_{\mathbf{a},\mathbf{c}}^{+}$ of smooth white noise functionals by:

$$(L^2)_{\mathbf{a},\mathbf{c}}^{+} := \bigoplus_n \sqrt{c_n\, n!}\; \mathcal{H}_{a_n}^{(n)}. \tag{3.4.a}$$

Thus, the norm of $\varphi \in (L^2)^+_{a,c}$, with $\mathcal{S}\varphi = (f^{(n)}, n \in \mathbb{N}_0)$, is given by

$$\| \varphi \|^2_{(L_2)^+_{a,c}} = \sum_{n=0}^{\infty} c_n\, n!\, \| f^{(n)} \|^2_{\mathcal{H}_{a_n}(\mathbb{R}^n)}. \tag{3.4.b}$$

The space $(L^2)^-_{a,c}$ of generalized white noise functionals is by definition the dual of $(L^2)^+_{a,c}$, and it is given by:

$$(L^2)^-_{a,c} := \bigoplus_n \sqrt{c_n^{-1}\, n!}\; \mathcal{H}^{(n)}_{-a_n}. \tag{3.5.a}$$

Therefore, $\Phi \in (L^2)^-_{a,c}$, if and only if it is in one-to-one correspondence with a sequence $(F^{(n)}, n \in \mathbb{N}_0)$, so that

$$\| \Phi \|^2_{(L^2)^-_{a,c}} = \sum_{n=0}^{\infty} c_n^{-1}\, n!\, \| F^{(n)} \|^2_{\mathcal{H}_{-a_n}}. \tag{3.5.b}$$

Thus we obtain a Gel'fand triple

$$(L^2)^-_{a,c} \supset (L^2) \supset (L^2)^+_{a,c}, \tag{3.6}$$

of spaces of functionals on $\mathcal{S}'(\mathbb{R})$.

Of course, the choice of the sequences **a**, **c** is a question of the application one has in the mind. In view of the calculus of differential operators (cf. Chapter 5), a suitable choice is given by $a_n = \frac{n+1}{2}$, $c_n = n$. In this case, let us simply write $(L^2)^{\pm}$. There are two reasons for this choice of **a** (both being related by duality):

i) By Sobolev's Lemma (e.g., Adams (1975)), the elements in $\hat{\mathcal{H}}_{\frac{n+1}{2}}(\mathbb{R}^n)$ have continuous representatives, and therefore they can be considered as pointwise defined functions. Hence functionals in $(L^2)^+$ have an integral representation with pointwise defined, continuous kernels. This allows for a convenient discussion of the differential operator ∂_t of Chapter 5, which requires the pointwise evaluation of the integral kernels, cf. also Chapter 5.

ii) Let δ_t be the Dirac distribution at $t\in\mathbb{R}$. Then $\delta_t^{\otimes n}$ is an element of $\hat{\mathcal{H}}_{-\frac{n+1}{2}}(\mathbb{R}^n)$. This fact provides a typical class of generalized white noise functionals, namely certain "polynomials" (in fact "Wick monomials") in the generalized random variable X_{δ_t}. The Wick monomials play in white noise analysis the role that monomials play in ordinary calculus.

Let us illustrate the second point above a little more. Let $\delta_t^\epsilon\in\mathcal{H}_1(\mathbb{R})$, $\epsilon>0$, $t\in\mathbb{R}$, be a smooth approximation of δ_t. For example,

$$\delta_t^\epsilon(u) = (2\pi\epsilon)^{-\frac{1}{2}} \exp(-\tfrac{1}{2\epsilon}(t-u)^2).$$

Then $(\delta_t^\epsilon)^{\otimes n}$ is in $\hat{\mathcal{H}}_{\frac{n+1}{2}}(\mathbb{R}^n)$ and converges to $\delta_t{}^{\otimes n}$ in $\hat{\mathcal{H}}_{-\frac{n+1}{2}}(\mathbb{R}^n)$ as ϵ tends to zero. For $\xi\in\mathcal{S}(\mathbb{R})$, form

$$(\mathcal{S}W_{t,\epsilon}^{(n)})(\xi) = \int_{\mathbb{R}^n} (\delta_t^\epsilon)^{\otimes n}(t_1, \dots ,t_n)\ \xi(t_1) \cdots \xi(t_n)\ d^n t. \tag{3.7}$$

Using Lemma 2.8, the image of the $\mathcal{S}(\mathbb{R})$–functional (3.7) under $\mathcal{S}^{-1}$ is the following element in $(L^2)^+$:

$$W_{t,\epsilon}^{(n)}(x) = :\ \langle x, \delta_t^\epsilon\rangle^n :, \quad x\in\mathcal{S}'(\mathbb{R}). \tag{3.8}$$

(Recall that $:\langle x, f\rangle^n:$ is the Hermite polynomial in $\langle x, f\rangle$ of order n, with parameter $\|f\|^2_{L^2(\mathbb{R})}$, and that it coincides with the projection of $\langle x, f\rangle^n$ onto $\mathcal{H}^{(n)}$.) By our preceding considerations, $W_{t,\epsilon}^{(n)}$ converges to some $W_t^{(n)}\in\mathcal{H}^{(n)}_{-\frac{n+1}{2}}$ as ϵ tends to zero. We denote this limit by

$$W_t^{(n)}(x) = \langle :x^{\otimes n}:, \delta_t^{\otimes n}\rangle \equiv\ :x(t)^n:, \quad x\in\mathcal{S}'(\mathbb{R}). \tag{3.9}$$

Of course, (3.9) is somewhat informal. However, for heuristic interpretations and also for calculations such expressions are very useful. Therefore we continue to use such notations whenever the rigorous interpretation of an object is clear. In this case, $W_t^{(n)} \in (L^2)^-$ has to be understood as acting on $\varphi \in (L^2)^+$ as follows. If φ has chaos decomposition given by $(f^{(n)}, n \in \mathbb{N}_0)$, $f^{(n)} \in H_{\frac{n+1}{2}}(\hat{\mathbb{R}}^n)$, then

$$\langle W_t^{(n)}, \varphi \rangle = n! \langle \delta_t^{\otimes n}, f^{(n)} \rangle = n! \, \tilde{f}^{(n)}(t, \ldots ,t),$$

where $\tilde{f}^{(n)}$ is the pointwise defined version of $f^{(n)}$.

Another remark concerning (3.9) is in order: in view of (3.8) and the definition of $:\langle x, f\rangle^n:$, one can interpret $:x(t)^n:$ as the n–th Hermite polynomial in x(t) 'with parameter $\delta_t(t) = \delta(0) = +\infty$', e.g., $:x(t)^2:$ '$= x(t)^2 - \delta(0)$'. In this sense we have subtracted an infinite constant from the ill–defined square of a distribution $x \in \mathcal{S}'(\mathbb{R})$, making their difference a well-defined generalized functional. I.e., we have performed a so–called renormalization[1].

It is clear, that in the same way as before (or by polarization) we find for $t = (t_1, \ldots ,t_n) \in \mathbb{R}^n$,

$$W_t^{(n)}(x) = :x(t_1) \cdots x(t_n):,$$

as an element in $\mathcal{H}^{(n)}_{-\frac{n+1}{2}}$.

In general we shall use the same notation as introduced in Chapter 2, namely

$$\Phi^{(n)}(x) = \sum_{n=0}^{\infty} \langle :x^{\otimes n}:, f^{(n)} \rangle, \tag{3.10}$$

and according to whether $f^{(n)} \in L^2(\hat{\mathbb{R}}^n)$ or $\mathcal{H}_{\pm\frac{n+1}{2}}(\hat{\mathbb{R}}^n)$, we have $\Phi \in (L^2)$ or $(L^2)^{\pm}$, respectively.

Because of the connection of the functional $\Phi^{(n)}$ in (3.10) to the multiple Wiener integral in case that $f^{(n)} \in L^2(\hat{\mathbb{R}}^n)$ (cf. Chapter 2), one could interpret

---

[1] It is not a renormalization in the strict sense that physicists use this notion, cf., e.g., Schweber (1966). However, we shall continue to call so any process of making an ill-defined object well-defined.

(3.10) for $f^{(n)} \in \hat{\mathcal{H}}_{-\frac{n+1}{2}}(\mathbb{R}^n)$ as a generalized multiple Wiener integral. Such an approach to generalized multiple Wiener integrals has been given in an article by Hida (1978).

In practice, one will have usually some informal expression of a white noise functional at hand and ask, whether this expression, or a renormalized version of it, belongs to $(L^2)^-_{\mathbf{a},\mathbf{c}}$ for some **a**, **c**. If a decomposition into homogeneous chaos components of the form (3.10) (from which one could read off the necessary properties of the kernels $f^{(n)}$) is not obvious, it will be useful to proceed as follows.

First one regularizes the expression 'by hand' in such a way that it becomes an element in (L^2). Then one computes the $\mathcal{S}$-transform of the regularized functional. Next, one has to renormalize the resulting functional on $\mathcal{S}(\mathbb{R})$ through subtraction or division (or both) of terms which diverge when the regularization is removed. Finally, one has to decompose the renormalized $\mathcal{S}(\mathbb{R})$-functional into homogeneous components in ξ, and to check whether the resulting kernels are in the Sobolev spaces $\hat{\mathcal{H}}_{-a_n}(\mathbb{R}^n)$, and if their norms are square summable with the weights $c_n^{-1} n!$ (cf. 3.5).

Let us make this concrete with several examples.

Example 3.1. Here we are going to review the generalized functional (3.9). To make things simple, let n = 2. The general case is then an easy exercise.

Let us start with the senseless expression

$$W^{(2)}_{t,un}(x) = x(t)^2,$$

where $t \in \mathbb{R}$, $x \in \mathcal{S}'(\mathbb{R})$. We regularize it by setting

$$W^{(2)}_{t,reg}(x) = \langle x, \delta_t^{\epsilon} \rangle^2,$$

with $\delta_t^{\epsilon} \in \mathcal{S}(\mathbb{R})$, $\epsilon > 0$, given above. Clearly $W^{(2)}_{t,reg} \in (L^2)$ as long as $\epsilon > 0$.

Then we compute the $\mathcal{S}$-transform of $W^{(2)}_{t,reg}$:

$$(\mathcal{S}W^{(2)}_{t,reg})(\xi) = \int (<x, \delta^{\epsilon}_{t}> + (\xi, \delta^{\epsilon}_{t}))^2 \, d\mu(x)$$

$$= (\xi, \delta^{\epsilon}_{t})^2 + \| \delta^{\epsilon}_{t} \|^2_{L^2(\mathbb{R},dt)}$$

$$= (\xi, \delta^{\epsilon}_{t})^2 + \frac{1}{2\sqrt{\pi\epsilon}}.$$

In the limit $\epsilon \downarrow 0$ the first term converges to $\xi(t)^2$, while the second diverges and has to be subtracted. Now we can take the limit and define

$$(\mathcal{S}W^{(2)}_{t,ren})(\xi) := \xi(t)^2,$$

which we write as

$$(\mathcal{S}W^{(2)}_{t,ren})(\xi) = <\delta_t^{\otimes 2}, \xi^{\otimes 2}>.$$

Since $\delta_t^{\otimes 2} \in \mathcal{H}_{-3/2}(\mathbb{R}^2)$, we have $W^{(2)}_{t,ren} \in (L^2)^-$. In fact, $W^{(2)}_{t,ren} \in \mathcal{H}^{(2)}_{-3/2}$. Now we drop the subscript 'ren' $W^{(2)}_{t,ren} \equiv W^{(2)}_{t}$, and the notation $W^{(2)}_{t}(x) = : x(t)^2:$ is used as in (3.9).

Our next example is less trivial. It plays an important role within the white noise approach to the Feynman integral, cf. the articles of Hida and Streit (1977), de Faria et al. (1990), and Chapter 12. Generalizations of this example will be discussed at the end of Chapter 4.A, and in Chapter 4.C.

Example 3.2. Consider the informal expression

$$\Phi_{un}(x) = \exp(c \int x(t)^2 \, dt), \quad x \in \mathcal{S}'(\mathbb{R}), \ c \in \mathbb{C}. \tag{3.11}$$

Let $(e_n, n \in \mathbb{N})$ be some (real) complete orthonormal system of $L^2(\mathbb{R},dt)$ which belongs to $\mathcal{S}(\mathbb{R})$ (e.g., the collection of Hermite functions). Then we may think of

$\int x(t)^2\, dt$ as $\sum_n <x, e_n>^2$ and define the regularized functional

$$\Phi_{reg}^N(x) = \exp(c\sum_{n=1}^{N} <x, e_n>^2), \quad x\in\mathcal{S}'(\mathbb{R}),$$

which is everywhere defined on $\mathcal{S}'(\mathbb{R})$. Moreover, Φ_{reg}^N belongs to (L^2) for all $N\in\mathbb{N}$ and all $c\in\mathbb{C}$ with Re $c<\frac{1}{4}$, as the evaluation of a simple integral shows. Thus, for the moment, let us put this restriction on $c\in\mathbb{C}$. (For the following it would actually be sufficient to assume Re $c<\frac{1}{2}$.)

Due to the independence of the random variables $<\cdot, e_n>$, $n\in\mathbb{N}_0$, it is easy to compute the $\mathcal{S}$–transform of Φ_{reg}^N:

$$(\mathcal{S}\Phi_{reg}^N)(\xi) = \prod_{n=1}^{N} e^{c\xi_n^2}\int e^{c(u^2+2\xi_n u)-\frac{1}{2}u^2}\,\frac{du}{\sqrt{2\pi}}$$

$$= (\prod_{n=1}^{N}(1-2c)^{-\frac{1}{2}})\exp\Big(\frac{c}{1-2c}\sum_{n=1}^{N}\xi_n^2\Big),$$

where $\xi_n = (\xi, e_n)$, $\xi\in\mathcal{S}(\mathbb{R})$. Observe that for $c \neq 0$, the sequence of products $\prod_{n=1}^{N}(1-2c)^{-\frac{1}{2}}$ either converges to zero or diverges, as N tends to infinity. Thus – as a renormalization – we divide by this factor. Note that this factor is equal to $\mathcal{S}\Phi_{reg}^N(0) = \mathbb{E}\Phi_{reg}^N$. We set

$$(\mathcal{S}\Phi_{ren})(\xi) := \lim_{N\to\infty}\frac{(\mathcal{S}\Phi_{reg}^N)(\xi)}{(\mathcal{S}\Phi_{reg}^N)(0)}$$

$$= \exp(\tilde{c}\int \xi(t)^2\, dt), \tag{3.12}$$

with the notation $\tilde{c} = \frac{c}{1-2c}$. For all $c\in\mathbb{C}$ with $c \neq \frac{1}{2}$, the functional $\mathcal{S}\Phi_{reg}$ is everywhere defined on $\mathcal{S}(\mathbb{R})$.

Next we have to investigate whether this functional is the $\mathcal{S}$–transform of a generalized functional in $(L^2)^-_{\mathbf{a,c}}$ for some choice of **a**, **c**. Write

$$(\mathcal{S}\Phi_{ren})(\xi) = \sum_{n=0}^{\infty} \frac{\tilde{c}^n}{n!} \int_{\mathbb{R}^n} \xi(t_1)^2 \cdots \xi(t_n)^2 \, d^n t$$

$$= \sum_{n=0}^{\infty} \frac{\tilde{c}^n}{n!} \int_{\mathbb{R}^{2n}} f^{(2n)}(t_1, \ldots, t_{2n}) \, \xi(t_1) \cdots \xi(t_{2n}) \, d^{2n} t,$$

with $f^{(2n)} = (f^{(2)})^{\hat{\otimes} n}$, $f^{(2)}(t_1, t_2) = \delta(t_1 - t_2)$. More rigorously,

$$(\mathcal{S}\Phi_{reg})(\xi) = \sum_{n=0}^{\infty} \frac{\tilde{c}^n}{n!} <\mathrm{Tr}^{\otimes n}, \xi^{\otimes 2n}>,$$

where $\mathrm{Tr} \in \mathcal{S}'(\mathbb{R}^2)$ is the distribution introduced in Chapter 2 (cf. (2.41)). In order that $\mathcal{S}\Phi_{reg}$ be the $\mathcal{S}$–transform of a functional in $(L^2)^-_{\mathbf{a},c}$, we would have to have that $\mathrm{Tr} \in \hat{\mathcal{H}}_{-a_2}(\mathbb{R}^2)$ for some $a_2 \in \mathbb{R}$. Clearly, this is not the case for any $a_2 \in \mathbb{R}$ and thus we either have to modify our functional, or we must change our definition of $(L^2)^-_{\mathbf{a},c}$. Here, we take the first choice. But we encourage the interested reader to construct a space of generalized functionals along our lines using instead of the triple (3.1) the following one

$$\hat{\mathcal{S}}_{-a_n}(\mathbb{R}^n) \supset \hat{L}^2(\mathbb{R}^n) \supset \hat{\mathcal{S}}_{a_n}(\mathbb{R}^n),$$

where $\hat{\mathcal{S}}_{a_n}(\mathbb{R}^n)$ is defined at the beginning of Chapter 4.A (cf. also Appendix A.5), and to determine a sequence $\mathbf{a}$ so that $\mathcal{S}\Phi_{ren}$ (3.12) has a pre–image in this space of generalized functionals. (Sufficient is the choice $a_n = p$, for all $n \in \mathbb{N}_0$, with $p > \frac{1}{4}$ such that $|\tilde{c}| \, 2^{-(4p-1)} \frac{4p}{4p-1} < 1$. See also Lemma 4.5.) Cf. Example 4.23 in Section 4.A.

We modify our functional (3.12) in the following way. For $h \in L^2(\mathbb{R})$, define

$$(\mathcal{S}\Phi_h)(\xi) := \exp(\tilde{c} \int h(t) \, \xi(t)^2 \, dt). \tag{3.13}$$

We claim that for all $c\in\mathbb{C}$, $c \neq \frac{1}{2}$, there is $\Phi_h\in(L^2)^-$ satisfying (3.13). As before we write

$$(\mathcal{S}\Phi_h)(\xi) = \sum_{n=0}^{\infty} \frac{\tilde{c}^n}{n!} \int_{\mathbb{R}^{2n}} f^{(2n)}(t_1, \dots, t_{2n})\ \xi(t_1) \cdots \xi(t_{2n})\ d^{2n}t,$$

this time with kernels

$$f^{(2n)} = (f^{(2)})^{\hat{\otimes} n}, \quad f^{(2)}(t_1,t_2) = h\left(\frac{t_1+t_2}{2}\right) \delta(t_1-t_2).$$

Now we set

$$\Phi_h(x) := \sum_{n=0}^{\infty} \frac{\tilde{c}^n}{n!} <: x^{\otimes n} :, f^{(2n)}>, \ x\in\mathcal{S}'(\mathbb{R}).$$

By construction

$$\| \Phi_h \|^2_{(L^2)^-} = \sum_{n=0}^{\infty} | \tilde{c} |^{2n} \frac{(2n)!}{2n(n!)^2} \| f^{(2n)} \|^2_{-\frac{(2n+1)}{2}},$$

and

$$\| f^{(2n)} \|^2_{-\frac{2n+1}{2}} = \int_{\mathbb{R}^{2n}} | (f^{(2n)})^{\sim}(p) |^2 (1+| p |^2)^{-\frac{2n+1}{2}} d^{2n}p$$

$$= \int_{\mathbb{R}^{2n}} | ((f^{(2)})^{\sim})^{\hat{\otimes} n}(p) |^2 (1+| p |^2)^{-\frac{2n+1}{2}} d^{2n}p,$$

where $\sim$ denotes Fourier transform. By Schwarz' inequality and the fact that

$$(f^{(2)})^{\sim}(p_1,p_2) = \frac{1}{\sqrt{2\pi}} \tilde{h}(p_1+p_2),$$

we find the estimate

$$\| f^{(2n)} \|^2_{-\frac{2n+1}{2}} \leq$$

$$\leq (2\pi)^{-n} \int_{\mathbb{R}^{2n}} | \tilde{h}(p_1+p_2) |^2 \cdots | \tilde{h}(p_{2n-1}+p_{2n} |^2 (1+| p |^2)^{-\frac{2n+1}{2}} d^{2n}p$$

$$\leq (2\pi)^{-n} \| h \|_2^{2n} \int_{\mathbb{R}^n} (1+| p |^2)^{-\frac{2n+1}{2}} d^n p,$$

using the Plancherel theorem. The last integral is easily estimated by $(1+\frac{1}{n+1})\sigma_n \leq 2\sigma_n$, where σ_n is the area of the unit sphere in n dimensions: $\sigma_n = 2\pi^{n/2}\Gamma(n/2)^{-1}$. Finally we obtain

$$\| f^{(2n)} \|^2_{-\frac{2n+1}{2}} \leq 4\,(4\pi)^{-\frac{n}{2}}\,\Gamma(n/2)^{-1} \| h \|_2^{2n},$$

and

$$\| \Phi_h \|^2_{(L^2)^-} \leq 4 \sum_{n=0}^{\infty} | \tilde{c} |^{2n} (4\pi)^{-\frac{n}{2}} \frac{(2n)!}{2n(n!)^2\Gamma(n/2)} \| h \|_2^{2n}.$$

By Stirling's theorem, the last sum converges for all $c \in \mathbb{C}$, $c \neq \frac{1}{2}$, and all $h \in L^2(\mathbb{R}, dt)$.

We shall also use the following notation for Φ_h:

$$\Phi_h(x) = \mathcal{N}\exp(c \int h(t)\, x(t)^2\, dt) \tag{3.14}$$

Typically h will be the indicator of an interval, and for the Feynman integral (cf. Chapter 12) the choice $c = \frac{1+i}{2}$ will be made.

Our next example is Donsker's δ-function which has been treated within the white noise framework in several papers: Hida (1983a), Kuo (1983a), Hida and Streit (1983), Potthoff and Streit (1991a), Shieh and Yokoi (1990). Also it appeared in an article by Fukushima (1979) in a framework using stochastic differential equations. Related articles are Kubo (1983), Watanabe (1983). This generalized functional is also discussed in a slightly different context in Chapter 7.

Example 3.3. (*Donsker's δ-function*) We shall follow essentially the methods of Hida (1983a) and Kuo (1983a).

Let $t>0$, $y\in\mathbb{R}$, and $B(t;\cdot) = \langle\cdot, 1_{[0,t)}\rangle$ be the Brownian motion at time t defined in Chapter 1. We want to construct a meaningful white noise functional corresponding to the informal expression

$$\delta_{t,y}(x) := \delta(y-B(t;x)), \quad x\in\mathcal{S}'(\mathbb{R}). \tag{3.15}$$

Still on an informal level, we make the following computation

$$\delta_{t,y} = \frac{1}{2\pi}\int e^{ip(y-B(t))}\,dp$$

$$= \frac{1}{2\pi}\int e^{ipy}\, e^{-ipB(t)+\frac{1}{2}p^2t}\, e^{-\frac{1}{2}p^2t}\,dp$$

$$= \frac{1}{2\pi}\sum_{n=0}^{\infty}\frac{(-i)^n}{n!} :B(t)^n: \int p^n\, e^{ipy-\frac{1}{2}p^2t}\,dp$$

$$= \frac{1}{\sqrt{2\pi t}}\, e^{-\frac{y^2}{2t}}\sum_{n=0}^{\infty}\frac{1}{n!}\, t^{-n} :y^n:_t :B(t)^n:. \tag{3.16}$$

In the preceding informal computation we made use of the fact that $G(p) = \exp(ipB(t)+\frac{1}{2}p^2t)$ is the generating function of the Hermite polynomials $:B(t)^n:$

with parameter t (cf. (A.1.13)), and we used (A.1.25) to compute the last integral.

Thus we define the following regularized functional in (L^2),

$$\delta^N_{t,y}(x) := p(t,y) \sum_{n=0}^{N} \frac{1}{n!} t^{-n} : y^n :_t : <x, 1_{[0,t)}>^n :,$$

with N∈ℕ, and $p(t,y) = (2\pi t)^{-1/2} \exp(-\frac{1}{2t}y^2)$. Intuitively it is clear, that the sequence $(\delta^N_{t,y}, N \in \mathbb{N})$ can not converge in (L^2), because the probability that Brownian motion hits a given point $y \in \mathbb{R}$ at a given time $t>0$ is zero. More technically,

$$\| \delta^N_{t,y} \|_2^2 = p(t,y)^2 \sum_{n=0}^{N} n! \frac{1}{(n!)^2 t^{2n}} (: y^n :_t)^2 \| 1^{\otimes n}_{[0,t)} \|^2_{L^2(\mathbb{R}^n)}$$

$$= p(t,y)^2 \sum_{n=0}^{N} \frac{1}{n!\, t^n} (: y^n :_t)^2$$

$$= \frac{1}{\sqrt{2\pi t}} p(t,y) \sum_{n=0}^{N} [\frac{1}{\sqrt{n!\, t^{-n}}} e^{-\frac{y^2}{4t}} : y^n :_t]^2.$$

According to estimate (A.1.11) in Appendix A.1 for the Hermite functions (which has to be rewritten for the Hermite functions built with Wick powers $: y^n :_t$, using (A.1.14)), the expression in square brackets is of the order $n^{-1/4}$, and the sum (3.16) does not converge in (L^2).

However, in $(L^2)^-$ we can get convergence due to the weights $c_n^{-1} = n^{-1}$ (or alternatively due to a decrease provided by the Sobolev norms):

$$\| \delta^N_{t,y} \|^2_{(L^2)^-} \leq \frac{1}{\sqrt{2\pi t}} p(t,y) \sum_{n=0}^{N} \frac{1}{n} \xi_n(y,t)^2,$$

where $\xi_n(y,t)$ is the function in square brackets above. Estimate (A.1.12) gives the bound

$$\max_{y\in\mathbb{R},t\in\mathbb{R}_+} |\xi_n(y,t)| = O(n^{-1/12}).$$

As a result we know that Donsker's δ-function exists as a generalized functional in $(L^2)^-$ and that it is represented by the series (3.16), which is convergent in $(L^2)^-$, uniformly in $y\in\mathbb{R}$, $t\in\mathbb{R}_+$.

Donsker's delta function will be reconsidered in Chapter 4, Examples 4.32 and 4.44. There we shall prove that $\delta_{t,y}$ defines a probability measure on $(\mathcal{S}'(\mathbb{R}), \mathcal{B})$ and compute its characteristic function. Based on this, we shall derive there the basic probabilistic properties of Brownian motion.

The following example is a generalization of the preceding one: we consider the composition of a tempered distribution $T\in\mathcal{S}'(\mathbb{R})$ with Brownian motion $B(t)$. This example is due to Kubo (1983) and has been reconsidered in an article by Kallianpur and Kuo (1984). Note that this kind of generalized functionals is a special kind of those which were constructed by Watanabe (1983) in his reformulation of the Malliavin calculus. (Cf. also Chapter 7 for a white noise version of this construction.)

Example 3.4. (Kallianpur and Kuo (1984)) Let $T\in\mathcal{S}'(\mathbb{R})$. We want to construct $T\circ B(t)$, $t>0$, as an element in $(L^2)^-$. To this end write informally

$$T\circ B(t;\, x) = \int T(y)\ \delta(y - \langle x, 1_{[0,t)}\rangle)\ dy.$$

For $\varphi\in(L^2)^+$:

$$\langle T\circ B(t), \varphi\rangle = \int T(y)\ \langle \delta(y-\langle\cdot, 1_{[0,t)}\rangle), \varphi(\cdot)\rangle\ dy$$

$$= \langle T(\cdot), \langle \delta_{t,\cdot}\, , \varphi\rangle\rangle, \tag{3.17}$$

where the exterior dual pairing is the one between $\mathcal{S}'(\mathbb{R})$ and $\mathcal{S}(\mathbb{R})$. Thus, if we show that for all $\varphi \in (L^2)^+$, $t>0$, the function $y \mapsto \langle \delta_{t,y}, \varphi\rangle$ belongs to $\mathcal{S}(\mathbb{R})$, then we can define $T \circ B(t)$ in $(L^2)^-$ by (3.17).

From Example 3.3 we know that

$$\begin{aligned}\langle \delta_{t,y}, \varphi\rangle &= p(t,y) \sum_{n=0}^{\infty} \frac{1}{n!} t^{-n} :y^n:_t \langle :B(t)^n:, \varphi\rangle \\ &= \pi^{1/4}(2\pi t)^{-1/2} e^{-\frac{y^2}{4t}} \sum_{n=0}^{\infty} \frac{1}{(n!)^{-1/2}} t^{-n/2}\, e_n\!\left(\frac{y}{\sqrt{2t}}\right)\langle :B(t)^n:, \varphi\rangle,\end{aligned} \tag{3.18}$$

where e_n is the n-th normalized Hermite function (cf. A.1.9). If we can prove that

$$a_n := \frac{1}{(n!)^{1/2}} t^{-n/2} \langle :B(t)^n:, \varphi\rangle,$$

is rapidly decreasing, i.e. for all $m \in \mathbb{N}$,

$$\sup_{n\in\mathbb{N}} |a_n| n^m < \infty,$$

then it follows by a standard result about Schwartz space $\mathcal{S}(\mathbb{R})$ (e.g., Reed and Simon (1972, p. 143)) that the function (of $y \in \mathbb{R}$) defined as the sum in (3.18) belongs to $\mathcal{S}(\mathbb{R})$, and therefore so does $\langle \delta_{t,y}, \varphi\rangle$. Now

$$|a_n| \le (n!)^{-1/2} t^{-n/2} \| :B(t)^n: \|_{(L^2)^-} \cdot \|\varphi\|_{(L^2)^+}.$$

For $\xi \in \mathcal{S}(\mathbb{R})$,

$$(\mathcal{S} :B(t)^n:)(\xi) = \int_{\mathbb{R}^n} (1_{[0,t)}^{\otimes n})(s_1, \dots, s_n)\, \xi(s_1) \cdots \xi(s_n)\, d^n s.$$

Hence

$$\| : B(t)^n : \|^2_{(L^2)^-} = n! \, \| 1^{\otimes n}_{[0,t)} \|^2_{\mathcal{H}_{-\frac{n+1}{2}}(\mathbb{R}^n)}$$

$$= n! \int_{\mathbb{R}^n} | (1^{\otimes n}_{[0,t)})^{\sim}(p) |^2 \, (1 + | p |^2)^{-\frac{n+1}{2}} \, d^n p$$

$$\leq n! \, \frac{t^{2n}}{(2\pi)^n} \int_{\mathbb{R}^n} (1 + | p |^2)^{-\frac{n+1}{2}} \, d^n p$$

$$\leq 4 \cdot 2^{-n} \, t^{2n} \, \pi^{-n/2} \, n! \, \Gamma(n/2)^{-1}$$

$$= 4 \cdot (Kt)^{2n} \, n! \, \Gamma(n/2)^{-1},$$

where K is some positive constant. Hence we find

$$| a_n | \leq c^n t^n \, \Gamma(n/2)^{-1/2} \, \| \varphi \|_{(L^2)^+},$$

which is indeed rapidly decreasing and so (3.17) makes sense.

Moreover, by insertion of the series representation (3.16) of $\delta_{t,y}$ into (3.17) one finds the following representation of $T \circ B(t)$:

$$T \circ B(t) = \sum_{n=0}^{\infty} \frac{1}{n!} \langle T, \eta_{n,t} \rangle \, : B(t)^n :, \tag{3.19}$$

where $\eta_{n,t}(y) = t^{-n} : y^n :_t \, (2\pi t)^{-1/2} \exp(-\frac{y^2}{2t})$. It is left as an exercise to the reader to prove that the sum (3.19) converges in $(L^2)^-$ (cf. also the article of Kallianpur and Kuo (1984)).

The examples of this section show that usually one will face the problem of controlling two different kinds of divergence: one comes from the fact that the integral kernels of certain functionals become singular, the other one is the problem of convergence of the sum over the homogeneous chaos'. Of course, there are functionals for which both types play a role. Also, as the last two examples show, sometimes the convergence of the sum over the chaos' can be controlled with the help of suitable norms of the kernels. These two aspects will be re–discussed in the following two sections in the context of different spaces of generalized and test functionals.

3.B Triples of Functional Spaces

Let us now return to the general situation of the Gaussian space $(\mathcal{N}^*,\mathcal{B},d\mu)$ induced by the continuous dense embedding $\mathcal{H}\supset\mathcal{N}$, as defined in Chapter 1. Our strategy for constructing spaces of generalized and test functionals over $\mathcal{N}^*$ will be very much in spirit of the preceding section: we shall construct triples of Fock spaces — typically induced by Gel'fand triples built on $\mathcal{H}\supset\mathcal{N}$ — and then use the unitary mapping $\mathcal{I}$ to obtain spaces of functionals over $\mathcal{N}^*$. For related works we refer the reader to the articles quoted at the beginning of this chapter.

It should be noted that the topological structures obtained from our constructions are standard ones, so that a large variety of results of the theory of topological vector spaces (cf., e.g., Gel'fand and Shilov (1968), Gel'fand and Vilenkin (1968), Trèves (1967)) are applicable.

As before we do not distinguish between $\mathcal{H}$ and its complexification for notational convenience. It will be clear from the context which of them is under consideration.

Consider the real nuclear separable pre–Hilbert space $(\mathcal{N},(\cdot\,,\,\cdot))$ and assume that we are given a countable system $\{(\cdot\,,\,\cdot)_p,\ p\in\mathbb{N}_0\}$ of compatible scalar products (cf. Chapter 1 and below). We put $(\cdot\,,\,\cdot)_0 \equiv (\cdot\,,\,\cdot)$, and suppose furthermore (and without loss of generality) that the norms $|\cdot|_p$, $p\in\mathbb{N}_0$, associated with the scalar products $(\cdot\,,\,\cdot)_p$ satisfy

$$|\cdot| \leq |\cdot|_1 \leq \cdots \leq |\cdot|_p \leq \cdots$$

$\mathcal{H}_p$ is the real Hilbert space obtained by completing $\mathcal{N}$ with respect to $|\cdot|_p$, and $\mathcal{H}_{-p}$ denotes the dual space of $\mathcal{H}_p$. Since $\mathcal{H}_{-p}$ is a Hilbert space, it admits a scalar product which we denote by $(\cdot\,,\,\cdot)_{-p}$. We identify $\mathcal{H}_0 \equiv \mathcal{H}$ with its dual $\mathcal{H}^*$, so that we have for every $p \in \mathbb{N}$ the continuous, dense embedding $\mathcal{H}_p \subset \mathcal{H} \subset \mathcal{H}_{-p}$. For $f \in \mathcal{H}$ considered as an element of $\mathcal{H}_{-p}$ the pairing $<f, \xi>$ with $\xi \in \mathcal{H}_p$ is equal to (f, ξ). We obtained the following chain of dense, continuous embeddings

$$\cdots \supset \mathcal{H}_{-p} \supset \cdots \supset \mathcal{H}_{-1} \supset \mathcal{H} \supset \mathcal{H}_1 \supset \cdots \supset \mathcal{H}_p \supset \cdots .$$

The assumption that the system of scalar products $\{(\cdot\,,\,\cdot)_p, p \in \mathbb{N}_0\}$ be compatible is very important (s. below), and we pause here to explain it a little further. Consider the Hilbert spaces $(\mathcal{H}_p, (\cdot\,,\,\cdot)_p)$, $(\mathcal{H}_q, (\cdot\,,\,\cdot)_q)$, and assume that $p \leq q$, so that $\mathcal{H}_q \subset \mathcal{H}_p$. One may now consider $(\cdot\,,\,\cdot)_q$ as a quadratic form with dense domain $\mathcal{H}_q$ on the Hilbert space $\mathcal{H}_p$. Then the compatibility of the scalar product is the same as the requirement that the quadratic form $((\cdot\,,\,\cdot)_q, \mathcal{H}_q)$ be closed on $\mathcal{H}_p$. (Cf., e.g., Kato (1976), or the discussion in Chapter 11). Another equivalent way of expressing compatibility is as follows. Denote by ι_p^q the continuous embedding of $\mathcal{H}_q$ into $\mathcal{H}_p$. Then the compatibility of $(\cdot\,,\,\cdot)_q$ and $(\cdot\,,\,\cdot)_p$ is equivalent to the statement that ι_p^q is injective. Compatibility of the scalar products is very important for the discussion of the projective limit of the above chain of embeddings which we consider next.

Let $\mathcal{H}_\infty$ be the projective limit of the chain constructed above of dense, continuous embeddings (cf., e.g., Gel'fand and Vilenkin (1968), Trèves (1967)). That is, as a set we define

$$\mathcal{H}_\infty := \bigcap_{p \in \mathbb{N}_0} \mathcal{H}_p .$$

Furthermore, $\mathcal{H}_\infty$ is equipped with the projective limit topology: a basis of open neighborhoods of zero $U_{p,\epsilon}$, is given by choice of $p\in\mathbb{N}_0$, and $\epsilon>0$, and the set $\{\xi\in\mathcal{H}_\infty, |\xi|_p < \epsilon\}$. A sequence $\{\xi_n, n\in\mathbb{N}\}$ converges to $\xi\in\mathcal{H}_\infty$ in this topology if and only if it converges to ξ with respect to every norm $|\cdot|_p$, $p\in\mathbb{N}_0$. As a consequence of the compatibility of the scalar products one obtains the fact that $\mathcal{H}_\infty$ is complete with respect to the projective limit topology (see, e.g., Gel'fand and Shilov (1968)). Thus $\mathcal{H}_\infty$ is a complete countably Hilbert metrizable topological vector space, and hence a locally convex Fréchet space.

It is a standard result (e.g., Gel'fand and Shilov (1968)) for such spaces that

$$\mathcal{H}_{-\infty} := \bigcup_{p\in\mathbb{Z}} \mathcal{H}_p,$$

is the dual of $\mathcal{H}_\infty$.

We now lift this structure to Fock spaces, and we refer the reader to Appendix 2 for more details on Fock spaces and the notation we shall employ.

Let $\Gamma(\mathcal{H}_p)$ be the symmetric Fock space over the complexification of $\mathcal{H}_p$, $p\in\mathbb{Z}$. First we have to show the (seemingly obvious) fact that $\Gamma(\mathcal{H}_q)$ is continuously embedded into $\Gamma(\mathcal{H}_p)$ whenever $q\geq p$. To this end we prove that $\Gamma(\iota_p^q)$ is the embedding of $\Gamma(\mathcal{H}_q)$ into $\Gamma(\mathcal{H}_p)$. Indeed, by (A.2.33) we find that $\Gamma(\iota_p^q)$ acts on a symmetric tensor $\xi_1\hat{\otimes} \cdots \hat{\otimes}\xi_n\in\Gamma^{(n)}(\mathcal{H}_q)$ of degree n, $\xi_j\in\mathcal{H}_q$, $n\in\mathbb{N}$, as

$$\Gamma(\iota_p^q)\, \xi_1\hat{\otimes} \cdots \hat{\otimes}\xi_n = \iota_p^q\xi_1\hat{\otimes} \cdots \hat{\otimes}\iota_p^q\xi_n.$$

The last tensor product is obviously the image of $\xi_1\hat{\otimes} \cdots \hat{\otimes}\xi_n$ in $\Gamma^{(n)}(\mathcal{H}_p)$. Now we use that fact that such tensor products are total in $\Gamma^{(n)}(\mathcal{H}_q)$, and that Proposition A.2.4 states that $\Gamma(\iota_p^q)$ is contraction from $\Gamma(\mathcal{H}_q)$ into $\Gamma(\mathcal{H}_p)$, and consequently continuous. Thus the restriction of $\Gamma(\iota_p^q)$ is the continuous embedding of $\Gamma^{(n)}(\mathcal{H}_q)$ into $\Gamma^{(n)}(\mathcal{H}_p)$ for every $n\in\mathbb{N}$. Using again the continuity of $\Gamma(\iota_p^q)$ we see that indeed $\Gamma(\iota_p^q)$ is the embedding of $\Gamma(\mathcal{H}_q)$ into $\Gamma(\mathcal{H}_p)$, and it is easy to see that this embedding is dense, too. Moreover, the fact that $\Gamma(\iota_p^q)$ is a contraction yields that the sequence of norms $\|\cdot\|_{\Gamma(\mathcal{H}_p)}$, $p\in\mathbb{Z}$, of the symmetric Fock spaces $\Gamma(\mathcal{H}_p)$ with scalar product $(\cdot\,,\cdot)_{\Gamma(\mathcal{H}_p)}$ satisfies

$$\|\cdot\|_{\Gamma(\mathcal{H}_p)} \leq \|\cdot\|_{\Gamma(\mathcal{H}_q)}, \quad p,q\in\mathbb{Z},\ p\leq q.$$

Thus have the following chain of dense continuous embeddings

$$\ldots \supset \Gamma(\mathcal{H}_{-p}) \supset \ldots \supset \Gamma(\mathcal{H}_{-1}) \supset \Gamma(\mathcal{H}) \supset \Gamma(\mathcal{H}_1) \supset \ldots \supset \Gamma(\mathcal{H}_p) \supset \ldots .$$

Next we give an argument that the scalar products $(\cdot\,,\cdot)_{\Gamma(\mathcal{H}_p)}$, $p\in\mathbb{N}_0$, are compatible. Consider $p,q\in\mathbb{N}_0$, $p\leq q$. Since the compatibility of the scalar products $(\cdot\,,\cdot)_p$, $(\cdot\,,\cdot)_q$ of $\mathcal{H}_p$, $\mathcal{H}_q$ respectively, is equivalent to the closedness of the densely defined quadratic form $((\cdot\,,\cdot)_q, \mathcal{H}_q)$ on the Hilbert space $\mathcal{H}_p$ (see above), we can apply the well-known representation theorem for closed quadratic forms (e.g., Kato (1976, VI.2)). This implies that there exists a positive, self-adjoint operator T_p^q on $\mathcal{H}_p$ with domain $\mathcal{H}_q$, so that for all ξ_1, $\xi_2\in\mathcal{H}_q$, we have

$$(\xi_1, \xi_2)_q = (T_p^q\xi_1, T_p^q\xi_2)_p,$$

where we have identified ξ_i, $i=1,2$, with $\iota_p^q\xi_i$ for notational simplicity. In particular, T_p^q is closed. With a similar argument as we used before to prove the embedding property of $\Gamma(\iota_p^q)$, one can now show that for the Fock spaces one has for all $F,G\in\Gamma(\mathcal{H}_q)$

$$(F, G)_{\Gamma(\mathcal{H}_q)} = (\Gamma(T_p^q)\,F, \Gamma(T_p^q)\,G)_{\Gamma(\mathcal{H}_p)},$$

(where we have suppressed the embedding $\Gamma(\iota_p^q)$). Now we use that fact that $\Gamma(T_p^q)$ is self-adjoint, and hence closed (cf., Appendix A.2). This implies immediately that the scalar products $(\cdot\,,\cdot)_{\Gamma(\mathcal{H}_q)}$ and $(\cdot\,,\cdot)_{\Gamma(\mathcal{H}_p)}$ are compatible.

As before we define $\Gamma(\mathcal{H}_\infty)$ to be the projective limit of the above constructed chain of Fock spaces. $\Gamma(\mathcal{H}_\infty)$ is a complete countably Hilbert space, and therefore a locally convex Fréchet space. Let $\Gamma(\mathcal{H}_{-\infty})$ denote the dual of $\Gamma(\mathcal{H}_\infty)$. Then

$$\Gamma(\mathcal{H}_{-\infty}) = \bigcup_{p\in\mathbb{Z}} \Gamma(\mathcal{H}_p).$$

Thus we have the following chain of dense, continuous inclusions

$$\Gamma(\mathcal{H}_{-\infty}) \supset \ldots \supset \Gamma(\mathcal{H}_{-p}) \supset \ldots \supset \Gamma(\mathcal{H}_{-1}) \supset \Gamma(\mathcal{H}) \supset$$
$$\supset \Gamma(\mathcal{H}_1) \supset \ldots \supset \Gamma(\mathcal{H}_p) \supset \ldots \Gamma(\mathcal{H}_\infty).$$

Now we construct from this chain of Fock spaces a chain of spaces of $\mathcal{N}^*$ functionals. As we have shown in Chapter 2, Theorem 2.7, there is a unitary mapping $\mathcal{I}$ from the space $(L^2) = L^2(\mathcal{N}^*, \mathcal{B}, \mu)$ onto $\Gamma(\mathcal{H})$, where μ is the Gaussian measure over $(\mathcal{N}^*,\mathcal{B})$ determined by the scalar product of $\mathcal{H}$. We identify (L^2) with its dual. Let $p\in\mathbb{N}$, and consider the following diagram

$$\begin{array}{ccccc} \Gamma(\mathcal{H}_{-p}) & \supset & \Gamma(\mathcal{H}) & \supset & \Gamma(\mathcal{H}_p) \\ \downarrow & & \downarrow \mathcal{I}^{-1} & & \downarrow \\ (L^2)_{-p} & \supset & (L^2) & \supset & (L^2)_p. \end{array}$$

Here the right arrow is given by the restriction of $\mathcal{I}^{-1}$ to the subspace $\Gamma(\mathcal{H}_p)$ of $\Gamma(\mathcal{H})$. The space $(L^2)_{-p}$ is by definition the dual space of $(L^2)_p$, and therefore isomorphic to $\Gamma(\mathcal{H}_{-p})$, and we extend the mapping $\mathcal{I}^{-1}$ to $\Gamma(\mathcal{H}_{-p})$ with image $(L^2)_{-p}$. Moreover, we equip $(L^2)_p$ and $(L^2)_{-p}$ with scalar products so that the exterior arrows become unitary, too. I.e., if $\varphi,\psi\in(L^2)_p$, and $\Phi\in(L^2)_{-p}$, then

$$(\varphi, \psi)_{2,p} = (\mathcal{I}\varphi, \mathcal{I}\psi)_{\Gamma(\mathcal{H}_p)},$$

and

$$<\Phi, \varphi> = <\mathcal{I}\Phi, \mathcal{I}\varphi>,$$

where we denoted the scalar product of $(L^2)_p$, $p\in\mathbb{Z}$, by $(\cdot\,,\,\cdot)_{2,p}$, and where the last dual pairing is the one between $\Gamma(\mathcal{H}_{-p})$ and $\Gamma(\mathcal{H}_p)$. This way we finally obtain the following chain of $\mathcal{N}^*$-functionals

$$(L^2)_{-\infty} \supset \ldots \supset (L^2)_{-p} \supset \ldots \supset (L^2) \supset \ldots \supset (L^2)_p \supset \ldots \supset (L^2)_\infty,$$

where $(L^2)_\infty$ is the pre-image of $\Gamma(\mathcal{H}_\infty)$ under $\mathcal{I}$, and $(L^2)_{-\infty}$ its dual. Again, we topologize $(L^2)_\infty$ so that $\mathcal{I}$: $\Gamma(\mathcal{H}_\infty) \rightarrow (L^2)_\infty$ becomes a homoeomorphism. Thus we have the following result.

Theorem 3.5. $(L^2)_\infty$ *is a complete countably Hilbert space, and hence a Fréchet space. The space* $(L^2)_{-\infty}$ *is the union of the Hilbert spaces* $(L^2)_p$, $p\in\mathbb{Z}$.

We shall consider the space $(L^2)_\infty$ as a space of *smooth*, $(L^2)_{-\infty}$ as a space of *generalized functionals* on $\mathcal{N}^*$.

According to our construction and Proposition 2.6, Theorem 2.7 we know that that $\varphi\in(L^2)_p$, if and only if, $\mathcal{I}\varphi = (f^{(n)}, n\in\mathbb{N}_0)$ with $f^{(n)}\in\Gamma^{(n)}(\mathcal{H}_p)\simeq\mathcal{H}_p^{\hat{\otimes}n}$, and

$$\| \varphi \|_{2,p}^2 = \sum_{n=0}^{\infty} n! \| f^{(n)} \|^2_{\mathcal{H}_p^{\otimes n}}, \tag{3.20}$$

where we denote by $\|f^{(n)}\|_{\mathcal{H}_p^{\otimes n}}$ the Hilbert–Schmidt norm (cf. Appendix A.2.) of $f^{(n)}$ as an n–multilinear form on $\mathcal{H}_{-p}$. Therefore $\varphi\in(L^2)_\infty$, if and only if it has a chaos decomposition given by a sequence $(f^{(n)}, n\in\mathbb{N}_0)$, and for every $n\in\mathbb{N}$, $f^{(n)}$ is a symmetric n–multilinear form on $\mathcal{H}_{-\infty}$, so that for all $p\in\mathbb{N}$, (3.20) is finite. A sequence $\{\varphi_n, n\in\mathbb{N}\}$ in $(L^2)_\infty$ converges in that space, if and only if it converges in every of the norms (3.20), $p\in\mathbb{N}$.

Similarly, $\Phi\in(L^2)_{-\infty}$ if and only if there exist $p\in\mathbb{N}$ so that $\Phi\in(L^2)_{-p}$, i.e., if and only if Φ is in one-to-one correspondence with a sequence $(F^{(n)}, n\in\mathbb{N}_0)\in\Gamma(\mathcal{H}_{-p})$, such that

$$\| \Phi \|_{2,-p}^2 = \sum_{n=0}^{\infty} n! \| F^{(n)} \|_{\mathcal{H}_{-p}^{\otimes n}}^2 < \infty. \tag{3.21}$$

Moreover, for all $\varphi \in (L^2)_p$ we have

$$|\langle \Phi, \varphi \rangle| \leq \| \Phi \|_{2,-p} \| \varphi \|_{2,p}, \tag{3.22}$$

The smallest such p is called the *order of* Φ.

Next we are going to prove a criterion for whether the above constructed spaces are nuclear (cf. Gel'fand and Vilenkin (1964) or Trèves (1967)). Let $\mathcal{H}_\infty$ be a complete countably Hilbert topological vector space $\mathcal{H}_\infty = \bigcap_p \mathcal{H}_p$ as above. Let us recall that $\mathcal{H}_\infty$ is called *nuclear*, if and only if for each $p \in \mathbb{N}$, there exists $q \in \mathbb{N}$ $(q \geq p)$ so that the embedding $\iota_p^q : \mathcal{H}_q \to \mathcal{H}_p$ is of Hilbert–Schmidt type, i.e. $\left\| \iota_p^q \right\|_{H.S.}^2 = \sum_k \left\| \iota_p^q e_k \right\|_{\mathcal{H}_p}^2$ is finite for some (and therefore every) complete orthonormal system $(e_k, k \in \mathbb{N})$ of $\mathcal{H}_q$.

The following results have essentially been given in Kondratiev (1980), and in Kubo and Takenaka (1980a).

Lemma 3.6. *Let* $\mathcal{H}_i$, i=1, 2, *be completions of* $\mathcal{N}$ *with respect to two compatible scalar products* $(\cdot,\cdot)_i$, i=1, 2. *Assume that* B *is a Hilbert–Schmidt operator from* $\mathcal{H}_1$ *into* $\mathcal{H}_2$ *with Hilbert-Schmidt norm* $\|B\|_{H.S.} < 1$. *Then* $\Gamma(B)$ *is Hilbert-Schmidt from* $\Gamma(\mathcal{H}_1)$ *into* $\Gamma(\mathcal{H}_2)$, *and*

$$\left\| \Gamma(B) \right\|_{H.S.} \leq \left(1 - \|B\|_{H.S.}^2\right)^{-1/2}. \tag{3.23}$$

Proof. It is a standard result (cf., e.g., Gel'fand and Vilenkin (1968)) that a Hilbert-Schmidt operator $B : \mathcal{H}_1 \to \mathcal{H}_2$, $\mathcal{H}_i$ separable, i=1,2, can be written as Bf $= \sum_k \lambda_k (f, e_k)_1 \cdot f_k$, $f \in \mathcal{H}_1$, $(e_k, k \in \mathbb{N})$, resp. $(f_k, k \in \mathbb{N})$, being CONS of $\mathcal{H}_1$, resp. $\mathcal{H}_2$, and $(\cdot\, , \cdot)_i$, i=1,2, is the scalar product of $\mathcal{H}_i$. Moreover, $\lambda_k \geq 0$ for all $k \in \mathbb{N}$

and $\sum_k \lambda_k^2 = \|B\|_{H.S.}^2 < \infty$.

From the CONS's (e_k) and (f_k) of $\mathcal{H}_1$, $\mathcal{H}_2$ resp., we form the CONS's $((n(\alpha)!)^{-1/2}\, e_\alpha,\ \alpha\in I^n,\ n\in\mathbb{N}_0)$, $((n(\alpha)!)^{-1/2}\, f_\alpha,\ \alpha\in I^n,\ n\in\mathbb{N}_0)$ of $\Gamma(\mathcal{H}_1)$, $\Gamma(\mathcal{H}_2)$ as in Appendix 2, A.2.12 to A.2.15. Then

$$\|\Gamma(B)\|_{H.S.}^2 = \sum_{n=0}^{\infty} \sum_{\alpha\in I^n} \left\|\Gamma(B)\ (n(\alpha)!)^{-1/2} e_\alpha\right\|_{\Gamma(\mathcal{H}_2)}^2$$

$$= \sum_{n=0}^{\infty} \sum_{\alpha\in I^n} (n(\alpha)!)^{-1}\ \lambda_{\alpha_1}^2 \cdots \lambda_{\alpha_n}^2\ \|f_\alpha\|_{\Gamma(\mathcal{H}_2)}^2$$

$$= \sum_{n=0}^{\infty} \sum_{\alpha\in I^n} \lambda_{\alpha_1}^2 \cdots \lambda_{\alpha_n}^2$$

$$\leq \sum_{n=0}^{\infty} \sum_{\alpha\in \mathbb{N}^n} \lambda_{\alpha_1}^2 \cdots \lambda_{\alpha_n}^2$$

$$= \sum_{n=0}^{\infty} \|B\|_{H.S.}^{2n}$$

$$= (1-\|B\|_{H.S.}^2)^{-1}$$

where we used (A.2.14). □

Theorem 3.7. *Suppose that the Hilbert–Schmidt norm of* ι_p^q *is strictly smaller than 1. Then* $(L^2)_\infty$ *is nuclear.*

Proof. It is clear that it is equivalent to prove that $\Gamma(\mathcal{H}_\infty)$ is nuclear. Fix $p\in\mathbb{N}_0$, and let $q\in\mathbb{N}_0$ with $q\geq p$. As argued above, the injection of $\Gamma(\mathcal{H}_q)$ into $\Gamma(\mathcal{H}_p)$ is given by $\Gamma(\iota_p^q)$. By hypothesis we can find $q\in\mathbb{N}_0$ such that $\left\|\iota_p^q\right\|_{H.S.}^2 < 1$. Therefore it follows from Lemma 3.6 that $\left\|\Gamma(\iota_p^q)\right\|_{H.S.}$ is finite and the theorem is proved. □

Many topologies can be introduced on $(L^2)_{-\infty}$ (cf., e.g., Appendix 5, and the literature quoted there), here we mention only the weak, strong and inductive limit topologies.

The *weak topology* σ on $(L^2)_{-\infty}$ is given by the set of neighborhoods of zero, which are defined by a finite numbers of elements $\varphi_1, \ldots, \varphi_n$ in $(L^2)_\infty$, and consist of all those $\Phi \in (L^2)_{-\infty}$ so that $|<\Phi, \varphi_i>| < 1$ for all $i=1, \ldots, n$. The convergence of a sequence $(\Phi_k, k \in \mathbb{N})$ in $(L^2)_{-\infty}$ to $\Phi \in (L^2)_{-\infty}$ in this topology (i.e., the *weak convergence*) is then given as follows: $\Phi_k \xrightarrow{w} \Phi$, as $k \to \infty$, if and only if $<\Phi_k, \varphi> \to <\Phi, \varphi>$ for every $\varphi \in (L^2)_\infty$. Since $(L^2)_\infty$ is a Fréchet space, a standard result implies the following important fact: $(L^2)_{-\infty}$ is sequentially complete in the weak topology, i.e., if for every $\varphi \in (L^2)_\infty$, $(<\Phi_k, \varphi>, k \in \mathbb{N})$ is a Cauchy sequence, then there exists a unique $\Phi \in (L^2)_{-\infty}$ so that $\Phi_k \xrightarrow{w} \Phi$.

For the definition of the strong topology we need the notion of bounded sets in $(L^2)_\infty$: a set $B \subset (L^2)_\infty$ is called *bounded*, if and only if for each neighborhood of zero U in $(L^2)_\infty$, there exists $\lambda > 0$ so that $\lambda B \subset U$. It follows that $B \subset (L^2)_\infty$ is bounded, if and only if it is bounded in every norm $\| \cdot \|_{2,p}$, $p \in \mathbb{N}_0$. An open neighborhood of zero in $(L^2)_{-\infty}$ of the *strong topology* β is defined by a bounded set $B \subset (L^2)_\infty$, and consists of those $\Phi \in (L^2)_{-\infty}$ such that

$$\sup_{\varphi \in B} |<\Phi, \varphi>| < 1.$$

A sequence $(\Phi_k, k \in \mathbb{N})$ in $(L^2)_{-\infty}$ is converging to Φ in the strong topology , $\Phi_k \xrightarrow{s} \Phi$, if and only if it converges uniformly on every bounded set in $(L^2)_\infty$. I.e., for all bounded $B \subset (L^2)_\infty$

$$\sup_{\varphi \in B} |<\Phi - \Phi_k, \varphi>| \to 0,$$

as $k \to \infty$. An analogous result as above for the weak topology holds for the strong topology because $(L^2)_\infty$ is Fréchet: $(L^2)_{-\infty}$ is sequentially complete in the strong topology.

The *inductive limit topology* τ_{ind} on $(L^2)_{-\infty}$ is defined as follows. A convex subset U of $(L^2)_{-\infty}$ is an open neighborhood of zero in τ_{ind}, if and only if for every $p \in \mathbb{N}_0$, $U \cap (L^2)_{-p}$ is an open neighborhood of zero in the Hilbert space $(L^2)_{-p}$. A sequence $(\Phi_k, k \in \mathbb{N})$ converges to Φ in $(L^2)_{-\infty}$ in this topology, if and only if there exists $p \in \mathbb{N}_0$ so that for all large enough k, $\Phi_k \in (L^2)_{-p}$, $\Phi \in (L^2)_{-p}$ and $\Phi_k \to \Phi$ in $(L^2)_{-p}$.

A priori we have the following inclusions of topologies: $\sigma \subseteq \beta \subseteq \tau_{ind}$, and the following implications of convergence of sequences: τ_{ind}-convergence is stronger than strong convergence, which in turn implies weak convergence. But it turns out that the strong and the inductive limit topologies coincide, $\tau_{ind} = \beta$. For an argument we refer the reader to Appendix 5.

3.C Constructions with Second Quantized Operators

In this section we are going to consider a special situation of the one treated in the previous section. Namely, we shall study the case where the scalar products $(\cdot\, , \cdot)_p$ of Section 3.B are given by the quadratic forms of the powers of certain self-adjoint operators on $\mathcal{H}$. At the same time, we shall construct the corresponding spaces $(L^2)_p$ somewhat more directly as subspaces of (L^2) using second quantized operators. In addition, we shall set up some formalism and results which will be helpful in subsequent chapters. In the final part of this section we shall use a different kind of second quantized operators, which will lead to spaces which have not been obtained in Section 3.B.

Let A be self-adjoint semibounded operator on $\mathcal{H} = \mathcal{H}_0$, and without loss of generality we may assume that $A \geq 1$. Suppose furthermore that $\mathcal{D}(A) \supset \mathcal{N}$, and that A leaves $\mathcal{N}$ invariant and that $\mathcal{N}$ is a core for A. Therefore $A^p \mathcal{N} \subset \mathcal{N}$ for all $p \in \mathbb{N}$. Then we can define a system of scalar products $\{(\cdot, \cdot)_p, p \in \mathbb{N}\}$ on $\mathcal{N}$ by

$$(f, g)_p := (A^p f, A^p g).$$

As before, $\mathcal{H}_p$, $p \in \mathbb{N}$, is the Hilbert space which is given by the completion of $\mathcal{N}$ with respect to the norm $|\cdot|_p$ which is defined by the scalar product $(\cdot\,,\cdot)_p$. (Thus, the operator T_p^q which we considered in Section 3.B is given by A^{q-p}.) Using the construction of Section 3.B, we obtain the corresponding spaces $(L^2)_{\pm\infty}$ of smooth and generalized functionals on $\mathcal{N}^*$. Typically $\mathcal{N}$ will be a function space and depending on the nature of A, the integral kernels of elements in $(L^2)_\infty$ will have certain analytic properties. It is clear that the choice of A will depend on the application one has in mind.

We shall now give the above announced, slightly more direct construction of the spaces $(L^2)_{\pm\infty}$.

Suppose that A is a self-adjoint operator on $\mathcal{H}$ with the above properties. We shall consider the so-called second quantization $\Gamma(A)$ of A on (L^2). It can be described as follows. Consider either of the dense subspaces $\mathcal{P}$ or $\mathcal{E}$, $\mathcal{E}_r$ of (L^2) (cf. Definition 1.8). An element $P \in \mathcal{P}$ can be written as $(x \in \mathcal{N}^*)$

$$\begin{aligned} P(x) &= \sum_n \sum_{\alpha \in \mathbb{N}^n} b_\alpha \langle x, \eta_{\alpha_1}\rangle \cdots \langle x, \eta_{\alpha_n}\rangle \\ &= \sum_n \sum_{\alpha \in \mathbb{N}^n} a_\alpha :\langle x, \xi_{\alpha_1}\rangle \cdots \langle x, \xi_{\alpha_n}\rangle:, \end{aligned} \tag{3.24}$$

where the sums extend only over a finite number of non-zero terms, and where a_α, $b_\alpha \in \mathbb{C}$, η_{α_k}, $\xi_{\alpha_k} \in \mathcal{N}$, $\alpha = (\alpha_1, \ldots \alpha_n) \in \mathbb{N}^n$, $k=1,2,\ldots,n$. For such $P \in \mathcal{P}$, set

$$\Gamma(A)P(x) = \sum_n \sum_{\alpha \in \mathbb{N}^n} a_\alpha :\langle x, A\xi_{\alpha_1}\rangle \cdots \langle x, A\xi_{\alpha_n}\rangle:. \tag{3.25}$$

Alternatively, if E belongs to $\mathcal{E}$ or $\mathcal{E}_r$, it is of the form

$$E = \sum_n a_n :e^{\iota\langle\cdot,\, \xi_n\rangle}:,$$

with $a_n \in \mathbb{C}$ being non-zero for only a finite number of $n \in \mathbb{N}$, and $\iota = i$ or 1, $\xi_n \in \mathcal{N}$. Then we set

$$\Gamma(A)E = \sum_n a_n : e^{\iota<\cdot, A\xi_n>} :. \tag{3.26}$$

Either way, we have a densely defined operator, and it is straightforward to check that $\Gamma(A)$ is symmetric. Computing the $\mathcal{S}$–transforms of (3.24) and (3.25), one immediately observes that $\Gamma(A)$ coincides with the inverse $\mathcal{S}$–transform of the second quantization of A on the symmetric Fock space $\Gamma(\mathcal{H})$, as described in Appendix A.2 (denoted there by the same symbol). It is well-known that $\Gamma(A)$ is essentially self–adjoint on $\mathcal{P}$ (and therefore also on $\mathcal{E}$) (cf., e.g., Cook (1953), Simon (1974)). In the sequel we shall identify $\Gamma(A)$ with its self—adjoint extension on (L^2). Also, we note that $A \geq 1$ implies $\Gamma(A) \geq 1$ (cf., e.g., Reed and Simon (1972, Chap. VIII)).

Note that by our hypotheses on A, $\Gamma(A)$ leaves the algebras $\mathcal{P}$, $\mathcal{E}_{(r)}$ and $\mathcal{A}$ (cf. Definition 1.8) invariant. Therefore, for $p \in \mathbb{N}_0$, we may define the following norm on $\mathcal{P}$, $\mathcal{E}$, $\mathcal{E}_r$ and $\mathcal{A}$:

$$\| \varphi \|_{2,p,A} = \| \Gamma(A)^p \varphi \|_2 = \| \Gamma(A^p)\varphi \|_2. \tag{3.27}$$

Usually, if there is no danger of confusion we shall drop the subscript A, in order to simplify the notation. Note that the norm (3.27) is a Hilbertian norm, so that the space $(L^2)_p$, $p \in \mathbb{N}_0$, defined as the completion of $\mathcal{P}$ with respect to this norm is a Hilbert space. By construction we have that $\varphi \in (L^2)_p$, if and only if φ is in one-to-one correspondence with a sequence $(f^{(n)}, n \in \mathbb{N}_0)$ of n–multilinear forms $f^{(n)}$ on $\mathcal{H}^* \equiv \mathcal{H}$, so that

$$\| \varphi \|_{2,p}^2 = \sum_{n=0}^{\infty} n! \| (A^{\otimes n})^p f^{(n)} \|_{H.S.}^2$$

$$= \sum_{n=0}^{\infty} n! \| f^{(n)} \|_{\mathcal{H}_p^{\otimes n}}^2,$$

is finite. Here, $\| \cdot \|_{H.S.}$ is the Hilbert–Schmidt norm of n–multilinear forms on $\mathcal{H}$, while $\|\cdot\|_{\mathcal{H}_p^{\otimes n}}$ is the Hilbert-Schmidt norm on $\mathcal{H}_{-p}$. Comparison with (3.21) (and the remark that $\mathcal{SP}$ is dense in $\Gamma(\mathcal{H})$) shows that the Hilbert spaces $(L^2)_p$ defined here and in the preceding section are the same. (In this case, the compatibility of the system of norms (and the associated scalar products) is obvious.) Therefore we obtain in this special situation the same spaces $(L^2)_{\pm\infty}$ of generalized and test functionals over $\mathcal{N}^*$.

The following result will be later be useful.

Proposition 3.8. $\mathcal{E}$, $\mathcal{E}_r$, *and* $\mathcal{A}$ *are dense in every* $(L^2)_p$, $p \in \mathbb{N}$, *and in* $(L^2)_\infty$.

Proof. As argued above, $\Gamma(A)$ leaves the algebras $\mathcal{E}$, $\mathcal{E}_r$ and $\mathcal{A}$ invariant. Because they are subspaces of (L^2) (cf. Propositions 1.6 and 1.7), it follows that they are subspaces of every $(L^2)_p$, and of $(L^2)_\infty$. Since $\mathcal{A} \supset \mathcal{E}_r$, we only need to show the density of $\mathcal{E}_{(r)}$. We shall consider only the case $\mathcal{E}$, density of $\mathcal{E}_r$ is proved by the same arguments.

Let $p \in \mathbb{N}$, and consider $\varphi \in (L^2)_p$. Then we have $\Gamma(A)^p \varphi \in (L^2)$. By Theorem 1.9, we can find a sequence $\{\psi_n, n \in \mathbb{N}\}$ in $\mathcal{E}$ which converges to $\Gamma(A)^p \varphi$ in (L^2). Set $\varphi_n = \Gamma(A^{-p})\psi_n = \Gamma(A)^{-p}\psi_n$, and note that because of (3.26) and the invertibility of A on $\mathcal{N}$, we have $\varphi_n \in \mathcal{E}$. Clearly, φ_n converges to φ in $(L^2)_p$.

Now we prove the density of $\mathcal{E}$ in $(L^2)_\infty$. Let $\Phi \in (L^2)_{-\infty}$ be such that $\langle \Phi, \varphi \rangle = 0$ for all $\varphi \in \mathcal{E}$. Since $\Phi \in (L^2)_{-p}$ for some $p \in \mathbb{N}_0$ and we have proved that $\mathcal{E}$ is dense in $(L^2)_p$, continuity of Φ on $(L^2)_p$ implies that $\Phi = 0$ on $(L^2)_p$, and in particular on $(L^2)_\infty$. Since $(L^2)_\infty$ is a Fréchet space, we may apply the Hahn–Banach theorem with the conclusion that $\mathcal{E}$ is dense in $(L^2)_\infty$. □

In quite a similar way, one proves the following result concerning $(L^2)_{-\infty}$ and we leave the details to the interested reader.

Proposition 3.9. *The spaces* $\mathcal{A}$, $\mathcal{P}$, $\mathcal{E}$, $(L^2)_{+\infty}$, (L^2) *are strongly sequentially dense in* $(L^2)_{-\infty}$.

Let us now turn to some examples. In all of them we choose white noise as the basic probability space. However, generalizations to white noise with higher dimensional time parameter or other Gaussian spaces based on function spaces $\mathcal{N} \subset \mathcal{H}$ are immediate.

Example 3.10. Consider the orthonormal basis of $L^2(\mathbb{R},dt)$ given by the Hermite functions e_k, $k \in \mathbb{N}_0$, cf. Appendix A.1. Note that for all $k \in \mathbb{N}_0$, $e_k \in \mathcal{S}(\mathbb{R})$. Consider the operator H, defined on $\mathcal{S}(\mathbb{R})$ by

$$Hf(u) := -f''(u)+(u^2+1)f(u), \quad u \in \mathbb{R} \tag{3.28}$$

H is essentially self–adjoint on $L^2(\mathbb{R},du)$ with $\mathcal{S}(\mathbb{R})$ being a core (e.g., Glimm and Jaffe (1981)) and we denote the self–adjoint extension of H by the same symbol. Obviously, H leaves $\mathcal{S}(\mathbb{R})$ invariant. The spectrum of H is the set $\{2k+2,\ k \in \mathbb{N}_0\}$ and the Hermite functions are the eigenfunctions of H:

$$H\, e_k = (2k+2)\, e_k. \tag{3.29}$$

(We have added 1 to the conventional Hamiltonian of the harmonic oscillator for reasons which shall become clear later).

H fulfills all our above requirements and we therefore obtain a chain of spaces of functionals on $\mathcal{S}'(\mathbb{R})$ as above. Because of the particular importance of this example the spaces $(L^2)_p$ will be denoted by $(\mathcal{S})_p$, $p \in \mathbb{Z}$. The projective limit space is denoted by $(\mathcal{S})$ and its dual by $(\mathcal{S})^*$. We call elements in $(\mathcal{S})$ *white noise test functionals*,

and elements in $(\mathcal{S})^*$ *generalized white noise functionals.*

The triple

$$(\mathcal{S})^* \supset (L^2) \supset (\mathcal{S}) \tag{3.30}$$

will be discussed in detail in the next chapter. It will re-appear throughout the whole book. Here we just state that the integral kernels (cf. chapter 2) $f^{(n)}$, $n \in \mathbb{N}_0$, of an element $\varphi \in (\mathcal{S})$ are test functions in the symmetric Schwartz space $\mathcal{S}(\hat{\mathbb{R}}^n)$. (More precisely, they have a representative in this space). The generalized 'integral' kernels of elements in $(\mathcal{S})^*$ are symmetric tempered distributions.

Example 3.11. Here we take the self-adjoint extension of the differential operator $A = -\frac{d^2}{du^2}+1$. Thus $\mathcal{H}_\infty$ is the intersection of all Sobolev spaces $\mathcal{H}_p(\mathbb{R})$, $p \in \mathbb{N}$. In this case the Sobolev embedding theorem (cf., e.g., Adams (1975)) implies that the integral kernels $f^{(n)}$, $n \in \mathbb{N}_0$, of the elements in $(L^2)_\infty$ (have representatives which) are C^∞-functions, which together with all their derivatives are bounded, and belong to all $L^p(\mathbb{R}^n)$, $p \geq 2$.

In the set-up of this section, where we define the spaces $(L^2)_{\pm\infty}$ by second quantized operators $\Gamma(A)$, nuclearity of $(L^2)_{+\infty}$ is specially convenient to prove. By Theorem 3.7 we find the following result.

Corollary 3.12. *Assume that* A *is a self-adjoint operator on* $\mathcal{H}$ *with* $\mathcal{N} \subset \mathcal{D}(A)$ *a core of* A, *and* $A : \mathcal{N} \to \mathcal{N}$. *If* $\inf \sigma(A) > 1$, *and for some* $k \in \mathbb{N}$ *the operator* A^{-k} *is Hilbert–Schmidt, then* $(L^2)_\infty$ *is nuclear.*

Proof. It is an easy calculation to show that for $q \geq p$, $q-p \geq k$, $\| \iota_p^q \|_{H.S.} = \| A^{p-q} \|_{H.S.}$. Moreover, $\inf \sigma(A) > 1$ implies that the operator norm of A^{-m} converges to zero as m tends to infinity. Therefore we only have to observe that $\| A^{-(k+m)} \|_{H.S.} \leq \| A^{-k} \|_{H.S.} \cdot \| A^{-1} \|^m$ to conclude the proof. □

In particular we have shown that $(\mathcal{S})$, constructed in Example 3.10, is a nuclear space.

Among the many nice properties of nuclear spaces and their duals, we want to mention here only the following. A bounded closed subset of a nuclear space is compact. This entails that in the dual of a nuclear space weak and strong convergence of *sequences* coincide. In particular, in $(\mathcal{S})^*$ all the three types of convergence of sequences discussed in Section 3.B are equivalent. For other properties we refer the reader to the standard literature.

Another interesting class of spaces arises if in the above construction of $(L^2)_\infty$ based on an operator A on $\mathcal{H}$ we replace the second quantized operator $\Gamma(A)$ by the operator $d\Gamma(A)$ which is defined as follows. Assume that A satisfies the above assumptions, and let $P \in \mathcal{P}$ be of the form:

$$P(x) = \sum_n \sum_{\alpha \in \mathbb{N}^n} a_\alpha :\langle \cdot, \xi_{\alpha_1}\rangle \cdots \langle \cdot, \xi_{\alpha_n}\rangle:.$$

Then we set

$$d\Gamma(A)P(x) = \sum_n \sum_{\alpha \in \mathbb{N}^n} a_\alpha \sum_{k=1}^{n} :\langle \cdot, \xi_{\alpha_1}\rangle \cdots \langle \cdot, A\xi_{\alpha_k}\rangle \cdots \langle \cdot, \xi_{\alpha_n}\rangle:.$$

We leave it as an exercise to the interested reader, to verify that this definition of $d\Gamma(A)$ leads to the operator with the same symbol defined on the symmetric Fock space in Appendix A.2. $d\Gamma(A)$ is essentially self-adjoint on $\mathcal{P}$ (e.g. Cook (1953), Simon (1974)), and we shall denote its self-adjoint extension to (L^2) by $d\Gamma(A)$, too. We remark in passing the following formula. For $\xi \in \mathcal{N}$, and $\lambda \in \mathbb{C}$,

$$d\Gamma(A): e^{\lambda\langle \cdot, \xi\rangle}: = (\lambda\langle \cdot, A\xi\rangle - \lambda^2(\xi, A\xi)) :e^{\lambda\langle \cdot, \xi\rangle}:.$$

For $k \in \mathbb{N}$ we set

$$\mathcal{D}_A^{2,k} = \mathcal{D}(d\Gamma(A)^{k/2}),$$

where $\mathcal{D}(d\Gamma(A)^r)$, $r \in \mathbb{R}_+$, is the (L^2)-domain of $d\Gamma(A)^r$. $\mathcal{D}_A^{2,k}$ is the same as the completion of $\mathcal{P}$ under the norm given by the scalar product

$$(\varphi, \psi)_{2,k} := (\varphi, (1 + d\Gamma(A))^k \psi)_2,$$

where $(\cdot\,,\,\cdot)_2$ denotes the scalar product of (L^2). As before, we obtain a chain of dense continuous embeddings

$$\ldots \supset \mathfrak{D}_A^{2,-k} \supset \ldots \supset (L^2) \supset \ldots \supset \mathfrak{D}_A^{2,k} \supset \ldots,$$

where $\mathfrak{D}_{A,}^{2,-k}$ denotes the dual of $\mathfrak{D}_A^{2,k}$. The projective limit of the chain will be denoted by $\mathfrak{D}_A^{2,\infty}$, while its dual is denoted by $\mathfrak{D}_A^{2,-\infty}$.

The analogues of Examples 3.10 and 3.11 are easily worked out, and are left as exercises. In fact, it is clear that the integral kernels of elements in $\mathfrak{D}^2_{A,\infty}$ for these cases have the same analytic properties: as in Examples 3.10 and 3.11 they must consist of functions from a space which is stable under any number of applications of A in each argument. However there is an essential difference concerning the properties of the *sequence* $(f^{(n)}, n\in\mathbb{N}_0)$ of integral kernels. Let us explain this for the example where we choose A as in Example 3.10, i.e., $A = H$, with H given in (3.57). Due to the fact that $A\geq 2$ we find that $\| \Gamma(A)^{-1}|_{\mathcal{H}^{(n)}} \| = 2^{-n}$. As we shall see in the next chapter, this implies that for $\varphi\in(\mathcal{S})$ having kernels $(f^{(n)}, n\in\mathbb{N}_0)$, and any $p\in\mathbb{N}_0$, we have for all $c>0$,

$$\sum_{n=0}^{\infty} c^n\, n!\, \| f^{(n)} \|_{2,p}^2 < \infty, \tag{3.31}$$

where $\| f^{(n)} \|_{2,p} \equiv \| (A^{\otimes n})^p f^{(n)} \|_2$. On the other hand, for $\psi\in\mathfrak{D}_A^{2,\infty}$ with chaos expansion given by $(g^{(n)}, n\in\mathbb{N}_0)$, we only find convergence of polynomial type in n. Namely for all $p, k\in\mathbb{N}$,

$$\sum_{n=0}^{\infty} n^k\, n!\, \| g^{(n)} \|_{2,p}^2 < \infty.$$

Note that we would not obtain any 'uniform decrease in n' as in (3.31) if we did not choose H in (3.28) in such a way that $H>1$. This is one of the reasons why we have shifted the spectrum of the usual Hamiltonian of the harmonic oscillator by

1. Moreover, this property provided the nuclearity of $(\mathcal{S})$. Also, in the next chapter we shall see that the exponential decrease in n in (3.31) leads to another important property of $(\mathcal{S})$: it is an *algebra under pointwise multiplication.* However, this is not true for $\mathcal{D}_A^{2,\infty}$. The following simple argument has been provided by Meyer (1983).

Consider the simplified situation given by the choice $\mathcal{N} = \mathcal{H} = \mathbb{R}$. Consequently $(L^2) = L^2(\mathbb{R}, d\mu)$ with $d\mu(u) = (2\pi)^{-1/2}\exp(-\frac{1}{2}u^2)du$ and $\Gamma(\mathcal{H}) = l^2$, the Hilbert space of square summable complex sequences (with weight n!): $\Gamma^{(n)}(\mathcal{H}) = \mathbb{C}$. An element $\varphi \in (L^2)$ has the decomposition

$$\varphi(u) = \sum_{n=0}^{\infty} a_n' : u^n :,$$

where $u \mapsto \, : u^n :$ is the usual Hermite polynomial of degree n, cf. Appendix A.1. Moreover the coefficients $a_n' \in \mathbb{C}$ satisfy

$$\sum_{n=0}^{\infty} n! \mid a_n' \mid^2 < \infty.$$

Put $a_n := \sqrt{n!}\, a_n'$. Now let A be the operator which is multiplication by $c \geq 1$ on $\mathbb{R}$. For φ in the domain of $d\Gamma(A)$, we have

$$d\Gamma(A)\, \varphi\, (u) = \sum_{n=0}^{\infty} (nc)\, a_n\, e_n(u),$$

where e_n is the normalized Hermite polynomial $e_n(u) = (n!)^{-1/2} : u^n :$. If we now require that for all $k \in \mathbb{N}$, $d\Gamma(A)^k \varphi$ belongs to (L^2), then this means that for all $k \in \mathbb{N}$, the sequence $(a_n,\, n \in \mathbb{N}_0) \in \Gamma(\mathcal{H})$ has the property

$$\sum_{n=0}^{\infty} n^k \mid a_n \mid^2 < \infty.$$

In other words $(a_n,\, n \in \mathbb{N}_0)$ is rapidly decreasing (cf., e.g., Reed and Simon (1972, Chapt. V)). Write

$$\varphi(u) = e^{\frac{1}{4}u^2} \sum_{n=0}^{\infty} a_n\, e_n(u)\, e^{-\frac{1}{4}u^2}$$

$$= e^{\frac{1}{4}u^2}\, \psi(u), \tag{3.32}$$

where ψ belongs to $\mathcal{S}(\mathbb{R})$ (complexified), because of the rapid decrease of (a_n). (For a proof see Reed and Simon (1972), and use relation (A.1.12) for the comparison between the slightly different conventions of writing the Hermite functions $e_n(u)\cdot\exp(-\frac{1}{4}u^2)$). In other words: in this simplified case $\varphi\in\mathcal{D}_A^{2,\infty}$, if and only if φ is of the form (3.32) with $\psi\in\mathcal{S}(\mathbb{R})$. In particular we may choose $\psi(u) = \exp(-\frac{1}{8}u^2)$. But then clearly φ^2 does not belong to $\mathcal{D}_A^{2,\infty}$ since it does not even belong to (L^2).

In Meyer's article (1983) it was also proposed how to cure the disadvantage that the above defined space is not an algebra: in our construction of the space of smooth functionals we have to take into account additionally all L^p–norms of $d\Gamma(A)^k\varphi$.

For simplicity, let us from now on assume in addition that A has a complete set $\{e_n,\ n\in\mathbb{N}\}$ of normalized eigenvectors e_n in $\mathcal{H}$, so that $e_n\in\mathcal{N}$, $n\in\mathbb{N}$. Note that the algebra $\mathcal{P}_A$ generated by the random variables $\{Xe_n = \langle\cdot\,,e_n\rangle,\ n\in\mathbb{N}\}$ is dense in all (L^p), $p\geq 1$ (cf. the remark after Theorem 1.9). It is also clear that for all $k\in\mathbb{Z}$, the operator $(1 + d\Gamma(A))^{k/2}$ maps $\mathcal{P}_A$ into itself. Thus we may define the Banach spaces $\mathcal{D}_A^{p,k}$, $p\geq 1$, $k\in\mathbb{Z}$, as the completion of $\mathcal{P}_A$ under the norms

$$|||\,\varphi\,|||_{A,p,k} := \left\| (1 + d\Gamma(A))^{k/2}\varphi \right\|_p,\quad \varphi\in\mathcal{P}_A.$$

In order to establish inclusion relations between the spaces $\mathcal{D}_A^{p,k}$, we consider the semigroup $\{P_{A,t},\ t\geq 0\}$ generated by $d\Gamma(A)$ on (L^2): $P_{A,t} := \exp(-t\,d\Gamma(A))$, $t\geq 0$. Note that we have $P_{A,t} = \Gamma(e^{-tA})$, $t\geq 0$ (cf. also Appendix

2). Many properties are known for such semigroups and we refer to Glimm and Jaffe (1970), Nelson (1973), Simon (1974), and references quoted there. Moreover, for the special case where A is the identity on $\mathcal{H} = L^2(\mathbb{R})$, a number of results are stated in Chapter 7.A. Here we only mention the following two facts: (i) the semigroup $\{P_{A,t}, t\in\geq 0\}$ is strongly continuous on (L^2), and (ii) for all $t\geq 0$, $P_{A,t}$ extends (from $\mathcal{P}_A$) to a contraction on all (L^p), $1\leq p\leq +\infty$. (i) follows from the fact that $\mathrm{spec}(A)\geq 0$ implies that $\mathrm{spec}(d\Gamma(A))\geq 0$. (ii) follows from the fact that $P_{A,t}$, $t\geq 0$, is positivity preserving, self–adjoint, and $P_{A,t}1 = 1$ (cf., e.g., Simon (1974) for a detailed proof). Note that for all $\lambda>0$ we have

$$\lambda^{-1/2} = \pi^{-1/2} \int_0^\infty e^{-s\lambda}\, s^{-1/2}\, ds.$$

Thus the spectral theorem entails that on (L^2) we get the representation

$$(1 + d\Gamma(A))^{-1/2} = \pi^{-1/2} \int_0^\infty e^{-s}\, P_{A,s}\, s^{-1/2}\, ds,$$

and on $\mathcal{P}_A$ this representation holds pointwise. Hence, for all $\varphi\in\mathcal{P}_A$, $p\geq 1$, we can estimate in the following way.

$$\| (1 + d\Gamma(A))^{-1/2}\varphi \|_p \leq \pi^{-1/2} \int_0^\infty e^{-s}\, \| P_{A,s}\varphi \|_p\, s^{-1/2}\, ds$$

$$\leq \| \varphi \|_p\, \pi^{-1/2} \int_0^\infty e^{-s} s^{-1/2}\, ds$$

$$= \| \varphi \|_p,$$

where we used the above mentioned fact that $P_{A,t}$ is a contraction on all (L^p). Thus also $(1 + d\Gamma(A))^{-1/2}$ extends to a contraction on all (L^p), $p\geq 1$. In particular, we find that for all $p\geq 1$, $k\in\mathbb{Z}$, $\varphi\in\mathcal{D}_A^{p,k}$,

$$|||\,\varphi\,|||_{p,k} \le |||\,\varphi\,|||_{p,k+1}.$$

On the other hand, for all $q \ge p$, $k \in \mathbb{Z}$,

$$|||\,\varphi\,|||_{p,k} \le |||\,\varphi\,|||_{q,k}, \quad \varphi \in \mathfrak{D}_A^{q,k},$$

by an application of Hölder's inequality. Thus we obtain the following system of dense, continuous embeddings.

$$\begin{array}{ccccccc}
(L^1) & \supset & (L^2) & \supset \ldots \supset & (L^p) & \supset \ldots & \\
\cup & & \cup & & \cup & & \\
\mathfrak{D}_A^{1,1} & \supset & \mathfrak{D}_A^{2,1} & \supset \ldots \supset & \mathfrak{D}_A^{p,1} & \supset \ldots & \\
\cup & & \cup & & \cup & & \\
\vdots & & \vdots & \ddots & \vdots & & \\
\cup & & \cup & & \cup & & \\
\mathfrak{D}_A^{1,k} & \supset & \mathfrak{D}_A^{2,k} & \supset \ldots \supset & \mathfrak{D}_A^{p,k} & \supset \ldots & \\
\cup & & \cup & & \cup & & \\
\vdots & & \vdots & & \vdots & \ddots &
\end{array} \tag{3.33}$$

It is obvious, that for $p>1$, $k \in \mathbb{Z}$, the dual $(\mathfrak{D}_A^{p,k})^*$ of $\mathfrak{D}_A^{p,k}$ is given by $\mathfrak{D}_A^{q,-k}$, where $q>1$ is such that $p^{-1}+q^{-1} = 1$.

We would like to define a space $\mathfrak{D}_A$ of test functionals as the projective limit of the system (3.33). To this end, we need to establish the compatibility of the system of norms $|||\cdot|||_{p,k}$. Since $(1 + d\Gamma(A))^{k/2}$, $k \in \mathbb{Z}$, is densely defined (on $\mathcal{P}_A$) on all (L^p), $p \ge 1$, and self-adjoint on (L^2), $(1 + d\Gamma(A))^{k/2}$ is closable on all (L^p), $p>1$. The closures will be denoted by the same symbol (there is no danger of confusion). Closedness of $(1 + d\Gamma(A))^{k/2}$ on (L^p), $p>1$, clearly entails the compatibility of the system $\{|||\cdot|||_{p,k}, p>1, k\in\mathbb{Z}\}$. Thus we define the space $\mathcal{D}_A$ as the projective limit of the family $\{\mathcal{D}_A^{p,k}, p>1, k\in\mathbb{Z}\}$. Obviously, $\mathcal{D}_A$ is a Fréchet space. $\mathcal{D}_A^*$ denotes the dual of $\mathcal{D}_A$, and we have (cf., e.g., Gel'fand and Shilov (1968))

$$\mathcal{D}_A^* = \bigcup_{p>1,k\in\mathbb{N}_0} \mathcal{D}_A^{p,-k}. \tag{3.34}$$

A particular role is played by the dual pair of spaces which arises by the choice A = 1 (identity on $\mathcal{H}$). In this case, $d\Gamma(A)$ is equal to the number operator N (or to the negative of the Ornstein–Uhlenbeck operator). For this choice, we simply omit the subscript "A". For example $\mathcal{D}^{p,k}$, $p>1$, $k\in\mathbb{Z}$, has the norm

$$|||\,\varphi\,|||_{p,k} = \|\,(1+N)^{k/2}\varphi\,\|_p, \quad \varphi\in\mathcal{D}^{p,k}. \tag{3.35}$$

In the white noise case, $\mathcal{N} = \mathcal{S}(\mathbb{R})$ and $\mathcal{H} = L^2(\mathbb{R},du)$, $\mathcal{D}$ is the white noise version of the space investigated by Meyer (1982, 1983). $\mathcal{D}$ and $\mathcal{D}^*$ have been used by Watanabe (1983, 1984) and Sugita (1985) for a reformulation of the Malliavin calculus in terms of generalized Brownian functionals. We shall give an account of the ideas of Meyer and Watanabe in Chapter 7. We shall prove there that $\mathcal{D}$ is an algebra. The method of Watanabe (1983, 1984) allows then to construct an important large class of elements in $\mathcal{D}^*$, generalizing Example 3.4. Therefore we omit examples of generalized functionals in $\mathcal{D}^*$ here.

4. The Spaces $(\mathcal{S})$ and $(\mathcal{S})^*$

In this chapter we analyze the spaces $(\mathcal{S})$, $(\mathcal{S})^*$ of Example 3.10. In particular, we review results from the articles of Hida, Potthoff and Streit (1988), Kubo and Yokoi (1989), Kubo and Takenaka (1980b), Kuo, Potthoff and Streit (1991), Kuo, Potthoff and Yan (1992), Lee (1991), Meyer and Yan (1988, 1989), Potthoff and Streit (1990, 1991a,b), Potthoff and Yan (1992), Yan (1990a,b,c,d), and Yokoi (1990).

This chapter contains the following sections. In Section 4.A we establish the basic properties of the spaces $(\mathcal{S})$ and $(\mathcal{S})^*$. Positive generalized functionals in $(\mathcal{S})^*$ are defined and investigated in Section 4.B. In Section 4.C we shall prove results which characterize the spaces $(\mathcal{S})$ and $(\mathcal{S})^*$ in terms of the $\mathcal{S}$– and $\mathcal{T}$–transforms of their elements. Also, we shall give a number of applications of these characterizations. In Section 4.D we shall review the results of Lee (1991) concerning the analytic version of elements in $(\mathcal{S})$ and its consequences.

4.A Basic Properties and Examples

Let us first recall some of the notation established in Chapter 3. The scalar product of $L^2(\mathbb{R}^n,du)$, $n\in\mathbb{N}$, where du denotes Lebesgue measure, is denoted by $(\cdot\,,\cdot)_2$, the corresponding norm by $|\cdot|_2$. H is the self–adjoint extension of the differential operator

$$-\frac{d^2}{du^2} + (1+u^2)$$

on $\mathcal{S}(\mathbb{R})$ to $L^2(\mathbb{R},du)$. By $|\cdot|_{2,p}$, $p\in\mathbb{R}$, we denote the norm given by $|H^p\cdot|_2$. More generally, for a function $f^{(n)}$ on $\mathbb{R}^n$, $n\in\mathbb{N}$, we denote by $|f^{(n)}|_{2,p}$, $p\in\mathbb{R}$, the norm $|(H^{\otimes n})^p f^{(n)}|_2$. For $p\geq 0$, $\mathcal{S}_p(\mathbb{R}^n)$ denotes the completion of $\mathcal{S}(\mathbb{R}^n)$ with respect to the norm $|\cdot|_{2,p}$, by $\mathcal{S}_{-p}(\mathbb{R}^n)$ we denote the dual of $\mathcal{S}_p(\mathbb{R}^n)$. The natural norm of $\mathcal{S}_{-p}(\mathbb{R}^n)$ is given by $|\cdot|_{2,-p}$. Note that $\mathcal{S}(\mathbb{R}^n)$ is the projective limit of the spaces $\mathcal{S}_p(\mathbb{R}^n)$, and that $\mathcal{S}'(\mathbb{R}^n)$ is the union of the spaces $\mathcal{S}_{-p}(\mathbb{R}^n)$ (cf. Appendix A.5).

If φ is a function on $\mathcal{S}'(\mathbb{R})$, and $p \in \mathbb{R}$, we set $\| \varphi \|_{2,p} = \| \Gamma(H)^p \varphi \|_2$, where $\| \cdot \|_2$ is the norm of (L^2). $(\mathcal{S})_p$ is by definition the completed (L^2)–domain of $\Gamma(H)^p$, and the space $(\mathcal{S})$ of white noise test functionals is the projective limit of the spaces $(\mathcal{S})_p$, $p \in \mathbb{N}$. $(\mathcal{S})^*$ is the dual of $(\mathcal{S})$, and it is equal to the union of the spaces $(\mathcal{S})_{-p}$, $p \in \mathbb{N}_0$, $(\mathcal{S})_{-p}$ being the dual of $(\mathcal{S})_p$.

The results about $(\mathcal{S})$ which are established in Chapter 3 are collected in the following theorem.

Theorem 4.1. *$(\mathcal{S})$ is a nuclear Fréchet space. Furthermore, $\mathcal{E}$, $\mathcal{E}_r$, $\mathcal{A}$ are dense subspaces of $(\mathcal{S})$.*

Also we recall the following property from Chapter 3.

Proposition 4.2. *Assume that φ has a chaos decomposition given by the sequence $(f^{(n)}, n \in \mathbb{N})$. Then φ belongs to $(\mathcal{S})_p$, $p \in \mathbb{R}$, if and only if*

$$\| \varphi \|_{2,p}^2 = \sum_{n=0}^{\infty} n! \, | f^{(n)} |_{2,p}^2 < \infty. \tag{4.1}$$

Our first result in this chapter characterizes the integral kernels $f^{(n)}$, $n \in \mathbb{N}_0$, of an element in $(\mathcal{S})$.

Proposition 4.3. *Let $\varphi \in (\mathcal{S})$. Then for each $n \in \mathbb{N}$, the integral kernel $f^{(n)}$ of φ has a representative which belongs to $\mathcal{S}(\hat{\mathbb{R}}^n)$.*

Remark. As before, we shall denote $L^2(\mathbb{R})$ and its complexification by the same symbol. As indicated in Proposition 4.3, we have a chaos decomposition of $\varphi \in (\mathcal{S})$ with integral kernels $f^{(n)}$ in the *complexified* spaces $\mathcal{S}(\hat{\mathbb{R}}^n)$. In particular, $f^{(1)}$ belongs to the complexified Schwartz space $\mathcal{S}(\mathbb{R})$. However, also in this case it will be convenient not to distinguish the spaces in notation: Elements $f^{(n)}$, $F^{(n)}$, etc. appearing in chaos decompositions will in general be complex. On the other hand, unless mentioned otherwise, the arguments $\xi, \eta, \ldots \in \mathcal{S}(\mathbb{R})$, and $x, y, \ldots \in \mathcal{S}'(\mathbb{R})$ of functions will be real.

Proof of Proposition 4.3. $\varphi\in(\mathcal{S})$ implies that $\varphi^{(n)}\in(\mathcal{S})_p$ for all p, $n\in\mathbb{N}$. We have shown in Chapter 2, that as an element of (L^2), $\varphi^{(n)}$ has the representation

$$\varphi^{(n)}(x) = \langle : x^{\otimes n} :, f^{(n)} \rangle,$$

with $f^{(n)}\in L^2(\hat{\mathbb{R}^n})$. We may expand $f^{(n)}$ in $L^2(\hat{\mathbb{R}^n},du)$ as follows (cf. also Chapter 2)

$$f^{(n)} = \sum_{\alpha\in I^n} a_\alpha^{(n)}\, e_\alpha,$$

where we choose as an orthonormal basis $(e_k,\ k\in\mathbb{N}_0)$ of $L^2(\mathbb{R},du)$ the Hermite functions e_k, which are eigenfunctions of H with eigenvalues $2(k+1)$ (cf. (3.28)). For $\alpha\in I^n$, e_α is defined as in (A.2.13). The square of the $(\mathcal{S})_p$–norm of $\varphi^{(n)}$ is equal to

$$\sum_{\alpha\in I^n} n(\alpha)!\ |\, a_\alpha^{(n)} |^2\ (2^n(\alpha+1))^{2p},$$

where $(\alpha+1) = (\alpha_1+1)(\alpha_2+1)\ \cdots\ (\alpha_n+1)$. Since our hypothesis implies that the last sum is finite for all $p\in\mathbb{N}$, we immediately find that for all $p\in\mathbb{N}$

$$\sum_{\alpha\in I^n} |\, a_\alpha^{(n)} |^2\ (\alpha+1)^{2p} < \infty.$$

In other words, for each $n\in\mathbb{N}$ the sequence $(a_\alpha^{(n)},\ \alpha\in I^n)$ is rapidly decreasing, cf. Reed and Simon (1972). Therefore (loc. cit.), $f^{(n)}$ has a (unique) version belonging to $\mathcal{S}(\mathbb{R}^n)$. □

In the sequel we shall usually identify $f^{(n)}$ with its version in $\mathcal{S}(\mathbb{R}^n)$, $n\in\mathbb{N}$.

Denote $N := d\Gamma(1)$, where 1 is the identity on $L(\mathbb{R}, du)$. N is called the *number* or *Ornstein–Uhlenbeck operator.* Note that N acts on $\mathcal{H}^{(n)}$, the homogeneous chaos of degree n, by multiplication with n (cf. Chapter 3.C).

Proposition 4.4. *For every* $\lambda \in \mathbb{R}$, $\exp(\lambda N)$ *maps* $(\mathcal{S})$ *continuously into itself.*

Proof. First we remark that we have the following well-known formula: for a bounded symmetric operator A on $\mathcal{H}$, $\exp(d\Gamma(A)) = \Gamma(\exp A)$ (cf., e.g., Nelson (1973), Simon (1974), and Appendix A.2). In particular we have $\exp(\lambda N) = \Gamma(e^\lambda)$.

We only have to consider the case $\lambda > 0$, since by standard theorems (cf. Nelson (1973) or Simon (1974)) it is known that for negative λ, $\exp(\lambda N) = \Gamma(e^\lambda)$ is a contraction on all (L^p), $1 \leq p \leq +\infty$. Since $\Gamma(e^\lambda)$ commutes with $\Gamma(H)$ on the domain of $\Gamma(H)$ it follows that for negative λ, $\exp(\lambda N)$ is also a contraction from every $(\mathcal{S})_p$ into itself.

Let $\lambda > 0$. Fix $p \in \mathbb{N}$ and choose $q \in \mathbb{N}$ large enough so that $e^\lambda\, 2^{-(q-p)} \leq 1$. Let $\varphi \in (\mathcal{S})$ with sequence of integral kernels $(f^{(n)})$. Then

$$\| \varphi \|_{2,q}^2 = \sum_{n=0}^{\infty} n! \left\| (H^{\otimes n})^q f^{(n)} \right\|_2^2$$

$$\geq \sum_{n=0}^{\infty} n!\, e^{2\lambda n}\, 2^{-2n(q-p)} \left\| (H^{\otimes n})^q f^{(n)} \right\|_2^2$$

$$\geq \sum_{n=0}^{\infty} n!\, e^{2\lambda n} \left\| (H^{\otimes n})^{-1} \right\|^{2(q-p)} \left\| (H^{\otimes n})^q f^{(n)} \right\|_2^2$$

$$\geq \sum_{n=0}^{\infty} n!\, e^{2\lambda n} \left\| (H^{\otimes n})^p f^{(n)} \right\|_2^2$$

$$= \| \exp(\lambda N) \varphi \|_{2,p}^2,$$

which is finite since $\varphi \in (\mathcal{S})_q$. (The last equality follows since N is closed.) Moreover, the above estimation shows the continuity of the operator $\exp(\lambda N)$ on $(\mathcal{S})$. □

Proposition 4.3 provides the following observation. If $\varphi \in (\mathcal{S})$, then all its integral kernels $f^{(n)}$ have a version in $\mathcal{S}(\hat{\mathbb{R}}^n)$. Let us denote the corresponding version of φ by $\widetilde{\varphi}$. Assume that φ has only a finite number of non-vanishing terms in its chaos expansion. Then $\widetilde{\varphi}$ is pointwise defined on $\mathcal{S}'(\mathbb{R})$ since for every $n \in \mathbb{N}$, $: x^{\otimes n} :$ belongs to $\mathcal{S}'(\hat{\mathbb{R}}^n)$. Moreover, $\widetilde{\varphi}$ is strongly continuous on $\mathcal{S}'(\mathbb{R})$: as is well-known (e.g., Trèves (1967), Theorem 41.1 and its Corollary 2) the operation of forming tensor products of tempered distributions is strongly continuous, and therefore for every $n \in \mathbb{N}$, the mapping $x \mapsto \, : x^{\otimes n} :$ is strongly continuous from $\mathcal{S}'(\mathbb{R})$ into $\mathcal{S}'(\hat{\mathbb{R}}^n)$. Consequently, if $f^{(n)} \in \mathcal{S}(\hat{\mathbb{R}}^n)$, then $x \mapsto \langle : x^{\otimes n} :, f^{(n)} \rangle$ is strongly continuous from $\mathcal{S}'(\mathbb{R})$ into $\mathbb{C}$, and so is $\widetilde{\varphi}$ becaus it is a finite sum of such terms. It is thus natural to ask whether a general element in $(\mathcal{S})$ has a pointwise defined, continuous version. The answer is affirmative, and this important result has been proved by Kubo and Yokoi (1989). We shall present this result now with a proof which essentially follows the one of Kubo and Yokoi (1989). First, we prepare two lemmas.

Lemma 4.5. *For every* $p > \frac{1}{4}$, $\mathrm{Tr} \equiv \int dt \;\, \delta_t^{\otimes 2} \in \mathcal{S}_{-p}(\hat{\mathbb{R}}^2)$, *and* $| \mathrm{Tr} |_{2,-p} = \| H^{-2p} \|_{H.S.}$.

Proof. Let $(e_k,\, k \in \mathbb{N}_0)$ be the CONS of $L^2(\mathbb{R})$ given by the Hermite functions. As an element of $\mathcal{S}'(\mathbb{R}^2)$, Tr has the expansion $\mathrm{Tr} = \sum_{k=0}^{\infty} (e_k \otimes e_k, \cdot\,)$, where the last scalar product is the one of $L^2(\mathbb{R}^2)$. Therefore, by (3.29) we have

$$| \mathrm{Tr} |_{2,-p}^2 = | (H^{-p})^{\otimes 2}\, \mathrm{Tr} |_2^2$$

$$= \sum_{k=0}^{\infty} (2k+2)^{-4p}$$

$$= \| H^{-2p} \|^2_{H.S.} < \infty,$$

if $p > \frac{1}{4}$. □

Lemma 4.6.

a) *Let* $x \in \mathcal{S}'(\mathbb{R})$, *and put* $p_0 = \min(\deg(x), 1)$, *where* $\deg(x)$ *is the smallest integer* q *so that* $x \in \mathcal{S}_{-q}(\mathbb{R})$. *Then for all* $n \in \mathbb{N}$, $p \in \mathbb{N}$, $p \geq p_0$, $: x^{\otimes n} : \in \mathcal{S}_{-p}(\hat{\mathbb{R}^n})$, *and the following estimate holds:*

$$| : x^{\otimes n} : |_{2,-p} \leq \sqrt{n!} \, (| x |_{2,-p} + \| H^{-2p} \|^{1/2}_{H.S.})^n. \tag{4.2}$$

b) *Assume that* $x \in \mathcal{S}_{-q}(\mathbb{R})$, $q \in \mathbb{N}$. *Let* $m \in \mathbb{N}$, $m \geq q$, $z_q, \ldots, z_m \in \mathcal{S}'(\mathbb{R})$, *with* $| z_p |_{2,-p} \leq 1$, $p = q, q+1, \ldots, m$. *Furthermore, let* $\lambda_q, \ldots, \lambda_m > 0$, *so that* $\sum_p \lambda_p = 1$. *Put* $y = x + \sum_p \lambda_p z_p$. *Then for all* $n \in \mathbb{N}$, $p = q, \ldots, m$, *there exists* $v_{n,p} \in \mathcal{S}_{-p}(\hat{\mathbb{R}^n})$, *so that*

$$: y^{\otimes n} : - : x^{\otimes n} : = \sum_{p=q}^{m} \lambda_p \, v_{n,p}, \tag{4.3a}$$

and the following estimate holds:

$$| v_{n,p} |_{2,-p} \leq n \sqrt{n!} \, | z_p |_{2,-p} \, (1 + | x |_{2,-q} + \| H^{-2q} \|^{1/2}_{H.S.})^{n-1}. \tag{4.3b}$$

Proof.

a) Let $x \in \mathcal{S}'(\mathbb{R})$. By Lemma 2.11, we have

$$: x^{\otimes n} : = \sum_{k=0}^{[\frac{n}{2}]} \binom{n}{2k} (2k-1)!! \, (-1)^k \, x^{\otimes (n-2k)} \hat{\otimes} \mathrm{Tr}^{\otimes k}.$$

Using Lemma 4.5, we obtain

$$|x^{\otimes(n-2k)} \hat{\otimes} \mathrm{Tr}^{\otimes k}|_{2,-p} = |x|_{2,-p}^{n-2k} |\mathrm{Tr}|_{2,-p}^{k}$$

$$= |x|_{2,-p}^{n-2k} \| H^{-2p} \|_{H.S.}^{k}.$$

Therefore we get the following estimate

$$|:x^{\otimes n}:|_{2,-p} \leq \sum_{k=0}^{[\frac{n}{2}]} \binom{n}{2k} (2k-1)!! \, |x|_{2,-p}^{n-2k} \| H^{-2p} \|_{H.S.}^{k}.$$

Clearly we have $(2k\text{-}1)!! \leq ((2k)!)^{\frac{1}{2}} \leq \sqrt{n!}$, for $k \leq [\frac{n}{2}]$, and hence we obtain the bound (4.2).

b) Let $n \in \mathbb{N}$. Then

$$y^{\otimes n} = \sum_{k=0}^{n} \binom{n}{k} \sum_{p_1,\ldots,p_k=q}^{m} \lambda_{p_1} \cdots \lambda_{p_k} \, x^{\otimes(n-k)} \hat{\otimes} z_{p_1} \hat{\otimes} \cdots \hat{\otimes} z_{p_k}$$

$$= \sum_{k=0}^{n} \binom{n}{k} \sum_{p=q}^{m} \sum_{\max(p_1,\ldots,p_k)=p} \lambda_{p_1} \cdots \lambda_{p_k} \, x^{\otimes(n-k)} \hat{\otimes} z_{p_1} \hat{\otimes} \cdots \hat{\otimes} z_{p_k}$$

$$= x^{\otimes n} + \sum_{p=q}^{m} \lambda_p \, w_{n,p},$$

with

$$w_{n,p} = \lambda_p^{-1} \sum_{k=1}^{n} \binom{n}{k} \sum_{\max(p_1,\ldots,p_k)=p} \lambda_{p_1} \cdots \lambda_{p_k} \, x^{\otimes(n-k)} \hat{\otimes} z_{p_1} \hat{\otimes} \cdots \hat{\otimes} z_{p_k}.$$

Now we estimate the $\mathcal{S}_{-p}(\mathbb{R}^n)$–norm of $w_{n,p}$ in the following way:

$$| w_{n,p} |_{2,-p} \leq$$

$$\leq \lambda_p^{-1} \sum_{k=1}^{n} \binom{n}{k} | x |_{2,-p}^{n-k} \sum_{\max(p_1,\ldots,p_k)=p} \lambda_{p_1} \cdots \lambda_{p_k} | z_{p_1} |_{2,-p} \cdots | z_{p_k} |_{2,-p}$$

$$\leq \lambda_p^{-1} | z_p |_{2,-p} \sum_{k=1}^{n} \binom{n}{k} | x |_{2,-q}^{n-k} \sum_{\max(p_1,\ldots,p_k)=p} \lambda_{p_1} \cdots \lambda_{p_k},$$

where we made use of the fact that under the above sum there is at least one index equal to p, and that furthermore we have $| z_{p_i} |_{2,-p} \leq | z_{p_i} |_{2,-p_i} \leq 1$, because $p_i \leq p$. Note that the sum of the products $\lambda_{p_1} \cdots \lambda_{p_k}$ over $(p_1, \ldots, p_k)$ so that $\max(p_1, \ldots, p_k) = p$, is the same as k times the sum over $(p_1, \ldots, p_k)$ with $p_1 = p$, and $p_2, \ldots, p_k$ ranging from q to p. Thus we get

$$\sum_{\max(p_1,\ldots,p_k)=p} \lambda_{p_1} \cdots \lambda_{p_k} = k\, \lambda_p \sum_{p_1,\ldots,p_{k-1}=q}^{p} \lambda_{p_1} \cdots \lambda_{p_{k-1}}$$

$$\leq k\, \lambda_p \, (\sum_{r=q}^{m} \lambda_r)^{k-1}$$

$$= k\, \lambda_p,$$

because $\sum_r \lambda_r = 1$. Thus we have

$$| w_{n,p} |_{2,-p} \leq | z_p |_{2,-p} \sum_{k=1}^{n} \binom{n}{k} k \, | x |_{2,-q}^{n-k}$$

$$= n \, | z_p |_{2,-p} \, (1 + | x |_{2,-q})^{n-1}.$$

By Lemma 2.11 we have the following formula

$$:y^{\otimes n}: - :x^{\otimes n}: = \sum_{p=q}^{m} \lambda_p \, v_{n,p},$$

where

$$v_{n,p} = \sum_{k=0}^{[\frac{n}{2}]} \binom{n}{2k} (2k-1)!!(-1)^k \, w_{n-2k,p} \, \hat{\otimes} \, \mathrm{Tr}^{\otimes k}.$$

Using the above bound on $| w_{n,p} |_{2,-p}$, and a similar method as in the proof of a), we obtain finally inequality (4.3b). □

Theorem 4.7. *Every* $\varphi \in (\mathcal{S})$ *has a unique pointwise defined, strongly continuous representative* $\widetilde{\varphi}$.

Proof. Let φ be an element in $(\mathcal{S})$. In (L^2), φ has the decomposition

$$\varphi(x) = \sum_{n=0}^{\infty} \langle :x^{\otimes n}: , f^{(n)} \rangle. \tag{4.4}$$

Now we choose for the kernel functions $f^{(n)}$ the representatives in $\mathcal{S}(\hat{\mathbb{R}}^n)$ (denoted by the same symbol) in accordance with Proposition 4.3. We shall show first that then (4.4) converges pointwise in $x \in \mathcal{S}'(\mathbb{R})$ (uniformly on strongly bounded sets). The limit will be denoted by $\widetilde{\varphi}$. Then we shall prove that $\widetilde{\varphi}$ is strongly continuous. From this we shall also obtain the uniqueness of $\widetilde{\varphi}$.

Let $N \in \mathbb{N}$, and set

$$\widetilde{\varphi}_N(x) := \sum_{n=0}^{N} \langle :x^{\otimes n}: , f^{(n)} \rangle.$$

By Lemma 4.6a we have for all $N, M \in \mathbb{N}$, $M \geq N$,

$$| \widetilde{\varphi}_N(x) - \widetilde{\varphi}_M(x) | \leq \sum_{n=N+1}^{M} | \langle :x^{\otimes n}:, f^{(n)} \rangle |$$

$$\leq \sum_{n=N+1}^{M} |:x^{\otimes n}:|_{2,-p} \, |f^{(n)}|_{2,p}$$

$$\leq \sum_{n=N+1}^{M} \sqrt{n!}\, (\sigma_p(x))^n \, |f^{(n)}|_{2,p},$$

where $\sigma_p(x) = (|x|_{2,-p} + \| H^{-2p} \|_{H.S.}^{1/2})$. Choosing p large enough, we may assume that $\sigma_p(x) < 1$. Schwarz' inequality gives the bound

$$|\tilde{\varphi}_N(x) - \tilde{\varphi}_M(x)| \leq \Big(\sum_{n=N+1}^{M} n!\, |f^{(n)}|_{2,p}^2\Big)^{1/2} \Big(\sum_{n=N+1}^{M} \sigma_p(x)^n\Big)^{1/2}$$

$$\leq \|\varphi\|_{2,p} \Big(\sum_{n=N+1}^{M} \sigma_p(x)^n\Big)^{1/2},$$

where we made use of (4.1). The sum in the last expression is the Cauchy remainder of a convergent sum. Therefore we have proved that sum in (4.4) converges pointwise for every $x \in \mathcal{S}'(\mathbb{R})$, if we choose for the $f^{(n)}$'s their versions in $\mathcal{S}(\hat{\mathbb{R}}^n)$. (The uniformity of this convergence on bounded sets of the inductive limit topology on $\mathcal{S}'(\mathbb{R})$ follows, because these are bounded in some $\mathcal{S}_{-p}(\mathbb{R})$, and on these we find a uniform estimate for σ_p. Furthermore, the strong and the inductive limit topology on $\mathcal{S}'(\mathbb{R})$ coincide, cf. Appendix 5.)

Now we prove that the mapping $x \mapsto \tilde{\varphi}(x)$ is strongly continuous from $\mathcal{S}'(\mathbb{R})$ into $\mathbb{C}$. It will be convenient to consider bases of neighborhoods of the inductive limit topology (which – as pointed out before – is equal to the strong topology). A base of neighborhoods of zero in $\mathcal{S}'(\mathbb{R})$ in this topology is given by the absolutely convex envelopes of sets of the form $\bigcup_p V_p$, p ranging over $\mathbb{N}$, and where the V_p range over a base of neighborhoods of zero in $\mathcal{S}_{-p}(\mathbb{R})$, cf., e.g., Robertson and Robertson (1973), or Appendix 5. Of course, we may choose V_p to be of the form $\{x \in \mathcal{S}_{-p}(\mathbb{R});\ |x|_{2,-p} < \gamma_p\}$, for certain $\gamma_p > 0$. Thus as a base of neighborhoods of zero is determined by a sequence $\{\gamma_p,\ p \in \mathbb{N}\}$, $0 < \gamma_p \leq 1$, and every point y in such a neighborhood is of the form

$$y=\sum_{p=1}^{m}\lambda_p z_p,$$

where $m\in\mathbb{N}$, $\lambda_p>0$, $\sum_p \lambda_p=1$, and $|z_p|_{2,-p}<\gamma_p$.

Let $x\in\mathcal{S}'(\mathbb{R})$, $\deg(x)=q$. A base of neighborhoods of x is obtained by translation of the above neighborhoods of zero, and without loss of generality we may restrict the summation over p above from q to m. In order to show continuity of $\widetilde{\varphi}$ at x, we have to prove that for every $\epsilon>0$, there exists a sequence $\{\gamma_p, p\in\mathbb{N}\}$, so that for all $m\in\mathbb{N}$, $\lambda_q,\dots,\lambda_m>0$, $\sum_p \lambda_p=1$, and $z_q,\dots,z_m\in\mathcal{S}'(\mathbb{R})$, with $|z_p|_{2,-p}<\gamma_p$, we have

$$|\widetilde{\varphi}(y)-\widetilde{\varphi}(x)|\le\epsilon,$$

whenever

$$y=x+\sum_{p=q}^{m}\lambda_p z_p.$$

By Lemma 4.6b we have

$$\widetilde{\varphi}(y)-\widetilde{\varphi}(x)=\sum_{n=1}^{\infty}<:y^{\otimes n}:-:x^{\otimes n}:,f^{(n)}>$$

$$=\sum_{p=q}^{m}\lambda_p\sum_{n=1}^{\infty}<v_{n,p},f^{(n)}>.$$

Using the fact that $\|H^{-1}\|=\frac{1}{2}$, and inequality (4.3b), we may estimate for large enough $r\ge 0$ as follows:

$$|\widetilde{\varphi}(y)-\widetilde{\varphi}(x)|\le\sum_{p=q}^{m}\lambda_p\sum_{n=1}^{\infty}|v_{n,p}|_{2,-p}|f^{(n)}|_{2,p}$$

$$\leq \sum_{p=q}^{m} \lambda_p \sum_{n=1}^{\infty} \frac{1}{\sqrt{n!}} 2^{-rn} | v_{n,p} |_{2,p} \sqrt{n!} \, | f^{(n)} |_{2,p+r}$$

$$\leq \Big(\sum_{n=1}^{\infty} n^2 \, 2^{-2rn} \, \rho(x)^{2(n-1)} \Big)^{1/2} \sum_{p=q}^{m} \lambda_p \, | z_p |_{2,-p} \, \| \varphi \|_{2,p+r} ,$$

where we denoted $\rho(x) := 1 + | x |_{2,-q} + \| H^{-2p} \|^{1/2}_{H.S.}$, and we used Schwarz' inequality in the last step. If we choose r large enough, depending only on x, we obtain

$$\sum_{n=1}^{\infty} n^2 \, 2^{-2rn} \, \rho(x)^{2(n-1)} \leq 1.$$

Thus we get the bound

$$| \widetilde{\varphi}(y) - \widetilde{\varphi}(x) | \leq \sum_{p=q}^{m} \lambda_p \, | z_p |_{2,-p} \, \| \varphi \|_{2,p+r},$$

for large enough r. For given $\epsilon > 0$, we now choose the sequence $\{\gamma_p, \, p \in \mathbb{N}\}$ by setting $\gamma_p = \epsilon \cdot \min(1, \| \varphi \|^{-1}_{2,p+r})$, and obtain for $| z_p |_{2,-p} < \gamma_p$, $p = q, \ldots, m$,

$$| \widetilde{\varphi}(y) - \widetilde{\varphi}(x) | \leq \epsilon,$$

since $\sum_p \lambda_p = 1$.

Finally we prove the uniqueness of the version $\widetilde{\varphi}$ by an argument which we learned from M. Röckner (private communication). It is based on the following observations: (i) the σ-algebra generated by the strong topology β on $\mathcal{S}'(\mathbb{R})$ is the same as $\mathfrak{B}$, the σ-algebra generated by the cylinder sets, i.e., by the weak topology (cf., e.g., Badrikian (1970), Exposé n^0 8), and (ii) the support of the white noise measure is full: $\mathrm{supp}\mu = \mathcal{S}'(\mathbb{R})$. This is due to the fact that the white noise measure is quasi-invariant with respect to the translations of the dense subspace (in the weak topology) $\mathcal{S}(\mathbb{R})$ of $\mathcal{S}'(\mathbb{R})$. In particular, every open set has a strictly positive measure. Now assume that $\varphi \in (\mathcal{S})$ has two different, strongly continuous versions $\widetilde{\varphi}_1$ and $\widetilde{\varphi}_2$. Then there is $x \in \mathcal{S}'(\mathbb{R})$ such that $\widetilde{\varphi}_1(x) \neq \widetilde{\varphi}_2(x)$, and by continuity there is an open set $U \in \beta$ containing x, so that $\widetilde{\varphi}_1$ and $\widetilde{\varphi}_2$ are different on all of U. But (i) and (ii) above imply that $\mu(U) > 0$ which is a contradiction. □

As a corollary of the above proof we have the following result.

Corollary 4.8. *For every* $x \in \mathcal{S}'(\mathbb{R})$, *the evaluation map* $\delta_x \colon \varphi \mapsto \tilde{\varphi}(x)$, $\varphi \in (\mathcal{S})$, *belongs to* $(\mathcal{S})^*$.

Proof. As in the proof of Theorem 4.7, we have for $x \in \mathcal{S}'(\mathbb{R})$, $\varphi \in \mathcal{P}$ and some $p \in \mathbb{N}$

$$|\tilde{\varphi}(x)| \leq C_p(x) \| \varphi \|_{2,p},$$

where $C_p(x)$ does not depend on φ. Thus the evaluation map extends continuously from $\mathcal{P}$ to $(\mathcal{S})$, mapping φ into $\tilde{\varphi}(x)$. □

Unless there is danger of confusion, we shall usually identify $\varphi \in (\mathcal{S})$ with its pointwise defined and continuous version $\tilde{\varphi}$ on $\mathcal{S}'(\mathbb{R})$.

The next result is very important. It has first been indicated in Kubo and Takenaka (1980b). Detailed proofs can be found in Hida, Potthoff, and Streit (1988), Potthoff and Yan (1992) , and in Yokoi (1990). Also, in Kuo (1992) and Lee (1991) one finds proofs using different arguments.

Theorem 4.9. *Pointwise multiplication of two elements in* $(\mathcal{S})$ *is a continuous mapping from* $(\mathcal{S})$ *into itself. In particular,* $(\mathcal{S})$ *is an algebra under pointwise multiplication.*

This theorem is a consequence of the following sharper result (cf. Potthoff and Yan (1992)).

Proposition 4.10. *Let* $\varphi \in (\mathcal{S})$. *Then for all* $p \in \mathbb{N}_0$, *(pointwise) multiplication by* φ *is a continuous mapping from* $(\mathcal{S})_{p+1}$ *into* $(\mathcal{S})_p$.

Proof. Let $\psi \in (\mathcal{S})_p$ for some $p \in \mathbb{N}_0$, with chaos decomposition

$$\psi(x) = \sum_{n=0}^{\infty} \psi^{(n)}(x)$$

$$= \sum_{n=0}^{\infty} \langle :x^{\otimes n}:, g^{(n)} \rangle .$$

If we assume that φ is given as in (4.3), then we have the formula (cf. (2.51))

$$\varphi^{(n)}\psi^{(m)}(x) = \sum_{k=0}^{n \wedge m} k! \binom{n}{k} \binom{m}{k} \langle :x^{\otimes(n+m-2k)}:, f^{(n)} \hat{\otimes}_k g^{(m)} \rangle .$$

Therefore

$$\left\| \varphi^{(n)}\psi^{(m)} \right\|_{2,p} \leq \sum_{k=0}^{m \wedge n} k! \binom{m}{k} \binom{n}{k} \sqrt{(m+n-2k)!}\ \left| f^{(n)} \hat{\otimes}_k g^{(m)} \right|_{2,p} .$$

Using Schwarz' inequality and the fact that $\| H^{-1} \| = \frac{1}{2}$, we can bound the last norm by $|f^{(n)}|_{2,p} \cdot |g^{(m)}|_{2,p}$, and this is equal to $(m!\, n!)^{-1/2} \|\varphi^{(n)}\|_{2,p} \|\psi^{(m)}\|_{2,p}$. We obtain the bound

$$\left\| \varphi^{(n)}\psi^{(m)} \right\|_{2,p} \leq \left\| \varphi^{(n)} \right\|_{2,p} \left\| \psi^{(m)} \right\|_{2,p} \sum_{k=0}^{m \wedge n} \left(\binom{m}{k} \binom{n}{k} \binom{m+n-2k}{n-k} \right)^{1/2} .$$

Now we estimate $\binom{m}{k}\binom{n}{k} \leq \binom{m+n}{2k}$, and $\binom{m+n-2k}{n-k} \leq 2^{m+n-2k}$. Therefore

$$\sum_{k=0}^{m\wedge n}\left(\binom{m}{k}\binom{n}{k}\binom{m+n-2k}{n-k}\right)^{1/2} \le 2^{\frac{m+n}{2}}\sum_{k=0}^{m\wedge n}\binom{m+n}{2k}^{1/2}2^{-k}$$

$$\le 2^{\frac{m+n}{2}}\sum_{k=0}^{m+n}\binom{m+n}{k}^{1/2}2^{-k/2}$$

$$\le 2^{\frac{m+n}{2}}\Big(\sum_{k=0}^{\infty}2^{-k/2}\Big)^{1/2}\times$$

$$\times\Big(\sum_{k=0}^{m+n}\binom{m+n}{k}2^{-k/2}\Big)^{1/2}$$

$$=\Big(\frac{\sqrt{2}}{\sqrt{2}-1}\Big)^{1/2}\Big(\sqrt{2+\sqrt{2}}\,\Big)^{m+n},$$

where we used Schwarz' inequality in the third step. Due to the fact that $\left\|(H^{-1})^{\otimes n}\right\| \le 2^{-n}$, we find the estimation

$$\|\varphi\psi\|_{2,p} \le \sum_{m,n=0}^{\infty}\left\|\varphi^{(n)}\psi^{(m)}\right\|_{2,p}$$

$$\le \text{const.}\sum_{m,n=0}^{\infty}\left(\tfrac{\sigma}{2}\right)^{m+n}\left\|\varphi^{(n)}\right\|_{2,p+1}\left\|\psi^{(m)}\right\|_{2,p+1}$$

$$\le \text{const.}\Big(\sum_{n=0}^{\infty}\left\|\varphi^{(n)}\right\|^2_{2,p+1}\Big)^{1/2}\Big(\sum_{n=0}^{\infty}\left\|\psi^{(n)}\right\|^2_{2,p+1}\Big)^{1/2}$$

$$= \text{const.}\ \|\varphi\|_{2,p+1}\,\|\psi\|_{2,p+1},$$

where we have put $\sigma = \sqrt{2+\sqrt{2}}$ and used Schwarz' inequality again (together with $\sigma/2 < 1$). □

Remark. In Potthoff and Yan (1992) the above estimation has been worked out a little more: let $q_0 = \frac{\ln(2+\sqrt{2})}{2\ln 2}$ and $\varphi, \psi \in (\mathcal{S})_p$, $p > q_0$, then $\varphi \cdot \psi \in (\mathcal{S})_{p-q}$ for any $q > q_0$.

Corollary 4.11. *Let* $\varphi \in (\mathcal{S})$, $\Phi \in (\mathcal{S})^*$. *Then* $\varphi \cdot \Phi$ *defined by* $<\varphi \cdot \Phi, \psi> = <\Phi, \bar{\varphi} \cdot \psi>$, $\psi \in (\mathcal{S})$, *belongs to* $(\mathcal{S})^*$.

By construction, the algebra $\mathcal{P}$ of smooth polynomials is a subspace of $(\mathcal{S})$. The question arises, whether other composites with the smooth coordinate processes X_ξ, $\xi \in \mathcal{S}(\mathbb{R})$, provide elements in $(\mathcal{S})$. The next proposition states a sufficient condition for this to be the case. Interestingly the proof shows that this is not always the case for composition with a function in $\mathcal{S}(\mathbb{R})$ (cf. the remark after Proposition 4.12).

Proposition 4.12. *Let* $\xi_1, \dots \xi_n \in \mathcal{S}(\mathbb{R})$, $n \in \mathbb{N}$, *and* $X_\xi = (X_{\xi_1}, \dots, X_{\xi_n})$, *where* $X_{\xi_i} = <\cdot, \xi_i>$. *If* f *is the Fourier transform of a distribution* $\tilde{f}$ *in* $\mathcal{S}'(\mathbb{R}^n)$ *with compact support, then* $f \circ X_\xi \in (\mathcal{S})$.

Proof. Let us assume first that $\tilde{f} \in C_0^\infty(\mathbb{R}^n)$, and write

$$f \circ X_\xi(x) = (2\pi)^{-n/2} \int_{\mathbb{R}^n} \tilde{f}(k)\, e^{ik \cdot X_\xi(x)}\, dk, \quad x \in \mathcal{S}'(\mathbb{R}).$$

where $\cdot$ denotes the Euclidean inner produckt in $\mathbb{R}^n$. Define the symmetric $n \times n$ matrix $h(\xi,q)$, $q \geq 0$, by

$$(h(\xi,q))_{ij} = (\xi_i, H^q \xi_j); \quad 1 \leq i,j \leq n.$$

Note that for every $k \in \mathbb{R}^n$, $k \cdot X_{H^p \xi}$ is a centered Gaussian random variable on

$(\mathcal{S}'(\mathbb{R}),\mathcal{B},\mu)$ with variance

$$\mathbb{E}\,(k\cdot X_{H^p\xi})^2 = k\cdot h(\xi,2p)k.$$

Therefore, we have

$$:\exp(ik\cdot X_{H^p\xi}): = \exp(ik\cdot X_{H^p\xi} - \tfrac{1}{2}k\cdot\, h(\xi,2p)\, k),$$

and we find (since $\Gamma(H^p)$ is closed)

$$\Gamma(H^p)\, f\circ X_\xi = (2\pi)^{-n/2} \int_{\mathbb{R}^n} \tilde{f}(k)\, \exp(ik\cdot X_{H^p\xi})\, \exp(\tfrac{1}{2}k\cdot\, (h(\xi,2p)-h(\xi,0))\, k)\, dk.$$

Thus

$$\left\| f\circ X_\xi \right\|^2_{2,p} = (2\pi)^{-n} \int_{\mathbb{R}^{2n}} \bar{\tilde{f}}(k)\, \tilde{f}(q)\, \exp(k\cdot\, (h(\xi,p)-h(\xi,0))\, q)\, dk\, dq,$$

which is finite for every $p\in\mathbb{N}$. The general case follows now from a simple limiting argument. □

Remark. Clearly, we can relax the condition on the decrease of $\tilde{f}$. However, note that a stronger decay than Gaussian is necessary, as an explicit computation with Gaussian $\tilde{f}$ in the above proof shows. This demonstrates also that in general the composite $f\circ X_\xi$ with $f\in\mathcal{S}(\mathbb{R}^n)$ does *not* belong to $(\mathcal{S})$.

Next we state a result about the relation of the space $\mathcal{D}$ in the case of white noise (cf. subsection 3.C) and $(\mathcal{S})$. This result has been proved independently in Korezlioglu and Ustunel (1990), and in Potthoff and Yan (1992).

Proposition 4.13. $(\mathcal{S})$ *is continuously and densely embedded into* $\mathfrak{D}$.

Proof. Let $\varphi \in \mathcal{P}$. It is sufficient to show that for every $k \in \mathbb{N}_0$ and $p \in \mathbb{N}$, there exists $q \in \mathbb{N}$ so that

$$\| N^k \varphi \|_p \leq \text{const.} \| \varphi \|_{2,q}. \tag{4.5}$$

Since $\| H^{-1} \| \leq \frac{1}{2}$, Nelson's hypercontractivity theorem (cf. Nelson (1973)) implies that for some $m \in \mathbb{N}$ we may estimate as follows:

$$\| N^k \varphi \|_p = \| \Gamma(H^{-m})\, \Gamma(H^m)\, N^k \varphi \|_p$$

$$\leq \| \Gamma(H^m)\, N^k\, \varphi \|_2.$$

Choose $q \in \mathbb{N}$ so that $q > m$. Then, as in the proof of Proposition 4.4, we find the bound

$$\| N^k \varphi \|_p \leq \text{const.} \| \varphi \|_{2,q}. \qquad \square$$

Corollary 4.14. $\mathfrak{D}^* \subset (\mathcal{S})^*$. *In particular, for every* $p \in (1,+\infty)$, $(L^p) \subset (\mathcal{S})^*$.

Remarks. Because of Corollary 4.14, the examples of generalized functionals in $\mathfrak{D}^*$ constructed by Watanabe (1983, 1984), namely composites of certain elements in $\mathfrak{D}$ with tempered distributions, are elements in $(\mathcal{S})^*$. Cf. also Chapter 7.C.

Consider the class of spaces $(L^2)^+_{\mathbf{a},\mathbf{c}}$ defined in Chapter 3.A, where $\mathbf{c}$ is a sequence of at most exponential growth. Then it is straightforward to check that $(\mathcal{S})$ is continuously and densely embedded into these spaces too. Therefore, all the examples of generalized functionals constructed in subsection 3.A are also elements in $(\mathcal{S})^*$.

The fact that the elements in $(\mathcal{S})$ have a pointwise defined version (Theorem 4.7) allows to discuss the behavior of these functions under 'coordinate transformations', i.e., under transformations on $\mathcal{S}'(\mathbb{R})$. For certain affine transformations, namely translations and scalings, this has been done in Potthoff and Yan (1992) . The result is, that $(\mathcal{S})$ is stable under the induced transformations. This has interesting consequences for $(\mathcal{S})^*$, and especially for the measures which are represented in $(\mathcal{S})^*$ (cf. Section 4.B). We give now an account of the results in the above quoted article, with an extension to complex scaling.

Consider $\varphi\in(\mathcal{S})$ and its pointwise defined, continuous version $\tilde{\varphi}$. Fix $y\in\mathcal{S}'(\mathbb{R})$, $\lambda\in\mathbb{R}$. Then we may define two other pointwise defined functions $(\tau_y\varphi)^{\sim}$ and $(\sigma_\lambda\varphi)^{\sim}$, by setting for $x\in\mathcal{S}'(\mathbb{R})$:

$$(\tau_y\varphi)^{\sim}(x) := \tilde{\varphi}(x+y);$$

$$(\sigma_\lambda\varphi)^{\sim}(x) := \tilde{\varphi}(\lambda\cdot x).$$

By $\tau_y\varphi$ and $\sigma_\lambda\varphi$ we denote the μ-classes of $(\tau_y\varphi)^{\sim}$ and $(\sigma_\lambda\varphi)^{\sim}$. The question arises whether $\tau_y\varphi$ and $\sigma_\lambda\varphi$ belong again to $(\mathcal{S})$.

Theorem 4.15. *For all* $y\in\mathcal{S}'(\mathbb{R})$, $\lambda\in\mathbb{R}$, *the mappings* τ_y: $\varphi\mapsto\tau_y\varphi$, *and* σ_λ: $\varphi\mapsto\sigma_\lambda\varphi$, *are continuous from* $(\mathcal{S})$ *into itself. The same is true for the analytic extension of* σ *to arbitrary* $\lambda\in\mathbb{C}$.

First we prove two useful formulae.

Lemma 4.16. *Let* x, $y\in\mathcal{S}'(\mathbb{R})$, *and* $\lambda\in\mathbb{R}$. *Then for all* $n\in\mathbb{N}_0$

$$:(x+y)^{\otimes n}: = \sum_{k=0}^{n}\binom{n}{k} :x^{\otimes(n-k)}: \otimes\, y^{\otimes k} \tag{4.6}$$

$$= \sum_{k=0}^{n} \binom{n}{k} : x^{\otimes k} : \otimes\, y^{\otimes n-k},$$

and

$$: (\lambda x)^{\otimes n} : = \sum_{k=0}^{[\frac{n}{2}]} \binom{n}{2k} (2k-1)!!\, (-1)^{k}\, \lambda^{n-2k}(1-\lambda^2)^{k} : x^{\otimes(n-2k)} : \hat{\otimes}\, \mathrm{Tr}^{\otimes k}. \quad (4.7)$$

Proof. Both formulae follow by usual manipulations with Hermite polynomials. Let us only prove (4.6), equation (4.7) is shown by a similar computation. Let $\xi \in \mathcal{S}(\mathbb{R})$, and recall that for $n \in \mathbb{N}$ and $z \in \mathcal{S}'(\mathbb{R})$, we have (cf. (2.47))

$$<: z^{\otimes n} :, \xi^{\otimes n}> = \frac{d^n}{dt^n} e^{t<z,\, \xi> - \frac{t^2}{2}|\xi|_2^2}\Big|_{t=0}.$$

If $z = x+y$, then we get by Leibniz' formula

$$<: (x+y)^{\otimes n} :, \xi^{\otimes n}> =$$

$$= \sum_{k=0}^{n} \binom{n}{k} \frac{d^{n-k}}{dt^{n-k}} e^{t<x,\, \xi> - \frac{t^2}{2}|\xi|_2^2}\Big|_{t=0} \cdot \frac{d^k}{dt^k} e^{t<y,\, \xi>}\Big|_{t=0}$$

$$= \sum_{k=0}^{n} \binom{n}{k} <: x^{\otimes(n-k)} :, \xi^{\otimes(n-k)}> <y^{\otimes k}, \xi^{\otimes k}>,$$

where we used (2.47) again in the last step. Thus formula (4.6) is true when evaluated on $\xi^{\otimes n}$, and by linearity and polarization we see that (4.7) holds when evaluated on $\xi_1 \hat{\otimes} \cdots \hat{\otimes} \xi_n$, $\xi_j \in \mathcal{S}(\mathbb{R})$. Again by linearity, (4.7) holds then on all complex linear combinations of such tensor products, and hence on a dense subspace of $\mathcal{S}(\hat{\mathbb{R}}^n)$. Now continuity implies the statement of the lemma. □

Proof of Theorem 4.15. First we prove the continuity of τ_y, for $y \in \mathcal{S}'(\mathbb{R})$. Let $\varphi \in \mathcal{P}$ have a chaos decomposition given by

$$\varphi(x) = \sum_{n=0}^{\infty} <: x^{\otimes n} :, f^{(n)}>.$$

Identify each $f^{(n)}$ with its representative in $\mathcal{S}(\hat{\mathbb{R}}^n)$. Then we have by definition

$$(\tau_y \varphi)^{\sim}(x) = \sum_{n=0}^{\infty} <: (x+y)^{\otimes n} :, f^{(n)}>.$$

Using (4.6) we obtain

$$\tau_y \varphi(x) = \sum_{n=0}^{\infty} \sum_{k=0}^{n} \binom{n}{k} <: x^{\otimes k} :, f_y^{(n,n-k)}>,$$

with the notation $f_y^{(n,k)} = <y^{\otimes k}, f^{(n)}>$. Thus

$$\Gamma(H^p)\ \tau_y \varphi(x) = \sum_{n=0}^{\infty} \sum_{k=0}^{n} \binom{n}{k} <: x^{\otimes k} :, (H^{\otimes k})^p\ f_y^{(n,n-k)}>$$

for every p. Let $p \in \mathbb{N}_0$, then

$$\| \tau_y \varphi \|_{2,p}^2 = \sum_{m,n=0}^{\infty} \sum_{k=0}^{n} k! \binom{n}{k} \binom{m}{k} \left((H^{\otimes k})^p\ f_y^{(n,n-k)}, (H^{\otimes k})^p\ f_y^{(m,m-k)} \right)_{L^2(\mathbb{R}^k)}$$

$$\le \sum_{m,n=0}^{\infty} \sum_{k=0}^{n} k! \binom{n}{k}\binom{m}{k} | f_y^{(n,n-k)} |_{2,p}\ | f_y^{(m,m-k)} |_{2,p}.$$

Schwarz' inequality and $\| H^{-1} \| \le 1$ imply that

$$| f_y^{(n,n-k)} |_{2,p} \le | y |_{2,-(p+q)}^{n-k}\ | f^{(n)} |_{2,(p+q)},$$

whenever q is chosen large enough so that $y \in \mathcal{S}_{-(p+q)}(\mathbb{R})$. Now we use $\binom{n}{k}\binom{m}{k} \le$

$\binom{m+n}{2k}$ and the estimate $k! \leq (n!\, m!)^{1/2}$ under the above sum. Then we obtain (with $r = p+q$)

$$\| \tau_y \varphi \|_{2,p}^2 \leq \sum_{m,n=0}^{\infty} (n!\, m!)^{1/2} \,|\, f^{(n)} |_{2,r} \,|\, f^{(m)} |_{2,r} \sum_{k=0}^{m \wedge n} \binom{n+m}{2k} |\, y\, |_{2,-r}^{n+m-2k}$$

$$\leq \Big(\sum_{n=0}^{\infty} (n!)^{1/2} (1 + |\, y\, |_{2,-r})^n \,|\, f^{(n)} |_{2,r} \Big)^2 .$$

Since $\left\| (H^{\otimes n})^{-s} \right\| \leq 2^{-sn}$ for all $n \in \mathbb{N}$, $s \geq 0$, an application of Schwarz' inequality gives for large enough s the bound

$$\| \tau_y \varphi \|_{2,p}^2 \leq \text{const.} \sum_{n=0}^{\infty} n! \,|\, f^{(n)} |_{2,r+s}^2$$

$$= \text{const.} \| \varphi \|_{2,p+q+s}^2 .$$

Since $\mathcal{P}$ is dense in $(\mathcal{S})$, we have shown the existence of a continuous extension of τ_y from $\mathcal{P}$ to $(\mathcal{S})$. That this extension of τ_y from $\mathcal{P}$ to $(\mathcal{S})$ coincides with the above definition follows from the remark that the evaluation map $\delta_x : \varphi \mapsto \tilde{\varphi}(x)$ is continuous on $(\mathcal{S})$ for every $x \in \mathcal{S}'(\mathbb{R})$, cf. Corollary 4.8.

The proof of the continuous extension of the scaling operator σ_λ, $\lambda \in \mathbb{R}$, is quite similar. Let $\varphi \in (\mathcal{S})$ as above. Using formula (4.7) we obtain for $\lambda \in \mathbb{R}$, $x \in \mathcal{S}'(\mathbb{R})$,

$$\sigma_\lambda \varphi(x) = \sum_{n=0}^{\infty} \sum_{k=0}^{[\frac{n}{2}]} \binom{n}{2k} (2k-1)!!\, (-1)^k \lambda^{n-2k} (1-\lambda^2)^k <: x^{\otimes(n-2k)} :, \mathrm{Tr}^{\otimes k} f^{(n)} > .$$

Let $p \in \mathbb{N}$. Then

$$\Gamma(H)^p \sigma_\lambda \varphi(x) = \sum_{n=0}^{\infty} \sum_{k=0}^{[\frac{n}{2}]} \binom{n}{2k} (2k-1)!! \, (-1)^k \, \lambda^{n-2k} (1-\lambda^2)^k \times$$

$$\times \langle : x^{\otimes (n-2k)} :, (H^{\otimes (n-2k)})^p \mathrm{Tr}^{\otimes k} f^{(n)} \rangle,$$

and therefore

$$\| \sigma_\lambda \varphi \|_{2,p} \le \sum_{n=0}^{\infty} \sum_{k=0}^{[\frac{n}{2}]} \binom{n}{2k} (2k-1)!! \, | \lambda^{n-2k} (1-\lambda^2)^k | \, ((n-2k)!)^{\frac{1}{2}} \, | \mathrm{Tr}^{\otimes k} f^{(n)} |_{2,p}.$$

We estimate first the last norm:

$$\begin{aligned} | \mathrm{Tr}^{\otimes k} f^{(n)} |_{2,p} &= | (H^{\otimes (n-2k)})^p \, \mathrm{Tr}^{\otimes k} f^{(n)} |_2 \\ &= | ((H^{\otimes 2})^{-p} \mathrm{Tr})^{\otimes k} \, (H^{\otimes n})^p f^{(n)} |_2 \\ &\le | \mathrm{Tr} |_{2,-p}^k \, | f^{(n)} |_{2,p} \\ &= \| H^{-2p} \|_{H.S.} \, | f^{(n)} |_{2,p}. \end{aligned}$$

In addition we bound $(2k-1)!!$ by $((2k)!)^{\frac{1}{2}}$, and $(n-2k)! \cdot (2k)!$ by $n!$. Then we obtain

$$\begin{aligned} \| \sigma_\lambda \varphi \|_{2,p} &\le \sum_{n=0}^{\infty} (n!)^{\frac{1}{2}} \, | f^{(n)} |_{2,p} \sum_{k=0}^{[\frac{n}{2}]} \binom{n}{2k} | \lambda^{n-2k} (1-\lambda^2)^k | \, \| H^{-2p} \|_{H.S.}^k \\ &\le \sum_{n=0}^{\infty} (n!)^{\frac{1}{2}} \, | f^{(n)} |_{2,p} \, [| \lambda | + (| 1-\lambda^2 | \, \| H^{-2p} \|_{H.S.})^{1/2}]^n. \end{aligned}$$

Choose q large enough so that $2^{-q} \cdot [| \lambda | + (| 1-\lambda^2 | \cdot \| H^{-2p} \|_{H.S.})^{1/2}] < 1$. Thus we arrive at the following inequality

$$\| \sigma_\lambda \varphi \|_{2,p} \leq K_{p,\lambda} \| \varphi \|_{2,p+q},$$

for some positive constant $K_{p,\lambda}$. The same reasoning applies if we extend the above power series to complex λ. Therefore we have also proved the existence of a continuous extension of the scaling operator σ_λ, $\lambda \in \mathbb{C}$, to $(\mathcal{S})$. □

Working the above estimates out a little more, one can show the following result (for the details we refer to the article by Kuo, Potthoff and Yan (1992)):

Proposition 4.17. *Let* $\varphi \in (\mathcal{S})$.

(i) *The mapping* $y \mapsto \tau_y \varphi$ *is strongly continuous from* $\mathcal{S}'(\mathbb{R})$ *into* $(\mathcal{S})$.

(ii) *The mapping* $\lambda \mapsto \sigma_\lambda \varphi$ *is continuous from* $\mathbb{C}$ *into* $(\mathcal{S})$.

Theorem 4.15 will have some interesting consequences for the transformation of measures which are represented by positive elements in $(\mathcal{S})^*_+$, see below. Also, in Chapter 5 we shall prove the convergence of the Taylor expansion of $\tau_y \varphi$ in $(\mathcal{S})$.

Next we shall discuss the space $(\mathcal{S})^*$ of generalized functionals on $\mathcal{S}'(\mathbb{R})$ in more detail.

Let $\Phi \in (\mathcal{S})^*$. By construction, we know that there exists $p \in \mathbb{N}_0$ so that $\Phi \in (\mathcal{S})_{-p}$, cf. Chapter 3. Therefore, we have for every $\varphi \in (\mathcal{S})$ the inequality

$$| \langle \Phi, \varphi \rangle | \leq \| \Phi \|_{2,-p} \| \varphi \|_{2,p}. \tag{4.8}$$

In particular, we may choose for $n \in \mathbb{N}$, and $f^{(n)} \in \mathcal{S}(\hat{\mathbb{R}}^n)$, φ of the form $\varphi(x) =$

$\varphi^{(n)}(x) = \langle :x^{\otimes n}:, f^{(n)}\rangle$, and obtain in this case

$$|\langle\Phi, \varphi^{(n)}\rangle| \le \sqrt{n!}\, \|\Phi\|_{2,-p}\, |f^{(n)}|_{2,p}.$$

This implies that the restriction $\Phi^{(n)}$ of Φ to $(\mathcal{S})\bigcap\mathcal{H}^{(n)}$ is in one-to-one correspondence with an element $F^{(n)}$ in $\mathcal{S}_{-p}(\hat{\mathbb{R}}^n)$, so that

$$\langle\Phi^{(n)}, \varphi^{(n)}\rangle = n!\, \langle F^{(n)}, f^{(n)}\rangle.$$

Furthermore we have

$$\|\Phi^{(n)}\|_{2,-p} = \sqrt{n!}\, |F^{(n)}|_{2,-p},$$

as is easily checked. Since a general $\varphi\in(\mathcal{S})$ is the sum of such elements, we see that $\Phi\in(\mathcal{S})_{-p}$ is in one-to-one correspondence with a sequence $(F^{(n)}, n\in\mathbb{N}_0)$, $F^{(n)}\in\mathcal{S}_{-p}(\hat{\mathbb{R}}^n)$, so that for all $\varphi\in(\mathcal{S})$ we have

$$\langle\Phi, \varphi\rangle = \sum_{n=0}^{\infty} n!\, \langle F^{(n)}, f^{(n)}\rangle, \tag{4.9a}$$

$$\|\Phi\|_{2,-p}^2 = \sum_{n=0}^{\infty} n!\, |F^{(n)}|_{2,-p}^2, \tag{4.9b}$$

where φ has chaos decomposition given by $(f^{(n)}, n\in\mathbb{N}_0)$, and where the dual pairing under the sum in (4.9a) is between $\mathcal{S}_{-p}(\mathbb{R}^n)$ and $\mathcal{S}_p(\mathbb{R}^n)$. Sometimes we shall also denote such Φ informally as $\Phi(x) = \sum_n \langle :x^{\otimes n}:, F^{(n)}\rangle$, its interpretation given by (4.9a) being understood.

By Proposition 3.8, we know that for all $\lambda\in\mathbb{C}$, $\xi\in\mathcal{S}(\mathbb{R})$, $\exp(\lambda\langle\cdot, \xi\rangle)$ belongs to $(\mathcal{S})$. Therefore, we can extend the $\mathcal{T}$- and $\mathcal{S}$-transforms continuously to $(\mathcal{S})^*$. For $\Phi\in(\mathcal{S})^*$, and $\xi\in\mathcal{S}(\mathbb{R})$, we define

$$\mathcal{S}\Phi(\xi) := \langle\bar{\Phi}, :e^{\langle\cdot, \xi\rangle}:\rangle = e^{-\frac{1}{2}|\xi|_2^2} \langle\bar{\Phi}, e^{\langle\cdot, \xi\rangle}\rangle, \tag{4.10}$$

$$\mathcal{T}\Phi(\xi) := \langle \overline{\Phi}, e^{i\langle \cdot, \xi \rangle} \rangle. \tag{4.11}$$

Note that $:e^{\langle \cdot, \xi \rangle}:$ has a chaos expansion given by $(\frac{1}{n!}\, \xi^{\otimes n}, n \in \mathbb{N}_0)$. By (4.9a) we find that $\Phi \in (\mathcal{S})^*$, with chaos expansion given by $(F^{(n)}, n \in \mathbb{N}_0)$, has an $\mathcal{S}$–transform given by

$$\mathcal{S}\Phi(\xi) = \sum_{n=0}^{\infty} \langle \overline{F^{(n)}}, \xi^{\otimes n} \rangle. \tag{4.12}$$

We summarize our preceding discussion in the next proposition.

Proposition 4.18. *Assume that* $\Phi \in (\mathcal{S})^*$. *Then there is a* $p \in \mathbb{N}$, *so that* Φ *is in one–to–one correspondence with a sequence* $(F^{(n)}, n \in \mathbb{N}_0)$, *so that* $F^{(n)} \in \mathcal{S}_{-p}(\hat{\mathbb{R}}^n)$ *for all* $n \in \mathbb{N}$, *and such that* (4.9) *holds. In particular, for every* $n \in \mathbb{N}$, $F^{(n)} \in \mathcal{S}'(\hat{\mathbb{R}}^n)$. *Moreover, the* $\mathcal{S}$*–transform of* Φ *is given by equation* (4.12).

Conversely we can show the following.

Lemma 4.19. *Assume that* $(F^{(n)}, n \in \mathbb{N}_0)$ *is a sequence of symmetric distributions* $F^{(n)} \in \mathcal{S}'(\mathbb{R}^n)$, $F^{(0)} \in \mathbb{C}$, *with the following properties:*

(i) *there exists* $p \in \mathbb{N}$, *so that for all* $n \in \mathbb{N}$, $F^{(n)} \in \mathcal{S}_{-p}(\mathbb{R}^n)$,

(ii) *there exist constants* $K_1, K_2 > 0$, *so that for almost all* $n \in \mathbb{N}$

$$| F^{(n)} |_{2,-p} \le K_1\, K_2^n\, (n!)^{-1/2}.$$

Then $(F^{(n)}, n \in \mathbb{N}_0)$ *defines a unique element in* $(\mathcal{S})^*$, *whose* $\mathcal{S}$*–transform is given by*

$$\mathcal{S}\Phi\ (\xi) = \sum_{n=0}^{\infty} <\overline{F^{(n)}}, \xi^{\hat{\otimes} n}>.$$

Proof. Informally, we set

$$\Phi(x) := \sum_{n=0}^{\infty} <: x^{\otimes n} :, F^{(n)}>.$$

By construction, $\Phi \in (\mathcal{S})^*$ if and only if there exists $q \in \mathbb{N}_0$, so that

$$\sum_{n=0}^{\infty} n! \mid F^{(n)} \mid_{2,-q}^{2} < \infty,$$

cf. (3.21). Let $p<q$. Using (ii), we have the estimate

$$\sum_{n=0}^{\infty} n! \mid F^{(n)} \mid_{2,-q}^{2} = \sum_{n=0}^{\infty} n! \mid (H^{\otimes n})^{-(q-p)} F^{(n)} \mid_{2,-p}^{2}$$

$$\leq K_1 \sum_{n=0}^{\infty} K_2^n\, 2^{-2n(q-p)},$$

which for large enough q is finite. □

Remark. In Section 4.C we shall prove a criterion for whether a sequence $(F^{(n)}, n \in \mathbb{N}_0)$ defines an element in $(\mathcal{S})^*$, which appears to be more convenient in applications.

Example 4.20. Let $y \in \mathcal{S}'(\mathbb{R})$ and consider the sequence of kernels given by $F^{(n)} := \frac{1}{n!} y^{\otimes n}$, $n \in \mathbb{N}_0$. Clearly, this sequence fulfills the hypothesis of Lemma 4.19, and we denote the corresponding element in $(\mathcal{S})^*$ by E_y. For $\varphi \in (\mathcal{S})$ with kernels $(f^{(n)}, n \in \mathbb{N}_0)$, we have by (4.9)

$$<E_y, \varphi> = \sum_{n=0}^{\infty} <y^{\otimes n}, \tilde{f}^{(n)}>,$$

where $\tilde{f}^{(n)}$ is the version of $f^{(n)}$ which belongs to $\mathcal{S}(\hat{\mathbb{R}}^n)$. In case that $y \in L^2(\mathbb{R})$, we have the formula $E_y = :\exp(<\cdot\,, y>):$, and we shall continue to denote E_y this way also in the general case. It is interesting to observe that – again in the case $y \in L^2(\mathbb{R})$ – E_y is the Radon-Nikodym derivative of the white noise measure translated by y with respect to μ: $E_y = \frac{d\mu(\,\cdot - y)}{d\mu}$. Therefore this example hints to the fact, that the translate of the white noise measure in a *general* direction y of $\mathcal{S}'(\mathbb{R})$ – so that in general $\mu(\cdot - y)$ and μ are singular – admits a *generalized Radon-Nikodym derivative* E_y as an element in $(\mathcal{S})^*$. Based on Theorem 4.14, this has first been worked out in Potthoff and Yan (1992) . We shall give an account of these ideas in Section 4.B. The generalization to arbitrary strongly continuous affine transformations can be found in Section 4.C (cf. also the article Potthoff and Streit (1991b)).

As in the finite dimensional situation, there is in general no sense in a 'pointwise' multiplication by two elements in $(\mathcal{S})^*$. (Note however, that multiplication by an element in $(\mathcal{S})$ is meaningful, cf. Corollary 4.11.) On the other hand, there are two multiplicative operations on $(\mathcal{S})^*$. One of them will be discussed now, the other in Section 4.C. The following multiplicative property of $(\mathcal{S})^*$ has been discovered by Meyer and Yan (1988).

Theorem 4.21. *Let* $\Phi, \Psi \in (\mathcal{S})^*$ *and define a functional on* $\mathcal{S}(\mathbb{R})$ *by setting* $\Xi\,(\xi) := \mathcal{S}\Phi(\xi)\cdot\mathcal{S}\Psi(\xi)$. *Then* Ξ *is the* $\mathcal{S}$*-transform of an element* $\Phi{:}\Psi$ *in* $(\mathcal{S})^*$, *called the* Wick product *of* Φ *and* Ψ:

$$\mathcal{S}\Phi{:}\Psi\,(\xi) = \mathcal{S}\Phi(\xi)\cdot\mathcal{S}\Psi(\xi)\,. \tag{4.13}$$

Proof. Assume that $\Phi \in (\mathcal{S})_{-p}$ corresponds to the sequence $(F^{(n)}, n \in \mathbb{N}_0)$, while $\Psi \in (\mathcal{S})_{-q}$ corresponds to $(G^{(n)}, n \in \mathbb{N}_0)$. Then

$$\Xi(\xi) = \sum_{n=0}^{\infty}\sum_{m=0}^{n} <\overline{F^{(n-m)}}, \xi^{\otimes(n-m)}><\overline{G^{(m)}}, \xi^{\otimes m}>$$

$$= \sum_{n=0}^{\infty} <\overline{H^{(n)}}, \xi^{\otimes n}>,$$

with

$$H^{(n)} = \sum_{m=0}^{n} F^{(n-m)} \hat{\otimes} G^{(m)}.$$

(Note that the tensor product of two tempered distributions is again a tempered distribution by the nuclear theorem). It remains to show that for some $s \in \mathbb{N}_0$, we have

$$\sum_{n=0}^{\infty} n! \mid H^{(n)} |_{2,-s}^2 < \infty.$$

Let $r = p \vee q$, and $s \geq r$. Then

$$| H^{(n)} |_{2,-s}^2 \leq \Big(\sum_{m=0}^{n} | F^{(n-m)} \hat{\otimes} G^{(m)} |_{2,-s} \Big)^2$$

$$= \Big(\sum_{m=0}^{n} | F^{(n-m)} |_{2,-s} \, | G^{(m)} |_{2,-s} \Big)^2$$

$$\leq (n+1) \sum_{m=0}^{n} | F^{(n-m)} |_{2,-s}^2 \, | G^{(m)} |_{2,-s}^2 .$$

Let $s\text{-}r > \frac{1}{2}$. Then

$$\sum_{n=0}^{\infty} n! \mid H^{(n)} |_{2,-s}^2 \leq \sum_{n=0}^{\infty} n! \, (n+1) \sum_{m=0}^{n} | F^{(n-m)} |_{2,-s}^2 \, | G^{(m)} |_{2,-s}^2$$

$$= \sum_{n=0}^{\infty} (n+1) \sum_{m=0}^{n} \binom{n}{m} (n-m)! \, | F^{(n-m)} |_{2,-s}^2 \, m! \, | G^{(m)} |_{2,-s}^2$$

$$\leq \sum_{n=0}^{\infty} (n+1) \, 2^n \sum_{m=0}^{n} (n-m)! \, | F^{(n-m)} |_{2,-s}^2 \, m! \, | G^{(m)} |_{2,-s}^2$$

$$\leq \text{const.} \sum_{n=0}^{\infty} \sum_{m=0}^{n} (n\text{-}m)! \, | \, F^{(n-m)} \, |_{2,-r}^{2} \; m! \, | \, G^{(m)} \, |_{2,-r}^{2}$$

$$= \text{const.} \, \| \, \Phi \, \|_{2,-r}^{2} \, \| \, \Psi \, \|_{2,-r}^{2} \qquad \square$$

As a corollary to the proof of the last theorem we have the following result.

Corollary 4.22. *Let* $\Phi \in (\mathcal{S})_{-p}$, and $\Psi \in (\mathcal{S})_{-q}$, *with chaos decompositions* $(F^{(n)}, n \in \mathbb{N}_0)$, $(G^{(n)}, n \in \mathbb{N}_0)$ *resp. Then for every* $\alpha > \frac{1}{2}$, $\Phi{:}\Psi$ *belongs to* $(\mathcal{S})_{-p\vee q-\alpha}$, *and its chaos decomposition is given by* $(\sum_{m=0}^{n} F^{(n-m)} \hat{\otimes} G^{(m)}, n \in \mathbb{N}_0)$. *Moreover,*

$$\| \, \Phi : \Psi \, \|_{2,-p\vee q-\alpha} \leq \text{const.} \, \| \, \Phi \, \|_{2,-p} \, \| \, \Psi \, \|_{2,-q}.$$

Remark. In Section 4.C we give a simple proof of Theorem 4.21 based on the characterization results there.

We conclude this section with the discussion of a class of examples of functionals in $(\mathcal{S})^*$. Namely, we shall consider Gauß kernels, i.e. generalizations of the functional in Example 3.2. Applications of these functionals will be discussed in Chapters 11 and 12.

Example 4.23. (*Gauß Kernels*). In the following let K be a self–adjoint operator on $L^2(\mathbb{R})$ with domain $\mathfrak{D}(K) \supset \mathcal{S}(\mathbb{R})$, such that

(i) K maps $\mathcal{S}(\mathbb{R})$ into itself,

(ii) $\inf \sigma(K) > -1$,

where $\sigma(K)$ is the spectrum of K. We are interested in constructing a meaningful version of a functional which informally is given by

$$\Phi(x) = \exp(-\tfrac{1}{2}<x,\, Kx>).$$

Let us assume for the moment that the operator K has a pure point spectrum. We denote its spectral decomposition by

$$K\cdot = \sum_{m=1}^{\infty} \kappa_m\, (e_m,\,\cdot\,)\, e_m,$$

where $(e_m,\, m\in\mathbb{N})$ are the eigenvectors of K, which form a complete orthonormal system of $L^2(\mathbb{R})$. Without loss of generality for the following, we may assume that $e_m\in\mathcal{S}(\mathbb{R})$ for all $m\in\mathbb{N}$. Note that due to condition (ii) above, there exists $\epsilon > 0$ so that for all $m\in\mathbb{N}$, $\kappa_m > -1+\epsilon$. Now we set for $m\in\mathbb{N}$

$$K_n\cdot = \sum_{m=1}^{n} \kappa_m\, (e_m,\,\cdot\,)\, e_m,$$

and form the well-defined random variable $(x\in\mathcal{S}'(\mathbb{R}))$

$$\Phi_{un,n}(x) := \exp\,(-\tfrac{1}{2}<x,\, K_n x>)$$

$$= \exp\,(\textstyle\sum_{m=1}^{n} -\tfrac{1}{2}\,\kappa_m <x,\, e_m>^2).$$

Due to the fact that for all m, $\kappa_m > -1+\epsilon$, it is easy to see that there exists $p > 1$, so that for all $n\in\mathbb{N}$, $\Phi_{un,n}\in(L^p)\subset(\mathcal{S})^*$. Therefore we may compute the $\mathcal{S}$-transform of $\Phi_{un,n}$, and an elementary calculation gives the following result

$$\mathcal{S}\Phi_{un,n}(\xi) = (\det(1+K_n))^{-1/2} \exp(-\tfrac{1}{2}(\xi,\, \widetilde{K}_n\xi))$$

$$= (\mathbb{E}\Phi_{un,n}) \exp(-\tfrac{1}{2}(\xi,\, \widetilde{K}_n\xi)).$$

Here we denoted by $\widetilde{\ }$ the mapping $u \mapsto \widetilde{u} = \frac{u}{1+u}$, $u > -1$. Now we divide by $\mathbb{E}\Phi_{un,n}$ in order to renormalize the functional $\Phi_{un,n}$:

$$\mathcal{S}\Phi_n(\xi) := \exp(-\tfrac{1}{2}(\xi,\widetilde{K}_n\xi)). \tag{4.14}$$

By linearity of $\mathcal{S}$ it is clear that in fact the left hand side of (4.14) is the $\mathcal{S}$–transform of the functional $\Phi_n = (\mathbb{E}\Phi_{un,n})^{-1}\,\Phi_{un,n}$, which is for every $n \in \mathbb{N}$ in (L^p).

Since $\|H^{-q}\| < 1$, $q > 0$, it follows that $\Gamma(H^{-q})$ extends to a contraction on every (L^p), $1 \le p \le +\infty$ (cf., e.g., Nelson (1973)). It is straightforward to compute the following formula

$$\mathcal{S}\Gamma(H^{-q})\Phi_n(\xi) = \exp(-\tfrac{1}{2}(\xi,\, H^{-q}\widetilde{K}_n H^{-q}\xi)). \tag{4.15}$$

Now we want to argue that from (4.15) we can deduce that for sufficiently large q, $\Gamma(H^{-q})\Phi_n$ converges to a limit in (L^2) as n tends to infinity.

Consider the functional

$$\Xi(\xi) := \exp(-\tfrac{1}{2}(\xi,\, B\xi)), \quad \xi \in \mathcal{S}(\mathbb{R}),$$

where B is a Hilbert–Schmidt operator on $L^2(\mathbb{R})$, with Hilbert–Schmidt norm strictly less than 1. Then one can view the functional Ξ as corresponding to the following element in the Fock space $\Gamma(L^2(\mathbb{R}))$: $(\Xi^{(n)}, n \in \mathbb{N}_0)$, with

$$\Xi^{(n)} = \begin{cases} 1 & n=0 \\ 0 & n \text{ odd} \\ (-\tfrac{1}{2})^m \frac{1}{m!}\, b_{2m} & n=2m,\ m \in \mathbb{N}, \end{cases}$$

where b_{2m} is the symmetric 2m-multilinear form defined by B on $L^2(\mathbb{R})$. I.e., it is the symmetrization of the mapping

$$(f_1, f_2, \dots, f_{2m}) \to (f_1, Bf_2) \cdots (f_{2m-1}, Bf_{2m}).$$

One easily checks that

$$\| b_{2m} \|_{H.S.} \le \| B \|_{H.S.}^m.$$

We find that $\Xi \in \Gamma(L^2(\mathbb{R}))$, since the square of its norm in $\Gamma(L^2(\mathbb{R}))$ is estimated as follows

$$\| \Xi \|^2 = \sum_{m=0}^{\infty} \frac{(2m)!}{2^{2m}(m!)^2} \| b_{2m} \|_{H.S.}^2$$

$$\le \text{const.} \sum_{m=0}^{\infty} \| B \|_{H.S.}^{2m}$$

$$= \text{const.} \, (1 - \| B \|_{H.S.}^2)^{-1}, \tag{4.16}$$

where we used Stirling's theorem.

Similarly one can show the following. Let B_1 and B_2 be two operators as above, and consider the functionals

$$\Xi_i(\xi) := \exp(-\tfrac{1}{2}(\xi, B_i \xi)), \quad i=1,2, \ \xi \in \mathcal{S}(\mathbb{R}).$$

Then one finds the following inequality

$$\| \Xi_1 - \Xi_2 \|^2 \le \text{const.} \| B_1 - B_2 \|_{H.S.}^2 \sum_{m=0}^{\infty} m^2 \beta^{2(m-1)}, \tag{4.17}$$

where $\beta = \max(\|B_1\|_{H.S.}, \|B_2\|_{H.S.})$, and the last sum is convergent.

Now let us return to the expression (4.15). Since the sequence $(\widetilde{K}_n, n \in \mathbb{N})$

of operators is uniformly bounded (by $1+\epsilon^{-1}$), and since H^{-1} is Hilbert–Schmidt, $\{\widetilde{K}_n H^{-1}, n\in\mathbb{N}\}$ is uniformly bounded in Hilbert–Schmidt norm. Furthermore, for large enough q we have $\| H^{-q}\widetilde{K}_n H^{-q} \|_{H.S.} < 1$, because $\| H^{-1} \| = \frac{1}{2}$. Therefore inequality (4.16) implies that for large enough q, the sequence $(\mathcal{S}\Gamma(H^{-q})\Phi_n, n\in\mathbb{N})$ belongs to $\Gamma(L^2(\mathbb{R}))$. By Theorem 2.7 we have that for every $n\in\mathbb{N}$, $\Gamma(H^{-q})\Phi_n \in (L^2)$.

Next we observe that K_n converges to K in strong resolvent sense. By standard theorems (e.g., Reed and Simon (1972), Theorem VIII.20) this implies that $\widetilde{K}_n$ converges strongly to $\widetilde{K}$ on $L^2(\mathbb{R})$. From this, the fact that H^{-1} is compact, so that it maps a bounded set in $L^2(\mathbb{R})$ into a pre-compact set, and Lemma III.3.7 of Kato (1976), it follows that $\widetilde{K}_n H^{-q}$ converges to $\widetilde{K}H^{-q}$ in norm. Therefore $H^{-q}\widetilde{K}_n H^{-q}$ converges to $H^{-q}\widetilde{K}H^{-q}$ in Hilbert–Schmidt norm, whenever $q\geq 1$. Thus we conclude from (4.17) (and the fact that $\mathcal{S}$ is unitary) that for large enough q, $\Gamma(H^{-q})\Phi_n$ converges in (L^2) to an element denoted by Ψ, whose $\mathcal{S}$–transform is given by $\exp(-\frac{1}{2}(\xi, H^{-q}\widetilde{K}H^{-q}\xi))$.

Let $k = [q]+1$ and let Φ be the element in $(\mathcal{S})_{-k}$ which is given by $\Gamma(H^q)\Psi$, i.e., for $\varphi\in(\mathcal{S})$ we set

$$<\Phi, \varphi> = (\Psi, \Gamma(H^q)\varphi)_{(L^2)}.$$

Then it is obvious that Φ_n converges to Φ in $(\mathcal{S})_{-k}$ and that Φ has $\mathcal{S}$–transform given by $\mathcal{S}\Phi(\xi) = \exp(-\frac{1}{2}(\xi, \widetilde{K}\xi))$.

Finally, let us remark that we may drop the assumption that K has pure point spectrum. By the von Neumann–Weyl theorem (e.g., Kato (1976)), there exists a sequence $(K_n, n\in\mathbb{N})$ of self-adjoint operators such that $K-K_n$ vanishes in Hilbert–Schmidt norm as n tends to infinity, and every K_n has pure point spectrum. Approximating K first by such operators and then repeating all the above arguments we have proven the following theorem.

Theorem 4.24. *Let* K *be a self–adjoint operator on* $L^2(\mathbb{R})$ *such that* $\mathcal{D}(K)\supset\mathcal{S}(\mathbb{R})$, K *leaves* $\mathcal{S}(\mathbb{R})$ *invariant, and* $\inf \sigma(K) > -1$. *Then there exists* $\Phi \in (\mathcal{S})^*$ *such that*

(i) $$\mathcal{T}\Phi(\xi) = \exp(-\tfrac{1}{2}(\xi, (K+1)^{-1}\xi)), \tag{4.18}$$

(ii) $$\mathcal{S}\Phi(\xi) = \exp(-\tfrac{1}{2}(\xi, \frac{K}{K+1}\xi)), \tag{4.19}$$

(iii) *there exists* $p > 1$, *and a sequence* $(K_n, n \in \mathbb{N})$ *of finite rank operators converging to* K *in strong resolvent sense, and*

$$\Phi_n(x) = Z_n^{-1} \exp(-\tfrac{1}{2}\langle x, K_n x\rangle), \tag{4.20a}$$

$$Z_n = \mathbb{E}(\exp(-\tfrac{1}{2}\langle \cdot, K_n \cdot\rangle)), \tag{4.20b}$$

belongs to (L^p) *for all* $n \in \mathbb{N}$, *and* Φ *is the strong limit of* Φ_n in $(\mathcal{S})^*$. *Moreover, the degree of* Φ *is equal to the smallest integer* k, *so that* $\| H^{-k} K H^{-k} \|_{H.S.} < 1$.

The generalized functional $\Phi \in (\mathcal{S})^*$ in Theorem 4.26 will also informally be denoted by $\Phi(x) = \mathcal{N}\exp(-\tfrac{1}{2}\langle x, Kx\rangle)$.

For the proof of (4.18), one only has to apply the relation (2.10), which extends from (L^2) to $(\mathcal{S})^*$ (cf. also Section 4.C).

Assume that we are given a Gaussian measure ν on the Borel σ-algebra $\mathcal{B}$ of $\mathcal{S}'(\mathbb{R})$, with characteristic functional equal to the right hand side of (4.18). Suppose furthermore that K satisfies the above conditions. From the preceding discussion it is clear that the expectations of this measure can be realized via the generalized functional Φ. The next two sections deal with related questions. In Section 4.B, we investigate the question whether positive elements in $(\mathcal{S})^*$ are given by measures on $(\mathcal{S}'(\mathbb{R}), \mathcal{B})$. As in the finite dimensional case the answer is positive, and this result is due to Kondratiev and Samoilenko (1976), and independently to Yokoi (1990). (For corresponding results in the case of the space $\mathcal{D}^*$ we also refer to Itô (1990), Potthoff (1987) and Sugita (1988).) In Section 4.C we answer the inverse question: when is a measure ν on $(\mathcal{S}'(\mathbb{R}), \mathcal{B})$ represented by an element in $(\mathcal{S})^*$? The answer will be given in terms of the characteristic functional of the measure ν. On the basis of this result, we shall discuss classes of non-Gaussian measures arising in quantum field theory in Chapter 11 (cf. also Albeverio, Hida, et al. (1990a), and Potthoff and Streit (1990)).

4.B Positive Generalized Functionals

In the case of the space $\mathcal{S}'(\mathbb{R}^d)$, $d \in \mathbb{N}$, one has the well-known theorem of Bochner–Schwartz, which states that every positive element in $\mathcal{S}'(\mathbb{R}^d)$ is given by a measure (with at most polynomial increase at infinity). In the recent years, this theorem found various generalizations to spaces of generalized functions in infinite dimensions. We refer to the articles of Itô (1990), Kondratiev and Samoilenko (1976), Potthoff (1987), Sugita (1988), and Yokoi (1990). In the case of the space $(\mathcal{S})^*$, the corresponding results have been proved independently by Kondratiev and Samoilenko (1976), and by Yokoi (1990).

We begin with a definition.

Definition 4.25. *An element* $\Phi \in (\mathcal{S})^*$ *with the property that for every* μ*–a.e. positive element* $\varphi \in (\mathcal{S})$ *we have* $\langle \Phi, \varphi \rangle \geq 0$, *is called* positive. *The cone of positive elements in* $(\mathcal{S})^*$ *is denoted by* $(\mathcal{S})^*_+$.

Remark. Yokoi (1990) defines positivity of $\Phi \in (\mathcal{S})^*$ by the following condition: Φ is positive if $\varphi \in (\mathcal{S})$ with continuous version $\widetilde{\varphi} \geq 0$ implies $\langle \Phi, \varphi \rangle \geq 0$. However, both definitions are equivalent as an argument analogous to the one used in the proof of the uniqueness of $\widetilde{\varphi}$ in Proposition 4.7 shows.

Theorem 4.26. *Let* $\Phi \in (\mathcal{S})^*_+$. *Then there exists a finite measure* ν_Φ *on* $\mathcal{B}$, *so that for all* $\varphi \in (\mathcal{S})$

$$\langle \Phi, \varphi \rangle = \int \widetilde{\varphi}(x)\, d\nu_\Phi(x), \tag{4.21}$$

where $\widetilde{\varphi}$ *is the pointwise defined, continuous version of* φ.

For the proof we prepare a lemma.

Lemma 4.27. *Let* $\Phi \in (\mathcal{S})^*$. *Then the map* $\xi \mapsto \mathcal{T}\Phi(\xi) := \langle\overline{\Phi}, \exp(i\langle\cdot, \xi\rangle)\rangle$ *is continuous from* $\mathcal{S}(\mathbb{R})$ *into* $\mathbb{C}$.

Proof. We claim that the mapping $\xi \mapsto \langle\cdot, \xi\rangle$ is continuous from $\mathcal{S}(\mathbb{R})$ into $(\mathcal{S})$. Namely, if $\xi, \eta \in \mathcal{S}(\mathbb{R})$, we find for every $p \in \mathbb{N}$,

$$\| \langle\cdot, \xi\rangle - \langle\cdot, \eta\rangle \|_{2,p} = \| \langle\cdot, H^p\xi - H^p\eta\rangle \|_2$$

$$= | H^p\xi - H^p\eta |_2$$

$$= | \xi - \eta |_{2,p}.$$

Since the standard topology on $\mathcal{S}(\mathbb{R})$ is equivalent to the topology generated by the norms $|\cdot|_{2,p}$, we get the claimed continuity. Note that the exponential series of $\exp(\lambda\langle\cdot, \xi\rangle)$ converges in $(\mathcal{S})$ for every $\lambda \in \mathbb{C}$, uniformly on bounded sets in $\mathcal{S}(\mathbb{R})$: this follows directly from Proposition 4.2 and a simple computation. Therefore, the fact that $(\mathcal{S})$ is an algebra with continuous (pointwise) multiplication (Theorem 4.9) implies that the mapping $\xi \mapsto \exp(i\langle\cdot, \xi\rangle)$ is continuous from $\mathcal{S}(\mathbb{R})$ into $(\mathcal{S})$. Continuity of Φ on $(\mathcal{S})$ concludes the proof. □

Proof of Theorem 4.26. Let $\Phi \in (\mathcal{S})^*_+$. Clearly, this entails that Φ is real: $\overline{\Phi} = \Phi$. If $\varphi \in \mathcal{E}$ then $\overline{\varphi}\cdot\varphi$ is positive and $\langle\Phi, \overline{\varphi}\cdot\varphi\rangle \geq 0$. This implies that $\xi \mapsto \mathcal{T}\Phi(\xi) = \langle\Phi, \exp(i\langle\cdot, \xi\rangle)\rangle$ is positive definite (cf. Appendix A.3). Due to Lemma 4.27 we may apply Minlos' theorem, Theorem 1.1, to the continuous $\mathcal{S}(\mathbb{R})$-functional $\xi \mapsto \mathcal{T}\Phi(\xi)$, and have the existence of a finite measure ν_Φ on $\mathcal{B}$ so that

$$\langle\Phi, \exp(i\langle\cdot, \xi\rangle)\rangle = \int_{\mathcal{S}'(\mathbb{R})} \exp(i\langle x, \xi\rangle)\, d\nu_\Phi(x).$$

By linearity this equality extends to the dense subspace $\mathcal{E}$ of $(\mathcal{S})$: for all $\varphi \in \mathcal{E}$ we have

$$<\Phi, \varphi> = \int \widetilde{\varphi}(x)\, d\nu_\Phi(x). \tag{4.22}$$

Now let $\varphi \in (\mathcal{S})$ (and identify φ with its continuous version). By Theorem 4.1 we know that there exists a sequence $(\varphi_n,\, n \in \mathbb{N})$ in $\mathcal{E}$ which converges to φ in $(\mathcal{S})$. Since multiplication is continuous in $(\mathcal{S})$ (Theorem 4.9), this implies that also $|\varphi_n - \varphi|^2$ converges to zero in $(\mathcal{S})$. But this shows that (φ_n) is Cauchy in $L^2(\mathcal{S}'(\mathbb{R}), \mathcal{B}, d\nu_\Phi)$, converging there to an element ψ. Selecting a subsequence which we denote again by $(\varphi_n,\, n \in \mathbb{N})$, we find that $\varphi_n \to \psi$, ν_Φ–a.e. On the other hand, Corollary 4.8 tells us that $\varphi_n \to \varphi$ pointwise. Thus $\varphi = \psi$, ν_Φ–a.e., and (4.24) holds for $\varphi \in (\mathcal{S})$. □

The next result has been shown in Yokoi (1990).

Theorem 4.28. *Let $\Phi \in (\mathcal{S})^*$. Then Φ is positive, if and only if $\mathcal{T}\Phi$ is positive definite.*

Proof. If Φ is positive it is obvious that $\mathcal{T}\Phi$ is positive definite. Assume that $\mathcal{T}\Phi$ is positive definite. Note that due to Lemma 4.27, $\xi \mapsto \mathcal{T}\Phi(\xi)$ is continuous from $\mathcal{S}(\mathbb{R})$ into $\mathbb{C}$. Therefore we may apply Minlos' theorem as before to obtain the positive measure ν_Φ with equality (4.22) for all $\varphi \in \mathcal{E}$. By the same arguments as in the proof of Theorem 4.26, this equality extends to all $\varphi \in (\mathcal{S})$. Clearly, if $\varphi \geq 0$, it has a pointwise positive representative (cf. the Remark after Definition 4.25). Thus we find $<\Phi, \varphi> \geq 0$. □

Here are a number of examples of positive elements in $(\mathcal{S})^*$.

Example 4.29. Consider the generalized functional $E_y = :e^{<\cdot, y>}:$, $y \in \mathcal{S}'(\mathbb{R})$, Example 4.20. It easy to compute its $\mathcal{T}$–transform: for $\xi \in \mathcal{S}(\mathbb{R})$, $\mathcal{T}E_y(\xi) =$

$e^{i<y,\,\xi>-\frac{1}{2}|\,\xi\,|_2^2}$. Since this is the product of two positive definite functionals, it is positive definite. Thus by Theorem 4.28, $E_y \in (\mathcal{S})^*_+$. Actually, $\mathcal{T}E_y$ is the Fourier transformation of the Gaussian measure on $\mathcal{S}'(\mathbb{R})$ with mean $y \in \mathcal{S}'(\mathbb{R})$, and covariance being given by the scalar product of $L^2(\mathbb{R})$. We shall use this below.

Example 4.30. Consider the class of generalized functionals in Example 4.23. By the convergence statement in Theorem 4.24.iii, it is clear that the Gauß kernel Φ of the operator K is positive, since it is the strong limit of the positive functionals Φ_n. However, we want to include a warning here. Consider the simplified case, where K is multiplication by a number $-2c > -1$. Then this example and Example 3.2 coincide (for the choice h=1). Note, that as in Example 3.2 one can actually prove that there exists an element $\Phi_c \in (\mathcal{S})^*$, with $\mathcal{S}$-transform $\mathcal{S}\Phi_c(\xi) = \exp(\hat{c}|\,\xi\,|_2^2)$, $\hat{c} = \frac{c}{1\text{-}2c}$, whenever $c \in \mathbb{C}$ with $c \neq \frac{1}{2}$. (For another proof we refer the reader to the next section.) In particular, c may be *any* complex number different from $\frac{1}{2}$. But then Φ_c will in general *not* be positive for real c! Actually, it is positive if and only if c is real, and $c < \frac{1}{2}$, as the following computation shows.

From the $\mathcal{S}$-transform of Φ_c, we can read off that its chaos decomposition is given by the sequence of kernels $(\frac{1}{n!}\,\hat{c}^n\,\mathrm{Tr}^{\otimes n},\ n \in \mathbb{N}_0)$. Let $\eta \in \mathcal{S}(\mathbb{R})$. Then

$$X_\eta^2 = <\cdot\,,\eta>^2 = :<\cdot\,,\eta>^2: + |\,\eta\,|_2^2,$$

is certainly a positive element in $(\mathcal{S})$. We find

$$<\Phi_c, X_\eta^2> = 2!\,\hat{c}\,<\mathrm{Tr},\eta^{\otimes 2}> + |\,\eta\,|_2^2$$

$$= \frac{1}{1\text{-}2c}\,|\,\eta\,|_2^2,$$

which is positive if and only if $c < \frac{1}{2}$.

Example 4.31. The evaluation mapping $\varphi \mapsto \varphi(x)$, $\varphi \in (\mathcal{S})$, $x \in \mathcal{S}'(\mathbb{R})$, of Corollary 4.8 is obviously a positive element in $(\mathcal{S})^*$.

Example 4.32. Consider Donsker's delta function $\delta_{t,y} = \delta_y \circ B(t)$ in Example 3.3, where δ_y is the Dirac distribution concentrated at $y \in \mathbb{R}$, B is the white noise version of Brownian motion, and $t>0$. Shieh and Yokoi (1990) have shown that $\delta_{t,y}$ is an element in $(\mathcal{S})^*_+$. Another simple proof that $\delta_{t,y} \in (\mathcal{S})^*$ can be found in Example 4.44, while Example 4.33 below gives an argument for the positivity of $\delta_{t,y}$. Let us study Donsker's delta function a little more.

According to Theorem 4.26, $\delta_{t,y}$ defines a positive finite measure on $(\mathcal{S}'(\mathbb{R}),\mathcal{B})$. The corresponding normalized measure will be denoted by $\mu_{t,y}$. We shall now compute the characteristic function of $\mu_{t,y}$ by using the series representation (3.16) of $\delta_{t,y}$. First we calculate the $\mathcal{T}$-transform of $\delta_{t,y}$. Let $\xi \in \mathcal{S}(\mathbb{R})$, then

$$\mathcal{T}\delta_{t,y}(\xi) = \langle\delta_{t,y}, e^{i\langle\cdot\,,\,\xi\rangle}\rangle$$

$$= e^{-\frac{1}{2}|\xi|_2^2} \langle\delta_{t,y}, :e^{i\langle\cdot\,,\,\xi\rangle}:\rangle.$$

Now we make use of (3.16), and denote $p_t(y) = (2\pi t)^{-1/2}\exp(-y^2/2t)$:

$$\langle\delta_{t,y}, :e^{i\langle\cdot\,,\,\xi\rangle}:\rangle = p_t(y) \sum_{n=0}^{\infty} \frac{1}{n!\,t^n} :y^n:_t \frac{i^n}{n!} \langle :B(t)^n:, :\langle\cdot\,,\,\xi\rangle^n:\rangle$$

$$= p_t(y) \sum_{n=0}^{\infty} \frac{i^n}{n!} :y^n:_t \Big(t^{-1}\int_0^t \xi(s)\,ds\Big)^n$$

$$= p_t(y) :\exp\Big(\frac{i\,y}{t}\int_0^t \xi(s)\,ds\Big):_t,$$

where we used the fact that

$$\sum_{n=0}^{\infty} \frac{\lambda^n}{n!} :y^n:_t = :e^{\lambda y}:_t = \exp(\lambda y - \tfrac{\lambda^2}{2}t), \quad y \in \mathbb{R}.$$

Thus we get the following formula

$$\mathcal{T}\delta_{t,y}(\xi) = p_t(y)\exp\left(\tfrac{i}{t}\,(1_{[0,t]},\,\xi) - \tfrac{1}{2}(\xi,\,K_t\xi)\right),$$

where – informally – K_t is given by the kernel

$$K_t(u,v) = \delta(u-v) - t^{-1}1_{[0,t]}(u)\,1_{[0,t]}(v).$$

More precisely, K_t is the orthogonal projection in $L^2(\mathbb{R})$ onto the subspace which is orthogonal to the space spanned by $1_{[0,t]}$, i.e., $K_t\xi = \xi - t^{-1}1_{[0,t]}\,(\xi,\,1_{[0,t]})$. It is obvious that $p_t(y) = \langle\delta_{t,y},\,1\rangle$ is the total mass of the measure corresponding to $\delta_{t,y}$. Therefore $\mu_{t,y}$ has the characteristic functional

$$\int_{\mathcal{S}'(\mathbb{R})} e^{i\langle x,\,\xi\rangle}\,d\mu_{t,y}(x) = \exp\left(\tfrac{i}{t}\,(1_{[0,t]},\,\xi) - \tfrac{1}{2}(\xi,\,K_t\xi)\right).$$

This shows that $\mu_{t,y}$ is the measure of the noise of a Brownian motion which is pinned at y at time $t>0$. In fact, obviously $\mu_{t,y}$ is a Gaussian measure, and therefore the random variables $B(s) = \langle\cdot\,,\,1_{[0,s)}\rangle$, $s>0$, are Gaussian. From the characteristic functional above we can easily read off their mean and covariance under $\mu_{t,y}$:

$$\int B(s)\,d\mu_{t,y} = y\,\frac{s\wedge t}{t},$$

$$\int B(s_1)\,B(s_2)\,d\mu_{t,y} = s_1\wedge s_2 - \frac{t\wedge s_1\cdot t\wedge s_2}{t}.$$

(We ask the interested reader to derive these two equations directly from (3.16) as an exercise.) These are just the mean and covariance of a Brownian motion pinned at y at time $t>0$.

Clearly we have informally

$$\mathbb{E}(\varphi \mid B(t) = y) = \frac{\langle\delta_{t,y},\,\varphi\rangle}{\langle\delta_{t,y},\,1\rangle} = \int \varphi(x)\,d\mu_{t,y}(x),$$

for suitable φ, where $\mathbb{E}(\cdot \mid \cdot)$ stands for conditional expectation. Now observe that the covariance of $B(s_1)$, $B(s_2)$ is zero when t is between s_1 and s_2. Thus the above computation shows that for $s \geq t$, B(s) is conditionally independent of the past up to time t, given B at time t: this is the *Markov property* of Brownian motion. The equation for the mean of B(s) under $\mu_{t,y}$ tells us that $\mathbb{E}(B(s) \mid B(t)) = B(t)$, for all $s \geq t$. This is the *martingale property* of Brownian motion. Furthermore, if s_1, $s_2 \geq t$, then the covariance of $B(s_1)$ and $B(s_2)$ under $\mu_{t,y}$ is equal to $s_1 \wedge s_2 - t$. This is the same as the covariance of a Brownian motion starting at y at time t. Also, both – B(s) under $\mu_{t,y}$ for $s \geq t$, and Brownian motion starting at y at time t – have the same mean, namely y. Thus Brownian motion *starts afresh* at every moment of time.

Example 4.33. We can generalize the preceding example, as in Example 3.4. Assume that $T \in \mathcal{S}'(\mathbb{R})$ is positive. Then we can find a sequence $(T_n, n \in \mathbb{N})$ of positive functions in $\mathcal{S}(\mathbb{R})$ which converges strongly to T in $\mathcal{S}'(\mathbb{R})$. For example, we can regularize T by convolution with a sequence of normalized Gauß functions with variance parameter tending to zero. It follows from the proof of Watanabe's theorem (cf. Watanabe (1983) and Chapter 7), that for $t > 0$, $T_n \circ B(t)$ converges strongly in $\mathcal{D}^*$ to an element denoted by $T \circ B(t)$. It follows then from Proposition 4.13 that $T_n \circ B(t)$ converges weakly (and therefore strongly) to $T \circ B(t)$ in $(\mathcal{S})^*$. It is clear that for every $n \in \mathbb{N}$, $T_n \circ B(t)$ is positive. Consequently, $T \circ B(t) \in (\mathcal{S})^*_+$.

A very important class of positive generalized white noise functionals is given by the Euclidean and vacuum measures of certain models in quantum field theory. They will be discussed in Chapter 11.

Now we come back to the idea mentioned in Example 4.20, cf. also Example 4.29 above. From Theorem 4.15 we know that for every $y \in \mathcal{S}'(\mathbb{R})$, the translation operator τ_y acts continuously on $(\mathcal{S})$. Note that $1 \in (\mathcal{S})^*$, i.e., the expectation belongs to $(\mathcal{S})^*$ and represents in the sense of (4.21) the measure μ on $(\mathcal{S}'(\mathbb{R}), \mathcal{B})$. Theorem 4.15 implies that for every $y \in \mathcal{S}'(\mathbb{R})$, there exists some $q \in \mathbb{N}_0$, so that

$$|<1, \tau_y\varphi>| = \left|\int \varphi(x)\, d\mu(x-y)\right|$$

$$\leq \text{const.} \, \| \varphi \|_{2,q}.$$

As before we have denoted $\tau_y\varphi = \varphi(\cdot +y)$. Therefore the mapping $\varphi \mapsto <1, \tau_y\varphi>$ is a linear continuous functional on $(\mathcal{S})$. Hence there exists an element E_y in $(\mathcal{S})^*$, so that $<1, \tau_y\varphi> = <E_y, \varphi>$. It is not hard to see that E_y is exactly the generalized functional discussed in Example 4.20. (For example, for $y \in L^2(\mathbb{R})$ this follows from the translation formula for the white noise measure and the general case is obtained by limiting argument.) Thus we have the formula

$$\int \varphi(x)\, d\mu(x-y) = <E_y, \varphi> \tag{4.23}$$

for all $\varphi \in (\mathcal{S})$ and all $y \in \mathcal{S}'(\mathbb{R})$. In case that $y \in L^2(\mathbb{R})$, μ and $\mu(\cdot\, -y)$ are absolutely continuous and in this case $E_y(x) = \exp(<x, y> - \frac{1}{2}|y|_2^2) = :\exp(<x, y>):$, $x \in \mathcal{S}'(\mathbb{R})$, is the Radon–Nikodym derivative. In the general case $y \in \mathcal{S}'(\mathbb{R})$, we consider E_y as a *generalized Radon-Nikodym derivative.* The preceding considerations have been generalized in Potthoff and Yan (1992) as in the next two theorems.

Theorem 4.34. *Let* ν *be a positive finite measure on* $(\mathcal{S}'(\mathbb{R}), \mathcal{B})$, *such that* ν *is represented by a positive generalized functional* Φ *in* $(\mathcal{S})^*$. *In this case write* $\Phi = \frac{d\nu}{d\mu}$. *Let* $y \in \mathcal{S}'(\mathbb{R})$, *and denote* $\nu(\cdot\, -y)$ *by* $\tau_y\nu$. *Then also* $\tau_y\nu$ *is represented by an element* $\frac{d\tau_y\nu}{d\mu} \in (\mathcal{S})^*$ *and the following formula holds*

$$\frac{d\tau_y\nu}{d\mu} = \frac{d\mu_y}{d\mu} : \frac{d\nu}{d\mu} = E_y : \frac{d\nu}{d\mu}. \tag{4.24}$$

Proof. Without loss of generality we may assume that $\frac{d\nu}{d\mu} \in (\mathcal{S})_{-p}$, $p \in \mathbb{N}_0$. Let $\varphi \in (\mathcal{S})$, $\widetilde{\varphi}$ denotes its pointwise defined, continuous version. Then

$$\int \widetilde{\varphi}(x)\, d\tau_y\nu(x) = <\frac{d\nu}{d\mu}, \tau_y\varphi>$$

and from Theorem 4.15 we know that for some $q\in\mathbb{N}_0$, $q>p$, the following estimate holds

$$\left| \int \widetilde{\varphi}(x)\, d\tau_y\nu(x) \right| \leq \left\| \frac{d\nu}{d\mu} \right\|_{2,-p} \left\| \varphi \right\|_{2,q} .$$

Thus the mapping $\varphi \mapsto \int \widetilde{\varphi}\, d\tau_y\nu(x)$ defines a continuous linear functional on $(\mathcal{S})$. Therefore $\frac{d\tau_y\nu}{d\mu}\in(\mathcal{S})_{-q}$ and we have

$$<\frac{d\tau_y\nu}{d\mu}, \varphi> = <\frac{d\nu}{d\mu}, \tau_y\varphi>,$$

for all $\varphi\in(\mathcal{S})$. Making the particular choice $\varphi(x) = :\exp(<x, \xi>):$, $\xi\in\mathcal{S}(\mathbb{R})$, we find

$$\begin{aligned} \mathcal{S}\frac{d\tau_y\nu}{d\mu}(\xi) &= <\frac{d\tau_y\nu}{d\mu}, :\exp<\cdot, \xi>:> \\ &= e^{<y, \xi>} <\frac{d\nu}{d\mu}, :\exp<\cdot, \xi>:> \\ &= \mathcal{S}E_y(\xi)\cdot \mathcal{S}\frac{d\nu}{d\mu}(\xi) \end{aligned}$$

which proves formula (4.24) (cf. Theorem 4.21). □

The scaling operator σ_λ, $\lambda\in\mathbb{R}$, can be discussed similarly. Assume that ν is a measure on $(\mathcal{S}'(\mathbb{R}), \mathcal{B})$, and for $\lambda\in\mathbb{R}\backslash\{0\}$ put $\sigma_\lambda\nu(E) = \nu(\lambda^{-1}E)$, $E\in\mathcal{B}$. Then the following result has been proved in Potthoff and Yan (1992) .

Theorem 4.35. *If* $\frac{d\nu}{d\mu}\in(\mathcal{S})^*$, *then for every* $\lambda\in\mathbb{R}\backslash\{0\}$, $\frac{d\sigma_\lambda\nu}{d\mu}\in(\mathcal{S})^*$, *and the following formula holds*

$$\frac{d\sigma_\lambda \nu}{d\mu} = \frac{d\sigma_\lambda \mu}{d\mu} : \Gamma(\lambda)\frac{d\nu}{d\mu}, \tag{4.25}$$

where λ *is identified with the operator of multiplication by* λ *on* $L^2(\mathbb{R})$. *Moreover, for every* $\xi \in \mathcal{S}(\mathbb{R})$,

$$\mathcal{S}\frac{d\sigma_\lambda \mu}{d\mu}(\xi) = e^{-\frac{1}{2}(1-\lambda^2)|\xi|_2^2}. \tag{4.26}$$

Proof. The proof of this theorem is very similar to the one of Theorem 4.24. Let $\varphi \in (\mathcal{S})$. We have

$$\int \widetilde{\varphi}(x)\, d\sigma_\lambda \nu(x) = \int \widetilde{\varphi}(\lambda x)\, d\nu(x)$$

$$= <\frac{d\nu}{d\mu}, \sigma_\lambda \varphi>.$$

$d\nu/d\mu \in (\mathcal{S})_{-p}$, for some $p \in \mathbb{N}$. By Theorem 4.15 we obtain the estimate

$$|\int \widetilde{\varphi}(x)\, d\sigma_\lambda \nu(x)| \leq \| \frac{d\nu}{d\mu} \|_{2,-p} \| \sigma_\lambda \varphi \|_{2,p}$$

$$\leq K_\lambda \| \frac{d\nu}{d\mu} \|_{2,-p} \| \varphi \|_{2,q},$$

for some $q \in \mathbb{N}$, and some constant K_λ depending only on $\lambda \in \mathbb{R}$. Therefore the mapping $\varphi \mapsto \int \widetilde{\varphi}(x)\, d\sigma_\lambda \nu(x)$ belongs to $(\mathcal{S})^*$. Let $\xi \in \mathcal{S}(\mathbb{R})$, and choose $\varphi(x) = :\exp(<\cdot, \xi>):$. Then we get

$$<\frac{d\sigma_\lambda \nu}{d\mu}, :e^{<\cdot, \xi>}:> = e^{-\frac{1}{2}|\xi|_2^2} \int e^{<x, \lambda\xi>}\, d\nu(x)$$

$$= e^{-\frac{1}{2}(1-\lambda^2)|\xi|_2^2} \int :e^{<x, \lambda\xi>}:\, d\nu(x)$$

$$= e^{-\frac{1}{2}(1-\lambda^2)|\,\xi\,|_2^2} <\Gamma(\lambda)\,\frac{d\nu}{d\mu},\, :e^{<\cdot,\,\xi>}:>.$$

In terms of $\mathcal{S}$–transforms this reads

$$\mathcal{S}\frac{d\sigma_\lambda\nu}{d\mu}\,(\xi) = e^{-\frac{1}{2}(1-\lambda^2)|\,\xi\,|_2^2}\cdot\mathcal{S}\Gamma(\lambda)\frac{d\nu}{d\mu}\,(\xi).$$

Noting equation (4.26), which is easily verified, we have proved (4.25). □

We conclude this section with the remark that we shall generalize Theorems 4.34 and 4.35 to arbitrary strongly continuous affine transformations on $\mathcal{S}'(\mathbb{R})$ in the next section.

4.C. Characterization Theorems and Their Applications

The main result of this section gives an answer to the converse question of the preceding section: when is a measure ν on $(\mathcal{S}'(\mathbb{R}),\mathfrak{B})$ represented by an element in $(\mathcal{S})_+^*$? The answer will be given in terms of the characteristic function of the measure ν. Actually, we will state necessary and sufficient conditions for whether a functional on $\mathcal{S}(\mathbb{R})$ is the $\mathcal{T}$– or the $\mathcal{S}$–transform of an element in $(\mathcal{S})^*$. A similar result will be given for the space $(\mathcal{S})$ too.

The results in this section have been proved in the articles Potthoff and Streit (1990, 1991a,b) and Kuo, Potthoff and Streit (1991). A result concerning the characterization of $(\mathcal{S})$ through its $\mathcal{S}$–transforms has also been obtained independently in Kondratiev (1980). Refinements of the results of this section have appeared in Yan (1990b), and Meyer and Yan (1990). Also, Lee (1991) proved a characterization of the spaces $(\mathcal{S})_p$, $p\in\mathbb{R}$, by relating their $\mathcal{S}$–transforms to Bargmann–Segal spaces.

The characterization of $(\mathcal{S})^*$ in this section provides a simple discussion of many examples of generalized white noise functionals. Also, it will give the above mentioned generalization of Theorems 3.34 and 3.35 to strongly continuous affine transformations on $\mathcal{S}'(\mathbb{R})$ (cf. Potthoff and Streit (1991b)). In Chapter 11 we shall use the characterization result to establish the fact that the vacuum measures of $P(\phi)_2$ and other models of quantum field theory are represented by elements in $(\mathcal{S})^*$.

Consider $\Phi \in (\mathcal{S})^*$. Recall (cf. Proposition 4.18) that for some $p \in \mathbb{N}_0$, $\Phi \in (\mathcal{S})_{-p}$, and that Φ is in one-to-one correspondence with a sequence $(F^{(n)}, n \in \mathbb{N}_0)$ such that

$$\| \Phi \|_{2,-p}^2 = \sum_{n=0}^{\infty} n! \, | F^{(n)} |_{2,-p}^2 < \infty.$$

Moreover, the $\mathcal{S}$-transform of Φ at $\xi \in \mathcal{S}(\mathbb{R})$ is given by

$$\mathcal{S}\Phi(\xi) = <\overline{\Phi}, :e^{<\cdot,\, \xi>}:>$$

$$= \sum_{n=0}^{\infty} <\overline{F^{(n)}}, \xi^{\otimes n}>.$$

Consider $\lambda \in \mathbb{R}$, and the function $\lambda \mapsto \mathcal{S}\Phi(\lambda\xi)$. We estimate as follows

$$\sum_{n=0}^{\infty} | <\overline{F^{(n)}}, (\lambda\xi)^{\otimes n}> | \le \sum_{n=0}^{\infty} | \lambda |^n \, | F^{(n)} |_{2,-p} \, | \xi |_{2,p}^n$$

$$\le \Big(\sum_{n=0}^{\infty} \frac{1}{n!} | \lambda |^{2n} \, | \xi |_{2,p}^{2n} \Big)^{\frac{1}{2}} \Big(\sum_{n=0}^{\infty} n! \, | F^{(n)} |_{2,-p}^2 \Big)^{\frac{1}{2}}$$

$$= \| \Phi \|_{2,-p} \, e^{\frac{1}{2} | \lambda |^2 | \xi |_{2,p}^2}.$$

Hence, for every $\xi \in \mathcal{S}(\mathbb{R})$, the mapping $\lambda \mapsto \mathcal{S}\Phi(\lambda\xi)$ has an entire analytic extension to $z \in \mathbb{C}$ (which we denote in the same way). Moreover, we found for this extension the bound

$$| \mathcal{S}\Phi(z\xi) | \le \| \Phi \|_{2,-p} \, e^{\frac{1}{2} | z |^2 | \xi |^2_{2,p}} , \quad z \in \mathbb{C}. \tag{4.27}$$

In particular, $z \mapsto \mathcal{S}\Phi(z\xi)$ is for every $\xi \in \mathcal{S}(\mathbb{R})$ entire analytic and of order 2 (cf., e.g., Titchmarsh (1939)), uniformly on the the unit ball of $\mathcal{S}_p(\mathbb{R})$.

We can strengthen the above derived analyticity result by the following observation. Let $\eta \in \mathcal{S}(\mathbb{R})$, so that $:\exp(<\cdot, \eta>): \cdot \Phi \equiv \Phi_\eta$ belongs to $(\mathcal{S})^*$ (cf. Corollary 4.11). Thus for all $\eta, \xi \in \mathcal{S}(\mathbb{R})$, $\mathcal{S}\Phi_\eta(\lambda\xi)$ has an entire extension in λ. On the other hand we have

$$\mathcal{S}\Phi(\eta + \lambda\xi) = e^{-\lambda(\eta, \xi)} \, \mathcal{S}\Phi_\eta(\lambda\xi).$$

Therefore, for all $\eta, \xi \in \mathcal{S}(\mathbb{R})$, the mapping $\lambda \mapsto \mathcal{S}\Phi(\eta+\lambda\xi)$ has an entire analytic extension. Usually we shall just write $z \mapsto \mathcal{S}\Phi(\eta+z\xi)$, $z \in \mathbb{C}$, for this extension. However we warn the reader that it is understood that the *analytic* extension of this function is chosen. For example, if $\Phi = :<\cdot, y>^2:$, $y \in \mathcal{S}'(\mathbb{R})$, then $\mathcal{S}\Phi(\lambda\xi) = <y, \lambda\xi>^2$. This function of λ has the following extensions to $z \in \mathbb{C}$: $| z |^2 <y, \xi>^2$, $\bar{z}^2 <y, \xi>^2$, and $z^2 <y, \xi>^2$, of which clearly only the last is analytic. Therefore in this case $\mathcal{S}\Phi(z\xi) = z^2 <y, \xi>^2$.

Note that the analyticity of $z \mapsto \mathcal{S}\Phi(z\xi)$ implies also that relation (2.3) stays true on $(\mathcal{S})^*$, i.e., we have

$$\mathcal{T}\Phi(\xi) = e^{-\frac{1}{2}| \xi |^2_2} \, \mathcal{S}\Phi(i\xi). \tag{4.28}$$

By (4.28) it is obvious that the above stated results about $\mathcal{S}\Phi$ carry over to $\mathcal{T}\Phi$. The main result of this section will be that also the converse of these statements is true. Let us first make two definitions.

Definition 4.36. *A complex valued function* F *on* $\mathcal{S}(\mathbb{R})$ *is called* ray entire *at* $\eta \in \mathcal{S}(\mathbb{R})$, *if and only if for every* $\xi \in \mathcal{S}(\mathbb{R})$, *the mapping* $\lambda \mapsto F(\eta+\lambda\xi)$, $\lambda \in \mathbb{R}$, *has* *an entire analytic extension. This extension is denoted by* $z \mapsto F(\eta+z\xi)$, $z\in\mathbb{C}$. *If* F *is ray entire at every* $\eta\in\mathscr{S}(\mathbb{R})$, *then it is called* ray entire on $\mathscr{S}(\mathbb{R})$. *Assume that* F *is ray entire at* 0, *and set*

$$M(R, \xi) := \sup_{z,|z|=R} |F(z\xi)|, \quad R>0, \xi\in\mathscr{S}(\mathbb{R}).$$

F *is said to be of* growth (ρ, τ), $\rho\geq 0$, $\tau>0$, on $\mathscr{S}_p(\mathbb{R})$, *if and only if there exists a constant* $C>0$, *so that for all* $\xi\in\mathscr{S}(\mathbb{R})$, *and all* $R>0$,

$$M(R, \xi) \leq C\, e^{\tau R^\rho |\xi|_{2,p}^2}. \tag{4.29}$$

ρ *is called the* order of F, τ *is called the* type of F. F *is said to be of* growth $(\rho, 0)$ on $\mathscr{S}_p(\mathbb{R})$, *if and only if for every* $\tau>0$, *there exists* $C>0$ *so that the bound* (4.29) *holds.*

Remark. In the theory of functions of complex variables (e.g., Titchmarsh (1939)), one usually defines order and growth of an analytic function by requiring that the bound (4.29) holds asymptotically as R tends to infinity. For our purposes however, there will be no loss of generality if we require the estimate (4.29) for all $R>0$.

Definition 4.37. *A complex–valued function* F *on* $\mathscr{S}(\mathbb{R})$ *is called a* U–functional, *if and only if* F *is ray entire on* $\mathscr{S}(\mathbb{R})$ *and of growth* $(2, \tau)$ *on* $\mathscr{S}_p(\mathbb{R})$, *for some* $\tau\geq 0$ *and some* $p\in\mathbb{N}_0$.

Remark. Equivalently, we have that F is a U–functional if and only if it is ray entire on $\mathscr{S}(\mathbb{R})$, and if there exist constants K_1, $K_2>0$, and $p\in\mathbb{N}_0$ so that for all $z\in\mathbb{C}$, $\xi\in\mathscr{S}(\mathbb{R})$,

$$|F(z\xi)| \leq K_1\, e^{K_2 |z|^2 |\xi|_{2,p}^2}.$$

Moreover, it useful to note that the norm $|\xi|_{2,p}$ in the last bound may be replaced by a standard Schwartz space semi–norm $|\xi|_{\alpha,\beta}$ where α, β depend on p (cf. also Appendix A.5).

Theorem 4.38. *If* $\Phi\in(\mathcal{S})^*$, *then* $\mathcal{S}\Phi$ *and* $\mathcal{T}\Phi$ *are U–functionals. If* F *is a U–functional, then there exist unique* Φ_S *and* Φ_T *in* $(\mathcal{S})^*$ *so that* $\mathcal{S}\Phi_S = \mathcal{T}\Phi_T = F$.

Proof. The first part of the statement of the theorem has been proved in the preceding discussion.

Assume that F is a U–functional, and set $f(z,\xi) := F(z\xi)$, $\xi\in\mathcal{S}(\mathbb{R})$, $z\in\mathbb{C}$. Consider the power series expansion of $f(\cdot, \xi)$:

$$f(z, \xi) = \sum_{n=0}^{\infty} z^n f^{(n)}(\xi),$$

where we denoted the Taylor coefficients by $f^{(n)}(\xi)$. Clearly, for $n\in\mathbb{N}$,

$$\begin{aligned} f^{(n)}(\xi) &= \frac{1}{n!}\frac{d^n}{dz^n} f(z, \xi)\Big|_{z=0} \\ &= \frac{1}{n!}(D^n_\xi F)(0), \end{aligned} \tag{4.30}$$

where D_ξ is the Gâteaux derivative in direction ξ on $\mathcal{S}(\mathbb{R})$ (cf. Appendix A.4). Because of the analyticity of f, F is everywhere Gâteaux differentiable in every direction and to all orders, as one easily checks. Moreover, this implies that $\xi \mapsto D_\xi F$ is linear. Therefore equation (4.30) implies that the mapping $\xi \mapsto f^{(n)}(\xi)$ is homogeneous of order n.

Assume for the moment that $|\xi|_{2,p}=1$. By Cauchy's theorem we have

$$f^{(n)}(\xi) = \frac{1}{2\pi i}\oint \frac{f(z, \xi)}{z^{n+1}}\, dz,$$

where we integrate over a closed contour around the origin. Choose a circle of radius $R>0$ with center at the origin. Then we get the bound

$$\begin{aligned} | f^{(n)}(\xi) | &\leq R^{-n} \sup_{z, |z|=R} | f(z, \xi) | \\ &\leq C R^{-n} e^{\tau R^2 |\xi|_{2,p}^2}, \end{aligned}$$

where we used the hypothesis (4.29). Choose $R = \sqrt{n}$. An application of Stirling's theorem gives the bound

$$| f^{(n)} (\xi) | \leq K^n (n!)^{-1/2},$$

for some constant K, independent of n and ξ (which will possibly change its value from step to step below). Since $f^{(n)}$ is homogeneous of degree n, we get for general $\xi \in \mathcal{S}(\mathbb{R})$ the estimate $| f^{(n)} (\xi) | \leq K^n (n!)^{-1/2} |\xi|_{2,p}^n$.

Now we use the polarization identity to construct from $f^{(n)}$ an n–multilinear form $F^{(n)}$ on $\mathcal{S}(\mathbb{R})$:

$$F^{(n)}(\xi_1, \dots, \xi_n) := \frac{1}{n!} \sum_{k=1}^{n} (-1)^{n-k} \sum_{l_1 < \cdots < l_k} f^{(n)}(\xi_{l_1} + \cdots + \xi_{l_k}). \quad (4.31)$$

Note that $F^{(n)}(\xi, \dots, \xi) = f^{(n)}(\xi)$. If $\xi_1, \dots, \xi_n$, have all $|\cdot|_{2,p}$–norm equal to 1, we obtain the following estimation

$$\begin{aligned} | F^{(n)}(\xi_1, \dots, \xi_n) | &\leq \frac{1}{n!} \sum_{k=1}^{n} \sum_{l_1 < \cdots < l_k} | f^{(n)}(\xi_{l_1} + \cdots + \xi_{l_k}) | \\ &\leq K^n (n!)^{-3/2} \sum_{k=1}^{n} \sum_{l_1 < \cdots < l_k} k^n \\ &\leq K^n (n!)^{-3/2} n^n \sum_{k=1}^{n} \sum_{l_1 < \cdots < l_k} 1. \end{aligned}$$

We bound the last sum by $(k!)^{-1}\, n^k$, and $\sum_{k=1}^{n}(k!)^{-1}\, n^k$ by e^n. Applying Stirling's theorem again, we get

$$|\, F^{(n)}(\xi_1, \dots, \xi_n) \,| \le K^n \, (n!)^{-1/2}.$$

For general $\xi_1, \dots, \xi_n \in \mathcal{S}(\mathbb{R})$ we make use of the multilinearity of $F^{(n)}$ and obtain

$$|\, F^{(n)}(\xi_1, \dots, \xi_n) \,| \le K^n \, (n!)^{-1/2} \prod_{k=1}^{n} |\, \xi_k \,|_{2,p}. \tag{4.32}$$

Clearly, this entails that $F^{(n)}$ belongs to $\mathcal{S}_{-p}(\hat{\mathbb{R}^n})$. Next we estimate the $\mathcal{S}_{-q}(\mathbb{R}^n)$–norm, $q \ge p$, of $F^{(n)}$. Let $(e_k,\, k \in \mathbb{N}_0)$ be the system of Hermite functions (normalized in $L^2(\mathbb{R})$). Then

$$|\, F^{(n)} \,|_{2,-q}^2 = \sum_{k_1,\dots,k_n} |\, F^{(n)}\, (H^{-q} e_{k_1}, \dots, H^{-q} e_{k_n}) \,|^2$$

$$\le (n!)^{-1} \, K^{2n} \sum_{k_1,\dots,k_n} \prod_{l=1}^{n} |\, H^{-q} e_{k_l} \,|_{2,p}^2$$

$$= (n!)^{-1} \, K^{2n} \left(\sum_{k=0}^{\infty} |\, H^{-(q-p)} e_k \,|_2^2 \right)^n$$

$$= (n!)^{-1} \left(K \, \| \, H^{-(q-p)} \, \|_{H.S.} \right)^{2n}.$$

Choose $q > p$ large enough so that $K \, \| \, H^{-(q-p)} \, \|_{H.S.} < 1$. Then we get $|\, F^{(n)} \,|_{2,-q} \le (n!)^{-1/2} \cdot \rho$, for some $\rho < 1$.

Now we define Φ_S as corresponding to the sequence $(F^{(n)},\, n \in \mathbb{N}_0)$, i.e., informally we set

$$\Phi_S(x) = \sum_{n=0}^{\infty} \langle F^{(n)}, :x^{\otimes n}: \rangle.$$

Obviously, this defines an element in $(\mathcal{S})_{-q}$, since $\sum_n n! \mid F^{(n)} \mid_{2,-q}^2$ is convergent. Moreover, we have (cf. Proposition 4.18, (4.12))

$$\mathcal{S}\Phi_S(\xi) = \sum_{n=0}^{\infty} <\overline{F^{(n)}}, \xi^{\otimes n}>$$

$$= \sum_{n=0}^{\infty} F^{(n)}(\xi, \dots, \xi)$$

$$= \sum_{n=0}^{\infty} f^{(n)}(\xi)$$

$$= f(1, \xi)$$

$$= F(\xi).$$

Therefore, we have constructed from the U–functional F an element in $(\mathcal{S})^*$ with $\mathcal{S}$–transform given by F. Since F is a U–functional, clearly also $e^{-\frac{1}{2}|\xi|_2^2} F(-i\xi)$ is a U–functional. Construct Φ_T as before with this U–functional. Then $\Phi_T \in (\mathcal{S})^*$, and by (4.28) we have $\mathcal{T}\Phi_T = F$.

Finally we show uniqueness of Φ_S, the argument for Φ_T is similar. Assume that Φ and Ψ have $\mathcal{S}$–transform given by F. Then $\mathcal{S}(\Phi - \Psi)(\xi) = <\Phi - \Psi, :\exp(<\cdot, \xi>) :> = 0$ for all $\xi \in \mathcal{S}(\mathbb{R})$. By Theorem 4.1 this means that $\Phi - \Psi$ is zero on a total set of $(\mathcal{S})$, and hence $\Phi = \Psi$. □

It is clear (and has been used already for the discussion of the $\mathcal{T}$–transform) that the product of two U–functionals is again a U–functional. As a consequence we get the following result.

Corollary 4.39. *Let* $\Phi, \Psi \in (\mathcal{S})^*$. *Then* $\Phi{:}\Psi$ *and* $\Phi * \Psi$, *defined by*

$$\mathcal{S}\Phi{:}\Psi(\xi) = \mathcal{S}\Phi(\xi)\cdot\mathcal{S}\Psi(\xi), \tag{4.33}$$

$$\mathcal{T}\Phi{*}\Psi(\xi) = \mathcal{T}\Phi(\xi)\cdot\mathcal{T}\Psi(\xi), \tag{4.34}$$

belong to $(\mathcal{S})^*$ *too.*

The Wick product (4.33) has already been discussed in Section 4.A. We call the product in (4.34) the *convolution* of Φ and Ψ. Clearly both, : and $*$, are commutative. Let Φ and Ψ be positive. Then the right hand side of (4.36) is equal to the product of the Fourier transform of the measures corresponding to Φ and Ψ. Thus the left hand side is the Fourier transform of the convolution of these two measures, and our terminology is in agreement with the standard one.

Another direct consequence of Theorem 4.38 is the following

Corollary 4.40. *Let* ν *be a measure on* $(\mathcal{S}'(\mathbb{R}),\mathfrak{B})$. *Then* ν *is represented by* $\frac{d\nu}{d\mu}\in(\mathcal{S})^*_+$, *if and only if the characteristic function of* ν *is a U–functional.*

We shall now use the above ideas to prove a convergence result which is slightly stronger than a similar one which can be found in Potthoff and Streit (1991a).

Theorem 4.41. *Assume that* $\{F_n,\ n\in\mathbb{N}\}$ *is a sequence of U–functionals with the following properties:*

(a) *for every* $\xi\in\mathcal{S}(\mathbb{R})$, $\{F_n(\xi),\ n\in\mathbb{N}\}$ *is a Cauchy sequence;*

(b) *there exists* K_1, $K_2>0$, $p\in\mathbb{N}_0$, *so that for all* $n\in\mathbb{N}$, $\xi\in\mathcal{S}(\mathbb{R})$, $z\in\mathbb{C}$,

$$|\,F_n(z\xi)\,| \le K_1 \exp(K_2\,|\,z\,|^2\,|\,\xi\,|^2_{2,p}\,).$$

Then there exist unique Φ_S *and* Φ_T *in* $(\mathcal{S})^*$, *so that* $\mathcal{S}^{-1}F_n$ *converges strongly to* Φ_S, *and* $\mathcal{T}^{-1}F_n$ *converges strongly to* Φ_T.

Proof. By Theorem 4.38, for every $n\in\mathbb{N}$ there exists $\Phi_{n,S}\in(\mathcal{S})^*$ so that $\mathcal{S}\Phi_{n,S} = F_n$. By hypothesis, the sequence given by

$$\{<\overline{\Phi_{n,S}}, :e^{<\cdot\,,\,\xi>}:> = F_n(\xi),\ n\in\mathbb{N}\},$$

is Cauchy for every $\xi\in\mathcal{S}(\mathbb{R})$. It follows that $\{\Phi_{n,S},\ n\in\mathbb{N}\}$ is Cauchy on a dense subset of $(\mathcal{S})$, because the set $\{: \exp(<\cdot\,,\xi>:,\ \xi\in\mathcal{S}(\mathbb{R})\}$ is total in $(\mathcal{S})$. On the other hand, the uniform bound (b) and the estimates in the proof of Theorem 4.38 show that the sequence $\{\Phi_{n,,S},\ n\in\mathbb{N}\}$ is bounded. Consequently it converges weakly to a unique $\Phi_S\in(\mathcal{S})^*$, since $(\mathcal{S})^*$ is weakly sequentially complete. Moreover, because $(\mathcal{S})$ is nuclear it is perfect, and weak sequential convergence is equivalent to strong sequential convergence. Thus for $\mathcal{S}^{-1}F_n$ the statement of the theorem is proven. For $\mathcal{T}^{-1}F_n$ the proof is similar. □

Let us now consider and review some examples. First we want to mention that Theorem 4.38 excludes the possibility that the measures of Lévy–Khintchin processes could be represented by elements in $(\mathcal{S})^*$, unless they are Gaussian. This is so, because non–Gaussian Lévy–Khintchin measures have characteristic functions of infinite order, while Theorem 4.38 requires the order to be less or equal to 2. In particular, the measure of the noise of the Poisson process (e.g., Hida (1970b)) can *not* be represented by an element in $(\mathcal{S})^*$.

Example 4.42. Here we reconsider Example 4.20. Consider $y\in\mathcal{S}'(\mathbb{R})$, and the U-functional $F(\xi) = \exp<y, \xi>$, $\xi\in\mathcal{S}(\mathbb{R})$. By Theorem 4.38 there exists an element $E_y\in(\mathcal{S})^*$ with $\mathcal{S}E_y = F$. Since for $y\in(L^2)$, the element is given by $: \exp(<\cdot, y>) :$, we continue to denote it by the same symbol for $y\in\mathcal{S}'(\mathbb{R})$.

Example 4.43. This is a generalization of Example 4.23 (Theorem 4.24). Let Q be a quadratic form on $\mathcal{S}(\mathbb{R})$, so that there exist $p\in\mathbb{N}_0$ and $C>0$ with $|Q(\xi)| \le C\cdot |\xi|_{2,p}^2$, for all $\xi\in\mathcal{S}(\mathbb{R})$. Then $\exp(Q(\xi))$ defines an element in $(\mathcal{S})^*$ of Gaussian type. (However, note the warning given in Example 4.30.) In particular, if K is an operator on $L^2(\mathbb{R})$ with -1 in its resolvent set, then we may consider the quadratic form $Q_K(\xi) = (\xi, K(K+1)^{-1}\xi)$. If the above bound holds for Q_K, we have an element Φ_K in $(\mathcal{S})^*$ which we shall also informally denote by $\Phi_K(x) = \mathcal{N}\exp(-\frac{1}{2}<x, Kx>)$, in accordance with the symbolic notation introduced after Theorem 4.24.

Example 4.44. Consider again Donsker's delta function $\delta_{t,y}$, Example 3.3. It is not hard to compute its $\mathcal{S}$-transform from the formulae given there, and the result is (cf. also Hida (1983a))

$$\mathcal{S}\delta_{t,y}(\xi) = (2\pi t)^{-1/2} \exp[-\frac{1}{2t}(y-\int_0^t \xi(s)\, ds)^2]. \tag{4.35}$$

Clearly, the last expression is a U-functional, and therefore $\delta_{t,y}$ belongs to $(\mathcal{S})^*$.

Example 4.45. The evaluation mapping of Corollary 4.8, $\varphi \mapsto \widetilde{\varphi}(x)$, $x\in\mathcal{S}'(\mathbb{R})$, has obviously an $\mathcal{S}$-transform given by $\exp(<x, \xi>)$ which is a U-functional. Therefore it is an element in $(\mathcal{S})^*$.

Example 4.46. Consider the algebras $\mathcal{E}$ and $\mathcal{E}_r$, generated by $\exp(i<\cdot, \xi>)$, and $\exp(<\cdot, \xi>)$, respectively, $\xi\in\mathcal{S}(\mathbb{R})$. We can show that they are dense in $(\mathcal{S})$ as follows (cf. also Proposition 3.8, Theorem 4.1). Let $\Phi\in(\mathcal{S})^*$, and assume that

$$<\Phi, \exp(\iota<\cdot, \xi>)> = 0$$

for all $\xi\in\mathcal{S}(\mathbb{R})$, $\iota = 1$ or i. Then it follows that for every real λ, $\mathcal{T}\Phi(\lambda\cdot)$ and $\mathcal{S}\Phi(\lambda\cdot)$ are identically zero. By analyticity, they are zero for all complex λ too. Thus the $f^{(n)}$ and the $F^{(n)}$ in the proof of Theorem 4.38 are zero, and consequently $\Phi = 0$. Since $(\mathcal{S})$ is Fréchet, the Hahn–Banach theorem applies, and

$\mathcal{E}$ and $\mathcal{E}_r$ must be dense in $(\mathcal{S})$.

Another simple application of Theorem 3.38 has been given in the article Potthoff and Streit (1991b). Assume that ν is a measure on $(\mathcal{S}'(\mathbb{R}),\mathcal{B})$ which is represented by $\frac{d\nu}{d\mu} \in (\mathcal{S})^*_+$. Let B^* be a strongly continuous linear mapping on $\mathcal{S}'(\mathbb{R})$. Then B^* is the adjoint of a strongly continuous mapping B on $\mathcal{S}(\mathbb{R})$, as a consequence of the fact that $\mathcal{S}(\mathbb{R})$ is reflexive. In particular, for every $p \in \mathbb{N}$ there exists $q \in \mathbb{N}$, and a constant $C_{p,q}$ so that for every $\xi \in \mathcal{S}(\mathbb{R})$, $|B\xi|_{2,p} \le C_{p,q} |\xi|_{2,q}$. Set $\nu_B(E) := \nu((B^*)^{-1}E)$, $E \in \mathcal{B}$. Compute the characteristic function of ν_B: for $\xi \in \mathcal{S}(\mathbb{R})$,

$$\int e^{i\langle x,\, \xi\rangle}\, d\nu_B(x) = \int e^{i\langle B^*x,\, \xi\rangle}\, d\nu(x)$$

$$= \int e^{i\langle x,\, B\xi\rangle}\, d\nu(x)$$

$$= \mathcal{T}(\tfrac{d\nu}{d\mu})(B\xi).$$

Since $\xi \mapsto \mathcal{T}(\frac{d\nu}{d\mu})(\xi)$ is a U–functional, so is $\xi \mapsto \mathcal{T}(\frac{d\nu}{d\mu})(B\xi)$, by continuity of B. Therefore ν_B is represented by an element in $(\mathcal{S})^*_+$, too. Moreover, one can easily compute the following formula as in the proof of Theorem 4.35.

$$\frac{d\nu_B}{d\mu} = \frac{d\mu_B}{d\mu} : \Gamma(B^*)\, \frac{d\nu}{d\mu},$$

where $\mu_B(E) = \mu(B^{*-1}E)$, $E \in \mathcal{B}$. If we combine this with Theorem 4.34, we obtain the result that the positive cone of measures which are represented by elements in $(\mathcal{S})^*$ is stable under strongly continuous affine transformations:

Theorem 4.47. *Assume that* B^* *is a strongly continuous mapping on* $\mathcal{S}'(\mathbb{R})$, *and that* $y \in \mathcal{S}'(\mathbb{R})$. *Consider a measure* ν *on* $(\mathcal{S}'(\mathbb{R}),\mathcal{B})$, *and set* $\nu_{B,y}(E) := \nu(B^{*-1}E-y)$, $E \in \mathcal{B}$. *If* ν *is such that* $d\nu/d\mu \in (\mathcal{S})^*$, *then also* $d\nu_{B,y}/d\mu \in (\mathcal{S})^*$. *Moreover, the following formula holds.*

$$\frac{d\nu_{B,y}}{d\mu} = \frac{d\mu_{B,0}}{d\mu} : \frac{d\mu_{id,y}}{d\mu} : \Gamma(B^*)\frac{d\nu}{d\mu}, \tag{4.36}$$

where $\mu_{B,y}(E) = \mu(B^{*-1}E - y)$, $E \in \mathfrak{B}$.

Example 4.48. Choose $B = (1-\Delta)^{\alpha}$, $\alpha \in \mathbb{R}$. It is straightforward to see that B is strongly continuous on $\mathcal{S}(\mathbb{R})$ (cf., e.g., Potthoff and Streit (1991b)). In particular, choose $\alpha = -\frac{1}{2}$. Then it follows from Theorem 4.47 that there exists a generalized white noise functional $d\mu_B/d\mu$ in $(\mathcal{S})^*$, which represents the measure μ_B. Its $\mathcal{T}$-transform is given by $\exp(-\frac{1}{2}(\xi, (1-\Delta)^{-1}\xi))$, $\xi \in \mathcal{S}(\mathbb{R})$. Note that $(1-\Delta)^{-1}$ has an integral kernel given by $(1-\Delta)^{-1}(t,s) = \frac{1}{2}\exp(-|t-s|)$, s, $t \in \mathbb{R}$. Therefore we obtain the fact that the measure of the Ornstein–Uhlenbeck process is represented by an element in $(\mathcal{S})^*$.

Other applications of Theorems 4.38 and 4.41 will be given in Chapters 8, 9, 11, and 12.

Next we shall present a result characterizing $(\mathcal{S})$ in terms of its $\mathcal{S}$-transforms. It was indicated in Kondratiev (1980), and has been shown independently in Kuo et al. (1991).

Theorem 4.49. *If* $\varphi \in (\mathcal{S})$, *then* $\mathcal{S}\varphi$ *is ray entire on* $\mathcal{S}(\mathbb{R})$ *and of growth* (2,0) *on* $\mathcal{S}_{-p}(\mathbb{R})$ *for every* $p \in \mathbb{N}_0$. *Conversely, if* F *is ray entire on* $\mathcal{S}(\mathbb{R})$ *and of growth* (2,0) *on* $\mathcal{S}_{-p}(\mathbb{R})$ *for every* $p \in \mathbb{N}_0$, *then* F *is the* $\mathcal{S}$*-transform of an element in* $(\mathcal{S})$.

The proof of this theorem is very similar to the proof of Theorem 4.38, and we indicate only the major steps. For more details we refer the reader to Kuo, Potthoff and Streit (1991), or Yan (1990b).

Sketch of the Proof of Theorem 4.49. Let $\varphi \in (\mathcal{S})$. Since $\varphi \in (\mathcal{S})$ implies that $\varphi \in (\mathcal{S})^*$, Theorem 4.40 implies that $\mathcal{S}\varphi$ is ray entire on $\mathcal{S}(\mathbb{R})$. Moreover, we have for all $z \in \mathbb{C}$, $\xi \in \mathcal{S}(\mathbb{R})$, $q \in \mathbb{N}_0$,

$$
\begin{aligned}
|\mathcal{S}\varphi(z\xi)| &= |(\Gamma(H)^q\,\overline{\varphi},\ \Gamma(H)^{-q} : e^{z<\cdot,\,\xi>} :)| \\
&= |(\Gamma(H)^q\,\overline{\varphi},\ : e^{z<\cdot,\,H^{-q}\xi>} :)| \\
&\le \|\varphi\|_{2,q}\ \|: e^{z<\cdot,\,H^{-q}\xi>} :\|_2.
\end{aligned}
$$

The last norm is easily computed, and we obtain

$$|\mathcal{S}\varphi(z\xi)| \le \|\varphi\|_{2,q}\ e^{\frac{1}{2}|z|^2|\xi|^2_{2,-q}}.$$

Given $p\in\mathbb{N}_0$ and $\tau>0$, choose q large enough so that $2^{-(q-p+1)}<\tau$. Then we get the bound $|\mathcal{S}\varphi(z\xi)| \le \|\varphi\|_{2,q}\exp(\tau|z|^2|\xi|^2_{2,-p})$, and the first part of the theorem is proven.

Now assume that F is ray entire on $\mathcal{S}(\mathbb{R})$ and of growth (2,0) on every $\mathcal{S}_{-p}(\mathbb{R})$, $p\in\mathbb{N}_0$. As in the proof of Theorem 4.38, we construct - from the power series expansion of the analytic extension - the symmetric n–multilinear forms $F^{(n)}$ on the spaces $\mathcal{S}_{-p}(\mathbb{R})$, $p\in\mathbb{N}_0$, i.e., $F^{(n)}$ is an n–multilinear form on their union $\mathcal{S}'(\mathbb{R})$. For the coefficients $f^{(n)}$ of the power series of $f(z,\xi)$ we find this time an estimate of the following form: for all $K>0$ there exists $C_K>0$, so that for all $p\in\mathbb{N}_0$, $\xi\in\mathcal{S}(\mathbb{R})$

$$|f^{(n)}(\xi)| \le C_K\,(n!)^{-1/2}\,e^{-Kn}\,|\xi|^n_{2,-p}.$$

This follows as in the proof of Theorem 4.38, by choosing the radius R as $n^{1/2}e^{K'}$, for large enough K', and $\tau>0$ small enough. We use this bound to estimate the $|\cdot|_{2,p}$–norm of $F^{(n)}$ as follows. For every $n\in\mathbb{N}$, $p\in\mathbb{N}_0$, and $K>0$, $|F^{(n)}|_{2,p} \le (n!)^{-1/2}\,e^{-Kn}$. Define φ as corresponding to the sequence $(F^{(n)},\ n\in\mathbb{N}_0)$. Then φ belongs to every $(\mathcal{S})_p$, $p\in\mathbb{N}_0$, and $\mathcal{S}\varphi(\xi) = F(\xi)$, $\xi\in\mathcal{S}(\mathbb{R})$. □

Kuo (1992) used Theorem 4.49 to give an alternative proof of the algebraic property of $(\mathcal{S})$ (Theorem 4.9).

In various contexts one would like to handle integrals over families of generalized functionals. Their existence in the sense of Bochner - see, e.g., Section V.5. of Yosida (1971) - can be established with the help of Theorem 4.38. To prepare for this we first present a suitable, simple norm estimate for generalized functionals; a stronger version can be found, e.g., in Kondratiev and Streit (1991a), the present one follows if one keeps track of the constants in the proof of Theorem 4.38.

Lemma 4.50. *Let F be a U-functional F obeying*

$$|F(z\xi)| \leq C\, e^{K\, |z|^2 |\xi|_{2,p}^2}.$$

for some C, K > 0 and $p \in \mathbb{N}_0$. Then for any $\epsilon>0$ there is a $q = q(K,p,\epsilon)$ such that

$$\Phi = S^{-1}\, F \in (S)_{-q}$$

and

$$\|\Phi\|_{2,-q} \leq (1+\epsilon)\, C$$

As an application of Lemma 4.49 we derive the following criterion for *Bochner-integrability* of white noise functionals.

Theorem 4.51. *Let $(\Omega,\mathfrak{B}_0,m)$ be a measure space, and $\Phi\colon \Omega \to (\mathcal{S})^*$ such that*

(1) $\omega \mapsto (\mathcal{S}\Phi_\omega)(\xi)$ *is measurable for all* $\xi \in \mathcal{S}(\mathbb{R})$ *and*

(2) $|\ \mathcal{S}\Phi_\omega(z\xi)\ | \leq C(\omega)\ e^{K(\omega)\ |z|^2\ |\xi|_{2,p}^2}$ *with* $C \in L^1(m)$ *and* $K \in L^\infty(m)$.

Then the Bochner integral $\int dm(\omega)\ \Phi_\omega$ *exists in* $(\mathcal{S})^*$, *and*

$$\mathcal{S}\Big(\int dm(\omega)\ \Phi_\omega\Big)(\xi) = \int dm(\omega)\ (\mathcal{S}\Phi_\omega)(\xi).$$

Proof. Replacing $K(\omega)$ by its essential supremum we can invoke Lemma 4.49 to conclude that there is a q such that

$$\|\ \Phi_\omega\ \|_{-q} \leq 2\ C(\omega)$$

m-almost everywhere. The right hand side is evidently m-integrable. To establish the integrability of Φ_ω in $(\mathcal{S})_{-q}$ it is thus sufficient (see Yosida (1971)) to demonstrate its strong measurability. In the present case, by Pettis' Theorem, this is equivalent to weak measurability which in turn follows by a limiting argument from the measurability of the $\mathcal{S}$-transform since this is a dual pairing with arbitrary elements from the fundamental set of exponential functionals. Integration may be interchanged with bounded linear maps, hence in particular with the $\mathcal{S}$-transform. □

Remarks.

1. While the Bochner integral has been constructed as an element of a suitable space $(\mathcal{S})_{-q} \subset (\mathcal{S})^*$, a glance at the exchange of $\mathcal{S}$-transform and integration formula makes it clear that the result does not depend on the particular choice of q.
2. In view of equation (4.28), both Lemma 4.50 and Theorem 4.51 remain true if we replace $\mathcal{S}$- by $\mathcal{T}$-transform consistently.

Example 4.52. Consider the integral

$$\frac{1}{2\pi}\int_{\mathbb{R}} d\omega\ e^{i(\langle x,\ \eta\rangle - a)\omega}$$

for non-zero $\eta \in L^2(\mathbb{R})$ and complex a. To apply Theorem 4.51 we must control the $\mathcal{S}$-transform of $\Phi_\omega(x) = \exp(i\langle x,\eta\rangle\omega - ia\omega)$ for which we find

$$(\mathcal{S}\Phi_\omega)(\xi) = e^{i\omega(\xi,\, \eta) \;-\; \frac{1}{2}\omega^2(\eta,\, \eta) \;-\; ia\omega}\ .$$

Consequently we have, for all ξ with $(\xi,\, \xi) \leq 1$, the bound

$$|\ \mathcal{S}\Phi_\omega(z\xi)\ | \leq\ e^{-\frac{1}{4}\omega^2(\eta,\, \eta) \;+\; |a\omega|}\, e^{|z|^2}.$$

Clearly, in this estimate, the required conditions on C, K are fulfilled for the integral to exist. Inspection of $\mathcal{S}$-transforms reveals that we have obtained an integral expression which generalizes Donsker's δ-function. The latter results if η is the indicator function of the interval [0,t], and a is chosen to be real, see Example 4.44. We shall write

$$\frac{1}{2\pi}\int_{\mathbb{R}} d\omega\ e^{i\langle x,\, \eta\rangle\omega \;-\; ia\omega} = \delta(\langle x,\, \eta\rangle \;-\; a)$$

for the resulting generalized functional.

Example 4.53. In the case of the integral

$$\int_{\mathbb{R}} dt\ :x^2(t):$$

we have the integrand $:x(t)^2: \;\in (\mathcal{S})^*$ with $\mathcal{S}$-transform

$$\mathcal{S}\!:\ \ :x(t)^2: \;\rightarrow \xi^2(t).$$

For the right hand side it is not hard to find a bound

$$|\ \xi^2(t)\ | \leq C_1(t)\ |\ \xi\ |_1^2,$$

with $C_1 \in L^1(\mathbb{R}, dt)$. Hence Theorem 4.51 applies, and the integral exists as a Bochner integral in $(\mathcal{S})^*$.

For more on such vector valued integrals in the context of stochastic integration see Chapter 8.

4.D. Analytic Version of $(\mathcal{S})$ and Growth Estimates

In this section we give an account of the description of $(\mathcal{S})$ which is found in the article by Lee (1991). However, we make also use of the formulation of Lee's results by Yan (1990).

The idea of Lee is the following. Consider again Theorem 4.7, stating that every element φ in $(\mathcal{S})$ has a pointwise defined, strongly continuous representative. In the proof, it was shown that if φ corresponds to $(f^{(n)}, n\in\mathbb{N}_0)$, the 'power series' expansion $\widetilde{\varphi}(x) = \sum_n <: x^{\otimes n} :, \widetilde{f^{(n)}}>$ converges pointwise absolutely, and uniformly on strongly bounded sets in $\mathcal{S}'(\mathbb{R})$. Here, $\widetilde{f^{(n)}}$ is the version of $f^{(n)}$ in $\mathcal{S}(\hat{\mathbb{R}}^n)$. From now we shall identify φ and $f^{(n)}$ with their pointwise defined, continuous versions for simplicity. In particular, the above series converges pointwise absolutely on every Hilbert space $\mathcal{S}_{-p}(\mathbb{R})$, uniformly on bounded sets. Therefore, it converges also pointwise absolutely (uniformly on bounded sets) on the complexified Hilbert spaces $\mathcal{S}_{-p,c}(\mathbb{R})$. In particular, this extension of φ to $\mathcal{S}_{-p,c}(\mathbb{R})$ is analytic in the sense of Hille and Phillips (1957) for every $p\in\mathbb{N}_0$.

The following is a slightly more explicit argument due to Yan (1990). Consider formula (4.6) (Lemma 4.16). In accordance with this formula we define for $x, y\in\mathcal{S}'(\mathbb{R})$, $n\in\mathbb{N}$, $\lambda\in\mathbb{C}$,

$$: (x+\lambda y)^{\otimes n} : \equiv \sum_{k=0}^{n} \binom{n}{k} \lambda^k : x^{\otimes(n-k)} : \hat{\otimes}\, y^{\otimes k}. \tag{4.37}$$

For fixed $x, y\in\mathcal{S}_{-p}(\mathbb{R})$, consider the function $\lambda \mapsto \varphi(x+\lambda y)$, $\lambda\in\mathbb{R}$, which is given by

$$\varphi(x+\lambda y) = \sum_{n=0}^{\infty} \sum_{k=0}^{n} \binom{n}{k} \lambda^k <: x^{\otimes(n-k)} :\hat{\otimes}\, y^{\otimes k}, f^{(n)}>. \tag{4.38}$$

It follows from the proof of Theorem 4.7 that this series converges absolutely, uniformly on bounded subsets of λ. Thus it is clear that $\lambda \mapsto \varphi(x+\lambda y)$ has an entire analytic extension – which we denote by the same symbol –, and that this extension is given by (4.38) with complex λ. Moreover, it follows from the continuity of φ, that it is bounded on bounded sets. But these two properties are exactly the definition of analyticity of φ given in Hille and Phillips (1957, p. 109 ff). Therefore we have shown the following theorem.

Theorem 4.54. *Every* $\varphi \in (\mathcal{S})$ *has a version which has an entire analytic extension to* $\mathcal{S}_{-p,c}(\mathbb{R})$ *for all* $p \in \mathbb{N}_0$.

In the following we shall use the notation $z = x+iy$ for a typical element in the complexified Schwartz space of tempered distributions $\mathcal{S}_c'(\mathbb{R})$, with $x, y \in \mathcal{S}'(\mathbb{R})$. $:z^{\otimes n}:$ is then understood as in (4.37) with $\lambda = i$. Moreover, we shall simply say that φ is analytic on $\mathcal{S}'(\mathbb{R})$, and shall drop the subscript 'c' which indicates complexification, unless there is danger of confusion.

We want to emphasize that in Hille and Phillips a number of interesting consequences for such functions are proved. In particular, the reader finds results concerning the theory of Gâteaux and Fréchet derivatives. However, we shall develop the theory of differential operators on $(\mathcal{S})$ in a different manner in Chapter 5.B.

On the basis of Theorem 4.54, Lee (1991) has proved a useful growth estimate for the elements $\varphi \in (\mathcal{S})$. Following Yan (1990), we prepare first the following lemma.

Lemma 4.55. *For all* $n \in \mathbb{N}$, $z \in \mathcal{S}'(\mathbb{R})$,

$$:z^{\otimes n}: = \int_{\mathcal{S}'(\mathbb{R})} (z+iy)^{\otimes n}\, d\mu(y), \tag{4.39}$$

in the sense of a Bochner–integral in $\mathcal{S}_{-p}(\mathbb{R})$, $p = \max(\deg z, 1)$.

Proof. As argued before Lemma 2.11, it is sufficient if we show that

$$\langle :z^{\otimes n}:, \xi^{\otimes n}\rangle = \int \langle (z+iy)^{\otimes n}, \xi^{\otimes n}\rangle \, d\mu(y),$$

holds for every $\xi \in \mathcal{S}(\mathbb{R})$. Without loss of generality we may assume that $|\xi|_2 = 1$. It follows from our definition of $:x^{\otimes n}:$ in Chapter 2, and (4.37) that

$$\langle :z^{\otimes n}:, \xi^{\otimes n}\rangle = \frac{d^n}{d\lambda^n} e^{\lambda\langle z, \xi\rangle - \lambda^2/2}\Big|_{\lambda=0}.$$

Write $\exp(-\lambda^2/2) = \int \exp(i\lambda\langle y, \xi\rangle)\, d\mu(y)$. Upon insertion and interchange of the integral and the derivatives (which is easily justified), we obtain the following equality:

$$\langle :z^{\otimes n}:, \xi^{\otimes n}\rangle = \int \frac{d^n}{d\lambda^n} e^{\lambda\langle z+iy, \xi\rangle}\Big|_{\lambda=0} d\mu(y),$$

and the right hand side is obviously equal to $\int \langle (z+iy)^{\otimes n}, \xi^{\otimes n}\rangle \, d\mu(y)$. □

Corollary 4.56. *For every* $\varphi \in (\mathcal{S})$ *and* $z \in \mathcal{S}_c'(\mathbb{R})$,

$$\varphi(z) = \sum_{n=0}^{\infty} \int \langle (z+iy)^{\otimes n}, f^{(n)}\rangle \, d\mu(y). \tag{4.40}$$

Theorem 4.57. *Let* $\varphi \in (\mathcal{S})$. *There exists a constant* $C>0$, *and for all* $\epsilon>0$, *there exists* $p_0 > \frac{1}{2}$, *such that for all* $p \in \mathbb{N}_0$ *and every* $z \in \mathcal{S}_{-p,c}(\mathbb{R})$,

$$|\varphi(z)| \le C \, \|\varphi\|_{2,p+p_0} \, e^{\frac{1}{2}(1+\frac{1}{\epsilon})\, |z|^2_{2,-(p+p_0)}}. \tag{4.41}$$

Proof. Fix $\beta > \frac{1}{2}$, and note that $(\mathcal{S}_{-\beta}(\mathbb{R}), L^2(\mathbb{R}), \mu)$ is an abstract Wiener space (e.g., Kuo (1975)). In particular, Fernique's theorem (loc.cit.) states that there exists $\alpha>0$ so that

$$C^2 = \int e^{\alpha |y|^2_{2,-\beta}}\, d\mu(y) < +\infty.$$

Let $p_0 \geq \beta$, and consider $z \in \mathcal{S}_{-p,c}(\mathbb{R})$, $p \in \mathbb{N}_0$. Set $q = p+p_0$. For $f^{(n)} \in \mathcal{S}(\hat{\mathbb{R}}^n)$ we estimate $<: z^{\otimes n} :, f^{(n)}>$ using the representation (4.39):

$$|<: z^{\otimes n} :, f^{(n)}>| \leq \int |<(z+iy)^{\otimes n}, f^{(n)}>|\, d\mu(y)$$

$$\leq |f^{(n)}|_{2,q} \int (|z|_{2,-q} + |y|_{2,-q})^n\, d\mu(y).$$

Let $\varphi \in (\mathcal{S})$ have chaos decomposition given by $\{f^{(n)}, n \in \mathbb{N}_0\}$. Then

$$|\varphi(z)| \leq \sum_{n=0}^{\infty} |f^{(n)}|_{2,q} \int (|z|_{2,-q} + |y|_{2,-q})^n\, d\mu(y)$$

$$= \sum_{n=0}^{\infty} \sqrt{n!}\, |f^{(n)}|_{2,q} \frac{1}{\sqrt{n!}} \int (|z|_{2,-q} + |y|_{2,-q})^n\, d\mu(y)$$

$$\leq \| \varphi \|_{2,q} \Big(\sum_{n=0}^{\infty} \frac{1}{n!} \int (|z|_{2,-q} + |y|_{2,-q})^{2n}\, d\mu(y)\Big)^{\frac{1}{2}}$$

$$= \| \varphi \|_{2,q} \Big(\int e^{(|z|_{2,-q} + |y|_{2,-q})^2}\, d\mu(y)\Big)^{\frac{1}{2}}.$$

Note that for all $\epsilon>0$, $(|z|_{2,-q} + |y|_{2,-q})^2 \leq (1+\frac{1}{\epsilon})|z|^2_{2,-q} + (1+\epsilon)|y|^2_{2,-q}$. Thus we get the bound

$$|\varphi(z)| \leq \| \varphi \|_{2,q}\, e^{\frac{1}{2}(1+\frac{1}{\epsilon})|z|^2_{2,-q}} \Big(\int e^{(1+\epsilon)|y|^2_{2,-q}}\, d\mu(y)\Big)^{\frac{1}{2}}.$$

Now choose p_0 large enough so that $(1+\epsilon)2^{-2(p_0-\beta)} \leq \alpha$, where α is as above. Then estimate

$$(1+\epsilon)\,|\,y\,|_{2,-q}^2 \leq (1+\epsilon)\,|\,y\,|_{2,-p_0}^2$$

$$\leq (1+\epsilon)\,2^{-2(p_0-\beta)}\,|\,y\,|_{2,-\beta}^2$$

$$\leq \alpha\,|\,y\,|_{2,-\beta}^2,$$

to conclude the proof of the theorem. □

For simplicity, we shall choose from now on the above ϵ as equal to 1. On the basis of Theorem 4.57, Lee (1991) defines the following chain of Banach spaces.

$\mathcal{A}_p$, $p \in \mathbb{N}_0$, is defined to be the space of all (strongly) continuous functions f on $\mathcal{S}'(\mathbb{R})$, so that the restriction of f to $\mathcal{S}_{-p}(\mathbb{R})$ is entire analytic (in the sense of Hille and Phillips (1957)), and f is finite in the following norm:

$$\|\,f\,\|_{\mathcal{A}_p} := \sup_{z \in \mathcal{S}_{-p}(\mathbb{R})} |\,f(z)\,|\,e^{-\frac{1}{2}|\,z\,|_{2,-p}^2}. \tag{4.42}$$

Let $p \in \mathbb{N}_0$, $z \in \mathcal{S}_{-p,c}(\mathbb{R})$, $\varphi \in (\mathcal{S})$, and consider the estimate (4.41) for the choice $\epsilon = 1$:

$$|\,\varphi(z)\,| \leq C\,\|\,\varphi\,\|_{2,p+p_0}\,e^{|\,z\,|_{2,-(p+p_0)}^2}.$$

Since $p_0 > \frac{1}{2}$, we get $|\,z\,|_{2,-(p+p_0)}^2 \leq \frac{1}{2}\,|\,z\,|_{2,-p}^2$ due to $\|\,H^{-p_0}\,\| = 2^{-p_0}$, and hence

$$|\,\varphi(z)\,|\,e^{-\frac{1}{2}|\,z\,|_{2,-p}^2} \leq C\,\|\,\varphi\,\|_{2,p+p_0},$$

for some constant C which does not depend on z. Thus we have the following corollary to Theorem 4.57.

Corollary 4.58. *There exists* $p_0>\frac{1}{2}$, *so that for all* $p\in\mathbb{N}_0$, $\varphi\in(\mathcal{S})_{p+p_0}$ *implies that* $\varphi\in\mathcal{A}_p$, *and for some constant* $C>0$,

$$\|\varphi\|_{\mathcal{A}_p}\leq C_p\|\varphi\|_{2,p+p_0}. \tag{4.43}$$

It is natural to ask whether the systems of norms $\{\|\cdot\|_{2,p},\ p\in\mathbb{N}_0\}$ and $\{\|\cdot\|_{\mathcal{A}_p},\ n\in\mathbb{N}_0\}$ are actually equivalent. The affirmative answer has been given in Lee (1991). The following estimate is proved there with a slightly sharper bound.

Proposition 4.59. *For every* $p\in\mathbb{N}_0$ *there exist* $q\in\mathbb{N}$, *and a constant* $C_p>0$, *so that for all* $\varphi\in(\mathcal{S})$,

$$\|\varphi\|_{2,p}\leq C_p\|\varphi\|_{\mathcal{A}_q}. \tag{4.44}$$

Proof. Let $\varphi\in(\mathcal{S})$, with chaos decomposition given by $(f^{(n)},\ n\in\mathbb{N}_0)$, and let $p\in\mathbb{N}_0$. Using Lemma 4.16, or Lemma 4.55, it is not hard to verify that for z, $y\in\mathcal{S}'(\mathbb{R})$, n, $m\in\mathbb{N}$

$$\frac{d^n}{d\lambda^n}:(z+\lambda y)^{\otimes m}:\Big|_{\lambda=0}=(m-n)!:z^{\otimes(m-n)}:\hat{\otimes}\,y^{\otimes n},$$

if $n\leq m$, and otherwise zero. It follows immediately that for every $n\in\mathbb{N}_0$, and $y\in\mathcal{S}'(\mathbb{R})$, we have (cf. also Chapter 5.B)

$$\begin{aligned} f^{(n)}(y,\ldots,y) &\equiv \langle y^{\otimes n}, f^{(n)}\rangle \\ &= \mathbb{E}\Big[\frac{d^n}{d\lambda^n}\varphi(\cdot+\lambda y)\Big|_{\lambda=0}\Big]. \end{aligned}$$

By polarization we get for $y_1, \ldots, y_n \in \mathcal{S}'(\mathbb{R})$,

$$f^{(n)}(y_1, \ldots, y_n) = \mathbb{E}\Big[\frac{\partial^n}{\partial\lambda_1 \cdots \partial\lambda_n} \varphi(\cdot + \lambda\cdot y) \Big|_{\lambda=0} \Big],$$

where we have used the notation $\lambda\cdot y = \sum_{k=1}^n \lambda_k y_k$. We use the last formula and an estimation similar to the one in the proof of Theorem 4.38, to derive a suitable bound for $| f^{(n)} |_{2,p}$. Using the analyticity of $\varphi\in(\mathcal{S})$ and Cauchy's theorem, we find for all $z\in\mathcal{S}'(\mathbb{R})$, and all $R>0$, an estimate of the following form

$$\Big| \frac{\partial^n}{\partial\lambda_1 \cdots \partial\lambda_n} \varphi(z+\lambda\cdot y) \Big|_{\lambda=0} \Big| \leq R^{-n} \sup_{\zeta} | \varphi(z+\zeta\cdot y) |,$$

where the supremum is over all $\zeta_1, \ldots, \zeta_n \in \mathbb{C}$, with $| \zeta_k | = R$. From now on let $z\in\mathcal{S}_{-r}$, $\frac{1}{2}<r\leq 1$. For every large enough $q\in\mathbb{N}$, we get the estimation

$$\Big| \frac{\partial^n}{\partial\lambda_1 \cdots \partial\lambda_n} \varphi(z+\lambda\cdot y) \Big|_{\lambda=0} \Big| \leq R^{-n} \, \| \varphi \|_{\mathcal{A}_q} \sup_{\zeta} e^{\frac{1}{2}| z+\zeta\cdot y |_{2,-q}^2}$$

$$\leq R^{-n} \, \| \varphi \|_{\mathcal{A}_q} \, e^{| z |_{2,-q}^2 + n R^2 \sum_{k=1}^n | y_k |_{2,-q}^2}$$

$$\leq R^{-n} \, \| \varphi \|_{\mathcal{A}_q} \, e^{\alpha| z |_{2,-r}^2} \, e^{n R^2 \sum_{k=1}^n | y |_{2,-q}^2},$$

where we choose α as in the proof of Theorem 4.57, namely in accordance with Fernique's theorem. Applying the latter, we obtain

$$| f^{(n)}(y_1, \ldots, y_n) | \leq K R^{-n} \, \| \varphi \|_{\mathcal{A}_q} \, e^{n R^2 \sum_{k=1}^n | y_k |_{2,-q}^2},$$

where K is some positive constant whose value may change in the following steps. Note that we have

$$| f^{(n)} |_{2,p}^2 = \sum_{k_1,\cdots,k_n=0}^{\infty} (2k_1+2)^{-2} \cdots (2k_n+2)^{-2} \, | f^{(n)}(H^{p+1}e_{k_1}, \ldots, H^{p+1}e_{k_n}) |^2,$$

where $(e_k, k \in \mathbb{N}_0)$ is the CONS in $L^2(\mathbb{R})$ consisting of the Hermite functions. Therefore we apply the last estimate to this choice of the y_k's. Choosing q large enough, we have for every $\epsilon > 0$, and all $k_1, \ldots, k_n \in \mathbb{N}_0$, the bound

$$| f^{(n)}(H^{p+1} e_{k_1}, \ldots, H^{p+1} e_{k_n}) | \leq K\, R^{-n} \, \| \varphi \|_{\mathcal{A}_q} \, e^{\frac{\epsilon}{2} n^2 R^2} .$$

Therefore, for all $p \in \mathbb{N}_0$, $\epsilon > 0$, there exist a constant $K > 0$, and $q \in \mathbb{N}$, so that for all $n \in \mathbb{N}$, the following estimate holds

$$| f^{(n)} |^2_{2,p} \leq K^{2n} \, R^{-2n} \, \| \varphi \|^2_{\mathcal{A}_q} \, e^{\epsilon\, n^2 R^2} .$$

Now choose $R = n^{-1} K_1$, where K_1 is such that $(e\, K\, K_1^{-1}) < 1$. Moreover, choose ϵ small enough (by making q large enough) so that $\epsilon K_1^2 \leq 1$. Then we obtain the following statement: for every $p \in \mathbb{N}_0$, there exist $q \in \mathbb{N}$, and $\rho \in (0,1)$, so that for all $n \in \mathbb{N}$,

$$| f^{(n)} |^2_{2,p} \leq \rho^n \, \frac{1}{n!} \, \| \varphi \|^2_{\mathcal{A}_q} \, .$$

Thus we obtain the statement of the proposition. □

Define the space $\mathcal{A}$ as being the projective limit of the spaces $\mathcal{A}_p$, $p \in \mathbb{N}_0$. Let $\mathcal{A}^*$ be the dual of $\mathcal{A}$.

Theorem 4.60. *$\mathcal{A}$ is homeomorphic to $(\mathcal{S})$. $\mathcal{A}^*$ is isomorphic to $(\mathcal{S})^*$. $\mathcal{A}^*$ and $(\mathcal{S})^*$ are homeomorphic if both are equipped with their strong topologies.*

The following interesting result of Lee (1991) is a consequence of Theorem 4.60. It characterizes the measures which are represented in $(\mathcal{S})^*$ by a Fernique type bound.

Theorem 4.61. *Suppose that* ν *is a measure on* $(\mathcal{S}'(\mathbb{R}), \mathcal{B})$. *Then* $\frac{d\nu}{d\mu}\in(\mathcal{S})^*$, *if and only if the following conditions hold:*

(i) *there exists* $p\in\mathbb{N}_0$, *so that* ν *is carried by* $\mathcal{S}_{-p}(\mathbb{R})$;

(ii) $$\int_{\mathcal{S}_{-p}(\mathbb{R})} e^{\frac{1}{2}| x |_{2,-p}^2}\, d\nu(x) < \infty.$$

Proof. Assume that $\frac{d\nu}{d\mu}\in(\mathcal{S})^*$. By Theorem 4.60, $\frac{d\nu}{d\mu}\in\mathcal{A}_{-p}\subset(\mathcal{S})_{-q}$, for some p, $q\in\mathbb{N}_0$. For $n\in\mathbb{N}$ define the following function g_n on $\mathcal{S}'(\mathbb{R})$

$$g_n(x) := \exp(\tfrac{1}{2}\sum_{k=0}^{n} (2k+2)^{-2p} <x, e_k>^2),\quad x\in\mathcal{S}'(\mathbb{R}),$$

where $(e_k,\ k\in\mathbb{N}_0)$ is the system of Hermite functions. It is obvious that $g_n\geq 0$, for all $n\in\mathbb{N}$, that $\| g_n \|_{\mathcal{A}_p}\leq 1$, and that g_n increases monotonically to the function $g(x) = \exp(\frac{1}{2}| x |_{2,-p}^2)$, $x\in\mathcal{S}'(\mathbb{R})$. Since for every $n\in\mathbb{N}$, $g_n\in(\mathcal{S})$, we may form

$$< \frac{d\nu}{d\mu}, g_n> = \int g_n(x)\, d\nu(x).$$

We get by Theorem 4.60 the estimate

$$| < \frac{d\nu}{d\mu}, g_n> | \leq \| \frac{d\nu}{d\mu} \|_{\mathcal{A}_{-p}} \| g_n \|_{\mathcal{A}_p}$$

$$\leq C_p \| \frac{d\nu}{d\mu} \|_{2,-q},$$

for some $q\in\mathbb{N}$. The monotone convergence theorem implies that g belongs to $L^1(\mathcal{S}'(\mathbb{R}), d\nu)$. The fact that ν is carried by $\mathcal{S}_{-p}(\mathbb{R})$ is obvious, since for every $x\in\mathcal{S}'(\mathbb{R})\backslash\mathcal{S}_{-p}(\mathbb{R})$, g(x) is infinite.

Now suppose that conditions (i) and (ii) of the theorem hold. Define on $(\mathcal{S})$ a linear mapping $\hat{\nu}$ by setting for $\varphi\in(\mathcal{S})$,

$$\hat{\nu}(\varphi) := \int_{\mathcal{S}_{-p}(\mathbb{R})} \widetilde{\varphi}(x)\, d\nu(x).$$

The last integral exists because of the following estimate:

$$\int_{\mathcal{S}_{-p}(\mathbb{R})} |\, \widetilde{\varphi}(x)\, |\, d\nu(x) \leq \|\, \varphi\, \|_{\mathcal{A}_p} \int_{\mathcal{S}_{-p}(\mathbb{R})} e^{\frac{1}{2}|\, x\, |_{2,-p}^2}\, d\nu(x)$$

$$\leq C_p \|\, \varphi\, \|_{2,q},$$

for some $q \in \mathbb{N}$. Moreover, the last inequality shows that $\hat{\nu}$ belongs to $(\mathcal{S})^*$. □

5. Calculus of Differential Operators

This chapter is devoted to the theory of differentiation of – and differential operators acting on – functionals of white noise. On one hand the ideas are very much inspired by the work of P. Lévy (in particular Lévy (1951)), and on the other hand by the theory of creation and annihilation operators on Fock space (e.g., Berezin (1966)). Background references for this chapter are, Hida (1975a), Kubo and Takenaka (1980b), Kuo (1983d), Potthoff and Yan (1992), and for the general theory of Gâteaux and Fréchet derivatives, e.g., the book of Schwartz (1969).

In Section 5.A we describe the pointwise differential calculus on $\mathcal{S}'(\mathbb{R})$, starting with the notion of Gâteaux derivative. We shall compute these derivatives on functionals belonging to the algebras $\mathcal{P}$ and $\mathcal{E}$ (cf. Chapter 2) and relate Gâteaux derivatives in direction of the Dirac distribution at $t \in \mathbb{R}$ on these algebras, to the derivative ∂_t introduced in Hida (1975a). The latter is defined via Fréchet functional derivatives and the $\mathcal{T}$– or $\mathcal{S}$–transformation (cf. also Kubo and Takenaka (1980b)).

In Section 5.B we show that the Gâteaux derivatives can be extended continuously to the space $(\mathcal{S})$ of test functionals of white noise. Based on this, we develop a theory of differential operators on the space $(\mathcal{S})$. Among other results, we find that $(\mathcal{S})$ is infinitely differentiable, and that the chain and product rules hold in $(\mathcal{S})$. Also, it will be proved that $(\mathcal{S})$ is C^{∞} in the sense of Fréchet. Finally, we shall show convergence (in $(\mathcal{S})$) of the Taylor series for elements of $(\mathcal{S})$. The results reported in this section were established in the papers of Albeverio, Hida, et al. (1990a,b), Hida et al. (1988a,b), Kuo, Potthoff and Yan (1992), Potthoff and Röckner (1989), Potthoff and Yan (1992).

Section 5.C is devoted to the L^2-theory of differential operators. Also, we give a discussion of differential operators acting on the space $(\mathcal{S})^*$ of generalized functionals of white noise.

Finally, Section 5.D contains a table of rules.

Many people have contributed to this circle of ideas from different points of view. In particular, we mention in addition the approaches of Albeverio and Høegh-Krohn (1976), Berezansky (1986), Bismut (1981a, 1981b), Gross (1967a, 1967b), M. Krée (1977), M. Krée and P. Krée (1983), P. Krée (1974), Kusuoka (1982), Lee (1989), Malliavin (1978, 1984), Ocone (1984), Shigekawa (1980), Stroock (1981) and their respective groups. We should emphasize the relation to the theory of creation and annihilation operators in quantum field theory as a source of partial results and structural intuition, cf., e.g., Berezin (1966), Cook (1953), Glimm and Jaffe (1981), Segal (1956, 1958) and Simon (1974) to name only a few.

5.A Pointwise Calculus of Differential Operators

In this section we study Gâteaux derivatives of functions on $\mathcal{S}'(\mathbb{R})$ (cf., e.g., Schwartz (1969)). We shall compute the action of these derivatives on the algebras $\mathcal{P}$ of polynomials, and $\mathcal{E}$ of trigonometric exponentials (cf. Chapter 1). In the special case where we differentiate (in the sense of Gâteaux) in direction of the Dirac distribution δ_t, $t \in \mathbb{R}$, we shall compare with the action of the derivative ∂_t, $t \in \mathbb{R}$, introduced in Hida (1975a). It turns out, that on $\mathcal{P}$ and $\mathcal{E}$ both operations coincide.

Let $y \in \mathcal{S}'(\mathbb{R})$ be fixed[1]. Assume that F is an everywhere defined mapping from $\mathcal{S}'(\mathbb{R})$ into $\mathbb{R}$ or $\mathbb{C}$ (the choice of the latter spaces is for simplicity of the presentation only – the generalizations to other classes of spaces are most of the time obvious). For $x \in \mathcal{S}'(\mathbb{R})$ consider the function $\lambda \mapsto F(x+\lambda y)$ on $\mathbb{R}$. If this function is differentiable at $\lambda = 0$, we say that F *is Gâteaux differentiable at* x *in direction* y and denote

$$D_y\, F(x) := \frac{d}{d\lambda}\, F(x+\lambda y)\Big|_{\lambda=0}. \tag{5.1}$$

[1] As in the preceding chapters, $x, y, \ldots, \xi, \eta, \ldots$ denote *real* elements.

$D_yF(x)$ is called the *Gâteaux derivative of* F *at* x *in direction* y. If for every $x \in \mathcal{S}'(\mathbb{R})$, F is Gâteaux differentiable in direction y, we simply call F *Gâteaux differentiable in direction* y. Of course, D_y acts linearly and admits the usual chain and product rules (cf., e.g., Schwartz (1969)).

Let $P \in \mathcal{P}$. Then P is of the following form:

$$P(x) = \sum_n \sum_{\alpha \in \mathbb{N}^n} a_\alpha <x, \xi_{\alpha_1,n}> \cdots <x, \xi_{\alpha_n,n}>, \tag{5.2}$$

both sums ranging over only a finite number of non-zero terms, $a_\alpha \in \mathbb{C}$, $\xi_{\alpha_j,n} \in \mathcal{S}(\mathbb{R})$. Rewriting the monomials on the right hand side of (5.2) in terms of Wick monomials (which is the same as writing P as an orthogonal sum over n in (L^2), cf. Chapter 2) we have another representation of P:

$$P(x) = \sum_n \sum_{\alpha \in \mathbb{N}^n} b_\alpha : <x, \eta_{\alpha_1,n}> \cdots <x, \eta_{\alpha_n,n}> :, \tag{5.3}$$

with certain $b_\alpha \in \mathbb{C}$, $\eta_{\alpha_j,n} \in \mathcal{S}(\mathbb{R})$. Furthermore, we may also write P in our standard representation of elements in (L^2):

$$P(x) = \sum_n <: x^{\otimes n} :, f^{(n)}>, \tag{5.4}$$

where $f^{(n)}$ is the element in $\mathcal{S}(\hat{\mathbb{R}}^n)$ given by

$$f^{(n)} = \sum_\alpha b_\alpha \, \eta_{\alpha_1,n} \hat{\otimes} \cdots \hat{\otimes} \eta_{\alpha_n,n}. \tag{5.5}$$

Note that all representations of P are everywhere defined, continuous functions on $\mathcal{S}'(\mathbb{R})$.

Now let us compute the action of D_y on P in the different forms given above. Since $D_y<\cdot, \xi>(x) = <y, \xi>$, linearity of D_y and the chain rule give immediately the following formula

$$D_y P(x) = \sum_n \sum_{\alpha \in \mathbb{N}^n} a_\alpha \sum_{k=1}^{n} <y, \xi_{\alpha_k,n}> \times \\ \times <x, \xi_{\alpha_1,n}> \cdots <x, \not{\xi}_{\alpha_k,n}> \cdots <x, \xi_{\alpha_n,n}>. \tag{5.6}$$

In order to compute the formula for the representation (5.3), we note that by linearity of D_y and the polarization identity (2.28) it suffices to calculate

$$D_y : <\cdot, \eta>^n : (x).$$

$: <x, \eta>^n :$ is the n-th Hermite polynomial in $<x, \eta>$ with parameter $|\eta|_2^2$. Since the derivative of $: u^n :_{|\eta|_2^2}$ is equal to $n \cdot : u^{n-1} :_{|\eta|_2^2}$ (cf. Appendix A.1, (A.1.20)), we find by the chain rule

$$D_y : <\cdot, \eta>^n : (x) = n <y, \eta> : <x, \eta>^{n-1} : .$$

Applying the polarization identity backwards, we obtain

$$D_y P(x) = \sum_n \sum_{\alpha \in \mathbb{N}^n} b_\alpha \sum_{k=1}^{n} <y, \eta_{\alpha_k,n}> \times \\ \times \; : <x, \eta_{\alpha_1,n}> \cdots <x, \not{\eta}_{\alpha_k,n}> \cdots <x, \eta_{\alpha_n,n}> :. \tag{5.7}$$

Alternatively, one can use Lemma 2.9 to prove (5.7) in a convenient way.

We can read off from (5.7) the formula for our third representation (5.4) of P:

$$D_y P(x) = \sum_n n <: x^{\otimes n-1} :, <y, f^{(n)}>>, \tag{5.8}$$

where $<y, f^{(n)}>$ means that the distribution y is evaluated on $f^{(n)}$ in the first argument. Note that $<y, f^{(n)}> \in \mathcal{S}(\mathbb{R}^{\hat{n}-1})$.

The choice $y = \delta_t$, $t \in \mathbb{R}$, will be of special importance. In this case

$$D_{\delta_t} P(x) = \sum_n \sum_{\alpha \in \mathbb{N}^n} a_\alpha \sum_{k=1}^{n} \xi_{\alpha_k,n}(t) \times \\ \times <x, \xi_{\alpha_1,n}> \cdots <x, \not{\xi}_{\alpha_k,n}> \cdots <x, \xi_{\alpha_n,n}> \tag{5.9}$$

$$= \sum_n \sum_{\alpha \in \mathbb{N}^n} b_\alpha \sum_{k=1}^{n} \eta_{\alpha_k,n}(t) \times \\ \times :<x, \eta_{\alpha_1,n}> \cdots <x, \not{\eta}_{\alpha_k,n}> \cdots <x, \eta_{\alpha_n,n}>: \tag{5.10}$$

$$= \sum_n n <: x^{\otimes n-1} :, f^{(n)}(t, \cdot)>. \tag{5.11}$$

In (5.11), $f^{(n)}(t, \cdot)$ means that the first argument of $f^{(n)}$ is fixed to t.

Now let us consider an element $E \in \mathcal{E}$. In general, it is of the form

$$E(x) = \sum_n a_n \exp(i<x, \xi_n>), \tag{5.12}$$

with complex numbers a_n, $\xi_n \in \mathcal{S}(\mathbb{R})$, and only a finite number of the a_n are non-zero. Linearity of D_y, $y \in \mathcal{S}'(\mathbb{R})$, and the chain rule imply

$$D_y E(x) = \sum_n a_n\, i<y, \xi_n> \exp(i<x, \xi_n>). \tag{5.13}$$

Therefore, we also have

$$D_y \sum_n a_n : \exp(i<\cdot, \xi_n>) : (x) = D_y \sum_n a_n \exp(i<\cdot, \xi_n> + \tfrac{1}{2}| \xi_n |_2^2)\,(x)$$

$$= \sum_n a_n\, i<y, \xi_n> : \exp(i<x, \xi_n>) :. \tag{5.14}$$

In particular,

$$D_{\delta_t} E(x) = \sum_n a_n\, i\xi_n(t) \exp(i<x, \xi_n>), \tag{5.15}$$

and

$$D_{\delta_t} \sum_n a_n : \exp(i\langle \cdot, \xi_n \rangle) : (x) = \sum_n a_n \, i\xi_n(t) : \exp(i\langle x , \xi_n \rangle) : . \quad (5.16)$$

Let us now consider the derivative ∂_t, $t \in \mathbb{R}$, introduced in Hida (1975a). We shall use the definition given in Kubo and Takenaka (1980b), which is shown there to be equivalent to the definition in Hida (1975). We denote

$$\partial_t \varphi(x) = \left(\mathcal{S}^{-1} \frac{\delta}{\delta\xi(t)} \mathcal{S}\varphi \right)(x), \quad (5.17)$$

where $\frac{\delta}{\delta\xi(t)}$ is the Fréchet functional derivative (acting on $\mathcal{S}\varphi$ as depending on $\xi \in \mathcal{S}(\mathbb{R})$), whenever the right hand side of (5.17) exists. For background on Fréchet derivatives and Fréchet functional derivatives on $\mathcal{S}(\mathbb{R})$, we refer to Appendix A.4. We note in passing that there is another definition of ∂_t which can be found in Kuo (1983d), which is also equivalent to (5.17).

Considering $\mathcal{P}$ and $\mathcal{E}$ as subspaces of (L^2), we now compute the action of ∂_t on P in the form (5.3), and on E given in (5.12).

The $\mathcal{S}$-transform of P at $\xi \in \mathcal{S}(\mathbb{R})$ is (cf. Chapter 2)

$$\mathcal{S}P(\xi) = \sum_n \sum_\alpha b_\alpha \, (\eta_{\alpha_1,n}, \xi) \cdots (\eta_{\alpha_n,n}, \xi),$$

and the Fréchet functional derivative of this functional is easily computed:

$$\frac{\delta}{\delta\xi(t)} \mathcal{S}P(\xi) = \sum_n \sum_\alpha b_\alpha \sum_{k=1}^{n} \eta_{\alpha_k,n}(t) \times \quad (5.18)$$
$$\times (\eta_{\alpha_1,n}, \xi) \cdots \widehat{(\eta_{\alpha_k,n}, \xi)} \cdots (\eta_{\alpha_n,n}, \xi).$$

Clearly, the inverse image of this functional under $\mathcal{S}$ is given by the following formula:

$$\partial_t P\,(x) = \sum_n \sum_{\alpha \in \mathbb{N}^n} b_\alpha \sum_{k=1}^{n} \eta_{\alpha_k, n}(t) \times$$
$$\times : \langle x, \eta_{\alpha_1,n}\rangle \cdots \langle x, \not\eta_{\alpha_k,n}\rangle \cdots \langle x, \eta_{\alpha_n,n}\rangle :. \tag{5.19}$$

We found $\partial_t P = D_{\delta_t} P$. *A priori*, this equality holds only μ–a.e. However, since both sides are pointwise defined and continuous, the equality extends to a pointwise equality for the (unique) pointwise defined continuous representatives. In the sequel we shall suppress such comments, whenever the choice of the representatives is obvious.

Next consider the functional $E \in \mathcal{E}$ given in (5.12). We write

$$E(x) = \sum_n c_n : \exp(i\langle x, \xi_n\rangle) :,$$

with $c_n = a_n \cdot \exp(-\frac{1}{2} | \xi_n |_2^2)$. Since

$$\mathcal{S}\!: \exp(i\langle \cdot, \xi_n\rangle) : (\xi) = \exp(i(\xi, \xi_n)),$$

we find immediately the formula

$$\partial_t E(x) = \sum_n a_n\, i\xi_n(t) \exp(i\langle x, \xi_n\rangle) \tag{5.20}$$
$$= D_{\delta_t} E(x).$$

Hence we have proven

Proposition 5.1. *Let* $t \in \mathbb{R}$. *Then* ∂_t *and the Gâteaux derivative in direction* δ_t *coincide on* $\mathcal{P}$ *and* $\mathcal{E}$.

As before, we consider $L^2(\mathbb{R})$ (real) as a subspace of $\mathcal{S}'(\mathbb{R})$, by identifying $h \in L^2(\mathbb{R})$ with the mapping $\xi \mapsto (h, \xi)$ on $\mathcal{S}(\mathbb{R})$. Considering then D_h, $h \in L^2(\mathbb{R})$, we have from (5.6),

$$D_h P(x) = \sum_n \sum_{\alpha \in \mathbb{N}^n} a_\alpha \sum_{k=1}^{n} (h, \xi_{\alpha_k,n}) \times \times <x, \xi_{\alpha_1,n}> \cdots <x, \not\xi_{\alpha_k,n}> \cdots <x, \xi_{\alpha_n,n}>. \tag{5.21}$$

(5.21) shows that for fixed $P \in \mathcal{P}$, $x \in \mathcal{S}'(\mathbb{R})$, the mapping $h \mapsto D_h P(x)$ is a continuous linear functional on $L^2(\mathbb{R})$. Therefore it is of the form

$$(h, \nabla P(x)) = \int_{\mathbb{R}} h(t)\, \nabla P(x)(t)\, dt, \tag{5.22}$$

for some $\nabla P(x) \in L^2(\mathbb{R})$. Upon comparison of (5.9), (5.21), (5.22), and by using Proposition 5.1 we find that

$$\nabla P(x)(t) = \partial_t P(x). \tag{5.23}$$

It is obvious that we can do the same with functionals in $\mathcal{E}$. Therefore we have the following result.

Proposition 5.2. *For all* $h \in L^2(\mathbb{R})$, *and all* $\varphi \in \mathcal{P}$, $\mathcal{E}$ *the following formula holds*

$$D_h \varphi(x) = (h, \nabla \varphi(x)) = \int_{\mathbb{R}} h(t)\, \partial_t \varphi(x)\, dt. \tag{5.24}$$

More generally, if $y \in \mathcal{S}'(\mathbb{R})$, $\varphi \in \mathcal{P}$, $\mathcal{E}$, *then*

$$D_y \varphi(x) = <y, \nabla \varphi>, \tag{5.25}$$

where the pairing on the right hand side of (5.25) *is the one between* $\mathcal{S}'(\mathbb{R})$ *and* $\mathcal{S}(\mathbb{R})$.

Note that the white noise measure μ is quasi-invariant in the directions of $L^2(\mathbb{R})$, i.e., for every $h \in L^2(\mathbb{R})$, $\mu(\cdot - h)$ is absolutely continuous with respect to μ (the Radon–Nikodym derivative is given by $\frac{d\mu(\cdot - h)}{d\mu} = :\exp(<\cdot, h>):$, cf. Chapter 4). Therefore, $L^2(\mathbb{R})$ can be considered in a natural way as the space of translations of the white noise measure space, i.e., as its *tangent space*. Therefore (5.24) justifies that we call $\nabla\cdot = (\partial_t \cdot, t \in \mathbb{R})$ the *gradient* of functions on $(\mathcal{S}'(\mathbb{R}), \mathcal{B}, \mu)$.

5.B Differential Operators on $(\mathcal{S})$

In finite dimensional analysis the Schwartz space of test functions $\mathcal{S}(\mathbb{R}^d)$, or the space $\mathcal{D}(\mathbb{R}^d)$ of C^∞ functions with compact support play important roles as 'intermediate spaces'. Typically, they are used as an *ad hoc* domain of definition for differential operators. In the analysis and functional analysis over the white noise space $(\mathcal{S}'(\mathbb{R}), \mathcal{B}, \mu)$, the space $(\mathcal{S})$ can be employed in a similar way. This section provides the corresponding tools.

First we extend the Gâteaux derivatives from $\mathcal{P}$ and $\mathcal{E}$ to $(\mathcal{S})$. It turns out that $(\mathcal{S})$ is C^∞ under D_y for every $y \in \mathcal{S}'(\mathbb{R})$. Also, we prove that the elements of $(\mathcal{S})$ are infinitely often Fréchet differentiable. The Fréchet derivative is given by the continuous extension of the gradient ∇. Finally, we shall prove the convergence of the Taylor expansion for elements in $(\mathcal{S})$.

For a different approach to some of the results of this section we refer the interested reader also to the article by Lee (1991).

Throughout this section we shall consider $\mathcal{S}'(\mathbb{R})$ as equipped with its strong topology (cf., e.g., Appendix A.5).

Theorem 5.3. *Let* $\varphi \in (\mathcal{S})$, *and* $y \in \mathcal{S}'(\mathbb{R})$. *Assume that* φ *has chaos decomposition given by* $(f^{(n)}, n \in \mathbb{N}_0)$. *Denote by* $\widetilde{\varphi}$ *and* $\widetilde{f^{(n)}}$, $n \in \mathbb{N}_0$, *their corresponding continuous versions. Then* $\widetilde{\varphi}$ *is everywhere Gâteaux differentiable in direction* y, *and the following formulae hold for* $x \in \mathcal{S}'(\mathbb{R})$:

$$(D_y\widetilde{\varphi})(x) = \sum_{n=1}^{\infty} n <: x^{\otimes(n-1)} : \hat{\otimes}\, y, \widetilde{f^{(n)}}> \\ = \sum_{n=1}^{\infty} n <: x^{\otimes(n-1)} :, <y, \widetilde{f^{(n)}}>>. \tag{5.26}$$

In (5.26) $<y, \widetilde{f^{(n)}}>$ *means that the distribution* $y \in \mathcal{S}'(\mathbb{R})$ *is evaluated in the first variable of* $\widetilde{f^{(n)}} \in \mathcal{S}(\hat{\mathbb{R}}^n)$, *so that* $<y, \widetilde{f^{(n)}}> \in \mathcal{S}(\hat{\mathbb{R}}^{n-1})$.

We first prove a Lemma.

Lemma 5.4. *Let* $\widetilde{f^{(n)}} \in \mathcal{S}(\hat{\mathbb{R}}^n)$, *and set* $\widetilde{\varphi^{(n)}}(x) = <: x^{\otimes n} :, \widetilde{f^{(n)}}>$, $x \in \mathcal{S}'(\mathbb{R})$. *Then for all* $x, y \in \mathcal{S}'(\mathbb{R})$, $\lambda \in \mathbb{R}$, *the following formula holds:*

$$\frac{d}{d\lambda}\, \widetilde{\varphi^{(n)}}(x+\lambda y) = n <: (x+\lambda y)^{\otimes(n-1)} : \hat{\otimes}\, y, \widetilde{f^{(n)}}>.$$

In particular, $D_y\widetilde{\varphi^{(n)}}$ *exists for all* $y \in \mathcal{S}'(\mathbb{R})$, *and is given by*

$$(D_y\widetilde{\varphi^{(n)}})\,(x) = n <: x^{\otimes(n-1)} : \hat{\otimes}\, y, \widetilde{f^{(n)}}>.$$

Proof. We use Lemma 4.16 to write for $\lambda \in \mathbb{R}$,

$$\widetilde{\varphi^{(n)}}(x+\lambda y) = \sum_{k=0}^{n} \binom{n}{k} \lambda^k <: x^{\otimes(n-k)} : \hat{\otimes}\, y^{\otimes k}, \widetilde{f^{(n)}}>.$$

Now the derivative with respect to λ is readily computed:

$$\frac{d}{d\lambda}\, \widetilde{\varphi^{(n)}}(x+\lambda y) = n <: (x+\lambda y)^{\otimes(n-1)} : \hat{\otimes}\, y, \widetilde{f^{(n)}}>. \qquad \square$$

Proof of Theorem 5.3. In view of Lemma 5.4, we only need to show that

$$\sum_{n=1}^{\infty} n <: (x+\lambda y)^{\otimes(n-1)} :, <y, \widetilde{f^{(n)}}>>$$

converges uniformly in λ in a neighborhood of $\lambda = 0$. But this is proved as in the first part of Theorem 4.15. □

It is worthwhile to extract the following rule from the preceding argument.

Rule 5.5. *For* $x, y \in \mathcal{S}'(\mathbb{R})$, $n \in \mathbb{N}$,

$$D_y :x^{\otimes n}: = n :x^{\otimes(n-1)}: \hat{\otimes}\, y. \tag{5.27}$$

As a consequence of Theorem 5.3 we have the following result.

Corollary 5.6. *Let* $\varphi \in (\mathcal{S})$, $t \in \mathbb{R}$. *Then*

$$\partial_t \widetilde{\varphi}(x) = \sum_{n=1}^{\infty} n \langle :x^{\otimes(n-1)}:, \widetilde{f^{(n)}}(t,\cdot)\rangle, \tag{5.28}$$

where $\partial_t = D_{\delta_t}$.

Let $\varphi \in (\mathcal{S})$, $y \in \mathcal{S}'(\mathbb{R})$, and denote by $D_y\varphi$ the μ-class of $D_y\widetilde{\varphi}$. Consider the mapping D_y: $\varphi \mapsto D_y\varphi$.

Theorem 5.7. *For every* $y \in \mathcal{S}'(\mathbb{R})$, D_y *is continuous from* $(\mathcal{S})$ *into itself. In particular,* $(\mathcal{S})$ *is infinitely often Gâteaux differentiable in every direction of* $\mathcal{S}'(\mathbb{R})$.

Theorem 5.7 is implied by the following sharper result.

Proposition 5.8. *Let* $y \in \mathcal{S}_{-q}(\mathbb{R})$, $q \geq 0$. *Then for all* $p \geq 0$, $\varphi \in (\mathcal{S})$, *the following holds:*

$$\| D_y\varphi \|_{2,p} \leq 2^{q-p} | y |_{2,-q} \| N^{1/2}\, 2^{-(p\vee q-p)N} \varphi \|_{2,p\vee q}, \tag{5.29}$$

where $p \vee q = \max(p, q)$. *In particular, for every* $p \geq 0$, $\varphi \in (\mathcal{S})$,

$$\| D_y\varphi \|_{2,p} \leq 2^{q-p} | y |_{2,-q} \| \varphi \|_{2,p\vee q+1}, \tag{5.30}$$

and therefore D_y *extends to a continuous map from* $(\mathcal{S})_{p\vee q+1}$ *into* $(\mathcal{S})_p$.

Proof. According to Theorem 5.3, $D_y\varphi$ has a chaos expansion given by the $(n\langle y, f^{(n)}\rangle(\cdot), n\in\mathbb{N})$, where $\langle y, f^{(n)}\rangle$ is the class of $\langle y, \widetilde{f^{(n)}}\rangle$ in $L^2(\mathbb{R}^{n-1})$. Therefore we can estimate in the following way[2]:

$$\| D_y\varphi \|_{2,p}^2 = \sum_{n=1}^{\infty} (n-1)!\, n^2 \,| (H^{\otimes(n-1)})^p \langle y, f^{(n)}\rangle |_2^2$$

$$= \sum_{n=1}^{\infty} n!\, n \,| (H^p)^{\otimes(n-1)} \langle H^{-q}y, H_1^q f^{(n)}\rangle |_2^2$$

$$= \sum_{n=1}^{\infty} n!\, n \,| \langle H^{-q}y, H^q\otimes(H^p)^{\otimes(n-1)} f^{(n)}\rangle |_2^2$$

$$\leq | y |_{2,-q}^2 \sum_{n=1}^{\infty} n!\, n \,| H^q\otimes(H^p)^{\otimes(n-1)} f^{(n)} |_2^2$$

$$\leq 2^{2(q-p)} | y |_{2,-q}^2 \sum_{n=0}^{\infty} n!\, n\, 2^{-2(p\vee q-p)n} | f^{(n)} |_{2,p\vee q}^2$$

$$= 2^{2(q-p)} | y |_{2,-q}^2 \left\| N^{1/2}\, 2^{-(p\vee q-p)N} \varphi \right\|_{2,p\vee q}^2,$$

where we made use of the fact that $\| H^{-1} \| = \frac{1}{2}$. □

---

[2]By H_1 we mean that this operator acts only in the first variable, i.e. $H_1 = H\otimes id\otimes \cdots \otimes id$ (n factors).

Corollary 5.9. *Let* $\varphi \in (\mathcal{S})$. *The mapping* $y \mapsto D_y\varphi$ *is strongly continuous from* $\mathcal{S}'(\mathbb{R})$ *into* $(\mathcal{S})$.

Proposition 5.8 states that D_y, $y \in \mathcal{S}'(\mathbb{R})$, is the continuous extension of the Gâteaux derivative acting on $\mathcal{P}$ or $\mathcal{E}$. Using furthermore the fact that $(\mathcal{S})$ is an algebra, we find that the standard rules of calculus carry over:

Theorem 5.10.

(i) *Assume that* $\varphi \in (\mathcal{S})$ *is real valued and let* $f \in C^1(\mathbb{R})$, *such that* $f \circ \varphi \in (\mathcal{S})$, *and* $f' \circ \varphi \in (\mathcal{S})$. *Then the chain rule holds*

$$D_y f \circ \varphi = f' \circ \varphi \cdot D_y \varphi. \tag{5.31}$$

(ii) On $(\mathcal{S})$ *the product rule holds for* D_y *for all* $y \in \mathcal{S}'(\mathbb{R})$.

Remark. The hypothesis in (i) that $\varphi \in (\mathcal{S})$ is real valued has only been made for simplicity of the formulation of the theorem. It is of course inessential, and can readily be replaced by more general assumptions.

As in the previous chapter, it will be convenient to identify φ, and $f^{(n)}$, $n \in \mathbb{N}_0$, with their continuous versions. Usually there will be no danger of confusion.

As before, we set for $t \in \mathbb{R}$, $\partial_t = D_{\delta_t}$, where δ_t is the Dirac distribution concentrated at t. This is justified since Proposition 5.8 implies that $\mathcal{S}^{-1} \frac{\delta}{\delta \xi(t)} \mathcal{S}$ and D_{δ_t} coincide on $(\mathcal{S})$, because they do so on the dense subspace $\mathcal{P}$, cf. Proposition 5.1. The following result is sometimes useful.

Corollary 5.11. *Let* $\varphi \in (\mathcal{S})$, $y_1, \dots, y_n \in \mathcal{S}'(\mathbb{R})$, $t_1, \dots, t_n \in \mathbb{R}$. *Then the following formulae hold.*

$$\langle y_1 \hat{\otimes} \cdots \hat{\otimes} y_n, f^{(n)} \rangle = \frac{1}{n!} \mathbb{E}(D_{y_1} \cdots D_{y_n} \varphi), \tag{5.32}$$

and in particular

$$f^{(n)}(t_1, \dots, t_n) = \frac{1}{n!} \mathbb{E}(\partial_{t_1} \cdots \partial_{t_n} \varphi). \tag{5.33}$$

Proof. This follows by iteration from (5.26), and the observation that for every $n \in \mathbb{N}$, $<: x^{\otimes n} :, g^{(n)}>$ is orthogonal to 1 in (L^2). □

The following is a corollary to Proposition 5.8.

Corollary 5.12. *Let* $t \in \mathbb{R}$. *Then for every* $q > \frac{1}{4}$, *and every* $p \geq 0$, ∂_t *is a continuous mapping from* $(\mathcal{S})_{p+q}$ into $(\mathcal{S})_p$.

Proof. We only need to check that for every $t \in \mathbb{R}$, $\delta_t \in \mathcal{S}_{-q}(\mathbb{R})$, for every $q > \frac{1}{4}$. In $\mathcal{S}'(\mathbb{R})$, δ_t has the representation

$$<\delta_t, \cdot> = \sum_{n=0}^{\infty} e_n(t) \, (e_n, \cdot),$$

where we may choose as the CONS $(e_n, n \in \mathbb{N}_0)$ of $L^2(\mathbb{R})$ the Hermite functions (cf. Appendix A.1). Then we get for $q > \frac{1}{4}$,

$$| H^{-q} \delta_t |_2^2 = 2^{-2q} \sum_{n=0}^{\infty} (n+1)^{-2q} \, e_n(t)^2,$$

where we used the fact that the Hermite functions are the eigenfunctions of H (cf. (A.1.10). Using the estimate (A.1.11), which states that for every $t \in \mathbb{R}$, $| e_n(t) | = O(n^{-1/4})$, we finish the proof. □

Fix $x \in \mathcal{S}'(\mathbb{R})$, and $\varphi \in (\mathcal{S})$. Then the mapping $y \mapsto (D_y \varphi)(x)$ is strongly continuous from $\mathcal{S}'(\mathbb{R})$ into $(\mathcal{S})$ by Corollary 5.9 and the continuity of the evaluation mapping (Corollary 4.8). This entails that there exists $\eta_{\varphi,x} \in \mathcal{S}(\mathbb{R})$, with

$$D_y \varphi(x) = <y, \eta_{\varphi,x}>,$$

because $\mathcal{S}(\mathbb{R})$ is reflexive. (Of course, the dual pairing in the last equality is between $\mathcal{S}'(\mathbb{R})$ and $\mathcal{S}(\mathbb{R})$.) We shall denote $\eta_{\varphi,x}(t)$ by $\nabla\varphi(t,x)$, $t\in\mathbb{R}$. Note that

$$\nabla\varphi(t,x) = <\delta_t, \eta_{\varphi,x}>$$

$$= D_{\delta_t}\varphi(x)$$

$$= \partial_t\varphi(x),$$

so that our notation is consistent with the one of Section 5.A. For fixed $t\in\mathbb{R}$, $\partial_t\varphi$ belongs to $(\mathcal{S})$ by Theorem 5.7, and for fixed $x\in\mathcal{S}'(\mathbb{R})$, $\nabla\varphi(x) = \partial.\varphi(x)$ belongs to $\mathcal{S}(\mathbb{R})$. It follows from the nuclear theorem (e.g., Trèves (1967)) that $\nabla\varphi\in\mathcal{S}(\mathbb{R})\otimes(\mathcal{S})$. We summarize our discussion as follows.

Theorem 5.13. ∇ *maps every* $\varphi\in(\mathcal{S})$ *continuously into* $\nabla\varphi\in\mathcal{S}(\mathbb{R})\otimes(\mathcal{S})$. *Moreover, for every* $y\in\mathcal{S}'(\mathbb{R})$,

$$D_y\varphi = <y, \nabla\varphi>. \tag{5.34}$$

In particular, if $h\in L^2(\mathbb{R})$, *then*

$$D_h\varphi = \int h(t)\ \partial_t\varphi\ dt, \tag{5.35}$$

where the last integral exists pointwise as a function on $S'(\mathbb{R})$ *(and in the sense of Bochner).*

Equation (5.34) indicates that $\nabla\varphi$ is the Fréchet derivative of $\varphi\in(\mathcal{S})$. We shall state and prove this as our next result.

Theorem 5.14. *Every* $\varphi\in(\mathcal{S})$ *is infinitely often differentiable in Fréchet sense. The* k-th *Fréchet derivative of* φ *is given by* $\nabla^k\varphi\in\mathcal{S}(\hat{\mathbb{R}^k})\otimes(\mathcal{S})$.

Proof. Let $\varphi\in(\mathcal{S})$, x, $y\in\mathcal{S}'(\mathbb{R})$. Write

$$\varphi(x+y) - \varphi(x) = \int_0^1 \frac{d}{d\lambda}\, \varphi(x+\lambda y)\, d\lambda.$$

This equation makes sense, because $\frac{d}{d\lambda}\,\varphi(x+\lambda y)$ is equal to $(D_y\varphi)(x+\lambda y)$ (which exists by Theorem 5.7), and $\lambda \mapsto \psi(x+\lambda y)$ is continuous for every $\psi \in (\mathcal{S})$, cf. Proposition 4.17. By Theorem 5.13 we find

$$\varphi(x+y) - \varphi(x) = \int_0^1 \langle y,\, \nabla\varphi(x+\lambda y)\rangle\, d\lambda.$$

Let p be large enough. Then we estimate as follows.

$$\begin{aligned}
&|\, \varphi(x+y) - \varphi(x) - \langle y,\, \nabla\varphi(x)\rangle \,| = \\
&= \Big|\int_0^1 \langle y,\, \nabla\varphi(x+\lambda y) - \nabla\varphi(x)\rangle\, d\lambda \Big| \\
&\le \sup_{\lambda\in[0,1]} |\,(D_y\varphi)(x+\lambda y) - (D_y\varphi)(x)\,| \\
&= \sup_{\lambda\in[0,1]} |\,(\tau_{\lambda y} D_y\varphi)(x) - (D_y\varphi)(x)\,| \\
&= \sup_{\lambda\in[0,1]} |\,(D_y(\tau_{\lambda y}\varphi - \varphi))(x)\,| \\
&= \sup_{\lambda\in[0,1]} |\,\langle \delta_x,\, D_y(\tau_{\lambda y}\varphi - \varphi)\rangle\,| \\
&\le \|\, \delta_x \,\|_{2,-p} \sup_{\lambda\in[0,1]} \|\, D_y(\tau_{\lambda y}\varphi - \varphi)\,\|_{2,p}
\end{aligned}$$

$$\leq \| \delta_x \|_{2,-p} | y |_{2,-p} \sup_{\lambda\in[0,1]} \| \tau_{\lambda y}\varphi - \varphi \|_{2,p+1},$$

where we made use of the fact that the translations and Gâteaux derivatives commute, and of estimate (5.30). Furthermore, we used Corollary 4.8, which states that the evaluation mapping δ_x is continuous on $(\mathcal{S})$. To get the statement of the theorem for $k = 1$, it remains to show that the last term above vanishes uniformly in $\lambda\in[0,1]$, as y tends strongly in $\mathcal{S}'(\mathbb{R})$ to zero. But this follows from Proposition 4.17, which states the strong continuity of $y \mapsto \tau_y\varphi$, which is uniform on strongly bounded sets of $\mathcal{S}'(\mathbb{R})$. The general case $k\in\mathbb{N}$ follows now by iteration. □

As the final result of the theory of differential operators on $(\mathcal{S})$, we prove now the convergence of the Taylor expansion of elements in $(\mathcal{S})$. We denote

$$\exp D_y\varphi = \sum_{n=0}^{\infty} \frac{1}{n!} D_y^n\varphi \tag{5.36}$$

whenever the right hand side makes sense. Recall that τ_y is the translation operator on $(\mathcal{S})$ (cf. Theorem 4.15).

Theorem 5.15. *Let* $y\in\mathcal{S}'(\mathbb{R})$ *and* $\varphi\in(\mathcal{S})$. *Then*

$$\tau_y\varphi = \exp D_y\varphi, \tag{5.37}$$

i.e., the sum on the right hand side of (5.36) *converges in* $(\mathcal{S})$ *to* $\tau_y\varphi$.

Proof. In Chapter 4 (cf. Lemma 4.16 and the proof of Theorem 4.15) we have shown that for φ corresponding to the sequence $(f^{(n)}, n\in\mathbb{N}_0)$ we have the formula

$$\tau_y\varphi(x) = \sum_{n=0}^{\infty}\sum_{k=0}^{n} \binom{n}{k} <: x^{\otimes k} :, <y^{\otimes(n-k)}, f^{(n)}>>.$$

From (5.26) we find be iteration

$$D_y^k\varphi(x) = \sum_n \frac{n!}{(n-k)!} <: x^{\otimes k} :, <y^{\otimes(n-k)}, f^{(n)}>>,$$

(with the convention $(n-m)!^{-1} = 0$ if $m>n$). Thus

$$\sum_{k=0}^{N} \frac{1}{k!} D_y^k\varphi(x) = \sum_{n=0}^{\infty}\sum_{k=0}^{N\wedge n} \binom{n}{k} <: x^{\otimes k} :, <y^{\otimes(n-k)}, f^{(n)}>>,$$

and

$$(\tau_y - \sum_{k=0}^{N} \frac{1}{k!} D_y{}^k)\varphi = \sum_{n=N+1}^{\infty}\sum_{k=0}^{N} \binom{n}{k} <: x^{\otimes k} :, <y^{\otimes(n-k)}, f^{(n)}>>.$$

However, in the proof of Theorem 4.15 we have shown that the last sum is the Cauchy remainder of a sum which converges in $(\mathcal{S})$, and therefore it converges in $(\mathcal{S})$ to zero as N tends to infinity. □

Corollary 5.16. *For every* x, $y\in\mathcal{S}'(\mathbb{R})$ *and* $\varphi\in(\mathcal{S})$ *the following formula holds*

$$\widetilde{\varphi}(x+y) = \sum_{n=0}^{\infty} \frac{1}{n!} (D_y^n\widetilde{\varphi})(x) \tag{5.38}$$

Proof. The evaluation map belongs to $(\mathcal{S})^*$, Corollary 4.8. □

Remark. The results presented in Chapter 4.D (cf. also the article of Lee (1991)) could have been based alternatively on the last two results.

5.C Differential Operators and Their Adjoints on (L^2) and $(\mathcal{S})^*$

In this section we discuss first the operator ∂_t, for fixed $t\in\mathbb{R}$, on other domains of (L^2). Also we shall introduce its adjoint ∂_t^*. Then we consider $\partial_t\varphi$ 'as a function of t', i.e., the gradient ∇ on (L^2). We shall prove the well-known fact that ∇ maps the (L^2)-domain of $N^{\frac{1}{2}}$ continuously into $L^2(\mathbb{R})\otimes(L^2)$, and study the adjoint ∇^* of ∇ (Cf. also Gross (1963), Krée and Krée (1983), Meyer (1983), Nualart and Zakai (1986), Ocone (1984), Shigekawa (1980), Sugita (1985), Watanabe (1984), and Chapter 7.) Finally we discuss the Gâteaux derivatives (in smooth directions), their adjoints, and the gradient on $(\mathcal{S})^*$.

Fix $t\in\mathbb{R}$, and consider the operator $\partial_t = D_{\delta_t}$ of Section 5.B on $(\mathcal{S})$. We have seen in Corollary 5.12 that for all $p\geq 0$, $q>\frac{1}{4}$, ∂_t has a continuous extension as a bounded linear operator from $(\mathcal{S})_{p+q}$ into $(\mathcal{S})_p$. The reason for this is essentially that $\delta_t\in\mathcal{S}_{-q}(\mathbb{R})$, for all $q>\frac{1}{4}$. Namely, this implies that the map which takes $f^{(n)}\in\mathcal{S}(\mathbb{R}^n)$ into its trace $f^{(n)}(t,\cdot)\in\mathcal{S}(\mathbb{R}^{n-1})$ on the hyperplane $\{(u_1,\dots,u_n)\in\mathbb{R}^n,\ u_1 = t\}$, has a continuous extension which maps $\mathcal{S}_{-q}(\mathbb{R}^n)$ into $L^2(\mathbb{R}^{n-1})$.

The last remark hints at a class of somewhat larger domains for ∂_t, $t\in\mathbb{R}$, by application of the Sobolev trace theorem (cf., e.g., Adams (1975), Folland (1976), or Lions and Magenès (1972)). The idea to use this theorem for the construction of a domain for ∂_t appeared first in Kuo and Russek (1988), and we shall also use it in Chapter 8. The basic observation is contained in the following result.

Lemma 5.17. *Let* $q>\frac{1}{2}$. *For all* $t\in\mathbb{R}$, $n\in\mathbb{N}$, *the mapping* $f^{(n)}\mapsto f^{(n)}(t,\cdot)$ *from* $\mathcal{S}(\mathbb{R}^n)$ *into* $\mathcal{S}(\mathbb{R}^{n-1})$, *has a unique continuous extension to a bounded linear mapping from the Sobolev space* $\mathcal{H}_q(\mathbb{R}^n)$ *into* $L^2(\mathbb{R}^{n-1})$. *Moreover, there exists a constant* $C_q>0$, *so that for all* $n\in\mathbb{N}$, $t\in\mathbb{R}$, *and all* $f^{(n)}\in\mathcal{H}_q(\mathbb{R}^n)$, *the following inequality holds*:

$$\| f^{(n)}(t,\cdot) \|_{L^2(\mathbb{R}^{n-1})} \leq C_q \| f^{(n)} \|_{\mathcal{H}_q}.$$

The proof of Lemma 5.17 can be found in the above quoted literature, and is omitted here.

If $\varphi\in(\mathcal{S})$, and φ has its chaos decomposition given by the sequence $(f^{(n)}, n\in\mathbb{N}_0)$, $f^{(n)}\in\mathcal{S}(\mathbb{R}^n)$, then $\partial_t\varphi$ corresponds to the sequence $((n+1)f^{(n+1)}(t,\cdot), n\in\mathbb{N}_0)$ (cf. Corollary 5.6). Thus we shall use the spaces $\mathcal{H}_q(\mathbb{R}^n)$, $q>\frac{1}{2}$, for the construction of a domain of ∂_t, and we only have to take care of the factor 'n'.

For q, k$\geq$0, define the norm $|\cdot|_{q,k}$ on $(\mathcal{S})$ by

$$|\varphi|_{q,k}^2 := \sum_{n=0}^{\infty} n!\, n^k \,\| f^{(n)} \|_{\mathcal{H}_q}^2, \tag{5.39}$$

where $\varphi\in(\mathcal{S})$ corresponds to the sequence $(f^{(n)}, n\in\mathbb{N}_0)$ of integral kernels. $(\mathcal{H})_{q,k}$ denotes the completion of $(\mathcal{S})$ under this norm. Since $\mathcal{S}(\mathbb{R}^n)$ is dense in $\mathcal{H}_q(\mathbb{R}^n)$, $q\geq 0$, it follows that $(\mathcal{H})_{q,k}$ is isomorphic to a (weighted) direct sum of the Hilbert spaces $\mathcal{H}_q(\hat{\mathbb{R}}^n)$. Let $(\mathcal{H})_{-q,-k}$ denote the dual of $(\mathcal{H})_{q,k}$, q, k$\geq$0. Then (upon the usual identification of (L^2) with its dual) we obtain a class of triples

$$(\mathcal{H})_{-q,-k} \supset (L^2) \supset (\mathcal{H})_{q,k}, \tag{5.40}$$

q, k$\geq$0. It is easy to see that the embeddings in (5.40) are dense and continuous.

Let $q>\frac{1}{2}$ from now on. If $\varphi\in(\mathcal{H})_{q,1}$ with chaos decomposition $(f^{(n)}, n\in\mathbb{N}_0)$, $f^{(n)}\in\mathcal{H}_q(\mathbb{R}^n)$, we define $\partial_t\varphi$ as corresponding to the sequence $(nf^{(n)}(t,\cdot), n\in\mathbb{N})$, where $f^{(n)}(t,\cdot)$ is the unique element in $L^2(\hat{\mathbb{R}}^{n-1})$ which is given in Lemma 5.17. Clearly, we have the following result.

Theorem 5.18. *Let* $t\in\mathbb{R}$, $q>\frac{1}{2}$. *Then* ∂_t *has a unique continuous extension from* $(\mathcal{S})$ *to* $(\mathcal{H})_{q,1}$, *so that* ∂_t *is a bounded linear mapping from* $(\mathcal{H})_{q,1}$ *into* (L^2).

Corollary 5.19. *Let* $t\in\mathbb{R}$. *Then* ∂_t *maps* $(L^2)^+$ *continuously into* (L^2).

Of course, one can think of extensions of ∂_t to even larger domains than the spaces $(\mathcal{H})_{q,1}$, $q>\frac{1}{2}$, above. An example can be found in the construction of a domain of the operator ∂_{t+} in Chapter 8. However, it should be noted that there is *no unique maximal domain* for ∂_t in (L^2), $t\in\mathbb{R}$. Naturally, this is due to the fact that the evaluation mapping $f^{(n)} \mapsto f^{(n)}(t, \cdot)$ depends in general on the choice of a version of $f^{(n)}$. In the above constructions, $f^{(n)}$ has a unique version which is continuous in its first argument. The choice of this version makes the trace $f^{(n)}(t, \cdot)$ well-defined, and 'unique'. For function spaces, where such a version does not exist, there is a large amount of ambiguity in the definition of $f^{(n)}(t, \cdot)$, and therefore of ∂_t.

Although we have defined now a number of different extensions of D_{δ_t} from $\mathcal{P}$, we shall denote all of them by ∂_t, and in the sequel add – if necessary – a remark in which sense we consider this operation. In particular, we have discussed ∂_t as a continuous linear mapping from $(\mathcal{S})$ into $(\mathcal{S})$, from $(\mathcal{S})_{p+q}$ into $(\mathcal{S})_q$, $q>\frac{1}{4}$, $p\geq 0$, and from $(\mathcal{H})_{q,1}$ into (L^2), $q>\frac{1}{2}$. In all these cases we have the following rule.

Rule 5.20. *Let* $t\in\mathbb{R}$, *then*

$$\partial_t : \, :x(u_1) \cdots x(u_n): \,\to \sum_{k=1}^{n} \delta(t-u_k) :x(u_1) \cdots \not{x}(u_k) \cdots x(u_n):, \qquad (5.41)$$

or, equivalently

$$\partial_t : \, (f^{(n)}, n\in\mathbb{N}_0) \to ((n+1)\, f^{(n+1)}(t, \cdot\,), n\in\mathbb{N}_0). \qquad (5.42)$$

Usually, for computations Rule 5.20 will be all what one needs, since it describes (in a somewhat informal manner, however cf. the remarks at the end of Chapter 2) the action of ∂_t on the Wick monomials in $x\in\mathcal{S}'(\mathbb{R})$ – which in turn describe the dependence of φ on $x\in\mathcal{S}'(\mathbb{R})$. Equivalently, (5.42) describes how the derivative ∂_t acts on the kernel functions of φ. Since the gradient ∇ is defined in terms of ∂_t (Theorem 5.13, and the arguments preceding it), one can compute $D_y\varphi$ for all $y\in\mathcal{S}'(\mathbb{R})$ with Rule 5.20 via (5.34).

By ∂_t^*, $t\in\mathbb{R}$, we denote the adjoint of the operator ∂_t with the various domains mentioned above: ∂_t^* can be considered as a linear continuous operator from $(\mathcal{S})^*$ into $(\mathcal{S})^*$ (with respect to the weak or strong topology), from $(\mathcal{S})_{-p}$ into $(\mathcal{S})_{-(p+q)}$, $p\geq 0$, $q>\frac{1}{4}$, and from (L^2) into $(\mathcal{H})_{-q,-1}$, $q>\frac{1}{2}$. To compute the action of ∂_t^*, assume that $\Phi\in(\mathcal{S})^*$ is characterized by the sequence $(T^{(n)}\in\mathcal{S}'(\hat{\mathbb{R}}^n),\ n\in\mathbb{N}_0)$ (cf. Chapter 4.A). I.e., if $\varphi\in(\mathcal{S})$ has kernels $(f^{(n)}\in\mathcal{S}(\hat{\mathbb{R}}^n),\ n\in\mathbb{N}_0)$, then

$$<\Phi, \varphi> = \sum_{n=0}^{\infty} n!\ <T^{(n)}, f^{(n)}>, \tag{5.43}$$

where the pairings under the last sum are those of $\mathcal{S}'(\mathbb{R}^n)$ and $\mathcal{S}(\mathbb{R}^n)$. Thus by definition,

$$<\partial_t^*\Phi, \varphi> = <\Phi, \partial_t\varphi>$$

$$= \sum_{n=0}^{\infty} n!\ (n+1)\ <T^{(n)}, \varphi^{(n+1)}(t, \cdot\ )>$$

$$= \sum_{n=0}^{\infty} (n+1)!\ <\delta_t\hat{\otimes}T^{(n)}, \varphi^{(n+1)}>.$$

Hence we find that $\partial_t^*\Phi$ has the kernels $(0,\ \delta_t\hat{\otimes}T^{(n)};\ n\in\mathbb{N}_0)$. (For the other domains, the preceding computation is of course the same.)

Rule 5.21. *Let* $t\in\mathbb{R}$. *Then*

$$\partial_t^* :\ :x(u_1)\cdots x(u_n): \ \to\ :x(t)\,x(u_1)\cdots x(u_n):, \tag{5.44}$$

or, equivalently

$$\partial_t^* : (T^{(n)}, n\in\mathbb{N}_0) \to (0, \delta_t \hat{\otimes} T^{(n-1)}, n\in\mathbb{N}). \tag{5.45}$$

As a by-product, we found the following result (cf. also the generalized functional $W_t^{(n)}$ in Chapter 3.A).

Rule 5.22. *Let* $u_1, \ldots, u_n \in \mathbb{R}$. *Then*

$$:x(u_1) \cdots x(u_n): = \partial_{u_1}^* \cdots \partial_{u_n}^* 1. \tag{5.46}$$

If we compare formulae (5.41), (5.44) with the definition (2.42) of the Wick monomials, we see that informally multiplication by x(t) coincides with the application of $\partial_t^* + \partial_t$. More precisely we can state the following result.

Proposition 5.23. *For every* $t\in\mathbb{R}$, $q>\frac{1}{2}$, $p>\frac{1}{4}$, *multiplication by* x(t) *is defined as a continuous mapping from* $(H)_{q,1}$ $((\mathcal{S})_p$, *resp.) into* $(H)_{-q,-1}$ $((\mathcal{S})_{-p}$, *resp.) through the formula:* $x(t)\cdot = \partial_t^* + \partial_t \cdot$.

We shall discuss the adjoint of D_y and multiplication by $X_y = \langle\cdot, y\rangle$, $y\in\mathcal{S}'(\mathbb{R})$, later in this section in a more general way.

Next we shall consider the gradient ∇ as an operator acting on (L^2). Recall that N is the number operator: if φ corresponds to $(f^{(n)}, n\in\mathbb{N}_0)$, then $N\varphi$ corresponds to $(n\cdot f^{(n)}, n\in\mathbb{N})$. Note that N is a self-adjoint operator on (L^2).

Let $r\geq 0$, and denote by $\mathcal{D}^{2,r}$ the completion of $\mathcal{P}$ with respect to the norm given by (cf. also (3.35))

$$|||\,\varphi\,|||_{2,r} := \left\| (1+N)^{r/2}\varphi \right\|_2. \tag{5.47}$$

Since N is diagonalized by the chaos expansion of (L^2), it is easy to see that $\mathcal{D}^{2,r}$

is the same as the completion of the (L^2)–domain $\mathcal{D}(N^{r/2})$ of $N^{r/2}$ in the graph topology. Note that for all $r \geq 0$, $\mathcal{D}^{2,r}$ is a Hilbert subspace of (L^2). The next two results have been shown in many articles. We refer the reader to the literature quoted at the beginning of this chapter.

Theorem 5.24. *For all* $r \geq 1$, *the gradient* ∇ *extends from* $\mathcal{P}$ *(or* $\mathcal{E}$ *or* $(\mathcal{S})$*) to a continuous linear operator from* $\mathcal{D}^{2,r}$ *into* $L^2(\mathbb{R})\otimes\mathcal{D}^{2,r-1} \simeq L^2(\mathbb{R};\, \mathcal{D}^{2,r-1})$. *In particular,* ∇ *maps* $\mathcal{D}^{2,1}$ *continuously into* $L^2(\mathbb{R})\otimes(L^2)$. *Moreover, for all* $r \geq 1$, $\varphi \in \mathcal{D}^{2,r}$,

$$\| \nabla\varphi \|_{L^2(\mathbb{R})\otimes\mathcal{D}^{2,r-1}} \leq ||| \varphi |||_{2,r}. \tag{5.48}$$

Proof. Consider $\varphi \in \mathcal{P}$, and observe the following commutation relation, which is easy to prove. For all $y \in \mathcal{S}'(\mathbb{R})$,

$$(1+N)\, D_y\, \varphi = D_y\, N\, \varphi. \tag{5.49}$$

In particular, we find for all $r \geq 0$, $\varphi \in \mathcal{P}$, $\mathrm{id}\otimes(1+N)^r\, \nabla\varphi = \nabla\, N^r\varphi$. Now we estimate in the following way:

$$\begin{aligned}
\| \nabla\varphi \|_{L^2(\mathbb{R})\otimes\mathcal{D}^{2,r-1}} &= \left\| \mathrm{id}\otimes(1+N)^{(r-1)/2}\nabla\varphi \right\|_{L^2(\mathbb{R})\otimes(L^2)} \\
&= \left\| \nabla N^{(r-1)/2}\varphi \right\|_{L^2(\mathbb{R})\otimes(L^2)} \\
&= \sup_{|h|_2=1} \left\| (h,\, \nabla N^{(r-1)/2}\varphi) \right\|_{(L^2)} \\
&= \sup_{|h|_2=1} \left\| D_h N^{(r-1)/2}\varphi \right\|_{(L^2)}
\end{aligned}$$

$$\leq \left\| N^{r/2}\varphi \right\|_{(L^2)}$$

$$\leq |||\, \varphi \,|||_{2,r},$$

where we used Proposition 5.2, the estimate of Proposition 5.8, and the spectral theorem. Since $\mathcal{P}$ is dense in $\mathcal{D}^{2,r}$ the proof is finished. □

It is clear, that we can also treat powers of the gradient in the same way:

Corollary 5.25. *For all* $k\in\mathbb{N}$, $r\geq k$, ∇^k *extends to a continuous operator from* $\mathcal{D}^{2,r}$ *into* $L^2(\mathbb{R}^k)\hat{\otimes}\mathcal{D}^{2-k} \simeq L^2(\mathbb{R}^k;\hat{\mathcal{D}}^{2,r-k})$.

Since we have for all (real) $h\in L^2(\mathbb{R})$ on $\mathcal{P}$ the formula $D_h = (h, \nabla\cdot)$, we obtain immediately the following result.

Corollary 5.26. *Let* $k\in\mathbb{N}$, $r\geq k$, $h_1, \ldots, h_k\in L^2(\mathbb{R})$. *Then* $D_{h_1} \cdots D_{h_k}$ *extends to a bounded operator from* $\mathcal{D}^{2,r}$ *into* $\mathcal{D}^{2,r-k}$.

According to Theorem 5.24, ∇ has an adjoint ∇^* which is a bounded linear operator from $L^2(\mathbb{R})\otimes(\mathcal{D}^{2,r})^*$ into $(\mathcal{D}^{2,r+1})^*$. However, the following result is often useful.

Theorem 5.27. *For every* $r\geq 1$, ∇^* *is a continuous mapping from* $L^2(\mathbb{R})\otimes\mathcal{D}^{2,r}$ into $\mathcal{D}^{2,r-1}$. *Moreover, for all* $r\geq 1$, $X\in L^2(\mathbb{R})\otimes\mathcal{D}^{2,r}$, *the following inequality holds:*

$$|||\, \nabla^* X \,|||_{2,r-1} \leq 2^{r/2} \|\, X \,\|_{L^2(\mathbb{R})\otimes\mathcal{D}^{2,r}}. \tag{5.50}$$

Proof. It suffices to show the estimate (5.50) for $X\in L^2(\mathbb{R})\otimes\mathcal{P}$. For such X we have the estimation

$$
\begin{aligned}
|||\, \nabla^* X \,|||_{2,r-1} &= \left\| (1+N)^{(r-1)/2} \nabla^* X \right\|_{(L^2)} \\
&= \sup_{\|\varphi\|_{(L^2)}=1} \left| (\varphi, (1+N)^{(r-1)/2} \nabla^* X) \right| \\
&= \sup_{\|\varphi\|_{(L^2)}=1} \left| (\nabla (1+N)^{(r-1)/2} \varphi, X)_{L^2(\mathbb{R}) \otimes (L^2)} \right| \\
&= \sup_{\|\varphi\|_{(L^2)}=1} \left| ((2+N)^{r/2} \nabla (1+N)^{-1/2} \varphi, X)_{L^2(\mathbb{R}) \otimes (L^2)} \right| \\
&\le \sup_{\|\varphi\|_{(L^2)}=1} \left\| \nabla (1+N)^{-1/2} \varphi \right\|_{L^2(\mathbb{R}) \otimes (L^2)} \times \\
&\qquad\qquad \times \left\| (2+N)^{r/2} X \right\|_{L^2(\mathbb{R}) \otimes (L^2)} \\
&\le 2^{r/2} \sup_{\|\varphi\|_{(L^2)}=1} |||\, (1+N)^{-1/2} \varphi \,|||_{2,1} \times \\
&\qquad\qquad \times \left\| (1+N)^{r/2} X \right\|_{L^2(\mathbb{R}) \otimes (L^2)} \\
&\le 2^{r/2} \| X \|_{L^2(\mathbb{R}) \otimes \mathcal{D}^{2,r}},
\end{aligned}
$$

where we used the commutation relation (5.49) and estimate (5.48). □

We remark in passing that ∇^* is closely related to Itô's stochastic integral of non–anticipating processes with respect to Brownian motion, and to the Hitsuda–Skorokhod integral. This is worked out in detail in Chapter 8.

On $\mathcal{D}^{2,1}$ we had set for $h \in L^2(\mathbb{R})$, $D_h = (h, \nabla)_{L^2(\mathbb{R})}$. Let D_h^* denote the adjoint of D_h. Then it is not hard to see, that $\psi \in \mathcal{D}^{2,1}$ with kernels $(g^{(n)}, n \in \mathbb{N}_0)$ is mapped by D_h^* into $D_h^* \psi \in (L^2)$, with kernels given by $(0, h \hat{\otimes} g^{(n-1)}, n \in \mathbb{N})$. It is straightforward to check that actually we have $D_h^* \psi = \nabla^* h \otimes \psi = \int dt\ h(t)\ \partial_t^* \psi$ (cf. also (5.45)). If $\psi \in \mathcal{D}^{2,r}$, $r \geq 1$, then one finds by a similar calculation as in the proof of Theorem 5.27 that $D_h^* \psi \in \mathcal{D}^{2,r-1}$. Thus the following considerations make sense.

Let $\varphi \in \mathcal{D}^{2,2}$, with chaos decomposition given by $(f^{(n)}, n \in \mathbb{N}_0)$. Then we find from (5.26) and the above remark about D_h^*, that for f, $g \in L^2(\mathbb{R})$, $D_f D_g^* \varphi$ has integral kernels given by $((f, g) f^{(0)}, (f, f^{(n)})(\cdot) \hat{\otimes} g + (f, g) \cdot f^{(n)}, n \in \mathbb{N})$, while $D_g^* D_f \varphi$ has kernels given by $((f, f^{(n)}) \hat{\otimes} g, n \in \mathbb{N})$. Thus we obtain the following result.

Theorem 5.28. *Let* f, $g \in L^2(\mathbb{R})$, *and* $\varphi \in \mathcal{D}^{2,2}$. *Then*

$$[D_f\ ,\ D_g^*]\ \varphi = (f, g)_{L^2(\mathbb{R})} \cdot \varphi, \tag{5.51}$$

where [,] *denotes the commutator.*

Symbolically, we state the following

Rule 5.29. *Let* $s, t \in \mathbb{R}$. *Then*

$$[\partial_s\ ,\ \partial_t^*] = \delta(s-t)\ \ ;\ \ [\partial_s\ ,\ \partial_t] = [\partial_s^*\ ,\ \partial_t^*] = 0. \tag{5.52}$$

In the physics literature the relations (5.52) are called the *canonical commutation relations*, or simply *CCR*. These relations show that ∂_t^* and ∂_t are the images of the *creation* and *annihilation operators* on the symmetric Fock space $\Gamma(L^2(\mathbb{R}))$ under the standard isomorphism (cf. also Appendix A.2).

We conclude this chapter by a discussion of differential operators and their adjoints on $(\mathcal{S})^*$. First, we state the following analog of Proposition 5.8.

Proposition 5.30. *Let* $\eta\in\mathcal{S}_q(\mathbb{R})$, $q\geq 0$, $\varphi\in\mathcal{P}$. *Then for the every* $p\geq 0$ *the following estimate holds*

$$\| D_\eta\varphi \|_{2,-p} \leq 2^{p-q} \mid \eta \mid_{2,q} \| N^{1/2} 2^{-(p-\min(p,q))N} \varphi \|_{2,-\min(p,q)}$$

$$\leq 2^{p-q} \mid \eta \mid_{2,q} \| \varphi \|_{2,-\min(p,q)+1}.$$

The proof is exactly along the lines of the proof of Proposition 5.8, and left as an exercise to the interested reader.

Since $\mathcal{P}$ is strongly sequentially dense in $(\mathcal{S})^*$ (cf. Proposition 3.9), we obtain from Proposition 5.30 the following result.

Theorem 5.31. *Assume that* $\eta\in\mathcal{S}_q(\mathbb{R})$, $q\geq 0$. *Then* D_η *has a continuous extension as a bounded linear operator from* $(\mathcal{S})_{-\min(p,q)+1}$ *into* $(\mathcal{S})_{-p}$. *In particular, if* $\eta\in\mathcal{S}(\mathbb{R})$, *then* D_η *extends to a strongly continuous operator from* $(\mathcal{S})^*$ *into itself.*

Now we are going to consider the adjoint of the Gâteaux derivative D_y, $y\in\mathcal{S}'(\mathbb{R})$ (real). Since Theorem 5.7 states that D_y acts continuously on $(\mathcal{S})$, D_y has an adjoint D_y^* mapping $(\mathcal{S})^*$ weakly continuously into itself. Let us compute its explicit action on $\Phi\in(\mathcal{S})^*$, which has chaos decomposition given by the sequence $(T^{(n)}, n\in\mathbb{N}_0)$ of symmetric tempered distributions. Let $\varphi\in(\mathcal{S})$ have a sequence $(f^{(n)}, n\in\mathbb{N}_0)$ of integral kernels. We use (5.26) to make the following calculation.

$$<D_y^*\Phi, \varphi> = <\Phi, D_y\varphi>$$

$$= \sum_{n=0}^{\infty} n!\,(n+1) <T^{(n)}, <y, f^{(n+1)}>>$$

$$= \sum_{n=0}^{\infty} (n+1)! <T^{(n)} \hat{\otimes} y, f^{(n+1)}>.$$

Using in addition Proposition 5.8, we obtain the following result.

Theorem 5.32. *If* $y \in \mathcal{S}_{-q}(\mathbb{R})$, $q \geq 0$, *then for every* $p \geq 0$, D_y^* *is a continuous mapping from* $(\mathcal{S})_{-p}$ into $(\mathcal{S})_{-((p \vee q)+1)}$. *In particular, for every* $y \in \mathcal{S}'(\mathbb{R})$, D_y^* *is a strongly continuous mapping of* $(\mathcal{S})^*$ *into itself. Moreover, if* $\Phi \in (\mathcal{S})^*$ *is given by the sequence* $(T^{(n)}, n \in \mathbb{N}_0)$, *then* $D_y^* \Phi$ *is given by* $(0, y \hat{\otimes} T^{(n-1)}, n \in \mathbb{N})$.

Let $y \in L^2(\mathbb{R})$ be real, and $f_1, \dots, f_n \in L^2(\mathbb{R})$ be complex valued. In Section 5.A we have shown (cf. (5.7)) that

$$D_y : <x, f_1> \cdots <x, f_n> : = \sum_{k=1}^{n} (y, f_k) : <x, f_1> \cdots \widehat{<x, f_k>} \cdots <x, f_n> :,$$

while in the derivation of Theorem 5.32, we proved that

$$D_y^* : <x, f_1> \cdots <x, f_n> : = : <x, y> <x, f_1> \cdots <x, f_n> :.$$

By formula (2.30), we find that the operator $(D_y + D_y^*)$ coincides with the multiplication by the random variable $X_y = <\cdot, y>$ on the Wick monomials. By linearity, and the continuity statements of Proposition 5.8 and Theorem 5.31, this fact extends for $y \in \mathcal{S}_{-q}(\mathbb{R})$, $q \geq 0$, to the following result, which generalizes Proposition 5.23.

Theorem 5.33. *Let* $y \in \mathcal{S}_{-q}(\mathbb{R})$, $q \geq 0$. *Multiplication by the generalized random variable* $X_y = <\cdot, y>$, *extends to a continuous mapping from* $(\mathcal{S})_q$ *into* $(\mathcal{S})_{-q-1}$. *Moreover, for all* $\varphi \in (\mathcal{S})_q$,

$$X_y\varphi = (D_y + D_y^*)\varphi. \tag{5.53}$$

In particular, for every $y\in\mathcal{S}'(\mathbb{R})$, *multiplication by* X_y *is a strongly continuous map from* $(\mathcal{S})$ *into* $(\mathcal{S})^*$, *so that for all* $\varphi\in(\mathcal{S})$ (5.53) *holds.*

We conclude this section by considering the gradient ∇ as an operator on $(\mathcal{S})^*$.

Let $\Phi\in(\mathcal{S})^*$. Consider the mapping $(\eta, \varphi) \mapsto \langle D_\eta\Phi, \varphi\rangle$, from $\mathcal{S}(\mathbb{R})\times(\mathcal{S})$ into $\mathbb{C}$. It follows from Proposition (5.30), that this mapping is continuous in η and in φ. By the nuclear theorem it is therefore jointly continuous, i.e., continuous on $\mathcal{S}(\mathbb{R})\otimes(\mathcal{S})$. Thus there exists an element in $\mathcal{S}'(\mathbb{R})\otimes(\mathcal{S})^*$ denoted by $\nabla\Phi$, so that for all $\eta\in\mathcal{S}(\mathbb{R})$, and all $\varphi\in(\mathcal{S})$ we have $\langle D_\eta\Phi, \varphi\rangle = \langle\nabla\Phi, \eta\otimes\varphi\rangle = \langle\langle\eta, \nabla\Phi\rangle, \varphi\rangle$, i.e., $D_\eta\Phi = \langle\eta, \nabla\Phi\rangle$ in $(\mathcal{S})^*$. We call $\nabla\Phi$ the *gradient of* Φ. The reader convinces himself easily, that for $\varphi\in(\mathcal{S})$ or $\mathcal{D}^{2,1}$ this notion of gradient coincides with the previous ones. Obviously, we may iterate the operator ∇ on $(\mathcal{S})^*$, and obtain the following result.

Theorem 5.34. $(\mathcal{S})^*$ *is infinitely often differentiable with respect to* ∇, *and for every* $k\in\mathbb{N}$, $\Phi\in(\mathcal{S})^*$, $\nabla^k\Phi\in\mathcal{S}'(\hat{\mathbb{R}}^k)\otimes(\mathcal{S})^*$. *Moreover, for all* $k\in\mathbb{N}$, $\eta_1, \ldots, \eta_k\in\mathcal{S}(\mathbb{R})$, $\Phi\in(\mathcal{S})^*$,

$$\langle\eta_1\otimes \cdots \otimes\eta_k, \nabla^k\Phi\rangle = D_{\eta_1} \cdots D_{\eta_k}\Phi, \tag{5.54}$$

as an equality in $(\mathcal{S})^*$.

Since in general $\Phi\in(\mathcal{S})^*$ can not be given as an explicit function on $\mathcal{S}'(\mathbb{R})$, it will be convenient for computational purposes to describe the action of differential operators in terms of the $\mathcal{S}$– and $\mathcal{T}$–transforms of Φ. This is done in the following two results.

Theorem 5.35. *Let* $\Phi \in (\mathcal{S})^*$, $\eta \in \mathcal{S}(\mathbb{R})$. D_η *denotes the Gâteaux derivative in direction* η *on* $\mathcal{S}'(\mathbb{R})$ *and on* $\mathcal{S}(\mathbb{R})$. *Let* $\xi \in \mathcal{S}(\mathbb{R})$. *Then the following holds.*

(i) $$(\mathcal{S}D_\eta\Phi)(\xi) = (D_\eta\mathcal{S}\Phi)(\xi); \tag{5.55a}$$

(ii) $$(\mathcal{S}D_\eta^*\Phi)(\xi) = (\eta, \xi)_{L^2(\mathbb{R})} (\mathcal{S}\Phi)(\xi); \tag{5.55b}$$

(iii) $$(\mathcal{S}<\cdot, \eta>\Phi)(\xi) = (D_\eta\mathcal{S}\Phi)(\xi) + (\eta, \xi)_{L^2(\mathbb{R})} \mathcal{S}\Phi(\xi). \tag{5.55c}$$

Proof. (i) By Theorem 5.31, $D_\eta\Phi \in (\mathcal{S})^*$, and therefore the following computation is justified:

$$\begin{aligned} \mathcal{S}D_\eta\Phi(\xi) &= <D_\eta\overline{\Phi}, :e^{<\cdot, \xi>}:> \\ &= \sum_{n=0}^{\infty} \frac{1}{n!} <D_\eta\overline{\Phi}, :<\cdot, \xi>^n:> \\ &= \sum_{n=0}^{\infty} \frac{1}{n!} <\overline{\Phi}, D_\eta^*:<\cdot, \xi>^n:> \\ &= \sum_{n=0}^{\infty} \frac{1}{n!} <\overline{\Phi}, :<\cdot, \eta><\cdot, \xi>^n:>. \end{aligned}$$

Here we made use of the fact that

$$\sum_{n=0}^{\infty} \frac{1}{n!} :<\cdot, \xi>^n:$$

converges to $:e^{<\cdot, \xi>}:$ in $(\mathcal{S})$ (exercise!), and that

$$D_\eta^*:<\cdot, \xi>^n: = :<\cdot, \eta><\cdot, \xi>^n:,$$

which is readily derived, similarly to Rule 5.21. Now let Φ have its chaos decomposition being given by the sequence $(F^{(n)}, n \in \mathbb{N}_0)$, $F^{(n)} \in \mathcal{S}'(\hat{\mathbb{R}}^n)$. Note also that the kernel of $:\langle\cdot,\eta\rangle\langle\cdot,\xi\rangle^n:$ is given by $\eta\hat{\otimes}\xi^{\otimes n}\in\mathcal{S}(\hat{\mathbb{R}}^n)$. Thus

$$\mathcal{S}D_\eta\Phi(\xi) = \sum_{n=0}^{\infty}\frac{(n+1)!}{n!}\langle\overline{F^{(n+1)}},\ \eta\hat{\otimes}\xi^{\otimes n}\rangle$$

$$= \sum_{n=1}^{\infty} n\langle\overline{F^{(n)}},\ \eta\hat{\otimes}\xi^{\otimes(n-1)}\rangle.$$

On the other hand,

$$\mathcal{S}\Phi(\xi) = \sum_{n=0}^{\infty}\langle\overline{F^{(n)}},\ \xi^{\otimes n}\rangle.$$

Therefore obviously $\mathcal{S}D_\eta\Phi = D_\eta\mathcal{S}\Phi$, and (i) is proved.

Now we prove (ii).

$$(\mathcal{S}D_\eta^*\Phi)(\xi) = \langle D_\eta^*\overline{\Phi},\ :e^{\langle\cdot,\ \xi\rangle}:\rangle$$

$$= \langle\overline{\Phi},\ D_\eta:e^{\langle\cdot,\ \xi\rangle}:\rangle$$

$$= (\eta,\ \xi)_{L^2(\mathbb{R})}\ \langle\overline{\Phi},\ :e^{\langle\cdot,\ \xi\rangle}:\rangle$$

$$= (\eta,\ \xi)_{L^2(\mathbb{R})}\ \mathcal{S}\Phi(\xi).$$

Since multiplication by $\langle\cdot,\eta\rangle$ is given by $D_\eta + D_\eta^*$, (iii) follows directly from (i) and (ii). □

In the same way, one can prove the following Theorem. (Another proof can be made by using formulae (5.55) and (4.30).)

Theorem 5.36. *Let* $\Phi\in(\mathcal{S})^*$, $\xi,\ \eta\in\mathcal{S}(\mathbb{R})$. *Then*

(i) $$(\mathcal{T}D_\eta\Phi)(\xi) = -\,iD_\eta\mathcal{T}\Phi(\xi) - i(\eta,\,\xi)_{L^2(R)}\mathcal{T}\Phi(\xi); \tag{5.56a}$$

(ii) $$(\mathcal{T}D_\eta^*\Phi)(\xi) = i(\eta,\,\xi)_{L^2(R)}\mathcal{T}\Phi(\xi); \tag{5.56b}$$

(iii) $$(\mathcal{T}<\cdot,\,\eta>\Phi)(\xi) = -\,iD_\eta\mathcal{T}\Phi(\xi). \tag{5.56c}$$

5.D Table Of Rules

In this section we summarize without further comments some formulae which are useful for practical calculations, from Chapter 2 and the preceding sections. Sometimes a rather informal notation is used. In case of doubt about the interpretation of a formula, we advice the reader to consult the corresponding sections for precise statements.

Wick Powers

Recursive definition:

$$:1: = 1 \tag{5.57a}$$

$$:x: = x \tag{5.57b}$$

$$:x^{\otimes n}: \doteq :x^{\otimes(n-1)}: \otimes x - (n-1) :x^{\otimes(n-2)}: \otimes \mathrm{Tr} \tag{5.57c}$$

Generating functional:

$$\sum_{n=0}^{\infty} \frac{\lambda^n}{n!} :x^{\otimes n}: = \exp{\otimes}(\lambda x - \frac{\lambda^2}{2} \mathrm{Tr}), \tag{5.58}$$

where for a distribution $y \in \mathcal{S}'(\mathbb{R})$, $\exp{\otimes} y$ means $\sum_{n=0}^{\infty} \frac{1}{n!} y^{\otimes n}$.

Explicit Formulae:

$$:x^{\otimes n}: = \sum_{k=0}^{[\frac{n}{2}]} \binom{n}{2k} (2k-1)!! \, (-1)^k \, x^{\otimes(n-2k)} \hat{\otimes} \mathrm{Tr}^{\otimes k} \tag{5.59}$$

$$x^{\otimes n} = \sum_{k=0}^{[\frac{n}{2}]} \binom{n}{2k} (2k-1)!! :x^{\otimes(n-2k)}: \hat{\otimes} \mathrm{Tr}^{\otimes k} \tag{5.60}$$

Informal Recursive Definition:

$$:1: = 1 \tag{5.61a}$$

$$:x(u_1): = x(u_1) \tag{5.61b}$$

$$:x(u_1)\cdots x(u_n): = :x(u_1)\cdots x(u_{n-1}):x(u_n) - \\ - \sum_{k=1}^{n-1} \delta(u_n - u_k) :x(u_1)\cdots x(\not{u}_k)\cdots x(u_{n-1}):. \tag{5.61c}$$

Informal Generating Functional:

$$:x(u_1)\cdots x(u_n): = \prod_{k=1}^{n} \frac{\delta}{\delta\eta(u_k)} \exp\Big(\int x(u)\eta(u)du - \int \eta(u)^2 du\Big)\Big|_{\eta=0} \tag{5.62}$$

Examples:

$$:x^{\otimes 2}: = x^{\otimes 2} - \mathrm{Tr}$$

$$:x^{\otimes 3}: = x^{\otimes 3} - 3\, x \hat{\otimes} \mathrm{Tr}$$

$$:x^{\otimes 4}: = x^{\otimes 4} - 6\, x^{\otimes 2} \hat{\otimes} \mathrm{Tr} + 3\, \mathrm{Tr}^{\otimes 2}$$

$$:x(u_1)\, x(u_2): = x(u_1)\, x(u_2) - \delta(u_1 - u_2)$$

$$:x(u_1)\, x(u_2)\, x(u_3): = x(u_1)\, x(u_2)\, x(u_3) - (\, x(u_1)\, \delta(u_2\text{-}u_3) + \text{cycl. perm.}\,)$$

$\mathfrak{S}$– and $\mathfrak{T}$–Transforms:

$$\mathfrak{S}\colon :x^{\otimes n}: \mapsto \xi^{\otimes n} \tag{5.63}$$

$$\mathcal{T}:\ :x^{\otimes n}: \mapsto i^n\, e^{-\frac{1}{2}|\,\xi\,|_2^2}\, \xi^{\otimes n} \tag{5.64}$$

$$\mathcal{S}:\ :x(u_1)\cdots x(u_n): \mapsto \xi(u_1)\cdots \xi(u_n) \tag{5.65}$$

$$\mathcal{T}:\ :x(u_1)\cdots x(u_n): \mapsto i^n\, e^{-\frac{1}{2}|\,\xi\,|_2^2}\, \xi(u_1)\cdots \xi(u_n) \tag{5.66}$$

Differential operators

Action on Wick Monomials and Kernels, General Formulae:

$$D_y:\ :x^{\otimes n}: \to n :x^{\otimes(n\text{-}1)}: \hat{\otimes}\, y \tag{5.67}$$

$$D_y:\ (f^{(n)}, n\in\mathbb{N}_0) \to (n < y, f^{(n)} >, n\in\mathbb{N}_0) \tag{5.68}$$

$$D_y = <y, \nabla> = <y, \partial.>,\ \partial_t = D_{\delta_t} \tag{5.69}$$

$$\partial_t\!: x(u_1)\cdots x(u_n): = \sum_{k=1}^{n} \delta(t-u_k) : x(u_1)\cdots x(\not{u}_k)\cdots x(u_n): \tag{5.70}$$

$$\partial_t : f^{(n)} \to n\, f^{(n)}(t, \cdot\,) \tag{5.71}$$

$$\partial_t^*\!: x(u_1)\cdots x(u_n): = : x(t)\, x(u_1)\cdots x(u_n): \tag{5.72}$$

$$\partial_t^* : f^{(n)} \to \delta_t\, \hat{\otimes}\, f^{(n)} \tag{5.73}$$

$$[\partial_t\,,\partial_s] = [\partial_t^*\,,\partial_s^*] = 0,\ \ [\partial_t\,,\partial_s^*] = \delta(t-s) \tag{5.74}$$

$$X_y \cdot = <\cdot, y>\cdot = D_y + D_y^* \tag{5.75}$$

$$x(t) \cdot = \partial_t^* + \partial_t \tag{5.76}$$

Product and Chain Rules:

$$D_y\ \varphi\cdot\psi = (D_y\varphi)\cdot\psi + \varphi\cdot(D_y\psi) \tag{5.77}$$

$$D_y\ f\circ\varphi = f'\circ\varphi \cdot D_y\varphi \tag{5.78}$$

$$D_y^*\ \varphi\cdot\psi = (D_y^*\varphi)\ \psi - \varphi\ (D_y\psi) = \varphi\ (D_y^*\psi) - (D_y\varphi)\ \psi \tag{5.79}$$

$$D_y^*\ f\circ\varphi = X_y\cdot\ f\circ\varphi - f'\circ\varphi \cdot D_y\varphi \tag{5.80}$$

Representation of the Number Operator N:

$$N = \int_{\mathbb{R}} dt\ \partial_t^*\partial_t = \sum_{n=1}^{\infty} D_{e_n}^*\ D_{e_n}, \tag{5.81}$$

where $(e_n,\ n\in\mathbb{N})$ is a CONS of $L^2(\mathbb{R})$.

Product and Chain Rule for N:

$$N\ \varphi\psi = (N\varphi)\ \psi + \varphi\ (N\psi) - 2(\nabla\overline{\varphi},\ \nabla\psi)_{L^2(\mathbb{R})} \tag{5.82}$$

$$N\ f\circ\varphi = f'\circ\varphi\ N\varphi - f''\circ\varphi\ (\nabla\overline{\varphi},\ \nabla\varphi)_{L^2(\mathbb{R})}. \tag{5.83}$$

$\mathcal{S}$– and $\mathcal{T}$–Transforms of ∂_t, ∂_t^*:

$$\mathcal{S}(\partial_t\varphi)(\xi) = \frac{\delta}{\delta\xi(t)}\, \mathcal{S}\varphi(\xi) \tag{5.84}$$

$$\mathcal{T}(\partial_t\varphi)(\xi) = -\,i\, \frac{\delta}{\delta\xi(t)}\, \mathcal{T}\varphi(\xi) - i\, \xi(t)\mathcal{T}\varphi(\xi) \tag{5.85}$$

$$\mathcal{S}(\partial_t^*\varphi)(\xi) = \xi(t)\, \mathcal{S}\varphi(\xi) \tag{5.86}$$

$$\mathcal{T}(\partial_t^*\varphi)(\xi) = i\, \xi(t)\, \mathcal{T}\varphi(\xi) \tag{5.87}$$

6. Laplacian Operators

In the previous chapter we have studied the differential operator ∂_t and its adjoint ∂_t^*. In this chapter we will study operators involving higher order differential operators and their adjoints, in particular, the various Laplacian operators. These Laplacian operators are closely connected with the Fourier transform, which will be discussed in Chapter 9. With different observations, several infinite dimensional Laplacian operators have been introduced. For example, when we consider the gradient ∇ and its adjoint ∇^* with respect to the Gaussian measure μ, then we obtain the number operator $N=\nabla^*\nabla$, which we have studied in Chapters 3 and 4. On the other hand, when we consider the Laplacian operator Δ on $\mathbb{R}^n$ as given by $\Delta = \partial^2/\partial x_1^2 +\cdots+ \partial^2/\partial x_n^2$, then the natural generalization is the Gross Laplacian Δ_G in an abstract Wiener space, see Gross (1967) and Kuo (1975). But if we take the average of second partial derivatives, i.e. $n^{-1}(\partial^2/\partial x_1^2 + \cdots+ \partial^2/\partial x_n^2)$, then as $n\to\infty$ we get the Lévy Laplacian Δ_L. The Lévy Laplacian vanishes, as one can easily guess, on the space (L^2). But it plays an important role in harmonic analysis. A related Laplacian, i.e. the Volterra Laplacian Δ_V, is a generalization of the Gross Laplacian to the space of generalized white noise functionals. These Laplacian operators have been studied by Gross (1967), Hida (1975a, 1985, 1989), Hida and Saitô (1988), Kubo and Takenaka (1982), Kuo (1975, 1986, 1988a, 1988b, 1990a), Kuo, Obata and Saitô (1990), Obata (1988, 1989, 1990), Piech (1975) and Saitô (1987, 1988), among others.

6.A White Noise Kernel Operators

The above Laplacian operators in white noise analysis lead naturally to the consideration of the following operator being written in an informal expression:

$$\Xi_{j,k}(\kappa) = \tag{6.1}$$

$$= \int_{\mathbb{R}^{j+k}} \kappa(s_1, \ldots, s_j, t_1, \ldots, t_k)\, \partial_{s_1}^* \cdots \partial_{s_j}^* \partial_{t_1} \cdots \partial_{t_k}\, ds_1 \cdots ds_j dt_1 \cdots dt_k,$$

with kernel $\kappa \in \mathcal{S}'(\mathbb{R}^{j+k})$. This operator has been studied by Krée (1988), Hida, Obata and Saitô (1991), among others. In this section we will present some of the results from Hida, Obata and Saitô (1991). L(X, Y) will denote the space of continuous linear operators from X into Y.

It follows from Theorem 5.7 that the white noise differentiation ∂_t is a continuous linear operator from $(\mathcal{S})$ into itself. Moreover, by Corollary 5.9, for fixed φ in $(\mathcal{S})$, the function $t \mapsto \partial_t\varphi$ is continuous from $\mathbb{R}$ into $(\mathcal{S})$. On the other hand, the adjoint operator ∂_t^* is strongly continuous from $(\mathcal{S})^*$ into itself and, for each $\Phi \in (\mathcal{S})^*$, the function $t \mapsto \partial_t^*\Phi$ is continuous. We will see below that in fact for any $\Phi \in (\mathcal{S})^*$ and $\varphi \in (\mathcal{S})$, the function $t \mapsto \langle \partial_t^*\Phi, \varphi \rangle$ belongs to $\mathcal{S}(\mathbb{R})$.

Lemma 6.1. *For φ and ψ in $(\mathcal{S})$, and $s_1, \ldots, s_j, t_1, \ldots, t_k \in \mathbb{R}$, let*

$$\eta_{\varphi,\psi}(s_1, \ldots, s_j, t_1, \ldots, t_k) = \langle \partial_{s_1}^* \cdots \partial_{s_j}^* \partial_{t_1} \cdots \partial_{t_k}\varphi, \psi \rangle.$$

Then $\eta_{\varphi,\psi} \in \mathcal{S}(\mathbb{R}^{j+k})$. In fact, for all $\varphi, \psi \in (\mathcal{S})$ and $p>0$,

$$|\eta_{\varphi,\psi}|_{2,p} \leq a_p\, |\varphi|_{2,p}\, |\psi|_{2,p}, \tag{6.2}$$

where the constant a_p is given by

$$a_p = \sup\Big(2^{-2pn} \frac{\sqrt{(j+n)!(k+n)!}}{n!}\ ;\ n \geq 0\Big).$$

Proof. Let φ and ψ be represented by

$$\varphi(x) = \sum_{n=0}^{\infty} \langle :x^{\otimes n}:, f_n \rangle,$$

$$\psi(x) = \sum_{n=0}^{\infty} <: x^{\otimes n} :, g_n>.$$

Then, by Rule 5.20, we have

$$\partial_{t_1} \cdots \partial_{t_k} \varphi(x) = \sum_{n=0}^{\infty} \frac{(k+n)!}{n!} <: x^{\otimes n} :, f_{k+n}(t_1, \ldots, t_k, \cdot)>,$$

$$\partial_{s_j} \cdots \partial_{s_1} \psi(x) = \sum_{n=0}^{\infty} \frac{(j+n)!}{n!} <: x^{\otimes n} :, g_{j+n}(s_1, \ldots, s_j, \cdot)>.$$

Hence we have

$$\begin{aligned} \eta_{\varphi,\psi}(s_1, \ldots, s_j, t_1, \ldots, t_k) &= <\partial_{t_1} \cdots \partial_{t_k} \varphi, \partial_{s_j} \cdots \partial_{s_1} \psi> \\ &= \sum_{n=0}^{\infty} \frac{(k+n)!(j+n)!}{n!} \eta_n(s_1, \ldots, s_j, t_1, \ldots, t_k), \end{aligned} \tag{6.3}$$

where η_n is given by

$$\eta_n(s_1, \ldots, s_j, t_1, \ldots, t_k) = \int_{\mathbb{R}^n} f_{k+n}(t_1, \ldots, t_k, u)\, g_{j+n}(s_1, \ldots, s_j, u)\, du.$$

Then, by the Schwarz inequality, we have for any p>0,

$$\begin{aligned} |\eta_n|_{2,p} &= |(A^p)^{\otimes(j+k)} \eta_n|_2 \\ &\le |((A^p)^{\otimes k} \otimes Id^{\otimes n})\, f_{k+n}|_2\, |((A^p)^{\otimes j} \otimes Id^{\otimes n})\, g_{j+n}|_2 \\ &= |(Id^{\otimes k} \otimes (A^{-p})^{\otimes n})((A^p)^{\otimes(k+n)})\, f_{k+n}|_2 \times \\ &\qquad \times |(Id^{\otimes j} \otimes (A^{-p})^{\otimes n})((A^p)^{\otimes(j+n)})\, g_{j+n}|_2 \\ &\le 2^{-2pn} |(A^p)^{\otimes(k+n)} f_{k+n}|\, |(A^p)^{\otimes(j+n)} g_{j+n}|_2 \end{aligned}$$

$$= 2^{-2pn}\, |f_{k+n}|_{2,p}\, |g_{j+n}|_{2,p}. \tag{6.4}$$

Hence it follows from (6.3) and (6.4) that

$$|\eta_{\varphi,\psi}|_{2,p} \le \sum_{n=0}^{\infty} \frac{(k+n)!\,(j+n)!}{n!}\, 2^{-2pn}\, |f_{k+n}|_{2,p}\, |g_{j+n}|_{2,p}$$

$$= \sum_{n=0}^{\infty} 2^{-2pn} \frac{\sqrt{(k+n)!\,(j+n)!}}{n!} \left(\sqrt{(k+n)!}\, |f_{k+n}|_{2,p}\right)\left(\sqrt{(j+n)!}\, |g_{j+n}|_{2,p}\right).$$

Note that for $p>0$, the number a_p defined by

$$a_p := \sup\left(2^{-2pn} \frac{\sqrt{(j+n)!(k+n)!}}{n!}\ ;\ n\ge 0\right)$$

is easily seen to be finite. Therefore, we obtain that for any $p>0$

$$|\eta_{\varphi,\psi}|_{2,p} \le a_p \sum_{n=0}^{\infty} \left(\sqrt{(k+n)!}\, |f_{k+n}|_{2,p}\right)\left(\sqrt{(j+n)!}\, |g_{j+n}|_{2,p}\right)$$

$$\le a_p \left(\sum_{n=0}^{\infty} (k+n)!\, |f_{k+n}|_{2,p}^2\right)^{\frac{1}{2}} \left(\sum_{n=0}^{\infty} (j+n)!\, |g_{j+n}|_{2,p}^2\right)^{\frac{1}{2}}$$

$$\le a_p\, |\varphi|_{2,p}\, |\psi|_{2,p}.$$

Thus $|\eta_{\varphi,\psi}|_{2,p} < \infty$ for all $p>0$ and so $\eta_{\varphi,\psi} \in \mathcal{S}(\mathbb{R}^{j+k})$. □

Theorem 6.2. *Let $\kappa \in \mathcal{S}'(\mathbb{R}^{j+k})$. Then there exists a unique continuous linear operator $\Xi_{j,k}(\kappa)$ from $(\mathcal{S})$ into $(\mathcal{S})^*$ such that for all φ, $\psi \in (\mathcal{S})$,*

$$\langle \Xi_{j,k}(\kappa)\varphi,\, \psi\rangle = \langle \kappa,\, \eta_{\varphi,\psi}\rangle,$$

where $\eta_{\varphi,\psi}$ is given as in Lemma 6.1, i.e.,

$$\eta_{\varphi,\psi}(s_1, \ldots, s_j, t_1, \ldots, t_k) = <\partial_{s_1}^* \cdots \partial_{s_j}^* \partial_{t_1} \cdots \partial_{t_k}\varphi, \psi>.$$

Moreover, if $\kappa \in \mathcal{S}_p'(\mathbb{R}^{j+k})$, *then*

$$|\Xi_{j,k}(\kappa)\varphi|_{2,-p} \leq a_p |\kappa|_{2,-p} |\varphi|_{2,p}, \quad \varphi \in (\mathcal{S}), \tag{6.5}$$

where a_p *is given as in Lemma* 6.1.

Proof. Suppose $\kappa \in \mathcal{S}'(\mathbb{R}^{j+k})$. Then there exists some p>0 such that $\kappa \in \mathcal{S}_p'(\mathbb{R}^{j+k})$. Consider the following sesquilinear form:

$$B(\varphi, \psi) = <\kappa, \eta_{\varphi,\psi}>, \quad \varphi, \psi \in (\mathcal{S}).$$

By (6.2) in Lemma 6.1, we have

$$\begin{aligned} |B(\varphi, \psi)| &\leq |\kappa|_{2,-p} |\eta_{\varphi,\psi}|_{2,p} \\ &\leq a_p |\kappa|_{2,-p} |\varphi|_{2,p} |\psi|_{2,p}. \end{aligned} \tag{6.6}$$

Therefore, B is a continuous sesquilinear form on $(\mathcal{S})$, which is a nuclear space. Hence, by the kernel theorem, there exists a unique continuous linear operator $\Xi_{j,k}(\kappa)$ from $(\mathcal{S})$ into $(\mathcal{S})^*$ such that

$$B(\varphi, \psi) = <\Xi_{j,k}(\kappa)\varphi, \psi>, \quad \varphi, \psi \in (\mathcal{S}). \tag{6.7}$$

Thus $<\Xi_{j,k}(\kappa)\varphi, \psi> = <\kappa, \eta_{\varphi,\psi}>$ for all φ and ψ in $(\mathcal{S})$. The inequality (6.5) follows from (6.6) and (6.7). □

Definition 6.3. *An operator* T *from* $(\mathcal{S})$ *into* $(\mathcal{S})^*$ *is called a* white noise kernel operator of order (j, k) *if it is given by* $T=\Xi_{j,k}(\kappa)$ *for some* $\kappa \in \mathcal{S}'(\mathbb{R}^{j+k})$.

Remark. Symbolically, we will write $\Xi_{j,k}(\kappa)$ in a more suggestive form as follows:

$$\Xi_{j,k}(\kappa) = \int_{\mathbb{R}^{j+k}} \kappa(s_1, \ldots, s_j, t_1, \ldots, t_k)\, \partial_{s_1}^* \cdots \partial_{s_j}^*\, \partial_{t_1} \cdots \partial_{t_k}\, ds dt.$$

Example 6.4. Let $y \in \mathcal{S}'(\mathbb{R})$. Then we have $\int_{\mathbb{R}} y(t)\partial_t\, dt = D_y$, when the Gâteaux differentiation D_y is regarded as an operator from $(\mathcal{S})$ into $(\mathcal{S})^*$. On the other hand, the operator $\int_{\mathbb{R}} y(s)\partial_s^*\, ds$ is equal to D_y^*, when D_y^* is restricted to $(\mathcal{S})$.

Note that given κ in $\mathcal{S}'(\mathbb{R}^{j+k})$, the associated white noise kernel operator $\Xi_{j,k}(\kappa)$ is, in general, a continuous linear operator from $(\mathcal{S})$ into $(\mathcal{S})^*$. But sometimes we need to know whether $\Xi_{j,k}(\kappa)$ is actually a continuous linear operator from $(\mathcal{S})$ into itself. For instance, the operator $\int_{\mathbb{R}} y(t)\partial_t\, dt$ is continuous from $(\mathcal{S})$ into itself (see Theorem 5.7 or Theorem 6.6 below.) It is also of interest to determine when $\Xi_{j,k}(\kappa)$ can be extended continuously to $(\mathcal{S})^*$ so that it is a continuous linear operator from $(\mathcal{S})^*$ into itself. For instance, the operator $\int_{\mathbb{R}} y(t)\partial_s^*\, ds$ can be extended continuously to $(\mathcal{S})^*$ to be a continuous linear operator from $(\mathcal{S})^*$ into itself (see Theorem 5.32 or Theorem 6.7 below.)

To answer the above questions, we need a lemma. But first note that

$$\mathcal{S}(\mathbb{R}^j) \otimes \mathcal{S}'(\mathbb{R}^k) \subset \mathcal{S}'(\mathbb{R}^j) \otimes \mathcal{S}'(\mathbb{R}^k) = \mathcal{S}'(\mathbb{R}^{j+k}).$$

Moreover, there is a one-to-one correspondence between $\mathcal{S}(\mathbb{R}^j) \otimes \mathcal{S}'(\mathbb{R}^k)$ and $L(\mathcal{S}(\mathbb{R}^k), \mathcal{S}(\mathbb{R}^j))$ given by: For $\kappa \in \mathcal{S}(\mathbb{R}^j) \otimes \mathcal{S}'(\mathbb{R}^k)$ corresponding to $K \in L(\mathcal{S}(\mathbb{R}^k), \mathcal{S}(\mathbb{R}^j))$,

$$\langle \kappa, \eta \otimes \zeta \rangle = \langle \overline{\eta}, K\zeta \rangle, \quad \eta \in \mathcal{S}(\mathbb{R}^j),\ \zeta \in \mathcal{S}(\mathbb{R}^k).$$

Lemma 6.5. *Suppose* $K \in L(\mathcal{S}(\mathbb{R}^k), \mathcal{S}(\mathbb{R}^j))$ *is such that for some* $p, q, c \geq 0$,

$$|K\eta|_{2,p} \leq c\,|\eta|_{2,p+q}, \quad \eta \in \mathcal{S}(\mathbb{R}^k).$$

Then the following holds for all $f \in \mathcal{S}(\mathbb{R}^{k+n})$

$$|(K \otimes \mathrm{Id}^{\otimes n})\, f|_{2,p} \leq c\, 2^{-qn}\, |f|_{2,p+q}.$$

Proof. To avoid confusion, we will use $|\cdot|_{(m)}$ to denote the $L^2(\mathbb{R}^m)$-norm. Note that

$$|(K \otimes \mathrm{Id}^{\otimes n})\, f|_{2,p} = |(A^p)^{\otimes (j+n)} (K \otimes \mathrm{Id}^{\otimes n})\, f|_{(j+n)}$$

$$= |(A^p)^{\otimes n}\, |(A^p)^{\otimes j} K\, f(\cdot)|_{(j)}\, |_{(n)}, \tag{6.8}$$

where $f(\cdot)$ indicates that the last n variables of f are held fixed when the operator $(A^p)^{\otimes j}K$ is applied to the first j variables of f. Now, by the assumption on K, we have

$$|(A^p)^{\otimes j} K\, f(\cdot)|_{(j)} = |Kf(\cdot)|_{2,p}$$

$$\leq c|f(\cdot)|_{2,p+q}$$

$$= c|(A^{p+q})^{\otimes k} f(\cdot)|_{(k)}. \tag{6.9}$$

Therefore, it follows from (6.8) and (6.9) that

$$|(K \otimes \mathrm{Id}^{\otimes n})\, f|_{2,p} \leq c|(A^p)^{\otimes n}\, |(A^{p+q})^{\otimes k} f(\cdot)|_{(k)}\, |_{(n)}$$

$$= c|(\mathrm{Id}^{\otimes k} \otimes (A^{-q})^{\otimes n})\, (A^{p+q})^{\otimes (k+n)}\, f|_{(k+n)}$$

$$\leq c 2^{-qn} |(A^{p+q})^{\otimes (k+n)}\, f|_{(k+n)}$$

$$= c2^{-qn}|f|_{2,p+q}. \qquad \square$$

Theorem 6.6. *Let* $\kappa \in \mathcal{S}'(\mathbb{R}^{j+k})$. *Then* $\Xi_{j,k}(\kappa)$ *is a continuous linear operator from* $(\mathcal{S})$ *into itself if and only if* $\kappa \in \mathcal{S}(\mathbb{R}^j) \otimes \mathcal{S}'(\mathbb{R}^k)$. *In this case, for any* p, q ≥ 0,

$$|\Xi_{j,k}(\kappa)\varphi|_{2,p} \leq a_{q/2}\, |(A^p)^{\otimes j} \otimes (A^{-(p+q)})^{\otimes k}\, \kappa|_{L^2(\mathbb{R}^{j+k})}\, |\varphi|_{2,p+q},$$

where the constant $a_{q/2}$ *is as given in Lemma* 6.1.

Proof. Suppose that $\kappa \in \mathcal{S}(\mathbb{R}^j) \otimes \mathcal{S}'(\mathbb{R}^k)$. Let $K \in L(\mathcal{S}(\mathbb{R}^k), \mathcal{S}(\mathbb{R}^j))$ be the corresponding operator of κ, i.e.

$$\langle \kappa, \eta \otimes \zeta \rangle = \langle \overline{\eta}, K\zeta \rangle, \quad \eta \in \mathcal{S}(\mathbb{R}^j),\ \zeta \in \mathcal{S}(\mathbb{R}^k). \tag{6.10}$$

Now, let φ, ψ, $\eta_{\varphi,\psi}$ and η_n be given as in the proof of Lemma 6.1. Then we have

$$\langle \Xi_{j,k}(\kappa)\varphi, \psi \rangle = \langle \kappa, \eta_{\varphi,\psi} \rangle$$

$$= \sum_{n=0}^{\infty} \frac{(k+n)!\,(j+n)!}{n!} \langle \kappa, \eta_n \rangle$$

$$= \sum_{n=0}^{\infty} \frac{(k+n)!\,(j+n)!}{n!} \langle \kappa, \langle f_{k+n}(\cdot), g_{j+n}(\cdot) \rangle_{L^2(\mathbb{R}^n)} \rangle.$$

Hence, by (6.10), we get that

$$\langle \Xi_{j,k}(\kappa)\varphi, \psi \rangle = \sum_{n=0}^{\infty} \frac{(k+n)!\,(j+n)!}{n!} \langle g_{j+n}, (K \otimes I^{\otimes n})\, f_{k+n} \rangle. \tag{6.11}$$

But K is a continuous linear operator from $\mathcal{S}(\mathbb{R}^k)$ into $\mathcal{S}(\mathbb{R}^j)$. Thus for any given $p\geq 0$, there exist $q>0$ and $C=C(p,q)\geq 0$ such that for all η in $\mathcal{S}(\mathbb{R}^k)$

$$|K\eta|_{2,p} \leq C|\eta|_{p+q}.$$

Therefore, by Lemma 6.5,

$$|<g_{j+n}, (K\otimes Id^{\otimes n})\, f_{k+n}>| \leq |g_{j+n}|_{2,-p}\, |(K\otimes Id^{\otimes n})\, f_{k+n}|_{2,p}$$

$$\leq C2^{-qn}\, |g_{j+n}|_{2,-p}\, |f_{k+n}|_{2,p+q}. \tag{6.12}$$

It follows from (6.11) and (6.12) that

$$|<\Xi_{j,k}(\kappa)\varphi,\, \psi>| \leq \sum_{n=0}^{\infty} \frac{(k+n)!\,(j+n)!}{n!}\, C2^{-qn}\, |f_{k+n}|_{2,p+q}\, |g_{j+n}|_{2,-p}$$

$$\leq Ca_{q/2} \sum_{n=0}^{\infty} \Big(\sqrt{(k+n)!}\, |f_{k+n}|_{2,p+q}\Big)\Big(\sqrt{(j+n)!}\, |g_{j+n}|_{2,-p}\Big),$$

where $a_{q/2}$ is defined as in Lemma 6.1. Therefore, by the Schwarz inequality, the following holds for all φ, $\psi\in(\mathcal{S})$

$$|<\Xi_{j,k}(\kappa)\varphi,\, \psi>| \leq Ca_{q/2}\, |\varphi|_{2,p+q}\, |\psi|_{2,-p}.$$

This implies that $|\Xi_{j,k}(\kappa)\varphi|_{2,p} \leq Ca_{q/2}\, |\varphi|_{2,p+q}$ for all φ in $(\mathcal{S})$. Hence $\Xi_{j,k}(\kappa)$ is a continuous linear operator from $(\mathcal{S})$ into itself.

Conversely, suppose $\Xi_{j,k}(\kappa)$ is a continuous linear operator from $(\mathcal{S})$ into itself. Then for any $p\geq 0$, there exist $q>0$ and $C=C(p,q)\geq 0$ such that for all $\varphi\in(\mathcal{S})$

$$|\Xi_{j,k}(\kappa)\varphi|_{2,p} \leq C|\varphi|_{2,p+q}.$$

Now, for any fixed $\zeta\in\mathcal{S}(\mathbb{R}^k)$ and $\eta\in\mathcal{S}(\mathbb{R}^j)$, consider φ and ψ defined by

$$\varphi(x) = <:x^{\otimes k}:, \zeta>,$$

$$\psi(x) = <:x^{\otimes j}:, \eta>.$$

It is easy to check that $\eta_{\varphi,\psi} = k!j!\eta\otimes\zeta$. Therefore,

$$\begin{aligned}|<\kappa, \eta\otimes\zeta>| &= \frac{1}{k!j!}|<\kappa, \eta_{\varphi,\psi}>| \\ &= \frac{1}{k!j!}|<\Xi_{j,k}(\kappa)\varphi, \psi>| \\ &\le \frac{1}{k!j!}|\Xi_{j,k}(\kappa)\varphi|_{2,p}|\psi|_{2,-p} \\ &\le \frac{1}{k!j!}C|\varphi|_{2,p+q}|\psi|_{2,-p} \\ &= \frac{1}{k!j!}C\sqrt{k!}|\zeta|_{2,p+q}\sqrt{j!}|\eta|_{2,-p} \\ &= \frac{C}{\sqrt{k!j!}}|\zeta|_{2,p+q}|\eta|_{2,-p}.\end{aligned}$$

This implies that $\kappa\in\mathcal{S}(\mathbb{R}^j)\otimes\mathcal{S}'(\mathbb{R}^k)$. □

Theorem 6.7. *Let $\kappa\in\mathcal{S}'(\mathbb{R}^j)\otimes\mathcal{S}(\mathbb{R}^k)$. Then the white noise kernel operator*

$$\Xi_{j,k}(\kappa) = \int_{\mathbb{R}^{j+k}} \kappa(s_1, \ldots, s_j, t_1, \ldots, t_k)\partial_{s_1}^*\cdots\partial_{s_j}^*\partial_{t_1}\cdots\partial_{t_k}\, ds_1\cdots ds_j dt_1\cdots dt_k$$

extends to be a continuous linear operator from $(\mathcal{S})^*$ into itself.

Proof. Define $\tilde{\kappa}$ by

$$\tilde{\kappa}(t_1, \ldots, t_k, s_1, \ldots, s_j) = \kappa(s_j, \ldots, s_1, t_k, \ldots, t_1).$$

Then $\tilde{\kappa}\in\mathcal{S}(\mathbb{R}^k)\otimes\mathcal{S}'(\mathbb{R}^j)$. By Theorem 6.6 the white noise kernel operator $\Xi_{k,j}(\tilde{\kappa})$ is continuous from $(\mathcal{S})$ into itself. Therefore, the adjoint $\Xi_{k,j}(\tilde{\kappa})^*$ is continuous from $(\mathcal{S})^*$ into itself. But it is easy to see that $\Xi_{k,j}(\tilde{\kappa})^*=\Xi_{j,k}(\kappa)$ and the conclusion follows immediately. □

Example 6.8. Let $\xi\in\mathcal{S}(\mathbb{R})$, $y\in\mathcal{S}'(\mathbb{R})$ and let $\kappa(s, t)=\xi(s)y(t)$. Then by Theorem 6.6 the operator $\int_{\mathbb{R}^2}\xi(s)y(t)\partial_s^*\partial_t dsdt$ is a continuous linear operator from $(\mathcal{S})$ into itself. In fact, it is obvious that $\int_{\mathbb{R}^2}\xi(s)y(t)\partial_s^*\partial_t dsdt = D_\xi^* D_y$. The continuity of the operator $D_\xi^* D_y$ from $(\mathcal{S})$ into itself follows from Theorems 5.7 and 5.30 and the fact that the restriction of D_ξ^* to $(\mathcal{S})$ is continuous from $(\mathcal{S})$ into itself.

Example 6.9. Let Tr denote the distribution in $\mathcal{S}'(\mathbb{R}^2)$ given by

$$\langle \mathrm{Tr}, f\rangle = \int_{\mathbb{R}^2} f(t, t)\, dt, \quad f\in\mathcal{S}(\mathbb{R}^2).$$

Actually, $\mathrm{Tr}\in\mathcal{S}(\mathbb{R})\otimes\mathcal{S}'(\mathbb{R})$. To see this, note that

$$\mathrm{Tr} = \sum_{n=0}^{\infty} (e_n\otimes e_n, \cdot),$$

where e_n is the Hermite function of order n. Recall that $Ae_n=(2n+2)e_n$. Hence

$$(A^p\otimes A^{-q})\,\mathrm{Tr} = \sum_{n=0}^{\infty} (2n+2)^p(2n+2)^{-q}\,(e_n\otimes e_n, \cdot).$$

Therefore, we have

$$|(A^p\otimes A^{-q})\,\mathrm{Tr}|_2^2 = \sum_{n=0}^{\infty} (2n+2)^{2p}(2n+2)^{-2q}.$$

Obviously, for any p, there exists q such that $|(A^p\otimes A^{-q})\,\mathrm{Tr}|_2<\infty$. Hence Tr belongs to $\mathcal{S}(\mathbb{R})\otimes\mathcal{S}'(\mathbb{R})$. In particular, when $q=p+1$, we get

$$|(A^p \otimes A^{-(p+1)}) \, \mathrm{Tr}|_2 = \Big(\sum_{n=0}^{\infty} (2n+2)^{-2}\Big)^{\frac{1}{2}} = \frac{\pi}{2\sqrt{6}}. \tag{6.13}$$

Therefore, by Theorem 6.6, the white noise kernel operator

$$\Xi_{1,1}(\mathrm{Tr}) = \int_{\mathbb{R}^2} \mathrm{Tr}(s, t) \, \partial_s^* \partial_t \, dsdt = \int_{\mathbb{R}} \partial_t^* \partial_t \, dt$$

is continuous from $(\mathcal{S})$ into itself. Moreover, by Theorem 6.7, this operator is also continuous from $(\mathcal{S})^*$ into itself. It follows from Theorem 6.6 and (6.13) that

$$|\Xi_{1,1}(\mathrm{Tr}) \, \varphi|_{2,p} \le a_{p+1} \frac{\pi}{2\sqrt{6}} |\varphi|_{2,p+1}, \quad \varphi \in (\mathcal{S}).$$

We will see in Section 6.B that this operator $\Xi_{1,1}(\mathrm{Tr})$ can be identified as the number operator N (cf. also (5.81)).

Example 6.10. For any $\kappa \in \mathcal{S}'(\mathbb{R}^k)$, the operator $\Xi_{0,k}(\kappa) = \int_{\mathbb{R}^k} \kappa(t_1, \ldots, t_k) \partial_{t_1} \cdots \partial_{t_k} dt_1 \cdots dt_k$ is a continuous linear operator from $(\mathcal{S})$ into itself by Theorem 6.6. In particular, when $\kappa = \mathrm{Tr}$, the operator

$$\Xi_{0,2}(\mathrm{Tr}) = \int_{\mathbb{R}^2} \mathrm{Tr}(s, t) \, \partial_s \partial_t \, dsdt = \int_{\mathbb{R}} \partial_t^2 \, dt$$

is continuous from $(\mathcal{S})$ into itself. Moreover, from the representation of Tr given in the previous example, we get easily that for any $p, q \ge 0$, $p+q > \frac{1}{4}$,

$$|(A^{-(p+q)})^{\otimes 2} \, \mathrm{Tr}|_2 = \Big(\sum_{n=0}^{\infty} (2n+2)^{-4(p+q)}\Big)^{\frac{1}{2}}.$$

Therefore, by Theorem 6.6, we have

$$|\Xi_{0,2}(\mathrm{Tr}) \, \varphi|_{2,p} \le a_{q/2} \Big(\sum_{n=0}^{\infty} (2n+2)^{-4(p+q)}\Big)^{\frac{1}{2}} |\varphi|_{2,p+q}, \quad \varphi \in (\mathcal{S}).$$

We will see in Section 6.C that this operator $\Xi_{0,2}(\mathrm{Tr})$ can be identified as the

Gross Laplacian Δ_G.

6.B The Number Operator

Recall the number operator N from Chapters 3 and 4. From (5.81) it can be represented by

$$N = \int_{\mathbb{R}} dt\ \partial_t^* \partial_t = \sum_{n=0}^{\infty} D_{e_n}^* D_{e_n},$$

where $(e_n, n \in \mathbb{N})$ is a CONS for $L^2(\mathbb{R})$. Moreover, if $\varphi \in (\mathcal{S})$ is given by

$$\varphi(x) = \sum_{n=0}^{\infty} \langle :x^{\otimes n}:, f_n \rangle,$$

then we have

$$N\varphi(x) = \sum_{n=1}^{\infty} n \langle :x^{\otimes n}:, f_n \rangle.$$

The number operator N is a continuous linear operator from $(\mathcal{S})$ into itself. Its adjoint N^* is a continuous linear operator from $(\mathcal{S})^*$ into itself. Moreover, we have $N^*|_{(\mathcal{S})} = N$. Thus we will use the same notation N to denote its extension as a continuous operator of $(\mathcal{S})^*$.

Theorem 6.11. *Let* $\Phi \in (\mathcal{S})^*$ *and* $F = \mathcal{S}\Phi$. *Then* F *is Fréchet differentiable and*

$$\mathcal{S}(N\Phi)(\xi) = \langle F'(\xi), \xi \rangle, \quad \xi \in \mathcal{S}(\mathbb{R}).$$

Proof. It is easy to check that F is Fréchet differentiable and

$$\langle F'(\xi), \eta \rangle = -\langle \xi, \eta \rangle F(\xi) + \langle \overline{\Phi}, \langle \cdot, \eta \rangle : e^{\langle \cdot, \xi \rangle} : \rangle, \quad \xi, \eta \in \mathcal{S}(\mathbb{R}).$$

In particular, we have

$$<F'(\xi), \xi> = -|\xi|_2 F(\xi) + <\overline{\Phi}, <\cdot, \xi> : e^{<\cdot, \xi>} :>, \quad \xi \in \mathcal{S}(\mathbb{R}). \tag{6.14}$$

On the other hand, it is easy to check that

$$N : e^{<\cdot, \xi>} : = (<\cdot, \xi> - |\xi|_2^2) : e^{<\cdot, \xi>} :, \quad \xi \in \mathcal{S}(\mathbb{R}).$$

Therefore, for $\xi \in \mathcal{S}(\mathbb{R})$, we have

$$\begin{aligned} \mathcal{S}(N\Phi)(\xi) &= <N\overline{\Phi}, : e^{<\cdot, \xi>} :> \\ &= <\overline{\Phi}, N^* : e^{<\cdot, \xi>} :> \\ &= <\overline{\Phi}, N : e^{<\cdot, \xi>} :> \\ &= <\overline{\Phi}, (<\cdot, \xi> - |\xi|_2^2) : e^{<\cdot, \xi>} :> \\ &= <\overline{\Phi}, <\cdot, \xi> : e^{<\cdot, \xi>} :> - |\xi|_2^2 F(\xi). \end{aligned} \tag{6.15}$$

It follows from (6.14) and (6.15) that $\mathcal{S}(N\Phi)(\xi) = <F'(\xi), \xi>$, $\xi \in \mathcal{S}(\mathbb{R})$. □

Now we will identify the number operator as a white noise kernel operator. Recall that from Example 6.9 the white noise kernel operator

$$\Xi_{1,1}(\mathrm{Tr}) = \int_{\mathbb{R}^2} \mathrm{Tr}(s, t)\, \partial_s^* \partial_t \, ds dt$$

is a continuous linear operator from $(\mathcal{S})$ into itself and from $(\mathcal{S})^*$ into itself.

Lemma 6.12. *Let* $\varphi_\xi = : e^{<\cdot, \xi>} :$, $\xi \in \mathcal{S}(\mathbb{R})$. *Then for any* ξ *and* ζ *in* $\mathcal{S}(\mathbb{R})$

$$<\Xi_{1,1}(\mathrm{Tr})\varphi_\xi, \varphi_\zeta> = <\xi, \zeta>\, e^{<\xi, \zeta>}.$$

Proof. Recall that from the definition of $\Xi_{1,1}(\kappa)$, we have for any φ and ψ in $(\mathcal{S})$

$$<\Xi_{1,1}(\mathrm{Tr})\varphi,\ \psi> = <\mathrm{Tr},\ \eta_{\varphi,\psi}>,$$

where $\eta_{\varphi,\psi}$ is given by

$$\eta_{\varphi,\psi}(s,\ t) = <\partial_s^*\partial_t\varphi,\ \psi>.$$

Now, it is easy to check that $\partial_t\varphi_\xi = \xi(t)\varphi_\xi$ and $<\varphi_\xi,\ \varphi_\zeta> = e^{<\xi,\ \zeta>}$. Therefore, we have

$$\begin{aligned}\eta_{\varphi_\xi,\varphi_\zeta}(s,\ t) &= <\partial_s^*\partial_t\varphi_\xi,\ \varphi_\zeta>\\ &= <\partial_t\varphi_\xi,\ \partial_s\varphi_\zeta>\\ &= \xi(t)\xi(s)\, e^{<\xi,\ \zeta>}.\end{aligned}$$

Therefore,

$$<\Xi_{1,1}(\mathrm{Tr})\varphi_\xi,\ \varphi_\zeta> = <\xi,\ \zeta>\, e^{<\xi,\ \zeta>}.$$ □

Theorem 6.13. *The number operator is a white noise kernel operator given by*

$$N = \Xi_{1,1}(\mathrm{Tr}) = \int_{\mathbb{R}^2} \mathrm{Tr}(s,\ t)\ \partial_s^*\partial_t\ ds dt.$$

Proof. Note that $\varphi_\xi = :e^{<\cdot,\ \xi>}:$ has the following series expansion:

$$\varphi_\xi = \sum_{n=0}^{\infty} \frac{1}{n!} <:x^{\otimes n}:,\ \xi^{\otimes n}>.$$

Therefore, we get

$$<N\varphi_\xi, \varphi_\zeta> = \sum_{n=1}^{\infty} n! <\frac{1}{(n-1)!}\,\xi^{\otimes n}, \frac{1}{n!}\,\zeta^{\otimes n}>$$

$$= \sum_{n=1}^{\infty} \frac{1}{(n-1)!} <\xi, \zeta>^n$$

$$= <\xi, \zeta>\, e^{<\xi, \zeta>}. \tag{6.16}$$

Since the linear span of the set $\{\varphi_\zeta;\ \zeta\in\mathcal{S}(\mathbb{R})\}$ is dense in $(\mathcal{S})$, it follows from Lemma 13 and (6.16) that $N\varphi_\xi = \Xi_{1,1}(\mathrm{Tr})\varphi_\xi$ for all $\xi\in\mathcal{S}(\mathbb{R})$. Now, observe that the linear span of the set $\{\varphi_\zeta;\ \zeta\in\mathcal{S}(\mathbb{R})\}$ is also dense in $(\mathcal{S})^*$ and recall that both N and $\Xi_{1,1}(\mathrm{Tr})$ are continuous linear operators from $(\mathcal{S})^*$ into itself. Hence $N=\Xi_{1,1}(\mathrm{Tr})$. □

6.C The Gross and Volterra Laplacians

The Gross Laplacian Δ_G was introduced by Gross (1967) for functions defined on an abstract Wiener space (H, B). Here H is a real separable Hilbert space and B is the completion of H with respect to a measurable norm. For details, see Gross (1967) or Kuo (1975). Suppose f is a twice H–differentiable function defined on B such that its second H–derivative f'' is a trace class operator of H. Then the *Gross Laplacian* $\Delta_G f$ of f is defined by

$$\Delta_G f = \mathrm{trace}_H\, f''.$$

Note that in our white noise space $(\mathcal{S}'(\mathbb{R}), \mu)$ the Gaussian measure is actually supported in the space $\mathcal{S}_{-p}(\mathbb{R})$ for any $p>\frac{1}{2}$, i.e., $(L^2(\mathbb{R}), \mathcal{S}_{-p}(\mathbb{R}))$ is an abstract Wiener space (see Gross (1965) or Kuo (1975)). Therefore, for a function φ defined on $\mathcal{S}'(\mathbb{R})$, we can consider its second $L^2(\mathbb{R})$–derivative and define $\Delta_G\varphi$ as

$$\Delta_G\varphi = \mathrm{trace}_{L^2(\mathbb{R})}\varphi''$$

if φ'' is a trace class operator of $L^2(\mathbb{R})$. In particular, we will show that $\Delta_G\varphi$ exists for φ in $\mathcal{P}$. And then we will extend Δ_G to $(\mathcal{S})$ by continuity.

Recall that in section 4.A we show that $\mathrm{Tr}\in\mathcal{S}_{-p}(\hat{\mathbb{R}^2})$ for any $p>\frac{1}{4}$ and, by Lemma 4.5, we have $|\mathrm{Tr}|_{2,-p} = \|\, H^{-2p}\,\|_{H.S.}$. For a function f in $\mathcal{S}(\hat{\mathbb{R}^n})$, we define the *trace* tr f of f by

$$\mathrm{tr}\, f = \langle \mathrm{Tr}, f(\cdot)\rangle = \int_{\mathbb{R}} f(t, t, \cdot)\, dt.$$

Note that $\mathrm{tr}\, f\in\mathcal{S}(\hat{\mathbb{R}^{n-2}})$ and the following holds for any $F\in\mathcal{S}'(\hat{\mathbb{R}^{n-2}})$ and $f\in\mathcal{S}(\hat{\mathbb{R}^n})$:

$$\langle F\hat{\otimes}\mathrm{Tr}, f\rangle = \langle F, \mathrm{tr}\, f\rangle. \tag{6.17}$$

Lemma 6.14. *Let* $p>\frac{1}{4}$. *Then for any* $f\in\mathcal{S}(\hat{\mathbb{R}^n})$

$$|\mathrm{tr}\, f|_{2,p} \leq \|\, H^{-2p}\,\|_{H.S.}\, |f|_{2,p}. \tag{6.18}$$

Proof. For any $F\in\mathcal{S}'(\hat{\mathbb{R}^{n-2}})$, by (6.17) and Lemma 4.5

$$\begin{aligned} |\langle F, \mathrm{tr}\, f\rangle| &= |\langle F\hat{\otimes}\mathrm{Tr}, f\rangle| \\ &\leq |F|_{2,-p}\, |\mathrm{Tr}|_{2,-p}\, |f|_{2,p} \\ &= |F|_{2,-p}\, \|\, H^{-2p}\|_{H.S.}\, |f|_{2,p}. \end{aligned}$$

This implies the inequality (6.18). □

Now, suppose that $\varphi(x) = \langle :x^{\otimes n}:, f\rangle$, $f\in\mathcal{S}(\hat{\mathbb{R}^n})$, $n\geq 2$. It follows from Proposition 5.2 and Lemma 5.4 that φ is twice $L^2(\mathbb{R})$-differentiable and

$$\langle \varphi''(x), h\otimes k\rangle = n(n-1)\langle :x^{\otimes(n-2)}:, \langle f(\cdot, \cdot), h\otimes k\rangle\rangle.$$

Obviously, $\varphi''(x)$ is a trace class operator of $L^2(\mathbb{R})$ and that

$$\mathrm{trace}_{L^2(\mathbb{R})}\ \varphi''(x) = n(n-1)\langle :x^{\otimes(n-2)}:, \mathrm{tr}\, f\rangle.$$

Therefore, the Gross Laplacian $\Delta_G\varphi$ exists and is given by

$$\Delta_G\varphi(x) = n(n-1)\langle :x^{\otimes(n-2)}:, \mathrm{tr}\, f\rangle. \tag{6.19}$$

Lemma 6.15. *Let* $\varphi\in\mathcal{P}$ *be given by* $\varphi(x)=\sum_{n=0}^{m}\langle :x^{\otimes n}:, f^{(n)}\rangle$, $f^{(n)}\in\mathcal{S}(\hat{\mathbb{R}}^n)$. *Then* $\Delta_G\varphi$ *exists and is given by*

$$\Delta_G\varphi(x) = \sum_{n=2}^{m} n(n-1)\langle :x^{\otimes(n-2)}:, \mathrm{tr}\, f^{(n)}\rangle.$$

Moreover, for any $p>\frac{1}{4}$, *the following holds for all* $\varphi\in\mathcal{P}$:

$$|\Delta_G\varphi|_{2,p} \le \| H^{-2p} \|_{H.S.}\, |\varphi|_{2,p+\frac{1}{2}}.$$

Proof. The first assertion follows from the above discussion and the obvious fact that $\Delta_G\varphi=0$ for $\varphi=c$, a constant, or $\varphi(x)=\langle x, f\rangle$, $f\in\mathcal{S}(\mathbb{R})$. On the other hand, by Lemma 6.14, we have

$$|\Delta_G\varphi|_{2,p}^2 = \sum_{n=2}^{m} (n-2)!\, n^2(n-1)^2\, |\mathrm{tr}\, f^{(n)}|_{2,p}^2$$

$$\le \sum_{n=2}^{m} n!\, n(n-1)\, \| H^{-2p} \|_{H.S.}^2\, |f^{(n)}|_{2,p}^2$$

$$\le \| H^{-2p} \|_{H.S.}^2 \sum_{n=2}^{m} n!\, n(n-1)\, 2^{-n}\, |f^{(n)}|_{2,p+\frac{1}{2}}^2$$

$$\leq \| H^{-2p} \|^2_{H.S.} \sum_{n=2}^{m} n! \, |f^{(n)}|^2_{2,p+\frac{1}{2}}$$

$$\leq \| H^{-2p} \|^2_{H.S.} \, |\varphi|^2_{2,p+\frac{1}{2}}. \qquad \square$$

Now, suppose $\varphi \in (\mathcal{S})$. Choose a Cauchy sequence $\{\varphi_n;\ n \geq 1\} \subset \mathcal{P}$ converging to φ in $(\mathcal{S})$. By Lemma 6.15, the sequence $\{\Delta_G \varphi_n;\ n \geq 1\}$ is Cauchy in $(\mathcal{S})$. Then we define the Gross Laplacian $\Delta_G \varphi$ of φ to be the limit, i.e., $\Delta_G \varphi = \lim_{n \to \infty} \Delta_G \varphi_n$. It is easy to see that $\Delta_G \varphi$ is well-defined.

Theorem 6.16. *The Gross Laplacian Δ_G is a continuous linear operator from $(\mathcal{S})$ into itself. Moreover, if $\varphi \in (\mathcal{S})$ is represented by*

$$\varphi(x) = \sum_{n=0}^{\infty} \langle :x^{\otimes n}:, f^{(n)} \rangle,$$

then $\Delta_G \varphi$ is given by

$$\Delta_G \varphi(x) = \sum_{n=2}^{\infty} n(n-1) \langle :x^{\otimes (n-2)}:, \operatorname{tr} f^{(n)} \rangle. \qquad (6.20)$$

Proof. The continuity of Δ_G follows from the above definition of $\Delta_G \varphi$ for φ in $(\mathcal{S})$ and Lemma 6.15. To prove (6.20), let

$$\varphi_m(x) = \sum_{n=0}^{m} \langle :x^{\otimes n}:, f^{(n)} \rangle$$

and

$$\psi_m(x) = \sum_{n=2}^{m} n(n-1) \langle :x^{\otimes (n-2)}:, \operatorname{tr} f^{(n)} \rangle.$$

Then $\varphi_m \to \varphi$ in $(\mathcal{S})$ as $m \to \infty$. Moreover, by the calculation in the proof of Lemma 6.15, ψ_m is Cauchy in $(\mathcal{S})$. Hence $\psi_m \to \psi$ in $(\mathcal{S})$ with

$$\psi(x) = \sum_{n=2}^{\infty} n(n-1) <: x^{\otimes(n-2)} :, \operatorname{tr} f^{(n)} >.$$

Now, by the continuity of Δ_G, $\Delta_G\varphi_m \to \Delta_G\varphi$ as $m\to\infty$. On the other hand, by Lemma 6.15, $\Delta_G\varphi_m = \psi_m$ and so $\Delta_G\varphi_m \to \psi$ as $m\to\infty$. Therefore, $\Delta_G\varphi=\psi$. □

Theorem 6.17. *Let* $\varphi\in(\mathcal{S})$ *and* $F=\mathcal{S}\varphi$. *Then for each* $\xi\in\mathcal{S}(\mathbb{R})$, *the second functional derivative* $F''(\xi;\cdot,\cdot)$ *belongs to* $\mathcal{S}(\hat{\mathbb{R}}^2)$. *Moreover,*

$$\mathcal{S}(\Delta_G\varphi)(\xi) = \int_{\mathbb{R}} F''(\xi; t, t)\, dt, \quad \xi\in\mathcal{S}(\mathbb{R}).$$

Proof. Suppose $\varphi\in(\mathcal{S})$ is represented by

$$\varphi(x) = \sum_{n=0}^{\infty} <: x^{\otimes n} :, f^{(n)}>.$$

Then we have

$$F(\xi) = \sum_{n=0}^{\infty} <f^{(n)}, \xi^{\otimes n}>.$$

F is easily seen to be Fréchet differentiable and its second functional derivative is given by

$$F''(\xi; s, t) = \sum_{n=2}^{\infty} n(n-1) <f^{(n)}(s, t, \cdot), \xi^{\otimes(n-2)}>.$$

Thus $F''(\xi;\cdot,\cdot)\in\mathcal{S}(\hat{\mathbb{R}}^2)$ and

$$\int_{\mathbb{R}} F''(\xi; t, t)\, dt = \sum_{n=2}^{\infty} n(n-1) <\operatorname{tr} f^{(n)}, \xi^{\otimes(n-2)}>. \tag{6.21}$$

On the other hand, by Theorem 6.16, we have

$$\Delta_G\varphi(x) = \sum_{n=2}^{\infty} n(n-1) <:x^{\otimes n}:, \operatorname{tr} f^{(n)}>.$$

Therefore,

$$\mathcal{S}(\Delta_G\varphi)(\xi) = \sum_{n=2}^{\infty} n(n-1) <\operatorname{tr} f^{(n)}, \xi^{\otimes(n-2)}>. \tag{6.22}$$

From (6.21) and (6.22), we get

$$\mathcal{S}(\Delta_G\varphi)(\xi) = \int_{\mathbb{R}} F''(\xi; t, t)\, dt, \quad \xi \in \mathcal{S}(\mathbb{R}). \qquad \square$$

Theorem 6.18. *For any φ and ψ in $(\mathcal{S})$, the following equalities hold:*

$$\Delta_G(\varphi\psi) = (\Delta_G\varphi)\psi + \varphi(\Delta_G\psi) + 2\int_{\mathbb{R}} (\partial_t\varphi)(\partial_t\psi)\, dt,$$

$$\Delta_G(\varphi:\psi) = (\Delta_G\varphi):\psi + \varphi:(\Delta_G\psi) + 2\int_{\mathbb{R}} (\partial_t\varphi):(\partial_t\psi)\, dt.$$

Proof. The first equality for $\varphi, \psi \in \mathcal{P}$ follows from the fact that

$$<(\varphi\psi)''(x), h\otimes k> = <\varphi''(x), h\otimes k>\psi(x) + \varphi(x)<\psi''(x), h\otimes k> +$$
$$+ D_h\varphi(x)D_k\psi(x) + D_k\varphi(x)D_h\psi(x).$$

Then we can use the same argument as in the proof of Theorem 6.16 to extend the equality for $\varphi, \psi \in (\mathcal{S})$. On the other hand, let F, G and H be the $\mathcal{S}$-transformations of φ, ψ and $\varphi:\psi$, respectively. Then $H(\xi)=F(\xi)G(\xi)$ and we have

$$H''(\xi; s, t) = F''(\xi; s, t)G(\xi) + F(\xi)G''(\xi; s, t) +$$
$$+ F'(\xi; s)G'(\xi; t) + F'(\xi; t)G'(\xi; s).$$

This yields the second equality in view of Theorem 6.17. □

Now we want to identify the Gross Laplacian Δ_G as a white noise kernel operator. Recall that from Example 6.10 the operator

$$\Xi_{0,2}(\mathrm{Tr}) = \int_{\mathbb{R}} \mathrm{Tr}(s, t)\, \partial_s \partial_t \, dsdt = \int_{\mathbb{R}} \partial_t^2 \, dt$$

is a continuous linear operator from $(\mathcal{S})$ into itself.

Theorem 6.19. *The Gross Laplacian is a white noise kernel operator given by*

$$\Delta_G = \Xi_{0,2}(\mathrm{Tr}) = \int_{\mathbb{R}^2} \mathrm{Tr}(s, t)\, \partial_s \partial_t \, dsdt.$$

Proof. Let $\varphi_\xi = :e^{<\cdot,\, \xi>}:$. By using (6.20) we can check easily that $\Delta_G \varphi_\xi = |\xi|_2^2 \varphi_\xi$. Hence for any ξ and ζ in $\mathcal{S}(\mathbb{R})$

$$\begin{aligned} <\Delta_G \varphi_\xi, \varphi_\zeta> &= |\xi|_2^2 <\varphi_\xi, \varphi_\zeta> \\ &= |\xi|_2^2 \, e^{<\xi,\, \zeta>}. \end{aligned} \tag{6.23}$$

On the other hand, from the definition of $\Xi_{0,2}(\mathrm{Tr})$, we have

$$<\Xi_{0,2}(\mathrm{Tr})\varphi_\xi, \varphi_\zeta> = <\mathrm{Tr}, \eta_{\varphi_\xi, \varphi_\zeta}>,$$

where $\eta_{\varphi_\xi, \varphi_\zeta}(s, t) = <\partial_s \partial_t \varphi_\xi, \varphi_\zeta>$. Note that $\partial_t \varphi_\xi = \xi(t) \varphi_\xi$. Therefore,

$$\eta_{\varphi_\xi, \varphi_\zeta}(s, t) = \xi(s)\xi(t)\, e^{<\xi,\, \zeta>}.$$

It follows immediately that for any ξ and ζ in $\mathcal{S}(\mathbb{R})$,

$$<\Xi_{0,2}(\mathrm{Tr})\varphi_\xi, \varphi_\zeta> = |\xi|_2^2 \, e^{<\xi,\, \zeta>}. \tag{6.24}$$

Since both Δ_G and $\Xi_{0,2}(\mathrm{Tr})$ are continuous linear operators from $(\mathcal{S})$ into itself, it follows from (6.23) and (6.24) that $\Delta_G = \Xi_{0,2}(\mathrm{Tr})$. □

Theorem 6.20. *The adjoint operator* Δ_G^* *of* Δ_G *is a white noise kernel operator given by*

$$\Delta_G^* = \Xi_{2,0}(\mathrm{Tr}) = \int_{\mathbb{R}^2} \mathrm{Tr}(s, t)\, \partial_s^* \partial_t^* \, ds dt.$$

Moreover, for any Φ *in* $(\mathcal{S})^*$, *we have*

$$\mathcal{S}(\Delta_G^* \Phi)(\xi) = |\xi|_2^2 \, \mathcal{S}\Phi(\xi), \quad \xi \in \mathcal{S}(\mathbb{R}).$$

Proof. Since Δ_G is continuous from $(\mathcal{S})$ into itself, its adjoint Δ_G^* is continuous from $(\mathcal{S})^*$ into itself. On the other hand, by Theorem 6.7, the operator $\Xi_{2,0}(\mathrm{Tr})$ extends to be continuous from $(\mathcal{S})^*$ into itself. Now, it is obvious from (6.23) and (6.24) that $\langle \Delta_G^* \varphi_\xi, \varphi_\zeta \rangle = \langle \Xi_{2,0}(\mathrm{Tr}) \varphi_\xi, \varphi_\zeta \rangle$ for any ξ and ζ in $\mathcal{S}(\mathbb{R})$. This implies immediately that $\Delta_G^* = \Xi_{2,0}(\mathrm{Tr})$. To find the $\mathcal{S}$-transformation of $\Delta_G^* \Phi$ for $\Phi \in (\mathcal{S})^*$, note that

$$\Delta_G : e^{\langle \cdot, \xi \rangle} : = |\xi|_2^2 : e^{\langle \cdot, \xi \rangle} :.$$

Hence, for any $\xi \in \mathcal{S}(\mathbb{R})$, we have

$$\mathcal{S}(\Delta_G^* \Phi)(\xi) = \langle \Delta_G^* \overline{\Phi}, : e^{\langle \cdot, \xi \rangle} : \rangle$$

$$= \langle \overline{\Phi}, \Delta_G : e^{\langle \cdot, \xi \rangle} : \rangle$$

$$= |\xi|_2^2 \langle \overline{\Phi}, : e^{\langle \cdot, \xi \rangle} : \rangle$$

$$= |\xi|_2^2 \, \mathcal{S}\Phi(\xi).$$

□

The number operator N and the Gross Laplacian Δ_G are related by the following well-known equation

$$N\varphi = -\Delta_G\varphi + \langle\cdot, \nabla\varphi\rangle, \quad \varphi\in(\mathcal{S}).$$

This equation can be checked easily by using the identities (5.75) and (5.81).

At the end of this section, we will study an extension of the Gross Laplacian Δ_G to the space $(\mathcal{S})^*$ of generalized white noise functionals. Obviously, (6.19) does not make sense for a general φ in $(\mathcal{S})^*$. However, we can use the $\mathcal{S}$-transformation to define a new Laplacian, i.e. the Volterra Laplacian Δ_V, which is indeed an extension of Δ_G.

Definition 6.21. *Let* $\Phi\in(\mathcal{S})^*$ *and* $F=\mathcal{S}\Phi$. *We will call* Φ *or* F a V–functional *if* F *has the second derivative of the form: for* ξ, η, $\zeta\in\mathcal{S}(\mathbb{R})$,

$$\langle F''(\xi), \eta\otimes\zeta\rangle = \langle f(\xi), \eta\zeta\rangle + \int_{\mathbb{R}^2} g(\xi; s, t)\eta(s)\zeta(t)\, dsdt, \tag{6.25}$$

where for each ξ, $f(\xi)\in\mathcal{S}'(\mathbb{R})$ *and* $g(\xi;\cdot,\cdot)$ *is symmetric and defines a trace class operator* $g(\xi)$ *of* $L^2(\mathbb{R})$, *and the function* $\mathrm{trace}_{L^2(\mathbb{R})}\, g(\xi)$, $\xi\in\mathcal{S}(\mathbb{R})$, *is a U–functional.*

We point out that the decomposition in (6.25) is unique. To see this, suppose $f\in\mathcal{S}'(\mathbb{R})$, and g defines a compact operator K of $L^2(\mathbb{R})$ such that for all η, $\zeta\in\mathcal{S}(\mathbb{R})$,

$$\langle f, \eta\zeta\rangle + \int_{\mathbb{R}^2} g(s, t)\eta(s)\zeta(t)\, dsdt = 0.$$

Let M_f denote the operator from $\mathcal{S}(\mathbb{R})$ into $\mathcal{S}'(\mathbb{R})$ defined by $M_f\xi=f\xi$. Then the above equality implies that $M_f\eta=-K\eta$ for all η in $\mathcal{S}(\mathbb{R})$. The fact that K is a bounded operator of $L^2(\mathbb{R})$ implies that f must be in $L^\infty(\mathbb{R})$. But since K is a compact operator of $L^2(\mathbb{R})$, we conclude that f=0. This implies that $K\eta=0$ for all

$\eta\in\mathcal{S}(\mathbb{R})$, and so g=0. Therefore, the decomposition in (6.25) is unique. This justifies the following definition of the Volterra Laplacian. We will use F''_V to denote the function g in (6.25), F''_V is called the *Volterra part of* F.

Definition 6.22. *Let Φ be a V–functional and $F=\mathcal{S}\Phi$. The* Volterra Laplacians $\Delta_V\Phi$ *and* $\Delta_V F$ *of Φ and F, respectively, are defined by*

$$\Delta_V\Phi = \mathcal{S}^{-1}(\mathrm{trace}_{L^2(\mathbb{R})} F''_V),$$

$$\Delta_V F = \mathrm{trace}_{L^2(\mathbb{R})} F''_V.$$

The domain $\mathcal{D}(\Delta_V)$ of the Volterra Laplacian Δ_V consists of all V–functionals Φ in $(\mathcal{S})^*$. We will also use the same notation $\mathcal{D}(\Delta_V)$ to denote the set of all V–functionals $F=\mathcal{S}\Phi$.

Example 6.23. Let $\Phi(x)=\langle :x^{\otimes n}:, \delta_r^{\otimes n}\rangle$, $n\geq 2$, and $r\in\mathbb{R}$. Then $F(\xi)=\xi(r)^n$ and it is easy to check that F has the second derivative given by

$$\langle F''(\xi), \eta\otimes\zeta\rangle = \langle n(n-1)\xi^{n-2}\delta_r, \eta\zeta\rangle.$$

Therefore, $\Phi\in\mathcal{D}(\Delta_V)$ and $\Delta_V\Phi=0$.

Example 6.24. Let $f\in\mathcal{S}'(\mathbb{R})$ and $F(\xi)=\int f(t)\xi(t)^n\,dt$. Then its second derivative is given by

$$\langle F''(\xi), \eta\otimes\zeta\rangle = \langle n(n-1)f\xi^{n-2}, \eta\zeta\rangle.$$

Therefore, $F\in\mathcal{D}(\Delta_V)$ and $\Delta_V F=0$.

Example 6.25. Let $f\in\mathcal{S}(\mathbb{R}^2)$ and

$$F(\xi) = \int_{\mathbb{R}^2} f(u, v)\xi(u)^2\xi(v)\, dudv.$$

Then its second derivative is given by

$$<F''(\xi), \eta\otimes\zeta> = \int_{\mathbb{R}}\Big(\int_{\mathbb{R}} 2f(t, v)\xi(v)dv\Big)\eta(t)\zeta(t)\, dt$$

$$+ \int_{\mathbb{R}^2} 2\Big(f(s, t)\xi(s) + f(t, s)\xi(t)\Big)\eta(s)\zeta(t)\, dsdt.$$

Obviously, F is a V–functional and

$$\text{trace}_{L^2(\mathbb{R})}\ F''_V(\xi) = 4\int_{\mathbb{R}} f(t, t)\xi(t)\, dt.$$

Therefore, $F\in\mathfrak{D}(\Delta_V)$ and $\Delta_V F(\xi) = 4<\tilde{f}, \xi>$. Here $\tilde{f}(u)=f(u, u)$.

Example 6.26. Let $\Phi=g_c=\mathcal{N}\exp(-\frac{1}{2c}\ |\cdot|_2^2)$, $c\neq-1$ (see Example 4.23). Then

$$F(\xi)=\exp(-\frac{1}{2(c+1)}\ |\xi|_2^2)$$

and its second derivative is given by

$$<F''(\xi), \eta\otimes\zeta> = \int_{\mathbb{R}} -(1+c)^{-1}F(\xi)\eta(t)\zeta(t)\, dt +$$

$$+ \int_{\mathbb{R}^2} (1+c)^{-2}F(\xi)\xi(s)\xi(t)\eta(s)\zeta(t)\, ds\, dt.$$

Note that $F''_V(\xi;\ s, t)=(1+c)^{-2}F(\xi)\xi(s)\xi(t)$ defines a trace class operator of $L^2(\mathbb{R})$

and that $\text{trace}_{L^2(\mathbb{R})} F''_V(\xi) = (1+c)^{-2}|\xi|_2^2\ F(\xi)$. Hence $g_c \in \mathfrak{D}(\Delta_V)$ and, by Theorem 6.20, we have $\Delta_V g_c = (1+c)^{-2}\Delta_G^* g_c$.

Example 6.27. Let $\theta:\mathbb{R}^+ \to \mathbb{R}$ be a C^2-function such that $|\theta''(u)| \le c_1\exp(c_2 u)$, c_1, $c_2 \ge 0$. Then $F(\xi) = \theta(|\xi|_2^2)$ is a U-functional. It is easy to check that

$$<F''(\xi),\ \eta\otimes\zeta> = \int_{\mathbb{R}} 2\theta'(|\xi|_2^2)\eta(t)\zeta(t)\ dt + \int_{\mathbb{R}^2} 4\theta''(|\xi|_2^2)\xi(s)\xi(t)\eta(s)\zeta(t)\ dsdt.$$

Therefore, $F \in \mathfrak{D}(\Delta_V)$ and $(\Delta_V F)(\xi) = 4|\xi|_2^2\ \theta''(|\xi|_2^2)$.

Example 6.28. Let θ: $\mathbb{R} \to \mathbb{R}$ be a C^2-function such that $|\theta''(u)| \le c_1\exp(c_2 u^2)$, c_1, $c_2 \ge 0$. Then $F(\xi) = \theta(<f,\ \xi>)$, $f \in L^2(\mathbb{R})$, is a U-functional and

$$<F''(\xi),\ \eta\otimes\zeta> = \theta''(<f,\ \xi>)<f,\ \eta><f,\ \zeta>.$$

Hence $F \in \mathfrak{D}(\Delta_V)$ and $\Delta_V F(\xi) = |f|_2^2\theta''(<f,\ \xi>)$.

Theorem 6.29. *The restriction of* Δ_V *to* $(\mathcal{S})$ *is* Δ_G, *i.e.*, $\Delta_V|_{(\mathcal{S})} = \Delta_G$.

Proof. We need to show that any $\varphi \in (\mathcal{S})$ is a V-functional and $\Delta_V\varphi = \Delta_G\varphi$. Suppose φ is represented by

$$\varphi(x) = \sum_{n=0}^{\infty} <:x^{\otimes n}:,\ f^{(n)}>.$$

Then its $\mathcal{S}$-transformation F is given by

$$F(\xi) = \sum_{n=0}^{\infty} <f^{(n)},\ \xi^{\otimes n}>.$$

We can check easily that

$$\langle F''(\xi), \eta\otimes\zeta\rangle = \sum_{n=2}^{\infty} n(n-1) \langle f^{(n)}, \xi^{\otimes(n-2)}\hat{\otimes}\eta\hat{\otimes}\zeta\rangle.$$

Therefore, $F''_V(\xi; s, t)$ is given by

$$F''_V(\xi; s, t) = \sum_{n=2}^{\infty} n(n-1) \langle f^{(n)}(s, t, \cdot), \xi^{\otimes(n-2)}\rangle.$$

Thus for each fixed ξ, $F''_V(\xi; \cdot, \cdot)$ defines a trace class operator of $L^2(\mathbb{R})$ and

$$\text{trace}_{L^2(\mathbb{R})} F''_V(\xi) = \sum_{n=2}^{\infty} n(n-1) \langle \text{tr } f^{(n)}, \xi^{\otimes(n-2)}\rangle.$$

Hence, we get

$$\Delta_V\varphi(x) = \sum_{n=2}^{\infty} n(n-1) \langle :x^{\otimes(n-2)}:, \text{tr } f^{(n)} \rangle.$$

In view of (6.20), we have $\Delta_V\varphi=\Delta_G\varphi$. □

Theorem 6.30. *Let* Φ *and* Ψ *be two* V*–functionals,* $F=\mathcal{S}\Phi$ *and* $G=\mathcal{S}\Psi$. *Assume that* $F'(\xi; \cdot)$, $G'(\xi; \cdot)\in L^2(\mathbb{R})$ *for each* ξ. *Then* $\Phi:\Psi$ *is also a* V*–functional and*

$$\Delta_V(\Phi:\Psi) = (\Delta_V\Phi):\Psi + \Phi:(\Delta_V\Psi) + 2\int_{\mathbb{R}} (\partial_t\Phi):(\partial_t\Psi)\, dt.$$

Proof. Let $H=\mathcal{S}(\Phi:\Psi)$. Then $H(\xi)=F(\xi)G(\xi)$. The conclusion follows by direct calculation. □

6.D The Lévy Laplacian

The Lévy Laplacian was introduced and investigated in Lévy (1922, 1925, 1951, 1956) for functions defined on the Hilbert space $H=L^2(0, 1)$. Let $\{e_n; n\geq1\}$ be an orthonormal basis for H. The Lévy Laplacian $\Delta_L f$ of a C^2–function f on H with respect to $\{e_n; n\geq1\}$ is given by

$$\Delta_L f(x) = \lim_{n\to\infty} \frac{1}{n} \sum_{k=1}^{n} <f''(x), e_k \otimes e_k>, \quad x \in H,$$

provided that the limit exists. This Laplacian is very different from the Gross Laplacian. For instance, consider the function $f(x)=|x|^2$, $|\cdot|$ being the H-norm. Then $<f''(x), h\otimes k>=2<h, k>$ for all x, or equivalently, $f''(x)=2\mathrm{Id}$. Hence $\Delta_G f(x)$ does not exists, while $\Delta_L f(x)=2$. This example reflects the fact that $|\cdot|$ is not a measurable norm on H (see Gross (1965) or Kuo (1975)). Moreover, it is obvious that if f is a C^2-function on H depending only on a finite dimensional subspace of H, then $\Delta_L f=0$. The Lévy Laplacian can be expressed in terms of the asymptotic spherical mean, see e.g. Lévy (1951), Obata (1989) and Kuo, Obata and Saitô (1990). On the other hand, it has been shown in Kuo, Obata and Saitô (1990) that $\Delta_L f$ is the integral of the singular part along the diagonal line of its second derivative f''. This is the rationale for our definition of the Lévy Laplacian below.

Before we define the Lévy Laplacian, let us take a closer look at the simple example $F(\xi)=|\xi|_2^2$, which is a U-functional. Its second derivative is given by

$$<F''(\xi), \eta\otimes\zeta> = 2<\eta, \zeta> = \int_{\mathbb{R}} 2\eta(t)\zeta(t)\, dt = \int_{\mathbb{R}} f(t)\eta(t)\zeta(t)\, dt.$$

Observe that the function $f(t)=2$ is not integrable on $\mathbb{R}$, but is locally integrable. This is the motivation for the following definition of L-functionals.

Definition 6.31. *Let* $\Phi\in(\mathcal{S})^*$ *and* $F=\mathcal{S}\Phi$. *We will call* Φ *or* F *an* L-functional *if* F *has first two derivatives of the forms: for* $\xi, \eta, \zeta\in\mathcal{S}(\mathbb{R})$,

$$<F'(\xi), \eta> = \int_{\mathbb{R}} F'(\xi; t)\eta(t)\, dt$$

$$<F''(\xi), \eta\otimes\zeta> = \int_{\mathbb{R}} f(\xi; t)\eta(t)\zeta(t)\, dt + \int_{\mathbb{R}^2} g(\xi; s, t)\eta(s)\zeta(t)\, ds dt, \tag{6.26}$$

where F', f *and* g *satisfy the following conditions:*

(1) *For each* ξ, $F'(\xi;\ \cdot)\in L^1_{loc}(\mathbb{R})$, $f(\xi;\ \cdot)\in L^1_{loc}(\mathbb{R})$ *and* $g(\xi;\ \cdot,\ \cdot)\in L^1_{loc}(\mathbb{R}^2)$ *and* $g(\xi;\cdot,\ \cdot)$ *is symmetric;*

(2) *For each* ξ, *there exist constants* a, $p\geq 0$ *depending on* ξ *such that*

$$\int_{|t|\geq a} |F'(\xi;\ t)|^2|t|^{-p}\ dt < +\infty;$$

$$\int_{|t|\geq a} |f(\xi;\ t)||t|^{-p}\ dt < +\infty;$$

$$\int_{|(s,t)|\geq a} |g(\xi;\ s,\ t)|^2|(s,t)|^{-p}\ ds\ dt < +\infty;$$

(3) *For any finite interval* T, $\int_T f(\cdot;\ t)\ dt$ *is a* U*–functional.*

First we show that the decomposition (6.26) is unique. Suppose f and g are locally integrable functions on $\mathbb{R}$ and $\mathbb{R}^2$, respectively, satisfying condition (2) such that for all η and ζ in $\mathcal{S}(\mathbb{R})$

$$\int_{\mathbb{R}} f(t)\eta(t)\zeta(t)\ dt + \int_{\mathbb{R}^2} g(s,\ t)\eta(s)\zeta(t)\ dsdt = 0.$$

Let T be any finite interval. By considering those η and ζ with support in T in the above equation, we see easily that for almost all $s\in T$

$$f(s) = -\int_T g(s,\ t)\ dt.$$

In particular, take $T=[0,\ u]$ and differentiate the above equation with respect to u, then we see that $g(s,\ u)=0$ a.e. on T^2 for any finite interval T. Therefore, $g=0$ a.e. and so $f=0$ a.e. This implies that the decomposition in (6.26) is unique.

Let $\mathfrak{D}(\Delta_L)$ denote the set of all L-functionals Φ. We will also use the same notation $\mathfrak{D}(\Delta_L)$ to denote the set of all L-functionals F. The function f in (6.26) is called the Lévy part of Φ or F and will be denoted by F_L''.

Definition 6.32. *Let* T *be a fixed finite interval on* $\mathbb{R}$. *Let* Φ *be an* L*-functional and* $F=\mathcal{S}\Phi$. *The* Lévy Laplacians $\Delta_L^T\Phi$ *and* $\Delta_L^T F$ *of* Φ *and* F, *respectively, with the interval* T *are defined by*

$$\Delta_L^T\Phi = \mathcal{S}^{-1}\Big(\frac{1}{|T|}\int_T F_L''(\cdot;\, t)\, dt\Big);$$

$$\Delta_L^T F = \frac{1}{|T|}\int_T F_L''(\cdot;\, t)\, dt.$$

Example 6.33. The function $F(\xi)=\int_{\mathbb{R}} f(t)\xi(t)^n dt$, $n\geq 2$, $f\in L^1(\mathbb{R})$, is an L-functional and $\Delta_L^T F(\xi) = n(n-1)|T|^{-1}\int_T f(t)\xi(t)^{n-2}dt$. Note that $\Delta_L^T F\to 0$ as $T\to\mathbb{R}$. When n=2, we have $\Delta_L^T F(\xi) = 2|T|^{-1}\int_T f(t)dt$, which indicates that Δ_L^T is the infinite dimensional Laplacian with respect to the coordinate system $\{\dot{B}(t);\, t\in T\}$.

Example 6.34. The function $F(\xi)=\int_{\mathbb{R}^2} f(u,\, v)\xi(u)^2\xi(v)dudv$, $f\in L^1(\mathbb{R}^2)$, is in $\mathfrak{D}(\Delta_L)$ and $\Delta_L^T F(\xi)=2|T|^{-1}\int_{\mathbb{R}^2} f(u,\, v)\xi(v)dudv$. We have $\Delta_L^T F\to 0$ as $T\to\mathbb{R}$.

Example 6.35. The Gaussian white noise functional $g_c=\mathcal{N}\exp(-\frac{1}{2c}\, |\cdot|_2^2)$ in Example 6.26 is an L-functional. We have $F(\xi) = \exp(-\frac{1}{2(c+1)}|\xi|_2^2)$ and $F_L''(\xi;\, t) = -(1+c)^{-1}F(\xi)$. Hence for any T, we have $\Delta_L^T F = -(1+c)^{-1}F$ and $\Delta_L^T\Phi = -(1+c)^{-1}\Phi$. Thus g_c is an eigenfunction of Δ_L^T with eigenvalue $-(1+c)^{-1}$.

Example 6.36. The function $F(\xi)=\theta(|\xi|_2^2)$ in Example 6.27 is an L-functional with $F_L''(\xi;\, t) = 2\theta'(|\xi|_2^2)$. Hence for any T, we have $\Delta_L^T F = 2\theta'(|\xi|_2^2)$. This indicates that Δ_L^T behaves like a first order differential operator in some sense.

Example 6.37. The function $F(\xi)=\theta(<f, \xi>)$ in Example 6.28 is an L–functional and $\Delta_L^T F=0$. Thus F is Δ_L^T–harmonic for any T.

An important class of L–functionals is the set of normal polynomials in Lévy's terminology. By a *normal polynomial* we mean a finite linear combination of functions of the form

$$\int_{A^n} f(t_1, \ldots, t_n)\, \xi(t_1)^{p_1} \cdots \xi(t_n)^{p_n}\, dt_1 \cdots dt_n,$$

where A is a finite interval in $\mathbb{R}$ and $f \in L^1(A^n)$ and $p_1, \ldots, p_n \geq 0$. It can be checked easily that a normal polynomial is an L–functional.

For the rest of this section T will be a fixed finite interval of $\mathbb{R}$. Thus for convenience, we will write Δ_L^T simply as Δ_L.

We will show that our definition of Δ_L is in consistent with Lévy's original definition. Then we will obtain some properties of Δ_L. In particular, we will show the first order differential operator nature of Δ_L, contrary to the fact that it is defined from the second derivative. We will also show that the Lévy Laplacian vanishes on the space (L^2). Thus Δ_L appears non-trivially only when we study generalized white noise functionals.

Definition 6.38. *An orthonormal basis* $\{e_n;\, n \geq 1\}$ *for* $L^2(T)$ *is called* equally dense *if the following holds for any* $f \in L^\infty(T)$:

$$\lim_{n \to \infty} \frac{1}{n} \sum_{k=1}^{n} \int_T f(t)\, e_n(t)^2\, dt = \frac{1}{|T|} \int_T f(t)\, dt. \qquad (6.27)$$

It is called uniformly bounded *if* $\sup\{\|e_n\|_\infty;\, n \geq 1\} < \infty$.

For the existence of an equally dense and uniformly bounded orthonormal basis for $L^2(T)$, see Hida (1980a). It is easy to check that if $\{e_n;\, n \geq 1\}$ is such a basis, then (6.27) holds for any $f \in L^1(T)$. Moreover, the following holds for any $g \in L^1(T^2)$

$$\lim_{n,m\to\infty} \int_{T^2} g(s, t)\, e_n(s)e_m(t)\, dsdt = 0. \tag{6.28}$$

Theorem 6.39. *Let* $F\in\mathfrak{D}(\Delta_L)$. *Then the following holds for any equally dense and uniformly bounded orthonormal basis* $\{e_n;\ n\geq 1\}$ *for* $L^2(T)$:

$$\Delta_L F(\xi) = \lim_{n\to\infty} \frac{1}{n}\sum_{k=1}^{n} <F''(\xi),\ e_k\otimes e_k>.$$

Proof. By assumption, we have

$$<F''(\xi),\ \eta\otimes\zeta> = \int_{\mathbb{R}} F_L''(\xi;\ t)\eta(t)\zeta(t)\, dt + \int_{\mathbb{R}^2} g(\xi;\ s,\ t)\eta(s)\zeta(t)\, dsdt, \tag{6.29}$$

where F_L'' and g satisfy the conditions in Definition 6.31. By (6.27) and the remark following it, we have

$$\lim_{n\to\infty} \frac{1}{n}\sum_{k=1}^{n} \int_T F_L''(\xi;\ t)\, e_k(t)^2\, dt = \frac{1}{|T|}\int_T F_L''(\xi;\ t)\, dt. \tag{6.30}$$

On the other hand, by (6.28), we have

$$\lim_{n\to\infty} \int_{T^2} g(\xi;\ s,\ t)\, e_n(s)e_n(t)\, dsdt = 0$$

and this implies that

$$\lim_{n\to\infty} \frac{1}{n}\sum_{k=1}^{n} \int_{T^2} g(\xi;\ s,\ t)\, e_k(s)e_k(t)\, dsdt = 0. \tag{6.31}$$

It follows from (6.29)–(6.31) that

$$\lim_{n\to\infty} \frac{1}{n}\sum_{k=1}^{n} <F''(\xi),\ e_k\otimes e_k> = \frac{1}{|T|}\int_T F_L''(\xi;\ t)\, dt.$$

But $\Delta_L F(\xi) = |T|^{-1}\int_T F_L''(\xi;\, t)dt$ and so we have shown the assertion of the theorem. □

Theorem 6.40. *Suppose either* (i) θ *is a polynomially bounded* C^2*-function on* $\mathbb{R}$, *and* F *is an* $\mathbb{R}$*-valued* L*-functional, or* (ii) θ *is a polynomially bounded analytic function on* $\mathbb{C}$, *and* F *is a* $\mathbb{C}$*-valued* L*-functional. Then* $G=\theta\circ F$ *is also an* L*-functional and*

$$\Delta_L G(\xi) = \theta'(F(\xi))\Delta_L F(\xi), \quad \xi\in\mathcal{S}(\mathbb{R}).$$

Proof. First note that G is a U–functional since θ is polynomially bounded. For any ξ, η and ζ in $\mathcal{S}(\mathbb{R})$, we have

$$\langle G'(\xi),\, \eta\rangle = \theta'(F(\xi))\langle F'(\xi),\, \eta\rangle,$$

$$\langle G''(\xi),\, \eta\otimes\zeta\rangle = \theta'(F(\xi))\langle F''(\xi),\, \eta\otimes\zeta\rangle + \theta''(F(\xi))\langle F'(\xi),\, \eta\rangle\langle F'(\xi),\, \zeta\rangle.$$

It is obvious that G is also an L–functional and $G_L''(\xi;\, t)$ is given by

$$G_L''(\xi;\, t) = \theta'(F(\xi))\, F_L''(\xi;\, t).$$

This implies that $\Delta_L G(\xi) = \theta'(F(\xi))\Delta_L F(\xi)$ for any $\xi\in\mathcal{S}(\mathbb{R})$. □

Theorem 6.41. *Let* Φ *and* Ψ *be in* $(\mathcal{S})^*$. *Suppose* $F=\mathcal{S}\Phi$ *and* $G=\mathcal{S}\Psi$ *are* L*-functionals. Then* FG *is also an* L*-functional, and*

$$\Delta_L(FG) = (\Delta_L F)G + F(\Delta_L G),$$

or equivalently,

$$\Delta_L(\Phi{:}\Psi) = (\Delta_L \Phi){:}\Psi + \Phi{:}(\Delta_L \Psi).$$

Remark. This property shows that the Lévy Laplacian Δ_L is a derivation. Thus Δ_L behaves like a first order differential operator. Note that on the other hand, it is defined from the second derivative of F.

Proof. Let H=FG. We can check easily that for ξ, η and ζ in $\mathcal{S}(\mathbb{R})$,

$$<H'(\xi), \eta> = <F'(\xi), \eta>G(\xi) + F(\xi)<G'(\xi), \eta>,$$

$$<H''(\xi), \eta\otimes\zeta> = <F''(\xi), \eta\otimes\zeta>G(\xi) + F(\xi)<G''(\xi), \eta\otimes\zeta>$$

$$+ <F'(\xi), \eta><G'(\xi), \zeta> + <F'(\xi), \zeta><G'(\xi), \eta>.$$

It is straightforward to check that H is an L–functional and

$$H''_L(\xi; t) = F''_L(\xi; t)G(\xi) + F(\xi)G''_L(\xi; t),$$

which implies that $\Delta_L H = (\Delta_L F)G + F(\Delta_L G)$. The other assertion follows immediately since $\mathcal{S}(\Phi:\Psi)=(\mathcal{S}\Phi)(\mathcal{S}\Psi)$. □

Theorem 6.42. *The Lévy Laplacian vanishes on* (L^2), *i.e.*, $\Delta_L\Phi=0$ *for all* $\Phi\in(L^2)$.

Proof. Let $\Phi\in(L^2)$ be represented by

$$\Phi(x) = \sum_{n=0}^{\infty} <: x^{\otimes n} :, f^{(n)}>, \ f^{(n)}\in L^2(\widehat{\mathbb{R}^n}).$$

Then $F=\mathcal{S}\Phi$ is given by

$$F(\xi) = \sum_{n=0}^{\infty} <f^{(n)}, \xi^{\otimes n}>.$$

It is easy to check that $F'(\xi; t)$ is given by

$$F'(\xi; t) = \sum_{n=1}^{\infty} n\langle f^{(n)}(t, \cdot), \xi^{\otimes(n-1)}\rangle.$$

By the Schwarz inequality, we get

$$|F'(\xi; t)|^2 \le \Big(\sum_{n=1}^{\infty} n! \int_{\mathbb{R}^{n-1}} |f^{(n)}(t, u)|^2 \, du\Big)\Big(\sum_{n=1}^{\infty} \frac{n^2}{n!} |\xi|_2^{2(n-1)}\Big)$$

$$= b(\xi)\Big(\sum_{n=1}^{\infty} n! \int_{\mathbb{R}^{n-1}} |f^{(n)}(t, u)|^2 \, du\Big),$$

where $b(\xi)=(1+|\xi|_2^2)\exp(|\xi|_2^2)$. Now, note that

$$|\Phi|_2^2 = \sum_{n=0}^{\infty} n! \, |f^{(n)}|^2_{L^2(\mathbb{R}^n)}.$$

Therefore, we get immediately that

$$\int_{\mathbb{R}} |F'(\xi; t)|^2 \, dt \le b(\xi)|\Phi|_2^2.$$

This implies conditions (1) and (2) for $F'(\xi; \cdot)$ in Definition 6.31. Furthermore, the second derivative F'' is given by

$$\langle F''(\xi), \eta\otimes\zeta\rangle = \sum_{n=2}^{\infty} n(n-1) \langle f^{(n)}, \xi^{\otimes(n-2)}\hat{\otimes}\eta\hat{\otimes}\zeta\rangle$$

$$= \int_{\mathbb{R}^2} g(\xi; s, t)\eta(s)\zeta(t) \, dsdt,$$

where $g(\xi; s, t)$ is defined as follows:

$$g(\xi; s, t) = \sum_{n=2}^{\infty} n(n-1) \langle f^{(n)}(s, t, \cdot), \xi^{\otimes(n-2)}\rangle.$$

By the Schwarz inequality, we get

$$|g(\xi;\, s,\, t)|^2 \le \Big(\sum_{n=2}^{\infty} n! \int_{\mathbb{R}^{n-2}} |f^{(n)}(s,\, t,\, u)|^2\, du\Big)\Big(\sum_{n=2}^{\infty} \frac{n^2(n-1)^2}{n!}\, |\xi|_2^{2(n-2)}\Big)$$

$$\le c(\xi)\Big(\sum_{n=2}^{\infty} n! \int_{\mathbb{R}^{n-2}} |f^{(n)}(s,\, t,\, u)|^2\, du\Big),$$

where $c(\xi)=(|\xi|_2^4 + 4|\xi|_2^2 + 2)\exp(|\xi|_2^2)$. Therefore, we have

$$\int_{\mathbb{R}^2} |g(\xi;\, s,\, t)|^2\, dsdt \le c(\xi)|\Phi|_2^2.$$

This implies that $g(\xi;\, \cdot,\, \cdot)$ satisfies conditions (1) and (2) in Definition 6.31 and the Lévy part of Φ is zero. Hence $\Delta_L \Phi=0$. □

Theorem 6.43. *Let* Φ *be an* L*–functional. Then for any* φ *in the linear span of* $\{: e^{<\cdot,\, \zeta>} :;\ \zeta\in\mathcal{S}(\mathbb{R})\}$, $\varphi\Phi$ *is also an* L*–functional and*

$$\Delta_L(\varphi\Phi) = \varphi(\Delta_L \Phi).$$

Proof. First recall that for any ξ and η in $\mathcal{S}(\mathbb{R})$

$$(: e^{<\cdot,\, \xi>} :)(: e^{<\cdot,\, \eta>} :) = : e^{<\cdot,\, \xi+\eta>} : e^{<\xi,\, \eta>}.$$

By using this identity we can check that for any ζ in $\mathcal{S}(\mathbb{R})$ and Φ in $(\mathcal{S})^*$

$$\mathcal{S}(: e^{<\cdot,\, \zeta>} :\Phi)(\xi) = \mathcal{S}\Phi(\xi+\zeta)e^{<\zeta,\, \xi>},\quad \xi\in\mathcal{S}(\mathbb{R}). \tag{6.32}$$

To prove the theorem it is sufficient to consider φ of the form $\varphi_\zeta =: e^{<\cdot,\, \zeta>} :$, $\zeta\in\mathcal{S}(\mathbb{R})$. Let F and G be the $\mathcal{S}$–transformations of Φ and $\varphi_\zeta\Phi$, respectively. Then by (6.32) we have $G(\xi)=F(\xi+\zeta)e^{<\zeta,\, \xi>}$, $\xi\in\mathcal{S}(\mathbb{R})$. Therefore, the first two derivatives of G are given by

$$<G'(\xi), \eta> = <F'(\xi+\zeta), \eta>e^{<\zeta, \xi>} + F(\xi+\zeta)<\zeta, \eta>e^{<\zeta, \xi>},$$

$$<G''(\xi), \eta\otimes\omega> = <F''(\xi+\zeta), \eta\otimes\omega>e^{<\zeta, \xi>} + <F'(\xi+\zeta), \eta><\zeta, \omega>e^{<\zeta, \xi>}$$

$$+ <F'(\xi+\zeta), \omega><\zeta, \eta>e^{<\zeta, \xi>} + F(\xi+\zeta)<\zeta, \eta><\zeta, \omega>e^{<\zeta, \xi>}.$$

It is straightforward to check that $\varphi_\zeta\Phi$ is an L–functional and

$$G''_L(\xi; t) = F''_L(\xi+\zeta; t)e^{<\zeta, \xi>}.$$

Therefore, we obtain that

$$\mathcal{S}(\Delta_L(\varphi_\zeta\Phi))(\xi) = \frac{1}{|T|}\int_T G''_L(\xi; t)\,dt = \frac{1}{|T|}\int_T F''_L(\xi+\zeta; t)e^{<\zeta, \xi>}\,dt. \quad (6.33)$$

On the other hand, by using (6.32), we have

$$\mathcal{S}(\varphi_\zeta(\Delta_L\Phi))(\xi) = \mathcal{S}(\Delta_L\Phi)(\xi+\zeta)e^{<\zeta, \xi>} = \frac{1}{|T|}\int_T F''_L(\xi+\zeta; t)e^{<\zeta, \xi>}\,dt. \quad (6.34)$$

It follows from (6.33) and (6.34) that $\Delta_L(\varphi_\zeta\Phi)=\varphi_\zeta(\Delta_L\Phi)$. □

Theorem 6.44. *The following holds for any* $\Phi\in\mathfrak{D}(\Delta_L)$,

$$\Delta_L\Delta_G^*\,\Phi = 2\Phi + \Delta_G^*(\Delta_L\Phi).$$

In particular, if $\Phi\in(L^2)$, *then* $\Delta_L\Delta_G^*\,\Phi = 2\Phi$.

Proof. Let $F=\mathcal{S}\Phi$. Then by Theorem 6.20, the $\mathcal{S}$–transformation $H(\xi)$ of $\Delta_G^*\Phi$ is given by $H(\xi)=|\xi|_2^2F(\xi)$. Let $G(\xi)=|\xi|_2^2$. We see easily that $\Delta_L G=2$. Hence by Theorem 6.41, we have $\Delta_L H(\xi) = 2F(\xi) + |\xi|_2^2(\Delta_L F)(\xi)$. This shows that

$$\Delta_L \Delta_G^* \Phi = 2\Phi + \Delta_G^*(\Delta_L \Phi).$$

In particular, when $\Phi\in(L^2)$, we have $\Delta_L \Phi=0$ by Theorem 6.42 and so $\Delta_L \Delta_G^* \Phi = 2\Phi$. □

At the end of this section we show that there is a relationship between the Gross Laplacian Δ_G and the Lévy Laplacian Δ_L, although they are very different. Intuitively, Δ_L is the limit of the average of finite dimensional Δ_G. For instance, consider the function $F(\xi)=|\xi|_2^2$. We have $F''(\xi)=2\mathrm{Id}$ and so $\Delta_G F$ does not exists. This is reflected by the fact that the Gross Laplacian acts on $(\mathcal{S})$ and there is no such functional $\varphi\in(\mathcal{S})$ such that $\mathcal{S}\varphi(\xi)=|\xi|_2^2$. On the other hand, this function F is the $\mathcal{S}$–transformation of a generalized white noise functional, i.e., $\Phi(x)=\int_{\mathbb{R}} :x(t)^2:dt$, and we have $\Delta_L F(\xi)=2$. Now, let P_n be an orthogonal projection of dimensional n such that $P_n\to \mathrm{Id}$ strongly as $n\to\infty$, and consider the function F_n defined by $F_n(\xi)=|P_n\xi|_2^2$. Then we have $\Delta_G F_n=2n$. Thus for this example, we have $\Delta_L F=\lim_{n\to\infty} n^{-1}\Delta_G F_n$. The following theorem shows that this is indeed true for any L–functional.

Theorem 6.45. *Let* $\{e_n;\ n\geq 1\}\subset\mathcal{S}(\mathbb{R})$ *be an equally dense and uniformly bounded orthonormal basis for* $L^2(T)$. *Let* P_n *be the projection onto the linear span of* $e_1, \ldots, e_n$. *Suppose* $F\in\mathfrak{D}(\Delta_L)$ *and assume that for any* $\epsilon>0$, *there exists* $\delta>0$ *such that*

$$|F_L''(\xi;t) - F_L''(\eta;t)| < \epsilon,\ \ \forall\ |\xi-\eta|_2<0,\ t\in T. \tag{6.35}$$

Then we have

$$\Delta_L F = \lim_{n\to\infty} \frac{1}{n}\, \Delta_G(F\circ P_n).$$

Proof. Let $F_n = F \circ P_n$. Then $F_n''(\xi) = P_n \circ F''(P_n\xi) \circ P_n$. Hence $F_n''(\xi)$ is a trace class operator of $L^2(T)$ and so $\Delta_G F_n$ exists and is given by

$$\Delta_G F_n = \sum_{k=1}^{\infty} \langle F''(P_n\xi), (P_n e_k)\otimes(P_n e_k)\rangle$$
$$= \sum_{k=1}^{n} \langle F''(P_n\xi), e_k \otimes e_k\rangle.$$

But by assumption $F''(\xi)$ is given as follows:

$$\langle F''(\xi), \eta\otimes\zeta\rangle = \int_{\mathbb{R}} F_L''(\xi; t)\eta(t)\zeta(t)\, dt + \int_{\mathbb{R}^2} g(\xi; s, t)\eta(s)\zeta(t)\, dsdt.$$

Therefore, we have

$$\frac{1}{n}\Delta_G F_n(\xi) = \frac{1}{n}\sum_{k=1}^{n} \int_T F_L''(P_n\xi; t)\, e_k(t)^2\, dt + \frac{1}{n}\sum_{k=1}^{n} \int_{T^2} g(P_n\xi; s, t)\, e_k(s)e_k(t)\, dsdt.$$

Thus with the assumption (6.35) and the same argument as in the proof of Theorem 6.39, we can show that

$$\lim_{n\to\infty} \frac{1}{n}\Delta_G F_n(\xi) = \frac{1}{|T|}\int_T F_L''(\xi; t)\, dt.$$

But from the definition of Δ_L, we have

$$\Delta_L F(\xi) = \frac{1}{|T|}\int_T F_L''(\xi; t)\, dt.$$

Hence we get $\Delta_L F = \lim_{n\to\infty} n^{-1}\, \Delta_G(F\circ P_n)$. □

A similar relationship between Δ_L and Δ_G has been obtained by Kuo, Obata and Saitô (1990). The following presentation is based on their idea.

For $\epsilon>0$, let K_ϵ be a C^∞-function on $\mathbb{R}$ satisfying the following conditions:

(1) $\text{supp}(K_\epsilon)\subset(-\frac{1}{2}, \frac{1}{2})$;

(2) $0\le K_\epsilon(t)\le\epsilon^{-1}$ for all $t\in\mathbb{R}$;

(3) $\int_{-\frac{1}{2}}^{\frac{1}{2}} K_\epsilon(t)dt=1$;

(4) $\lim_{\epsilon\downarrow 0}\int_{-\frac{1}{2}}^{\frac{1}{2}} \varphi(t)K_\epsilon(t)dt=\varphi(0)$ for all $\varphi\in C(\mathbb{R})$;

(5) $\varliminf_{\epsilon\downarrow 0} \epsilon\alpha(\epsilon) > 0$, where $\alpha(\epsilon) = \int_{-\frac{1}{2}}^{\frac{1}{2}} K_\epsilon(t)^2dt$.

Observe that $\alpha(\epsilon)\to\infty$ as $\epsilon\downarrow 0$ in view of condition (5). Recall that T is a fixed finite interval in $\mathbb{R}$. Define an integral operator Q_ϵ by

$$Q_\epsilon\xi(t) = \int_T K_\epsilon(t-s)\xi(s)\, ds.$$

Note that Q_ϵ is a Hilbert–Schmidt operator of $L^2(T)$. In the following T_1 will denote the interval $\{x;\ \text{dist}(x, T)\le\frac{1}{2}\}$. It is obvious that if $\xi\in L^2(T)$, then $Q_\epsilon\xi$ is a C^∞-function with support contained in T_1 and so $Q_\epsilon\xi\in\mathcal{S}(\mathbb{R})$. In fact, Q_ϵ is continuous from $L^2(T)$ into $\mathcal{S}(\mathbb{R})$. Thus it is also continuous from $\mathcal{S}(\mathbb{R})$ into itself.

Theorem 6.46. *Let* $F\in\mathfrak{D}(\Delta_L)$ *have the second derivative given by*

$$<F''(\xi), \eta\otimes\zeta> = \int_{\mathbb{R}} F_L''(\xi;\, t)\eta(t)\zeta(t)\, dt + \int_{\mathbb{R}^2} g(\xi;\, s, t)\eta(s)\zeta(t)\, dsdt. \quad (6.36)$$

Assume that $F_L''(\xi;\, t)$ *and* $g(\xi;\, s, t)$ *are continuous in* ξ, s *and* t, *and there exist constants* c, c_1, $c_2>0$ *and* p, $q\geq 0$ *such that*

(i) $\sup\{|F''_L(\xi;\, t)|;\, t\in T_1\}\leq c_1\exp(c_2|\xi|^2_{2,p})$ *for all* ξ *in* $\mathcal{S}(\mathbb{R})$;

(ii) $\sup\{|g(\xi;\, s,\, t)|;\, s,\, t\in T_1\}\leq c_1\exp(c_2|\xi|^2_{2,p})$ *for all* ξ *in* $\mathcal{S}(\mathbb{R})$;

(iii) $\|H^p Q_\epsilon H^{-q}\|_{L^2(\mathbb{R})}\leq c$ *for all* $\epsilon>0$.

Then the following holds

$$\Delta_L F = \lim_{\epsilon\downarrow 0} \frac{1}{\alpha(\epsilon)|T|}\, \Delta_G(F\circ Q_\epsilon).$$

Proof. Let $F_\epsilon=F\circ Q_\epsilon$. Then $F''_\epsilon=Q_\epsilon\circ F''(Q_\epsilon)\circ Q_\epsilon$. From the discussion before the statement of the theorem, we know that $Q_\epsilon\eta$ vanishes outside T_1 for any η in $L^2(T)$. Therefore, from (6.36), the following holds for any η, $\zeta\in L^2(T)$:

$$\langle F''_\epsilon(\xi),\, \eta\otimes\zeta\rangle = \int_{T_1} F''_L(Q_\epsilon\xi;\, t)(Q_\epsilon\eta(t))(Q_\epsilon\zeta(t))\, dt$$
$$+ \int_{T_1^2} g(Q_\epsilon\xi;\, s,\, t)(Q_\epsilon\eta(s))(Q_\epsilon\zeta(t))\, dsdt. \tag{6.37}$$

By using the kernel function K_ϵ, we can easily rewrite (6.37) as follows:

$$\langle F''_\epsilon(\xi),\, \eta\otimes\zeta\rangle = \int_{T^2} A_\epsilon(\xi;\, s_1,\, s_2)\eta(s_1)\zeta(s_2)\, ds_1ds_2$$
$$+ \int_{T^2} B_\epsilon(\xi;\, s_1,\, s_2)\eta(s_1)\zeta(s_2)\, ds_1ds_2, \tag{6.38}$$

where A_ϵ and B_ϵ are given by

$$A_\epsilon(\xi;\, s_1,\, s_2) = \int_{T_1} F_L''(Q_\epsilon\xi;\, t)K_\epsilon(t-s_1)K_\epsilon(t-s_2)\, dt, \tag{6.39}$$

$$B_\epsilon(\xi;\, s_1,\, s_2) = \int_{{T_1}^2} g(Q_\epsilon\xi;\, s,\, t)K_\epsilon(s-s_1)K_\epsilon(t-s_2)\, dsdt. \tag{6.40}$$

Hence from (6.38) we obtain that

$$\Delta_G F_\epsilon(\xi) = \int_T A_\epsilon(\xi;\, s,\, s)\, ds + \int_T B_\epsilon(\xi;\, s,\, s)\, ds. \tag{6.41}$$

Now, by assumptions (ii) and (iii), we see that for all $\epsilon>0$ and $\xi\in\mathcal{S}(\mathbb{R})$

$$\sup\{|g(Q_\epsilon\xi;\, s,\, t)|;\, s,\, t\in T_1\}\le c_1\exp(cc_2|\xi|_{2,q}^2). \tag{6.42}$$

It follows from condition (3), (6.40) and (6.42) that

$$\int_T |B_\epsilon(\xi;\, s,\, s)|\, ds \le c_1\exp(cc_2|\xi|_{2,q}^2)\int_T\Big(\int_{T_1^2} K_\epsilon(t_1-s)K_\epsilon(t_2-s)\, dt_1dt_2\Big)ds$$

$$= c_1\exp(cc_2|\xi|_{2,q}^2)\int_T\Big(\int_{[-\frac{1}{2},\, \frac{1}{2}]^2} K_\epsilon(u)K_\epsilon(v)\, dudv\Big)ds$$

$$= c_1|T|\exp(cc_2|\xi|_{2,q}^2).$$

Recall that $\alpha(\epsilon)\to\infty$ as $\epsilon\downarrow 0$ and so

$$\lim_{\epsilon\downarrow 0}\frac{1}{\alpha(\epsilon)|T|}\int_T B_\epsilon(\xi;\, s,\, s)\, ds = 0. \tag{6.43}$$

On the other hand, from conditions (1) and (5), and (6.39), we get

$$\frac{1}{\alpha(\epsilon)|T|}\int_T A_\epsilon(\xi;\, s,\, s)\, ds - \frac{1}{|T|}\int_T F_L''(\xi;\, s)\, ds =$$

$$= \frac{1}{\alpha(\epsilon)|T|} \int_T \Big(A_\epsilon(\xi;\, s, s) - F_L''(\xi;\, s) \int_{T_1} K_\epsilon(t-s)^2\, dt \Big)\, ds$$

$$= \frac{1}{\alpha(\epsilon)|T|} \int_T \Big(\int_{T_1} (F_L''(Q_\epsilon \xi;\, t) - F_L''(\xi;\, s))\, K_\epsilon(t-s)^2\, dt \Big) ds.$$

Hence by condition (2),

$$\Big|\frac{1}{\alpha(\epsilon)|T|} \int_T A_\epsilon(\xi;\, s, s)\, ds - \frac{1}{|T|} \int_T F_L''(\xi;\, s)\, ds\Big|$$

$$\leq \frac{1}{\epsilon\alpha(\epsilon)|T|} \int_T \Big(\int_{T_1} |F_L''(Q_\epsilon \xi;\, t) - F_L''(\xi;\, s)|\, K_\epsilon(t-s)\, dt \Big) ds. \tag{6.44}$$

Now, note that by condition (3), we get

$$\int_T \Big(\int_{T_1} |F_L''(Q_\epsilon \xi;\, t) - F_L''(\xi;\, t)|\, K_\epsilon(t-s)\, dt \Big) ds$$

$$\leq \int_{T_1} |F_L''(Q_\epsilon \xi;\, t) - F_L''(\xi;\, t)|\, dt.$$

Therefore, by assumption (i) and the Lebesgue dominated convergence theorem,

$$\lim_{\epsilon \downarrow 0} \int_T \Big(\int_{T_1} |F_L''(Q_\epsilon \xi;\, t) - F_L''(\xi;\, t)|\, K_\epsilon(t-s)\, dt \Big) ds = 0. \tag{6.45}$$

Moreover, by using condition (1), we derive easily that

$$\int_T \Big(\int_{T_1} |F_L''(\xi;\, t) - F_L''(\xi;\, s)|\, K_\epsilon(t-s)\, dt \Big) ds$$

$$= \int_T \Big(\int_{T_1 - s} |F_L''(\xi;\, u+s) - F_L''(\xi;\, s)|\, K_\epsilon(u)\, du \Big)\, ds$$

$$= \int_T \left(\int_{-\frac{1}{2}}^{\frac{1}{2}} |F_L''(\xi; u+s) - F_L''(\xi; s)| \, K_\epsilon(u) \, du \right) ds.$$

Hence, by condition (4),

$$\lim_{\epsilon \downarrow 0} \int_T \left(\int_{T_1} |F_L''(\xi; t) - F_L''(\xi; s)| \, K_\epsilon(t-s) \, dt \right) ds =$$

$$= \int_T |F_L''(\xi; s) - F_L''(\xi; s)| \, ds$$

$$= 0. \tag{6.46}$$

It follows from condition (5) and (6.44)–(6.46) that

$$\lim_{\epsilon \downarrow 0} \frac{1}{\alpha(\epsilon)|T|} \int_T A_\epsilon(\xi; s, s) \, ds = \frac{1}{|T|} \int_T F_L''(\xi; s) \, ds. \tag{6.47}$$

The conclusion of the theorem follows from (6.41), (6.43) and (6.47). □

6.E The Commutation Relations

In the previous sections we have studied the operators N, Δ_G, Δ_G^*, Δ_L, and Δ_V. The commutation relations among them are given as below.

(1) $[N, \Delta_G] = -2\Delta_G$ on $(\mathcal{S})$.

(2) $[N, \Delta_G^*] = 2\Delta_G^*$ on $(\mathcal{S})^*$.

(3) $[N, \Delta_L] = -2\Delta_L$ on $\mathcal{D}(\Delta_L)$.

(4) $[N, \Delta_V] = -2\Delta_V$ on $\mathcal{D}(\Delta_V)$.

(5) $[\Delta_G, \Delta_L] = 0$ on $(\mathcal{S})$; in fact, $\Delta_G \Delta_L = \Delta_L \Delta_G = 0$ on $(\mathcal{S})$.

(6) $[\Delta_G, \Delta_V] = 0$ on $(\mathcal{S})$; in fact, $\Delta_G \Delta_V = \Delta_V \Delta_G = \Delta_G^2$ on $(\mathcal{S})$.

(7) $[\Delta_L, \Delta_G^*] = 2\mathrm{Id}$ on $\mathcal{D}(\Delta_L)$.

(8) $[\Delta_V, \Delta_G^*] = 4N$ on $\mathcal{D}(\Delta_V)$.

Remark. The commutator $[\Delta_G, \Delta_G^*]$ is not well–defined. It should be replaced by $[\Delta_V, \Delta_G^*]$ in view of Theorem 6.29. We do not know the commutation relation between Δ_L and Δ_V.

Proof of (1). Let $\varphi\in(\mathcal{S})$ be represented by

$$\varphi(x) = \sum_{n=0}^{\infty} <: x^{\otimes n} :, f^{(n)}>, \ f^{(n)}\in\mathcal{S}(\hat{\mathbb{R}}^n).$$

Then by Theorem 6.16, we have

$$\Delta_G\varphi(x) = \sum_{n=2}^{\infty} n(n-1) <: x^{\otimes(n-2)} :, \mathrm{tr}\ f^{(n)} >.$$

And then by the definition of the number operator N, we get

$$N\Delta_G\varphi(x) = \sum_{n=2}^{\infty} n(n-1)(n-2) <: x^{\otimes(n-2)} :, \mathrm{tr}\ f^{(n)} >. \tag{6.48}$$

Similarly, from the definition of N and then Theorem 6.16, we have

$$\Delta_G N\varphi(x) = \sum_{n=2}^{\infty} n^2(n-1) <: x^{\otimes(n-2)} :, \mathrm{tr}\ f^{(n)} >. \tag{6.49}$$

Therefore, from (6.48) and (6.49), we see that

$$\begin{aligned}(N\Delta_G-\Delta_G N)\varphi(x) &= -2\sum_{n=2}^{\infty} n(n-1) <: x^{\otimes(n-2)} :, \mathrm{tr}\ f^{(n)} > \\ &= -2\Delta_G\varphi(x).\end{aligned}$$

□

Proof of (2). Simply take the adjoint of (1) and recall that $N^*=N$. □

Proof of (3). Let $\Phi\in\mathfrak{D}(\Delta_L)$, $F=\mathcal{S}\Phi$ and $G=\mathcal{S}N\Phi$. Then by Theorem 6.11, we have $G(\xi)=<F'(\xi), \xi>$. Obviously,

$$<G''(\xi), \eta\otimes\zeta> = 2<F''(\xi), \eta\otimes\zeta> + <F'''(\xi), \xi\otimes\eta\otimes\zeta>. \tag{6.50}$$

It is easy to see from this equality that $G_L''(\xi; t)$ is given by

$$G_L''(\xi; t) = 2F_L''(\xi; t) + <(F_L''(\cdot; t))'(\xi), \xi>.$$

Therefore, we obtain that

$$\mathcal{S}\Delta_L(N\Phi)(\xi) = 2\mathcal{S}\Delta_L\Phi(\xi) + \mathcal{S}N(\Delta_L\Phi)(\xi),$$

which yields the equality (3). □

Proof of (4). This follows from (6.50) by taking the Volterra parts of both sides. □

Proof of (5). We have $\Delta_G\Delta_L=\Delta_L\Delta_G=0$ on $(\mathcal{S})$ since Δ_G is a continuous linear operator from $(\mathcal{S})$ into itself and $\Delta_L\varphi=0$ for all $\varphi\in(\mathcal{S})$ by Theorem 6.42. □

Proof of (6). We have $\Delta_G\Delta_V=\Delta_V\Delta_G=\Delta_G^2$ on $(\mathcal{S})$ since $\Delta_V\varphi=\Delta_G\varphi$ for all $\varphi\in(\mathcal{S})$ by Theorem 6.29. □

Proof of (7). This is a restatement of Theorem 6.44. □

Proof of (8). Let $\Phi\in\mathcal{D}(\Delta_V)$ and $F=\mathcal{S}\Phi$. Let $H=\mathcal{S}(\Delta_G^*\Phi)$. Then $H(\xi) = |\xi|_2^2\, F(\xi)$ by Theorem 6.20. We can check easily that for ξ, η and ζ in $\mathcal{S}(\mathbb{R})$,

$$\begin{aligned}<H''(\xi), \eta\otimes\zeta> &= 2<\eta, \zeta>F(\xi) + 2<\xi, \eta><F'(\xi), \zeta> \\ &\quad + 2<\xi, \zeta><F'(\xi), \eta> + |\xi|_2^2\, <F''(\xi), \eta\otimes\zeta>.\end{aligned}$$

Thus $H_V''(\xi; s, t)$ is given by

$$H_V''(\xi;\, s,\, t) = 2\xi(s)F'(\xi;\, t) + 2F'(\xi;\, s)\xi(t) + |\xi|_2^2\, F_V''(\xi;\, s,\, t).$$

Therefore, for each $\xi \in \mathcal{S}(\mathbb{R})$, $H_V''(\xi;\, \cdot,\, \cdot)$ defines a trace class operator of $L^2(\mathbb{R})$ and

$$\operatorname{trace}_{L^2(\mathbb{R})} H_V''(\xi) = 4\langle F'(\xi),\, \xi\rangle + |\xi|_2^2\, \operatorname{trace}_{L^2(\mathbb{R})} F_V''(\xi).$$

Note that $\mathcal{S}(N\Phi)(\xi) = \langle F'(\xi),\, \xi\rangle$ by Theorem 6.11 and $\mathcal{S}(\Delta_G^*\Phi)(\xi) = |\xi|_2^2\, \mathcal{S}\Phi(\xi)$ by Theorem 6.20. Hence the above equality yields that

$$\Delta_V(\Delta_G^*\Phi) = 4N\Phi + \Delta_G^*(\Delta_V\Phi). \qquad \square$$

7. The Spaces $\mathfrak{D}$ and $\mathfrak{D}^*$

This chapter presents some of the basic properties of the spaces $\mathfrak{D}$ and $\mathfrak{D}^*$ which were introduced at the end of Chapter 3. The most important results are the theorem of Meyer stating the algebraic property of $\mathfrak{D}$, and the theorem of Watanabe which says that composites of tempered distributions with non-degenerate elements in $\mathfrak{D}$ belong to $\mathfrak{D}^*$. Both results are based on the equivalence of the (L^p)–norms of $N^{1/2}$ and $|\nabla \cdot|_2$, which has been established in Meyer (1982, 1983) (cf. also Bakry (1985), Feyel and de la Pradelle (1989,1991), Gundy (1989), Krée and Krée (1983), Shigekawa (1990), and literature cited in these articles). The results of Meyer (1982, 1983) have been worked out further in Sugita (1985a) and Watanabe (1983, 1984). The necessity of estimating the (L^p)–norms of $|\nabla \cdot|_2$ by those of the number operator N, when proving that $\mathfrak{D}$ is an algebra, comes in because of the product rule for N: $N\varphi\psi = (N\varphi) + \varphi N\psi - 2(\nabla\overline{\varphi}, \nabla\psi)$, cf. (5.80).

The above mentioned theorem of Watanabe about the composition of tempered distributions and elements in $\mathfrak{D}$ entails an alternative proof of the well-known result of Malliavin (1978) on the existence of smooth densities (with respect to Lebesgue measure) of laws of certain random variables (in particular of non–degenerate diffusions). Without aiming at a complete list of references we want to mention also the following works in this direction: Albeverio and Høegh–Krohn (1976, 1977a), Bismut (1981), Itô (1990), Kusuoka and Stroock (1982, 1984, 1985), Malliavin (1984), Shigekawa (1980), Stroock (1981), Sugita (1985b), Watanabe (1990), Zakai (1985).

Meyer's equivalence of norms is based on certain Littlewood–Paley–Stein inequalities. Such inequalities have been proven in Stein (1970b), and have been developed further in Meyer (1976, 1982) and subsequently in Bakry (1985). Based on the results of Meyer and Bakry, Shigekawa and Yosida (1990) have proven rather general Littlewood–Paley–Stein inequalities for symmetric diffusion processes. Moreover, Shigekawa (1990) proves the corresponding generalization of

Meyer's equivalence of norms. The methods of Bakry and Meyer make use of deep results in martingale theory. In this chapter we shall follow the formulation of the results of Meyer and Stein as given in Potthoff (1988a). It is a purely functional analytic approach, which is actually an adoption of the proof of the Littlewood–Paley–Stein inequalities on compact Lie groups as found in Stein (1970b) for the Littlewood– Paley functions considered in Meyer (1982). As a preparation, some more or less well-known results about the Ornstein–Uhlenbeck semigroup are collected in Section 7.A. In Section 7.B we prove the Littlewood–Paley–Stein inequalities. Finally, in Section 7.C we show Meyer's equivalence of norms, and sketch the above mentioned theorem by Watanabe, following their work very closely.

For the sake of simplicity, we shall work all the time with the number operator (i.e., the Ornstein–Uhlenbeck semigroup), and the white noise probability space. Most of the results in Section 7.A and 7.B have obvious generalizations to other second quantized operators which are left to the interested reader. All the results of this chapter can be formulated for general Gaussian spaces. (We refer to the above mentioned literature.) Throughout this chapter we shall consider $\mathcal{S}'(\mathbb{R})$ as being equipped with the weak topology, unless stated otherwise.

7.A The Ornstein–Uhlenbeck Semigroup

In this section we collect some facts about the semigroup generated by the number operator $N = d\Gamma(1)$. Most of the material is very well-known, and has been borrowed from Glimm and Jaffe (1970), Simon and Høegh–Krohn (1972), Meyer (1982, 1983), Nelson (1972, 1973), Simon (1974), Stein (1970b). We also want to mention in this context the papers of Gross (1967), Kuo and Piech (1973), Piech (1975). For general background on the theory of semigroups on Banach spaces we refer the reader to Hille and Philipps (1957), Kato (1976) and Yoshida (1951).

Recall from Chapter 5 and 6 that the number operator N can be written in the form $N = \int \partial_t^* \partial_t \, dt$. In this chapter it will be more convenient to represent N by

$$N = \sum_{k=1}^{\infty} \partial_k^* \partial_k = \sum_{k=1}^{\infty} (<\cdot, e_k> \partial_k - \partial_k^2), \tag{7.1}$$

where we have set $\partial_k \equiv D_{e_k}$, the Gâteaux derivative in direction e_k, and $\{e_k, k \in \mathbb{N}\}$ is a CONS of $L^2(\mathbb{R})$ which lies in $\mathcal{S}(\mathbb{R})$. Similarly, we identify the gradient $\nabla\varphi \in L^2(\mathbb{R}) \otimes (L^2)$ of $\varphi \in \mathfrak{D}^{2,1}$ with the sequence $\{\partial_k \varphi, k \in \mathbb{N}\}$ in $l^2 \otimes (L^2)$. For $\varphi \in \mathfrak{D}^{2,1}$ we have

$$| \nabla\varphi |_2^2 = \sum_{k=1}^{\infty} | \partial_k \varphi |^2 \in (L^1).$$

Consider the semigroup $\{P_t = \exp(-tN), t \geq 0\}$ generated by N on (L^2). $\{P_t, t \geq 0\}$ is called *Ornstein–Uhlenbeck semigroup.* Because of the formula $\exp(-tN) = \Gamma(e^{-t})$, it is immediate to find the action of P_t on $\mathcal{P}$ and $\mathcal{E}$.

Let $\varphi \in \mathcal{E}$ be of the form

$$\varphi(x) = \sum_n a_n : e^{i<\cdot, \xi_n>} :, \quad x \in \mathcal{S}'(\mathbb{R}), \tag{7.2}$$

where $a_n \in \mathbb{C}$, $\xi_n \in \mathcal{S}(\mathbb{R})$, and the sum in (7.2) has only a finite number of non-trivial terms. Then

$$P_t \varphi(x) = \sum_n a_n : e^{i<\cdot, e^{-t}\xi_n>} :, \quad t \geq 0. \tag{7.3}$$

For $\psi \in \mathcal{P}$ of the form

$$\psi(x) = \sum_n \sum_\alpha a_\alpha : <x, \xi_{\alpha_1}> \cdots <x, \xi_{\alpha_n}> :, \quad x \in \mathcal{S}'(\mathbb{R}), \tag{7.4}$$

we find

$$P_t\psi(x) = \sum_n \sum_\alpha a_\alpha :\langle x, e^{-t}\xi_{\alpha_1}\rangle \cdots \langle x, e^{-t}\xi_{\alpha_n}\rangle:, \quad x\in\mathcal{S}'(\mathbb{R}). \tag{7.5}$$

Clearly, in terms of the chaos expansion $(\varphi^{(n)}, n\in\mathbb{N}_0)$ of a general element $\varphi\in(L^2)$, we have

$$P_t\varphi = \sum_{n=0}^{\infty} e^{-tn}\varphi^{(n)}. \tag{7.6}$$

The following theorem summarizes results which can be found in the works quoted at the beginning of this section.

Theorem 7.1.

(i) $\{P_t, t\geq 0\}$ *is a strongly continuous contraction semigroup on* (L^2);

(ii) $\{P_t, t\geq 0\}$ *extends to a positivity preserving contraction semigroup on all* (L^p), $1\leq p\leq +\infty$.

(iii) $\{P_t, t\geq 0\}$ *is hypercontractive: if* $e^{-t} \leq \sqrt{\frac{p-1}{q-1}}$, $1<p, q<+\infty$, $p\leq q$, *then* P_t *is a contraction from* (L^p) *into* (L^q).

Consider $\varphi\in\mathcal{E}$ as in (7.2). Note that $\mathbb{E}(:\exp(i\langle\cdot, \eta\rangle):) = 1$, for all $\eta\in\mathcal{S}(\mathbb{R})$. Thus (7.3) shows that for every $x\in\mathcal{S}'(\mathbb{R})$,

$$\lim_{t\to\infty} P_t\varphi(x) = \mathbb{E}\varphi.$$

On the other hand, it follows from Theorem 7.1.ii that for μ–a.e. $x\in\mathcal{S}'(\mathbb{R})$,

$$|P_t\varphi(x)| \leq \| P_t\varphi \|_\infty \leq \| \varphi \|_\infty.$$

Thus we have $|P_t\varphi(x) - \mathbb{E}\varphi| \leq 2\|\varphi\|_\infty$, and an application of the dominated convergence theorem implies that $P_t\varphi \to \mathbb{E}\varphi$, in every (L^p), $1\leq p\leq +\infty$, as t →

$+\infty$. Now let $\varphi \in (L^p)$, $1 \le p \le +\infty$, and consider a sequence $(\varphi_n, n \in \mathbb{N})$ in $\mathcal{E}$ converging to φ in (L^p). Then we may estimate in the following way:

$$\| P_t\varphi - \mathbb{E}\varphi \|_p \le \| P_t(\varphi - \varphi_n) \|_p + \| P_t\varphi_n - \mathbb{E}\varphi_n \|_p + \mathbb{E}(| \varphi - \varphi_n |)$$

$$\le 2 \| \varphi - \varphi_n \|_p + \| P_t\varphi_n - \mathbb{E}\varphi_n \|_p,$$

where we used Theorem 7.1.ii and Hölder's inequality. The second of the last two terms converges to zero as t tends to infinity, as argued above. Therefore we have shown the following result.

Theorem 7.2. *Let* $p \in [1,+\infty]$, $\varphi \in (L^p)$. *Then* $P_t\varphi$ *converges in* (L^p) *to* $\mathbb{E}\varphi$ *as* t *tends to infinity.*

Equations (7.3) and (7.5) will now be used to prove the powerful Mehler formula for P_t. Let $\xi \in \mathcal{S}(\mathbb{R})$, and consider the following computation.

$$P_t \colon e^{i\langle \cdot, \xi \rangle} \colon (x) = \colon e^{i\langle x, e^{-t}\xi \rangle} \colon$$

$$= e^{i\langle x, e^{-t}\xi \rangle + \frac{1}{2}e^{-2t}| \xi |_2^2}$$

$$= e^{i\langle x, e^{-t}\xi \rangle + \frac{1}{2}| \xi |_2^2 - \frac{1}{2}(1-e^{-2t})| \xi |_2^2}$$

$$= \colon e^{i\langle e^{-t}x, \xi \rangle} \colon \int e^{i\langle (1-e^{-2t})^{1/2}y, \xi \rangle} d\mu(y)$$

$$= \int \colon e^{i\langle e^{-t}x + (1-e^{-2t})^{1/2}y, \xi \rangle} \colon d\mu(y).$$

By linearity, we obtain the following formula for all $\varphi \in \mathcal{E}$:

$$P_t\varphi(x) = \int \varphi(e^{-t}x+(1-e^{-2t})^{1/2}y)\, d\mu(y), \quad x\in\mathcal{S}'(\mathbb{R}),\ t\geq 0. \tag{7.7}$$

Either by differentiation, or by direct computation using (7.5), one proves that (7.7) holds for $\varphi\in\mathcal{P}$ too. Equation (7.7) is called *Mehler's formula.*

Theorem 7.3. *For all* $\varphi\in(L^\infty)$, *Mehler's formula* (7.7) *holds pointwise* μ–a.e.

Proof. Let $\varphi\in(L^\infty)$ and let $(\varphi_n,\ n\in\mathbb{N})$ be a sequence in $\mathcal{E}$ converging in (L^∞) to φ. $\varphi_n \to \varphi$ in (L^∞) implies that $P_t\varphi_n \to P_t\varphi$ in (L^∞) for all $t\geq 0$, since P_t is a contraction on (L^∞). Therefore, in order to prove the first statement we only need to show that

$$\left\| \int \left(\varphi_n(e^{-t}\cdot +(1-e^{-2t})^{1/2}y) - \varphi(e^{-t}\cdot +(1-e^{-2t})^{1/2}y)\right) d\mu(y) \right\|_\infty$$

converges to zero as n tends to infinity. But obviously we may bound the last expression by

$$\int \| \varphi_n(e^{-t}\cdot +(1-e^{-2t})^{1/2}y) - \varphi(e^{-t}\cdot +(1-e^{-2t})^{1/2}y) \|_\infty\, d\mu(y) =$$

$$= \| \varphi_n - \varphi \|_\infty,$$

which vanishes as $n \to \infty$. □

In the following sections it will be useful to work also with two other spaces of functions on $\mathcal{S}'(\mathbb{R})$: C_b and $\mathcal{F}C_b$. C_b is the space of continuous and bounded functions on $\mathcal{S}'(\mathbb{R})$, while $\mathcal{F}C_b$ consists of finitely based bounded continuous functions, i.e., $\varphi\in\mathcal{F}C_b$, if and only if there exists $n\in\mathbb{N}$, $\xi_1, \dots, \xi_n\in\mathcal{S}(\mathbb{R})$, and a bounded continuous functions f on $\mathbb{R}^n$ so that for all $x\in\mathcal{S}'(\mathbb{R})$,

$$\varphi(x) = f(<x, \xi_1>, \dots, <x, \xi_n>). \tag{7.8}$$

Without loss of generality we shall usually assume that the ξ's above are orthonormal in $L^2(\mathbb{R})$. It is clear that both – C_b and $\mathcal{F}C_b$ – are dense in all (L^p), $1 \leq p \leq +\infty$, because $\mathcal{E} \subset \mathcal{F}C_b \subset C_b$.

Let $\varphi \in C_b$, and consider φ as an element in (L^∞). Then (7.7) holds pointwise a.e., but note that the right hand side is well-defined for every $x \in \mathcal{S}'(\mathbb{R})$. Let us show that the right hand side is continuous in x. If $(x_\alpha, \alpha \in I)$ is a net which converges to x in $\mathcal{S}'(\mathbb{R})$, then for all $y \in \mathcal{S}'(\mathbb{R})$, $t \geq 0$,

$$\varphi(e^{-t}x_\alpha + (1-e^{-2t})^{1/2}y) \to \varphi(e^{-t}x + (1-e^{-2t})^{1/2}y).$$

Moreover, $|\varphi(e^{-t}x_\alpha + (1-e^{-2t})^{1/2}y)| \leq \| \varphi \|_\infty$, and we can apply the dominated convergence theorem to conclude that the right hand side of (7.7) is continuous. Therefore, $P_t\varphi$ has a version which is continuous, and we identify $P_t\varphi$ with this version. Thus P_t maps C_b into itself, and (7.7) holds pointwise.

Let $\varphi \in C_b$, $x \in \mathcal{S}'(\mathbb{R})$, and consider $t \mapsto P_t\varphi(x)$. Then it is immediate from formula (7.7) and the dominated convergence theorem that for all $t>0$, $t \mapsto P_t\varphi(x)$ is continuous. Moreover, one can read off that

$$\lim_{t \downarrow 0} P_t\varphi(x) = \varphi(x), \quad \lim_{t \to +\infty} P_t\varphi(x) = \mathbb{E}\varphi. \tag{7.9}$$

Theorem 7.4. P_t, $t \geq 0$, *maps* C_b *into itself, and for all* $\varphi \in C_b$, *formula* (7.7) *holds for all* $x \in \mathcal{S}'(\mathbb{R})$. *Moreover, for all* $\varphi \in C_b$, $x \in \mathcal{S}'(\mathbb{R})$, *the mapping* $t \to P_t\varphi(x)$ *is continuous on* $[0,+\infty)$, *and* (7.9) *holds.*

Let us now look at Mehler's formula on $\mathcal{F}C_b$ a bit more in detail. Consider $\varphi \in \mathcal{F}C_b$ of the form (7.8) (with orthonormal ξ's). For $u, v \in \mathbb{R}^n$, $t>0$, define the kernel

$$K_t(u,v) := (2\pi(1-e^{-2t}))^{-n/2} \exp\{-\frac{1}{2(1-e^{-2t})} \mid v-e^{-t}u \mid^2\}.$$

For $f \in C_b(\mathbb{R}^n)$ set

$$p_t f(u) := \int K_t(u,v)\, f(v)\, dv, \quad u \in \mathbb{R}^n.$$

Then it is a straightforward computation to show that

$$P_t\varphi(x) = (p_t f)(\langle x, \xi_1\rangle, \ldots, \langle x, \xi_n\rangle). \tag{7.10}$$

It is obvious that $(t,u) \mapsto p_t f(u)$ is infinitely often differentiable on $(0,+\infty)\times\mathbb{R}^n$ (and all derivatives are polynomially bounded). By the chain rule this implies that for every $x \in \mathcal{S}'(\mathbb{R})$, $t \mapsto P_t\varphi(x)$ is infinitely often differentiable on $(0,+\infty)$, and that for every $t>0$, $P_t\varphi$ is infinitely often Gâteaux differentiable in every direction of $L^2(\mathbb{R})$. Moreover, all derivatives (being polynomially bounded) belong to all (L^p), $p\geq 1$. Furthermore, it easy to see that for all $t>0$, $x\in\mathcal{S}'(\mathbb{R})$, $\mid \nabla P_t\varphi(x) \mid_2$ and (using (7.1)) $NP_t\varphi(x)$ exist. Finally another simple explicit calculation leads to the following result.

Lemma 7.5. *Let* $\varphi \in \mathcal{F}C_b$. *Then for every* $x\in\mathcal{S}'(\mathbb{R})$, $t \mapsto P_t\varphi(x)$ *is infinitely often differentiable on* $(0,+\infty)$, *and* $P_t\varphi$ *is infinitely often Gâteaux differentiable in every direction of* $L^2(\mathbb{R})$. *Moreover, for all* $x\in\mathcal{S}'(\mathbb{R})$, $t>0$,

$$(\frac{\partial}{\partial t} + N)P_t\varphi(x) = 0. \tag{7.11}$$

Remark. Clearly, we have (7.11) pointwise μ–a.e. for all $\varphi\in(L^2)$. We have stated this result here, because pointwise equality will be useful in Section 7.B. Also, at the end of this section we shall show (L^p)–smoothing properties of P_t.

Later in this chapter we shall make also use of the semigroup generated by $N^{1/2}$, and we denote it by $\{Q_t, t\geq 0\}$:

$$Q_t := e^{-tN^{1/2}}, \; t\geq 0.$$

Note the following formula for $\lambda\geq 0$, $t>0$:

$$e^{-t\lambda^{1/2}} = \int_0^\infty e^{-s\lambda} \, \nu(t,s) \, ds,$$

where

$$\nu(t,s) = (4\pi)^{-1/2} \, ts^{-3/2} \, e^{-t^2/4s}.$$

By the spectral theorem we find that

$$Q_t = \int_0^\infty P_s \, \nu(t,s) \, ds. \tag{7.12}$$

From this formula and the preceding discussion we can read off the following result.

Theorem 7.6. $\{Q_t, t\geq 0\}$ *is a positivity preserving contraction semigroup on all* (L^p), $1\leq p\leq +\infty$, *strongly continuous on* (L^2). *Moreover, for all* $\varphi\in(L^\infty)$ *the following formula holds for* μ–a.e. $x\in\mathcal{S}'(\mathbb{R})$,

$$Q_t\varphi(x) = \int_0^\infty \int \varphi(e^{-s}x + (1-e^{-2s})^{1/2}y) \, d\mu(y) \, \nu(t,s) \, ds. \tag{7.13}$$

Q_t, $t\geq 0$, *maps* C_b *into itself, and for* $\varphi\in C_b$ (7.13) *holds for all* $x\in\mathcal{S}'(\mathbb{R})$.

Theorem 7.1 and Theorem 7.6 show that Stein's maximal theorem, cf. Stein (1970b), Chapter III.§3, holds for P_t and Q_t:

Theorem 7.7. *Let* $p>1$, $\varphi\in C_b$. *Define*

$$\varphi^* := \sup_{t>0} | P_t\varphi |, \quad \varphi^{**} := \sup_{t>0} | Q_t\varphi |.$$

Then

$$\| \varphi^* \|_p \le K_p \| \varphi \|_p, \tag{7.14}$$

and

$$\| \varphi^{**} \|_p \le K_p' \| \varphi \|_p, \tag{7.15}$$

for some constants K_p, $K_p'>0$.

Note that

$$Q_t\varphi = \int_0^\infty P_{t^2s}\varphi \, \nu(1,s) \, ds.$$

From this, the fact that P_t is an (L^p)–contraction, the dominated convergence theorem, and Theorem 7.2, it follows that $Q_t\varphi \to \mathbb{E}\varphi$ in (L^p). Moreover, Theorem 7.4 implies the following result.

Theorem 7.8. *For all* $\varphi\in C_b$, $(t,x) \mapsto Q_t\varphi(x)$ *is continuous on* $[0,+\infty)\times\mathcal{S}'(\mathbb{R})$, *and*

$$\lim_{t\downarrow 0} Q_t\varphi(x) = \varphi(x), \quad \lim_{t\to+\infty} Q_t\varphi(x) = \mathbb{E}\varphi, \quad x\in\mathcal{S}'(\mathbb{R}). \tag{7.16}$$

Moreover, the analog of the first statement of Lemma 7.5 holds, and for all $\varphi\in\mathcal{F}C_b$, $t>0$, $NQ_t\varphi(x)$, $| \nabla Q_t\varphi(x) |_2$ exist pointwise for all $x\in\mathcal{S}'(\mathbb{R})$.

Lemma 7.9. *Assume that* $\varphi\in\mathcal{F}C_b$. *Then for every* $x\in\mathcal{S}'(\mathbb{R})$, $t \mapsto Q_t\varphi(x)$ *is infinitely often differentiable on* $(0,+\infty)$, *and* $Q_t\varphi$ *is infinitely often Gâteaux differentiable in every direction of* $L^2(\mathbb{R})$. *Moreover, for all* $x\in\mathcal{S}'(\mathbb{R})$, $t>0$,

$$(\frac{\partial^2}{\partial t^2} - N)Q_t\varphi(x) = 0. \tag{7.17}$$

Proof. Observe that for all s,t>0,

$$(\frac{\partial^2}{\partial t^2} - \frac{\partial}{\partial s})\nu(t,s) = 0.$$

Thus, if t>0

$$\frac{\partial^2}{\partial t^2}Q_t\varphi(x) = \frac{\partial^2}{\partial t^2}\int_0^\infty P_s\varphi(x)\ \nu(t,s)\ ds$$

$$= \int_0^\infty P_s\varphi(x)\ \frac{\partial}{\partial s}\nu(t,s)\ ds$$

$$= \lim_{s\to\infty} P_s\varphi(x)\ \nu(t,s) - \lim_{s\downarrow 0} P_s\varphi(x)\ \nu(t,s) +$$

$$+ \int_0^\infty NP_s\varphi(x)\ \nu(t,s)\ ds.$$

Now,

$$\lim_{s\to\infty} P_s\varphi(x)\ \nu(t,s) = \mathbb{E}\varphi \lim_{s\to\infty} \nu(t,s) = 0,$$

and

$$\lim_{s\downarrow 0} P_s\varphi(x)\ \nu(t,s) = \varphi(x) \lim_{s\downarrow 0} \nu(t,s) = 0.$$

Therefore we obtain

$$\frac{\partial^2}{\partial t^2}Q_t\varphi(x) = N\int_0^\infty P_s\varphi(x)\ \nu(t,s)\ ds$$

$$= NQ_t\varphi(x).$$

The interchanging of integrals and derivatives above is easily justified, and this is left as an exercise to the interested reader. □

Next we establish commutation relations of P_t with differential operators, and some useful inequalities. We recall from Chapter 3, that $\mathfrak{D}^{2,r}$, $r\geq 0$, is the Hilbert space which is the completion of $\mathcal{P}$ (or $\mathcal{S}$) under the norm $|||\cdot|||_{2,k} = \|(1+N)^{r/2}\cdot\|_2$. Moreover, Theorems 5.24 and 5.27 show that for every $h\in L^2(\mathbb{R})$, the (extended) Gâteaux derivative D_h in direction h, and its adjoint D_h^* map $\mathfrak{D}^{2,1}$ into (L^2).

Lemma 7.10. *Let* $\varphi\in\mathfrak{D}^{2,1}$, $h\in L^2(\mathbb{R})$. *Then for all* $t\geq 0$,

$$D_hP_t\varphi = e^{-t}P_tD_h\varphi, \tag{7.18}$$

and

$$D_h^*P_t\varphi = e^{t}P_tD_h^*\varphi. \tag{7.19}$$

Proof. Let φ have chaos expansion $\varphi = \sum_n \varphi^{(n)}$, $\varphi^{(n)}\in\mathcal{H}^{(n)}$. Note that $D_h\varphi^{(n)}\in\mathcal{H}^{(n-1)}$, $n\in\mathbb{N}$, and $D_h^*\varphi^{(n)}\in\mathcal{H}^{(n+1)}$, $n\in\mathbb{N}_0$. Therefore

$$D_hP_t\varphi = D_h\sum_{n=0}^{\infty} e^{-tn}\varphi^{(n)}$$

$$= \sum_{n=1}^{\infty} e^{-tn}D_h\varphi^{(n)}$$

$$= e^{-t}\sum_{n=1}^{\infty} e^{-t(n-1)}D_h\varphi^{(n)}$$

$$= e^{-t} P_t D_h \varphi,$$

where the interchange of D_h and the sum is justified because of the continuity of D_h. Relation (7.19) is proved in the same way. □

Remark. If we assume that $\varphi \in \mathcal{E}$, $\mathcal{P}$, or $(\mathcal{S})$, then relations (7.18) and (7.19) are true for $h \in \mathcal{S}'(\mathbb{R})$.

The following inequalities are very important for the derivation of the Littlewood–Paley–Stein inequalities in Section 7.B. The articles of Bakry (1985) show that they are actually at the heart of the matter.

Lemma 7.11. *Let* $\varphi \in \mathcal{F}C_b$, $t>0$, $u \in [0,t)$. *Then for all* $x \in \mathcal{S}'(\mathbb{R})$ *the following inequalities hold:*

$$|\nabla P_t \varphi(x)|_2^2 \le e^{-2u} (P_u |\nabla P_{t-u} \varphi|_2(x))^2 \tag{7.20}$$

$$\le e^{-2u} P_u |\nabla P_{t-u} \varphi|_2^2(x). \tag{7.21}$$

If $\varphi \in \mathcal{P}$ *or* $\mathcal{E}$, *then* (7.20), (7.21) *hold for all* $t \ge 0$, $u \in [0,t]$.

Proof. First note that by the discussion preceding Lemma 7.5, all terms in (7.20) and (7.21) make sense. By Lemma 7.10 and the semigroup property of P_t we have

$$\partial_k P_t \varphi = \partial_k P_u P_{t-u} \varphi$$

$$= e^{-u} P_u \partial_k P_{t-u} \varphi,$$

and we obtain

$$|\nabla P_t\varphi|_2^2(x) = e^{-2u}\sum_{k=1}^{\infty} |P_u\partial_k P_{t-u}\varphi(x)|^2.$$

Note that $\varphi\in\mathfrak{F}C_b$ implies that also $\partial_k P_{t-u}\varphi\in\mathfrak{F}C_b$, and therefore we may apply Mehler's formula (7.7):

$$|(P_u\partial_k P_{t-u}\varphi)(x)| = \left|\int \partial_k P_{t-u}\varphi(\gamma_{u,x}(y))\, d\mu(y)\right|$$

$$\leq \int |\partial_k P_{t-u}\varphi(\gamma_{u,x}(y))|\, d\mu(y), \tag{7.22}$$

where we used the notation $\gamma_{t,x}(y) = e^{-t}x + (1-e^{-2t})^{1/2}y$. Thus we get

$$\sum_{k=1}^{\infty} |P_u\partial_k P_{t-u}\varphi(x)|^2 \leq \left\| \int |\partial_{.} P_{t-u}\varphi(\gamma_{u,x}(y))|\, d\mu(y)\right\|_{l^2}^2$$

$$\leq \left(\int \| \partial_{.} P_{t-u}\varphi(\gamma_{u,x}(y))\|_{l^2}\, d\mu(y)\right)^2$$

$$= \left(\int |\nabla P_{t-u}\varphi(\gamma_{t,x}(y))|_2\, d\mu(y)\right)^2. \tag{7.23}$$

Since $\varphi\in\mathfrak{F}C_b$ implies that $|\nabla P_{t-u}\varphi|_2$ belongs to $\mathfrak{F}C_b$ too, we may use (7.7) in reverse and obtain the following inequality:

$$\sum_{k=1}^{\infty} |P_u\partial_k P_{t-u}\varphi(x)|^2 \leq (P_u|\nabla P_{t-u}\varphi(x)|_2)^2,$$

which proves (7.20). Using Schwarz' inequality in (7.23) and then (7.7), we obtain (7.21). □

Corollary 7.12. *For all* $\varphi\in\mathfrak{F}C_b$, $t>0$, $u\in[0,t)$, $x\in\mathcal{S}'(\mathbb{R})$,

$$| \nabla Q_t\varphi(x) |_2^2 \leq (Q_u| \nabla Q_{t-u}\varphi |_2(x))^2 \tag{7.24}$$

$$\leq Q_u| \nabla Q_{t-u}\varphi|_2^2(x). \tag{7.25}$$

Proof.

$$| \nabla Q_t\varphi(x) |_2^2 = | \nabla Q_u Q_{t-u}\varphi(x) |_2^2$$

$$= \sum_{k=1}^{\infty} | \partial_k \int_0^{\infty} P_s Q_{t-u}\varphi(x)\ \nu(u,s)ds |^2$$

$$= \sum_{k=1}^{\infty} | \int_0^{\infty} e^{-s}\ P_s \partial_k Q_{t-u}\varphi(x)\ \nu(u,s)ds |^2$$

$$\leq \sum_{k=1}^{\infty} \Big(\int_0^{\infty} P_s| \partial_k Q_{t-u}\varphi(x) |\ \nu(u,s)ds \Big)^2$$

$$\leq \Big(\int_0^{\infty} P_s| \nabla Q_{t-u}\varphi(x) |_2\ \nu(u,s)ds \Big)^2 \tag{7.26}$$

$$= (Q_u| \nabla Q_{t-u}\varphi |_2(x))^2,$$

where we used the subordination formula (7.12), the fact that P_t is positivity preserving to estimate $| P_t\psi(x) | \leq P_t|\cdot\psi |(x)$ above (cf. also Simon (1974), Lemma I.14), and Lemma 7.11. Thus (7.24) is proved. By (7.14) and (7.26) we have

$$| \nabla Q_t\varphi(x) |_2^2 \leq \Big(\int_0^{\infty} P_s| \nabla Q_{t-u}\varphi(x) |_2\ \nu(u,s)ds \Big)^2$$

$$= \Big(\int_0^\infty\int \big| \nabla Q_{t-u}\varphi(\gamma_{s,x}(y)) \big|_2 \, \nu(u,s)ds \, d\mu(y)\Big)^2$$

$$\leq \int_0^\infty\int \big| \nabla Q_{t-u}\varphi(\gamma_{s,x}(y)) \big|_2^2 \, \nu(u,s)ds \, d\mu(y),$$

where we used the same notation as in the proof of Lemma 7.11, and Schwarz' inequality for the normalized measure $\nu(u,s)ds\otimes d\mu(y)$. Using Mehler's formula again we obtain

$$\big| \nabla Q_t\varphi(x) \big|_2^2 \leq \int_0^\infty P_s\big| \nabla Q_{t-u}\varphi(x) \big|_2^2 \, \nu(u,s)ds$$

$$= Q_u\big| \nabla Q_{t-u}\varphi \big|_2^2(x).$$ □

We conclude this section with some remarks about the smoothing properties of the semigroups P_t, Q_t, $t\geq 0$, in (L^p)–sense. It is clear that for every $t>0$, P_t and Q_t map (L^2) into $\mathfrak{D}^{2,k}$, for all $k\in\mathbb{N}$. Therefore, if $\varphi\in(L^2)$, then $P_t\varphi$ and $Q_t\varphi$ admit infinitely many differentiations by the gradient ∇, and for all $k\in\mathbb{N}$, $\nabla^k P_t\varphi$, $\nabla^k Q_t\varphi\in L^2(\hat{\mathbb{R}}^k)\otimes(L^2)$. Consequently, $P_t\varphi$ and $Q_t\varphi$ are infinitely often Gâteaux differentiable (in (L^2)) in every direction in $L^2(\mathbb{R})$. It is not hard to show that Gâteaux differentiability in the $L^2(\mathbb{R})$–directions continues to hold in (L^p)–sense, $p>1$. In the following let $h\in L^2(\mathbb{R})$, $t>0$, $\varphi\in\mathcal{E}$ or $\mathcal{P}$. Compute as follows.

$$D_hP_t\varphi(x) = \frac{d}{d\lambda}\int \varphi(e^{-t}(x+\lambda h)+(1-e^{-2t})^{1/2}y) \, d\mu(x)\Big|_{\lambda=0}$$

$$= \frac{d}{d\lambda}\int \varphi(e^{-t}x+(1-e^{-2t})^{1/2}(y+\lambda\sigma(t)h)) \, d\mu(x)\Big|_{\lambda=0}$$

$$= \frac{d}{d\lambda} \int \varphi(\gamma_{t,x}(y)) : e^{<y,\, \lambda\sigma(t)\, h>} : d\mu(y) \Big|_{\lambda=0}$$

$$= \sigma(t) \int \varphi(\gamma_{t,x}(y)) <y, h> d\mu(y),$$

where we have set $\sigma(t) = e^{-t}(1-e^{-2t})^{-1/2}$, and $\gamma_{t,x}(y) = e^{-t}x+(1-e^{-2t})^{1/2}y$. (The interchanging of the integral and the λ-derivative above is easily justified – for example, by an application of the dominated convergence theorem.) Now use Hölder's inequality to obtain the following estimate for $p>1$:

$$| D_h P_t \varphi(x) | \leq C_p\, \sigma(t) |h|_2 \left(\int | \varphi(\gamma_{t,x}(y)) |^p\, d\mu(y) \right)^{1/p}$$

$$= C_p\, \sigma(t) |h|_2 (P_t | \varphi |^p(x))^{1/p},$$

where C_p is a constant depending only on p. Thus

$$\| D_h P_t \varphi \|_p \leq C_p\, \sigma(t) |h|_2 \| P_t | \varphi |^p \|_1^{1/p}$$

$$\leq C_p\, \sigma(t) |h|_2 \| | \varphi |^p \|_1^{1/p}$$

$$= C_p\, \sigma(t) |h|_2 \| \varphi \|_p,$$

because P_t is an (L^1)–contraction. Density of $\mathcal{E}$ and $\mathcal{P}$ imply the following result.

Proposition 7.13. *Let* $t>0$, $h \in L^2(\mathbb{R})$, $\varphi \in (L^p)$, $p>1$. *Then* $P_t\varphi$ *is differentiable in direction* h, *and* $D_h P_t \varphi \in (L^p)$. *Moreover,*

$$\| D_h P_t \varphi \|_p \leq C_p \frac{e^{-t}}{(1-e^{-2t})^{1/2}} |h|_2 \| \varphi \|_p. \qquad (7.27)$$

Remark. A similar result can be proved for Q_t, and we leave this as an exercise to the reader.

Now the following heuristic remark is in order. As t tends to zero, the right hand side of (7.17) develops a singularity of the type $t^{-1/2}$ (in the case of Q_t it is of the order t^{-1}). On the other hand, the function $n \mapsto n^{1/2}e^{-tn}$, $n \geq 0$, has a maximum of the order $t^{-1/2}$. Therefore one might guess that for normalized h, $N^{1/2}P_t$ behaves like D_hP_t on (L^p), or that the (L^p) norms of $D_h\varphi$ and $N^{1/2}\varphi$ are equivalent. (Of course, this is true on (L^2).) This is indeed so and this is the contents of the Meyer inequalities which will be proved in the next sections.

7.B Littlewood–Paley–Stein Inequalities

In this section we shall follow the method of Stein (1970b), Chapter II, quite closely to establish the Littlewood–Paley–Stein (LPS) inequalities found in Meyer (1976, 1982). On $\mathcal{F}C_b$ we introduce the following Littlewood–Paley functions G_i, i=1,2,3. Let $\varphi \in \mathcal{F}C_b$, $x \in \mathcal{S}'(\mathbb{R})$, and set

$$G_1(\varphi)(x) := \Big(\int_0^\infty (P_t |\nabla P_t\varphi|_2(x))^2 \, dt\Big)^{1/2}, \tag{7.28}$$

$$G_2(\varphi)(x) := \Big(\int_0^\infty P_t |\nabla P_t\varphi|_2^2(x) \, dt\Big)^{1/2}, \tag{7.29}$$

$$G_3(\varphi)(x) := \Big(\int_0^\infty t[|\nabla Q_t\varphi(x)|_2^2 + |\tfrac{\partial}{\partial t}Q_t\varphi(x)|^2] \, dt\Big)^{1/2}. \tag{7.30}$$

The aim of this section is to derive the LPS inequalities which estimate the (L^p)–norm of $G_i(\varphi)$ in terms of the (L^p)–norm of φ. In the cases i = 1,2 we shall call the inequalities *parabolic*, while for i = 3 *elliptic*, because $P_t\varphi$ solves the parabolic equation $(\frac{\partial}{\partial t}+N)P_t\varphi = 0$, and $Q_t\varphi$ the elliptic equation $(\frac{\partial^2}{\partial t^2}-N)Q_t\varphi = 0$.

First we give sense to the functions G_i, i = 1,2,3, by showing that their squares are well-defined for μ–a.e. $x \in \mathcal{S}'(\mathbb{R})$ as (L^1)–Bochner integrals. In the following let $\varphi \in \mathfrak{F}C_b$, unless stated otherwise. Then – according to our discussion in Section 7.A – the integrands in (7.28), (7.29), (7.30) make sense for $t>0$ and every $x \in \mathcal{S}'(\mathbb{R})$. Note that by Mehler's formula and Schwarz' inequality the integrand in (7.28) is dominated pointwise by the one in (7.29), and so we only need to consider G_2 and G_3.

Let $t>0$. Then (since P_t is positivity preserving)

$$\Big\| P_t | \nabla P_t \varphi |_2^2 \Big\|_1 = \int P_t | \nabla P_t \varphi |_2^2(x) \, d\mu(x)$$

$$= \int | \nabla P_t \varphi |_2^2(x) \, d\mu(x)$$

$$= (P_t \varphi, N P_t \varphi),$$

where we used the self–adjointness of P_t and (7.1). Thus we find

$$\Big\| P_t | \nabla P_t \varphi |_2^2 \Big\|_1 = -\tfrac{1}{2} \frac{d}{dt} \| P_t \varphi \|_2^2,$$

and therefore

$$\int_0^\infty \Big\| P_t | \nabla P_t \varphi |_2^2 \Big\|_1 dt = \tfrac{1}{2}\big(\lim_{t \downarrow 0} \| P_t \varphi \|_2^2 - \lim_{t \to \infty} \| P_t \varphi \|_2^2\big)$$

$$= \tfrac{1}{2}\big(\| \varphi \|_2^2 - | \mathbb{E} \varphi |^2\big).$$

Now consider G_3. The (L^1)–norm of its integrand is equal to

$$2t(Q_t \varphi, N Q_t \varphi) = \tfrac{1}{2} \, t \frac{d^2}{dt^2} \| Q_t \varphi \|_2^2.$$

Integration by parts leads to

$$\int_0^\infty t \left\| \,|\nabla Q_t\varphi|_2^2 + |\tfrac{\partial}{\partial t}Q_t\varphi|^2 \right\|_1 dt =$$

$$= \tfrac{1}{2}\{\lim_{t\to\infty} t\tfrac{\partial}{\partial t}\| Q_t\varphi \|_2^2 - \lim_{t\downarrow 0} t\tfrac{\partial}{\partial t}\| Q_t\varphi \|_2^2$$

$$+ \lim_{t\downarrow 0}\| Q_t\varphi \|_2^2 - \lim_{t\to\infty}\| Q_t\varphi \|_2^2\}$$

$$= \tfrac{1}{2}\{\| \varphi \|_2^2 - |\mathbb{E}\varphi|^2\},$$

because the first two limits vanish as we are going to show now. $\varphi\in\mathcal{F}C_b$ implies that $\varphi\in(L^2)$, and in (L^2) φ has a chaos decomposition given by $\varphi = \sum_n \varphi^{(n)}$, $\varphi^{(n)}\in\mathcal{H}^{(n)}$. Then

$$t\frac{\partial}{\partial t}\| Q_t\varphi \|_2^2 = \sum_{n=1}^{\infty} tn^{1/2}\exp(-2tn^{1/2})\,\| \varphi^{(n)} \|_2^2.$$

As $t\downarrow 0$, and as $t\to+\infty$, each summand converges to zero. Since $tn^{1/2}\exp(-2tn^{1/2})$ is bounded, the sum converges uniformly in $t\geq 0$, and hence it vanishes in the limits.

We summarize our discussion in the following proposition.

Proposition 7.14. *Let* $\varphi\in\mathcal{F}C_b$. *Then* $G_i(\varphi)^2$, $i = 1,2,3$, *is an* (L^1)–*Bochner integral, and* $G_i(\varphi)\in(L^2)$. *Moreover*

$$G_1(\varphi) \leq G_2(\varphi), \quad \mu\text{-a.e.,} \tag{7.31}$$

and

$$\| G_i(\varphi) \|_2^2 = \tfrac{1}{2}(\| \varphi \|_2^2 - |\mathbb{E}\varphi|^2) \leq \tfrac{1}{2}\| \varphi \|_2^2, \; i = 2,3. \tag{7.32}$$

Lemma 7.15. G_i, $i = 1,2,3$, *is subadditive on* $\mathcal{F}C_b$.

Proof. We only give the proof for G_2. The other cases are similar (and somewhat easier). Let φ, $\psi \in \mathfrak{F}C_b$.

$$G_2(\varphi+\psi)^2 = \int_0^\infty P_t\{|\nabla P_t\varphi|_2^2 + |\nabla P_t\psi|_2^2 + 2\mathrm{Re}(\nabla P_t\varphi, \nabla P_t\psi)_{L^2(\mathbb{R})}\}\, dt$$

$$\leq G_2(\varphi)^2 + G_2(\psi)^2 + 2\int_0^\infty P_t|\nabla P_t\varphi|_2 \, |\nabla P_t\psi|_2 \, dt,$$

where we use Schwarz' inequality on $L^2(\mathbb{R})$. Since $|\nabla P_t\varphi|_2$ and $|\nabla P_t\psi|_2$ belong to C_b (actually to $\mathfrak{F}C_b$), we may use Mehler's formula to write

$$\int_0^\infty (P_t|\nabla P_t\varphi|_2 \, |\nabla P_t\psi|_2)(x)\, dt =$$

$$\leq \int_0^\infty\int |\nabla P_t\varphi(\gamma_{t,x}(y))|_2 \, |\nabla P_t\psi(\gamma_{t,x}(y))|_2 \, d\mu(y)\, dt$$

$$\leq \Big(\int_0^\infty\int |\nabla P_t\varphi(\gamma_{t,x}(y))|_2^2 \, d\mu(y)\, dt\Big)^{1/2} \times$$

$$\times \Big(\int_0^\infty\int |\nabla P_t\psi(\gamma_{t,x}(y))|_2^2 \, d\mu(y)\, dt\Big)^{1/2}$$

$$= \Big(\int_0^\infty P_t|\nabla P_t\varphi|_2^2(x)\, dt\Big)^{1/2}\Big(\int_0^\infty P_t|\nabla P_t\psi|_2^2(x)\, dt\Big)^{1/2}$$

$$= G_2(\varphi)(x)\, G_2(\psi)(x),$$

where we have used the notation $\gamma_{t,x}(y) = e^{-t}x + (1-e^{-2t})y$. Thus we get

$$G_2(\varphi+\psi) \leq G_2(\varphi) + G_2(\psi). \qquad \square$$

Corollary 7.16. G_i *extends continuously from* $\mathcal{F}C_b$ *to* (L^2). *Moreover, for all* $\varphi \in (L^2)$,

$$\| G_i(\varphi) \|_2 \leq 2^{-1/2} \| \varphi \|_2, \quad i = 1,2,3. \tag{7.33}$$

Proof. Since G_i is subadditive, we obtain

$$| G_i(\varphi) - G_i(\psi) | \leq G_i(\varphi - \psi). \tag{7.34}$$

Let $\varphi \in (L^2)$, and $(\varphi_n, n \in \mathbb{N})$ be a sequence in $\mathcal{F}C_b$ approximating φ in (L^2). Then (7.34) and Proposition 7.14 imply that $(G_i(\varphi_n), n \in \mathbb{N})$ is Cauchy in (L^2). The limit is denoted in the same way, and it is clear that (7.33) holds. □

The following two formulae will play a crucial role in this section.

Lemma 7.17. *Let* $\varphi \in \mathcal{F}C_b$, $\varphi \geq 0$, $p \geq 1$. *Then for all* $t > 0$,

$$-(\frac{\partial}{\partial t} + N)(P_t\varphi)^p = p(p-1)\,(P_t\varphi)^{p-2} | \nabla P_t\varphi |_2^2, \tag{7.35}$$

$$(\frac{\partial^2}{\partial t^2} - N)(Q_t\varphi)^p = p(p-1)\,(Q_t\varphi)^{p-2}[(\frac{\partial}{\partial t} Q_t\varphi)^2 + | \nabla Q_t\varphi |_2^2], \tag{7.36}$$

hold pointwise on $\mathcal{S}'(\mathbb{R})$.

Proof. Lemma 7.5 and 7.9 show that all involved terms are pointwise defined, and that $(\frac{\partial}{\partial t} + N)P_t\varphi = 0$, $(\frac{\partial^2}{\partial t^2} - N)Q_t\varphi = 0$ hold. Thus (7.35) and (7.36) follow by the chain rules for $\frac{\partial}{\partial t}$ and N (cf. (5.81) or use (7.1)). □

Remark. (7.35) and (7.36) show that, for positive φ, $-(\frac{\partial}{\partial t} + N)(P_t\varphi)^p$ and $(\frac{\partial^2}{\partial t^2} - N)(Q_t\varphi)^p$ are positive.

Next we show that the functions on the right hand sides of (7.35) and (7.36) are (L^1)–Bochner integrable in $L^1(\mathbb{R}_+, dt)$, $L^1(\mathbb{R}_+, t\,dt)$ respectively. Since for positive φ these functions are positive, it follows that

$$\| -(\frac{\partial}{\partial t} + N)(P_t\varphi)^p \|_1 = - \int (\frac{\partial}{\partial t} + N)(P_t\varphi)^p(x)\, d\mu(x)$$

$$= - \int \frac{\partial}{\partial t} (P_t\varphi)^p(x)\, d\mu(x),$$

and similarly

$$\| (\frac{\partial^2}{\partial t^2} - N)(Q_t\varphi)^p \|_1 = \int \frac{\partial^2}{\partial t^2} (Q_t\varphi)^p(x)\, d\mu(x).$$

Using Fubini's theorem we obtain

$$\int_0^\infty \| -(\frac{\partial}{\partial t} + N)(P_t\varphi)^p \|_1 dt = - \int (\int_0^\infty \frac{\partial}{\partial t} (P_t\varphi)^p(x)\, dt)\, d\mu(x)$$

$$= \int \Big(\lim_{t\downarrow 0} (P_t\varphi)^p(x) - \lim_{t\to\infty} (P_t\varphi)^p(x) \Big)\, d\mu(x)$$

$$= \int (\varphi(x)^p - (\mathbb{E}\varphi)^p)\, d\mu(x)$$

$$\leq \| \varphi \|_p^p,$$

which is finite. The right hand side of (7.36) can be treated in a similar way (using the limits in the proof of Lemma 7.9), and this is omitted here. We have proved the following result.

Lemma 7.18. *For every positive* $\varphi \in \mathfrak{F}C_b$, $p \geq 1$,

$$- \int_0^\infty (\frac{\partial}{\partial t} + N)(P_t\varphi)^p\, dt \geq 0, \tag{7.37}$$

and

$$\int_0^\infty t\,(\frac{\partial^2}{\partial t^2} - N)(Q_t\varphi)^p \, dt \geq 0, \tag{7.38}$$

exist as (L^1)*–Bochner integrals, and*

$$\| \int_0^\infty (\frac{\partial}{\partial t} + N)(P_t\varphi)^p \, dt \|_1 \leq \| \varphi \|_p^p, \tag{7.39}$$

and

$$\| \int_0^\infty t(\frac{\partial^2}{\partial t^2} - N)(Q_t\varphi)^p \, dt \|_1 \leq \| \varphi \|_p^p. \tag{7.40}$$

Now we are ready to prove the first LPS inequality. From now on we shall use the following convenient notation: if there exists a constant $C_{p,q}$, depending only on $p,\ q \geq 1$, so that $\| \varphi \|_p \leq C_{p,q} \| \psi \|_q$, then we shall write $\| \varphi \|_p \lesssim \| \psi \|_q$.

Proposition 7.19. *Let* $\varphi \in (L^p)$, $1<p\leq 2$. *Then also* $G_1(\varphi) \in (L^p)$ *and*

$$\| G_1(\varphi) \|_p \lesssim \| \varphi \|_p. \tag{7.41}$$

Proof. Due to Corollary 7.16 we only need to consider the case $p<2$. Moreover, by the subadditivity of G_1 (Lemma 7.15), it is sufficient to prove (7.41) for $\varphi \in \mathfrak{F}C_b$. Assume that we have shown (7.41) for positive $\varphi \in \mathfrak{F}C_b$. Then (7.41) follows for arbitrary real $\varphi \in \mathfrak{F}C_b$ by the following argument. Decompose $\varphi = \varphi_+ - \varphi_-$ into its positive and real parts $\varphi_\pm \geq 0$. Then (7.41) holds for φ_+ and φ_-, and subadditivity of G_1 gives

$$\| G_1(\varphi_+ - \varphi_-) \|_p \leq \| G_1(\varphi_+) \|_p + \| G_1(\varphi_-) \|_p$$

$$\lesssim \| \varphi_+ \|_p + \| \varphi_- \|_p$$

$$\lesssim \| \varphi \|_p .$$

For complex $\varphi \in \mathfrak{F}C_b$, we now get (7.41) by the same argument. From now on assume that $\varphi \in \mathfrak{F}C_b$ is positive, and $\varphi \not\equiv 0$. It follows from Mehler's formula (7.7) that $P_t\varphi > 0$. Thus we may apply Lemma 7.16 to write

$$P_t| \nabla P_t\varphi |_2(x) = (p(p-1))^{-1/2}\, P_t\{(P_t\varphi)^{1-p/2}(-(\frac{\partial}{\partial t} + N)(P_t\varphi)^p)^{1/2}\}(x).$$

Now estimate in the following way by using (7.7) and Schwarz' inequality:

$$P_t| \nabla P_t\varphi |_2(x) \le$$

$$\le (p(p-1))^{-1/2}\, (P_t(P_t\varphi)^{2-p}(x))^{1/2}(-P_t(\frac{\partial}{\partial t} + N)(P_t\varphi)(x))^{1/2}$$

$$\le (p(p-1))^{-1/2}(((\varphi^*)^{2-p})^*)^{1/2}(x)(-P_t(\frac{\partial}{\partial t} + N)(P_t\varphi)(x))^{1/2},$$

where we used the notation of Theorem 7.7. Applying Hölder's inequality (with exponents $\frac{2}{p}, \frac{2}{2-p}$) and Theorem 7.7, we get the following estimate

$$\| G_1(\varphi) \|_p^p \lesssim \| \varphi \|_p^{p(1-p/2)} \| \int_0^\infty P_t(\frac{\partial}{\partial t} + N)(P_t\varphi)^p \,dt \|_1^{p/2}.$$

Finally we use the estimate (7.39) and obtain (7.41). □

Our next step is the following inequality.

Lemma 7.20. *Let $\varphi, \psi \in \mathfrak{F}C_b$, $\psi \ge 0$. Then*

$$\int G_2(\varphi)^2\psi \,d\mu \le K\{\int | \varphi |^2\psi \,d\mu + \int \varphi^* G_2(\varphi)\, G_1(\psi)\, d\mu\}, \qquad (7.42)$$

for some constant K>0.

Proof. Using the subadditivity of G_2, and the fact that $G_2(\mathrm{Im}\varphi) \le G_2(\varphi)$, and $G_2(\mathrm{Re}\varphi) \le G_2(\varphi)$, it is easy to see that it is sufficient to prove (7.42) for positive $\varphi \in \mathcal{F}C_b$. Choose p = 2 in (7.35) and compute as follows.

$$\int G_2(\varphi)^2 \, \psi \, d\mu = \int \Big(\int_0^\infty P_t | \nabla P_t \varphi |_2^2 \, dt \Big) \psi \, d\mu$$

$$= -\tfrac{1}{2} \int \Big(\int_0^\infty P_t \big(\tfrac{\partial}{\partial t} + N\big)(P_t\varphi)^2 \, dt \Big) \, \psi \, d\mu$$

$$= -\tfrac{1}{2} \int_0^\infty \int [(\tfrac{\partial}{\partial t} + N)(P_t\varphi)^2] \, P_t\psi \, d\mu \, dt,$$

where we used Fubini's theorem and self–adjointness of P_t. For real differentiable functions f, g on $\mathbb{R}_+ \times \mathcal{S}'(\mathbb{R})$ we have by the product rules for $\frac{\partial}{\partial t}$ and N (cf. (5.82))

$$(\tfrac{\partial}{\partial t} + N) \, f \cdot g = ((\tfrac{\partial}{\partial t} + N) f) \cdot g + f \cdot (\tfrac{\partial}{\partial t} + N) g - 2(\nabla f, \nabla g)_{L^2(\mathbb{R})}.$$

Thus

$$\int G_2(\varphi)^2 \, \psi \, d\mu =$$

$$= - \int \int_0^\infty (\tfrac{\partial}{\partial t} + N)((P_t\varphi)^2 \, P_t\psi) \, dt \, d\mu + \int \int_0^\infty (\nabla (P_t\varphi)^2, \nabla P_t\psi) \, dt \, d\mu,$$

where we used $(\frac{\partial}{\partial t} + N) P_t\psi = 0$. For the first of the last two expressions we use self–adjointness of N, N1 = 0 and the fundamental theorem of calculus, and obtain

$$- \int \int_0^\infty (\tfrac{\partial}{\partial t} + N)((P_t\varphi)^2 \, P_t\psi) \, dt \, d\mu = \tfrac{1}{2} \int (\varphi^2\psi - (\mathbb{E}\varphi)^2 \mathbb{E}\psi) \, d\mu$$

$$\le \tfrac{1}{2} \int \varphi^2 \psi \, d\mu.$$

Consider the second of the above terms. Use $\nabla(P_t\varphi)^2 = 2\, P_t\varphi\, \nabla P_t\varphi$, and Schwarz' inequality on $L^2(\mathbb{R})$ to estimate as follows:

$$\left| \int\int_0^\infty (\nabla(P_t\varphi)^2, \nabla P_t\psi) \, dt \, d\mu \right| \le 2 \int \varphi^* \int_0^\infty | \nabla P_t\varphi |_2 \, | \nabla P_t\psi |_2 \, dt \, d\mu.$$

By Lemma 7.11 we get

$$\left| \int\int_0^\infty (\nabla(P_t\varphi)^2, \nabla P_t\psi) \, dt \, d\mu \right| \le$$

$$\le 2 \int \varphi^* \int_0^\infty (P_{t/2} | \nabla P_{t/2}\varphi |_2^2)^{1/2} (P_{t/2} \, | \nabla P_{t/2}\psi |_2) \, dt \, d\mu$$

$$\le 4 \int \varphi^* \, G_2(\varphi) \, G_1(\psi) \, d\mu,$$

where we used Schwarz' inequality for the dt–integration in the last step. □

Now we are ready to prove the first main result: the parabolic LPS inequalities.

Theorem 7.21.

(i) *Let* $\varphi \in (L^p)$, $1 < p < +\infty$. *Then* $G_1(\varphi)$ *belongs to* (L^p) *too, and*

$$\| G_1(\varphi) \|_p \lesssim \| \varphi \|_p. \tag{7.43}$$

(ii) *If* $2 \le p < +\infty$, *and* $\varphi \in (L^p)$, *then also* $G_2(\varphi) \in (L^p)$, *and*

$$\| G_2(\varphi) \|_p \lesssim \| \varphi \|_p \tag{7.44}$$

(iii) *Let* $\varphi \in (L^p)$, $1<p\leq 2$, *with* $\mathbb{E}\varphi = 0$. *Then*

$$\| \varphi \|_p \lesssim \| G_2(\varphi) \|_p . \tag{7.45}$$

Proof. For $p\in(1,2]$, (7.43) was already proved in Proposition 7.19. Because of inequality (7.31), (7.43) is proved, if we show (7.44). Note that for $p = 2$ inequality (7.44) is true, cf. Corollary 7.16. We shall show (7.44) for $p\geq 4$, so that by an application of the Marcinkiewicz interpolation theorem (e.g. Stein (1970a), p.272), we obtain the inequality for all $p\geq 2$. Also, it is sufficient to prove (7.44) for $\varphi\in\mathfrak{F}C_b$.

Let $p\geq 4$, $\varphi\in\mathfrak{F}C_b$. Then

$$\begin{aligned}\| G_2(\varphi) \|_p^2 &= \| G_2(\varphi) \|_{p/2} \\ &= \sup_{\psi} \int G_2(\varphi)^2 \, \psi \, d\mu,\end{aligned}$$

where the sup is over all $\psi\in\mathfrak{F}C_b$, $\psi\geq 0$, so that $\|\psi\|_q \leq 1$ with q conjugate to p/2. Lemma 7.20 and Hölder's inequality provide the following estimate

$$\begin{aligned}\| G_2(\varphi) \|_p^2 &\leq K \sup_{\psi}\{\int | \varphi |^2 \, \psi \, d\mu + \int \varphi^* \, G_2(\varphi) \, G_1(\psi) \, d\mu\} \\ &\leq K \{\| \varphi \|_p^2 + \sup_{\psi} \| \varphi^* \, G_2(\varphi) \|_{p/2} \| G_1(\psi) \|_q\} \\ &\leq K \{\| \varphi \|_p^2 + \sup_{\psi} \| \varphi^* \|_p \| G_2(\varphi) \|_p \| G_1(\psi) \|_q,\end{aligned}$$

where K is some constant, and q is conjugate to $\frac{p}{2}$. Note that $q\leq 2$, so that we apply Proposition 7.19 for the last norm. Also, we make use of Theorem 7.7. We obtain

$$\| G_2(\varphi) \|_p^2 \lesssim \| \varphi \|_p (\| \varphi \|_p + \| G_2(\varphi) \|_p),$$

which implies inequality (7.44).

Now we prove inequality (7.45). Let $p \in (1,2]$. It is sufficient to consider real $\varphi \in \mathfrak{F}C_b$, with $\mathbb{E}\varphi = 0$. Note that by Proposition 7.14, we have the equality $\| G_2(\varphi) \|_2^2 = \frac{1}{2} \| \varphi \|_2^2$. Either by polarizing this equality, or by a direct computation like in the proof of Proposition 7.14, one finds for real $\psi \in \mathfrak{F}C_b$

$$\int \varphi \, \psi \, d\mu = 2 \int \int_0^\infty P_t | \nabla P_t \varphi |_2 \, | \nabla P_t \psi |_2 \, dt \, d\mu.$$

Using Mehler's formula (7.7) and Schwarz' inequality we get the estimate

$$| \int \varphi \, \psi \, d\mu | \leq 2 \int \int_0^\infty (P_t | \nabla P_t \varphi |_2^2)^{1/2} \, (P_t | \nabla P_t \psi |_2^2)^{1/2} \, dt \, d\mu$$

$$\leq 2 \int G_2(\varphi) \, G_2(\psi) \, d\mu.$$

Let q be conjugate to $p \in (1,2]$, so that $q \geq 2$. Then

$$\| \varphi \|_p = \sup_\psi | \int \varphi \, \psi \, d\mu |$$

$$\leq 2 \sup_\psi \int G_2(\varphi) \, G_2(\psi) \, d\mu$$

$$\lesssim \sup_\psi \| G_2(\varphi) \|_p \, \| G_2(\psi) \|_q$$

$$\lesssim \sup_\psi \| \psi \|_q \, \| G_2(\varphi) \|_p$$

$$\leq \| G_2(\varphi) \|_p,$$

where the sup was over all $\psi \in \mathfrak{F}C_b$, with $\| \psi \|_q \leq 1$, and we used (7.44). □

Next we derive the elliptic LPS inequalities for $G_3(\varphi)$. All the arguments will be very similar (and sometimes easier) to those used for the parabolic case. Therefore we only indicate the major steps.

First one uses equation (7.36), estimate (7.40), Theorem (7.7) and a computation like in the proof of Proposition 7.19 to prove the following result.

Proposition 7.22. *Assume that* $\varphi \in (L^p)$, $1<p\leq 2$. *Then* $G_3(\varphi) \in (L^p)$, *and*

$$\| G_3(\varphi) \|_p \lesssim \| \varphi \|_p. \tag{7.46}$$

The proof of the analog of Lemma 7.20 for $G_3(\varphi)$ follows along the same line, and one obtains

Lemma 7.23. *Let* $\varphi, \psi \in \mathfrak{F}C_b$, $\psi \geq 0$. *Denote* $\varphi^{**} = \sup_{t>0} Q_t \varphi$. *Then*

$$\int G_3(\varphi)^2 \, \psi \, d\mu \leq K\{\int | \varphi |^2 \, \psi \, d\mu + \int \varphi^{**} \, G_3(\varphi) \, G_3(\psi) \, d\mu\}, \tag{7.47}$$

for some constant $K>0$.

Now one uses the subadditivity of G_3, the Marcinkiewicz interpolation theorem, Proposition 7.22, Lemma 7.22, and equality (7.32) to prove the second major result of this section:

Theorem 7.24.

(i) *Let* $\varphi \in (L^p)$, $1<p<+\infty$. *Then* $G_3(\varphi) \in (L^p)$, *and*

$$\| G_3(\varphi) \|_p \lesssim \| \varphi \|_p. \tag{7.48}$$

(ii) *If* $\varphi \in (L^p)$, $1<p<+\infty$, with $\mathbb{E}\varphi = 0$, *then*

$$\| \varphi \|_p \lesssim \| G_3(\varphi) \|_p. \tag{7.49}$$

Finally we state some results which follow from Theorem 7.24 and which will be useful in the next section.

Define

$$G_4(\varphi) := \left(\int_0^\infty t \left| \frac{\partial}{\partial t} Q_t \varphi \right|^2 dt\right)^{1/2}, \tag{7.50}$$

$$G_5(\varphi) := \left(\int_0^\infty t \left| \nabla Q_t \varphi \right|_2^2 dt\right)^{1/2}. \tag{7.51}$$

As before we find that for $\varphi \in (L^2)$, with $\mathbb{E}\varphi = 0$, $\| G_4(\varphi) \|_2 = \| G_5(\varphi) \|_2 = \frac{1}{2} \| \varphi \|_2$. The first statement of the following Theorem follows directly from Theorem 7.24. The second one is proved by the preceding equalities, and an argument like in the proof of Theorem 7.21.iii.

Theorem 7.25.

(i) *Let* $\varphi \in (L^p)$, $1<p<+\infty$. *Then* $G_i(\varphi) \in (L^p)$, $i = 4,5$, *and*

$$\| G_i(\varphi) \|_p \lesssim \| \varphi \|_p, \quad i = 4,5. \tag{7.52}$$

(ii) *If* $\varphi \in (L^p)$, $1<p<+\infty$, *and* $\mathbb{E}\varphi = 0$, *then*

$$\| \varphi \|_p \lesssim \| G_i(\varphi) \|_p, \quad i = 4,5. \tag{7.53}$$

The following generalization of Theorem 7.25 to sequences $(\varphi_k, k\in\mathbb{N})$ can be proved by using Rademacher functions and the Khintchin inequalities. For a proof we refer to Meyer (1983).

Theorem 7.26.

(i) *Let* $\varphi = (\varphi_k, k\in\mathbb{N})$ *be a sequence of functions in* (L^p), $1<p<+\infty$. *Suppose that* $\| \varphi \|_{l^2} \in (L^p)$. *Then* $(\sum_k G_4(\varphi_k)^2)^{1/2} \in (L^p)$, *and*

$$\| (\sum_{k=1}^{\infty} G_4(\varphi_k)^2)^{1/2} \|_p \lesssim \| (\sum_{k=1}^{\infty} | \varphi_k |^2)^{1/2} \|_p. \tag{7.54}$$

(ii) *If in addition for all* $k\in\mathbb{N}$, $\mathbb{E}\varphi_k = 0$, *then*

$$\| (\sum_{k=1}^{\infty} | \varphi_k |^2)^{1/2} \|_p \lesssim \| (\sum_{k=1}^{\infty} G_4(\varphi_k)^2)^{1/2} \|_p. \tag{7.55}$$

We conclude this section with the following remark. It is clear that except for the properties of the semigroups P_t, Q_t, $t\geq 0$, that we have derived from Mehler's formula, no particular feature of our underlying white noise probability space entered the proofs of this section. Moreover, many of the statements about the semigroups hold for very general classes of semigroups (cf. the literature mentioned at the beginning of this chapter and of Section 7.A). Therefore, the results of this section hold in much more general frameworks. Apart from the book of Stein (1970b), we refer the interested reader in particular to the articles of Bakry (1985) for such general formulations.

7.C Meyer's Inequalities and Their Applications

In this section we establish the main results of this chapter, and we follow essentially the articles of Meyer (1982, 1983), Sugita (1985a) and Watanabe (1983) (cf. also Potthoff (1988b)). Namely, we first prove the equivalence of the (L^p)–norms of $| \nabla \cdot |_2$ and $N^{1/2}$ which is due to Meyer (1983). This equivalence allows to show that $\mathfrak{D}$ is an algebra with continuous (pointwise) multiplication

(Meyer (1983), Sugita (1985a)). Finally, we shall present the proof of Watanabe (1983) that composites of tempered distributions with random variables in $\mathfrak{D}$, satisfying a certain nondegeneracy condition, are elements in $\mathfrak{D}^*$ (and therefore also in $(\mathcal{S})^*$, cf. Corollary 4.14). This result can be used conveniently to provide a criterion for whether $\mathbb{R}^d$-valued random variables have a law admitting a smooth density with respect to Lebesgue measure.

We shall continue to use the notation of the previous sections. First we quote two results from Sugita (1985a) which enable us to handle functions of the number operator N on (L^p)-spaces.

Lemma 7.27.

(i) *Let* $n\in\mathbb{N}_0$, *and let* J_n *denote the orthogonal projection onto* $\mathcal{H}^{(n)}$ *in* (L^2). *Then for all* $n\in\mathbb{N}_0$, $p\in(1,+\infty)$, J_n *extends to a bounded linear operator on* (L^p).

(ii) *For all* $n\in\mathbb{N}_0$, $p\in(1,+\infty)$, *there exists a constant* $C_{p,n}>0$, *so that for all* $\varphi\in(L^p)$, $t\geq 0$,

$$\| P_t(1-\sum_{m=0}^{n-1} J_m)\, \varphi \|_p \leq C_{p,n}\, e^{-tn} \| \varphi \|_p. \tag{7.56}$$

Lemma 7.27 is essentially a consequence of Nelson's hypercontractivity theorem for P_t (cf. Nelson (1973c), Theorem 7.1.iii). Based on this lemma, the following important result is proved in Sugita (1985a).

Theorem 7.28. *Let* f *be a function on* $\mathbb{C}$, *which is real on the real line and analytic in a neighborhood of zero. Let* $\alpha>0$, *and consider the mapping* T_f *from* $\mathcal{P}$ *into itself given by*

$$T_f\varphi = \sum_{n=1}^{\infty} f(n^{-\alpha})\, J_n\varphi, \quad \varphi\in\mathcal{P}.$$

Then for every $p\in(1,+\infty)$, T_f *extends to a unique bounded linear operator from*

(L^p) *into itself.*

Now we can start proving Meyer's equivalence of norms following Meyer (1982).

Lemma 7.29. *Denote* $\mathcal{P}_2 := \mathcal{P} \bigcap \oplus_{n \le 2} \mathcal{H}^{(n)} = (1-J_0-J_1)\mathcal{P}$. *Let* $\varphi \in \mathcal{P}_2$, *and set* $\psi = (N-1)^{1/2}\varphi$. *Then for all* $p \in (1,+\infty)$

$$\| \, | \nabla\varphi |_2 \, \|_p \lesssim \| \, (\int_0^\infty e^{2t} \, | \nabla P_t \psi |_2^2 \, dt)^{1/2} \, \|_p .$$

Proof. Note that we have on $\mathcal{P}_2$ the following commutation relation

$$N^{1/2}\partial_k = \partial_k(N-1)^{1/2}. \tag{7.57}$$

Let $\varphi \in \mathcal{P}_2$, and compute as follows.

$$\begin{aligned} \frac{\partial}{\partial t} Q_t \partial_k \varphi &= - \, Q_t N^{1/2} \partial_k \varphi \\ &= - \, Q_t \partial_k (N-1)^{1/2} \varphi \\ &= - \int_0^\infty P_s \partial_k (N-1)^{1/2} \varphi \; \nu(t,s) \, ds \\ &= - \int_0^\infty e^s \, \partial_k P_s \varphi \; \nu(t,s) \, ds, \end{aligned}$$

where we used (7.12) and (7.18). Thus

$$\sum_{k=1}^\infty | \frac{\partial}{\partial t} Q_t \partial_k \varphi |_2^2 = | \int_0^\infty e^s \, \nabla P_s \psi \; \nu(t,s) \, ds |_2^2$$

$$\leq \Big(\int_0^\infty e^s \mid \nabla P_s \psi \mid_2 \nu(t,s)\, ds\Big)^{1/2}$$

$$\leq \int_0^\infty e^{2s} \mid \nabla P_s \psi \mid_2^2 \nu(t,s)\, ds,$$

where we made use of Schwarz' inequality for the normalized measure $\nu(t,s)$ ds in the last step. Integrate the last relation with respect to t·dt, use Fubini's theorem and

$$\int_0^\infty t\, \nu(t,s)\, dt = 1,$$

to get (cf. equation (7.50))

$$\sum_{k=1}^{\infty} G_4(\partial_k \varphi)^2 \leq \int_0^\infty e^{2s} \mid \nabla P_s \psi \mid_2^2 ds.$$

Note that by hypothesis we have $\mathbb{E}\varphi = 0$. Thus we obtain from Theorem 7.26.ii,

$$\| \mid \nabla \varphi \mid_2 \|_p \lesssim \| (\sum_{k=1}^{\infty} G_4(\partial_k \varphi)^2)^{1/2} \|_p$$

$$\leq \| (\int_0^\infty e^{2t} \mid \nabla P_t \psi \mid_2^2 dt)^{1/2} \|_p. \qquad \square$$

Remark. If ∂_k and N would commute, we would not have the factor e^{2t} under the integral in the last expression. Then we could use Lemma 7.11, Theorem 7.21.i and Theorem 7.28 to estimate in the following way:

$$\| (\int_0^\infty \mid \nabla P_t \psi \mid_2^2 dt)^{1/2} \|_p \leq \| (\int_0^\infty (P_{t/2} \mid \nabla P_{t/2} \psi \mid_2)^2 dt)^{1/2} \|_p$$

$$= 2^{1/2} \| G_1(\psi) \|_p$$

$$\lesssim \| \psi \|_p$$

$$= \| (N-1)^{1/2} \varphi \|_p$$

$$\lesssim \| N^{1/2} \varphi \|_{p,}$$

This estimation is parallel to the proof for the situations discussed in Stein (1970) (e.g., the classical Riesz transform). The appearance of the factor e^{2t} due to the non-commutativity of ∂_k and N was treated in Meyer (1982) by help of the following lemma.

Lemma 7.30. *Let* $\psi \in \mathcal{P}$, $\mathbb{E}\psi = 0$, $p \in (1,+\infty)$. *Then*

$$\| (\int_0^\infty e^{2t} | \nabla P_t \psi |_2^2 \, dt)^{1/2} \|_p \lesssim \| \psi \|_p + \int_0^\infty e^t \| P_t \psi \|_p \, dt.$$

Proof. Remark first that the condition $\mathbb{E}\psi = 0$, and estimate (7.56) show that the last integral exists. From Lemma 7.11 we have for $\eta \in \mathcal{P}$,

$$e^t | \nabla P_t \eta |_2^2 \le (P_{t/2} | \nabla P_{t/2} \eta |_2)^2, \quad t \ge 0.$$

Thus

$$\int_0^\infty e^t | \nabla P_t \eta |_2^2 \, dt \le 2 \int_0^\infty (P_t | \nabla P_t \eta |_2)^2 \, dt$$

$$= 2 \, G_1(\eta)^2,$$

and by Theorem 7.21.i we get for all $p \in (1,+\infty)$ the inequality

$$\| \Big(\int_0^\infty e^t \, | \nabla P_t \eta |_2^2 \, dt\Big)^{1/2} \|_p \lesssim \| \eta \|_p. \tag{7.58}$$

Make the choice $\eta = P_u \psi$, for $u \geq 0$. Then (7.58) reads as follows

$$\| \Big(\int_u^\infty e^t \, | \nabla P_t \psi |_2^2 \, dt\Big)^{1/2} \|_p \lesssim e^{u/2} \, \| P_u \psi \|_p. \tag{7.59}$$

Define

$$\zeta(t,u) = e^{t/2} \, | \nabla P_t \psi |_2 \, 1_{[u,\infty)}(t),$$

and integrate (7.59) over $\mathbb{R}_+$ with $e^{u/2} du$. Then we get

$$\int_0^\infty e^{u/2} \, \| \, \| \zeta(\cdot\, ,u) \|_{L^2(R_+)} \|_p \, du \lesssim \int_0^\infty e^u \, \| P_u \psi \|_p \, du.$$

On the other hand

$$\| \, \| \int_0^\infty e^{u/2} \, \zeta(\cdot\, ,u) \, du \, \|_{L^2(R_+)} \|_p \leq \int_0^\infty e^{u/2} \, \| \, \| \zeta(\cdot\, ,u) \|_{L^2(R_+)} \|_p \, du,$$

so that we get the following bound

$$\| \, \| \int_0^\infty e^{u/2} \, \zeta(\cdot\, ,u) \, du \, \|_{L^2(R_+)} \|_p \leq \int_0^\infty e^u \, \| P_u \psi \|_p \, du.$$

An elementary computation gives

$$\| \int_0^\infty e^{u/2} \, \zeta(\cdot\, ,u) \, du \, \|_{L^2(R_+)} = 2 \Big(\int_0^\infty e^t (e^{t/2} - 1)^2 \, | \nabla P_t \psi |_2^2 \, dt\Big)^{1/2},$$

and therefore we have obtained

$$\| \Big(\int_0^\infty e^t (e^{t/2} - 1)^2 \, | \nabla P_t \psi |_2^2 \, dt\Big)^{1/2} \|_p \lesssim \int_0^\infty e^u \, \| P_u \psi \|_p \, du.$$

In particular we get

$$\| \left(\int_1^\infty e^t (e^{t/2} - 1)^2 \mid \nabla P_t \psi \mid_2^2 dt \right)^{1/2} \|_p \lesssim \int_0^\infty e^u \| P_u \psi \|_p \, du.$$

Note that for large enough K we have $e^t \le K\,(e^{t/2} - 1)^2$, for all $t \ge 1$. Thus

$$\| \left(\int_1^\infty e^{2t} \mid \nabla P_t \psi \mid_2^2 dt \right)^{1/2} \|_p \lesssim \int_0^\infty e^u \| P_u \psi \|_p \, du.$$

Finally, the proof of the lemma is finished by estimating in the following way:

$$\| \left(\int_0^1 e^{2t} \mid \nabla P_t \psi \mid_2^2 dt \right)^{1/2} \|_p \lesssim \| \left(\int_0^\infty e^t \mid \nabla P_t \psi \mid_2^2 dt \right)^{1/2} \|_p$$

$$\lesssim \| \psi \|_p,$$

where we used inequality (7.58). □

Now we are ready to prove Meyer's equivalence of norms.

Theorem 7.31. *Let* $\varphi \in \mathcal{D}^{p,1}$, $1 < p < +\infty$. *Then*

$$\| \mid \nabla \varphi \mid_2 \|_p \lesssim \| N^{1/2} \varphi \|_p \lesssim \| \mid \nabla \varphi \mid_2 \|_p. \tag{7.60}$$

Proof. It is sufficient to prove (7.60) for $\varphi \in \mathcal{P}$. Assume first that $\varphi \in \mathcal{P}_2$. By Lemma 7.29 and Lemma 7.30 we have

$$\| \mid \nabla \varphi \mid_2 \|_p \lesssim \| (N-1)^{1/2} \varphi \|_p + \int_0^\infty e^t \| P_t (N-1)^{1/2} \varphi \|_p \, dt.$$

Observe that $(N-1)^{1/2} \varphi \in \mathcal{P}_2$, "so that Lemma 7.27.ii gives the bound

$$\int_0^\infty e^t \| P_t (N-1)^{1/2} \varphi \|_p \, dt \lesssim \| (N-1)^{1/2} \varphi \|_p,$$

whence

$$\| \, | \nabla\varphi |_2 \|_p \lesssim \| (N-1)^{1/2} \varphi \|_p$$

$$= \| (1-N^{-1})^{1/2} (1-J_0) N^{1/2} \varphi \|_p$$

$$\lesssim \| N^{1/2} \varphi \|_p,$$

because by Theorem 7.28 $(1-N^{-1})^{1/2}(1-J_0)$ is bounded on (L^p). Now consider a general $\varphi \in \mathcal{P}$, and decompose $\varphi = \varphi^{(0)} + \varphi^{(1)} + \varphi'$, with $\varphi^{(0)} = J_0\varphi$, $\varphi^{(1)} = J_1\varphi$. Then $\varphi^{(1)}$ is of the form $\varphi^{(1)} = \langle \cdot \, , \xi \rangle$, $\xi \in \mathcal{S}(\mathbb{R})$, and $| \nabla\varphi^{(1)} |_2 = | \xi |_2$, $N^{1/2}\varphi^{(1)} = \varphi^{(1)}$. Moreover, $\| \varphi^{(1)} \|_p = C_p | \xi |_2$, for some constant C_p. Thus (7.60) holds for $\varphi^{(1)}$, and by our preceding consideration for φ'. Therefore

$$\| \, | \nabla\varphi |_2 \|_p = \| \, | \nabla\varphi^{(1)} + \nabla\varphi' |_2 \|_p$$

$$\lesssim \| N^{1/2} \varphi^{(1)} \|_p + \| N^{1/2} \varphi' \|_p$$

$$= \| J_1 N^{1/2} \varphi \|_p + \| (1-J_0-J_1) N^{1/2} \varphi \|_p$$

$$\lesssim \| N^{1/2} \varphi \|_p,$$

and the first of the inequalities in (7.60) is proved.

To prove the second note first that for all $\varphi, \psi \in \mathfrak{D}^{2,1}$, we have

$$(\varphi, N\, \psi) = (\nabla\varphi, \nabla\psi)_{L^2(\mathbb{R}) \otimes (L^2)},$$

which can be seen by using (7.1). Now let $\varphi \in \mathcal{P}$, $p \in (1,+\infty)$, and q be conjugate to p. Then

$$\| N^{1/2} \varphi \|_p = \sup_{\psi} | (\psi, N^{1/2} \varphi) |$$

$$= \sup_{\psi} | ((1-J_0)\psi, N^{1/2}\varphi) |$$

$$= \sup_{\psi} | (N^{-1/2}(1-J_0)\psi, N\varphi) |$$

$$= \sup_{\psi} | (\nabla N^{-1/2}(1-J_0)\psi, \nabla\varphi)_{L^2(\mathbb{R})\otimes(L^2)} |$$

$$\leq \sup_{\psi} \| \, | \nabla N^{-1/2}(1-J_0)\psi |_2 \|_q \| \, | \nabla\varphi |_2 \|_p$$

$$\lesssim \sup_{\psi} \| (1-J_0)\psi \|_q \| \, | \nabla\varphi |_2 \|_p$$

$$\lesssim \| \, | \nabla\varphi |_2 \|_p,$$

where the sup is over all $\psi \in \mathcal{P}$ with $\| \psi \|_q \leq 1$, and we used the first of the inequalities (7.60), and Lemma 7.27.i. □

Now we are going to discuss the proof of the result of Meyer (1983) that $\mathcal{D}$ is an algebra under pointwise multiplication. Actually, it will follow from Theorem 7.31 that the multiplication of two elements in $\mathcal{D}$ is continuous. It is clear that this is proved if we can show that whenever $\varphi, \psi \in \mathcal{D}$, for all $k \in \mathbb{N}$, $p > 1$, then

$$\| N^k \varphi\psi \|_p \lesssim ||| \varphi |||_{2p,2k} \, ||| \psi |||_{2p,2k}. \tag{7.61}$$

(Cf. (3.35) for the definition of the norms $||| \cdot |||_{p,k}$.)

Let us first consider the case $k = 1$. Using the product rule (5.80) for N, we get

$$\| N\varphi\psi \|_p \lesssim \| (N\varphi)\psi \|_p + \| \varphi N\psi \|_p + 2 \| (\nabla\overline{\varphi}, \nabla\psi)_{L^2(\mathbb{R})} \|_p.$$

Application of Schwarz', Hölder's and Meyer's inequalities (7.60) gives

$$\| N\varphi\psi \|_p \lesssim \| N\varphi \|_{2p} \| \psi \|_{2p} + \| \varphi \|_{2p} \| N\psi \|_{2p} + \| N^{1/2}\varphi \|_{2p} \| N^{1/2}\psi \|_{2p}.$$

Finally, applying Theorem 7.28, we obtain the desired inequality (7.61) for $k = 1$. For general $k\in\mathbb{N}$, the product rule for N produces k-th order gradients $\nabla^k\varphi$, mixed terms and so forth. In order to treat this efficiently, Sugita (1985a) derived inequalities like (7.60) for random variables which take values in a separable Hilbert space $\mathcal{K}$. Then one can for example consider $\nabla^{k-1}\varphi$ as a random variable with values in $L(\mathbb{R}^{k-1})$, and apply the above mentioned generalization. The idea of proof of Sugita's generalization of (7.60) for Hilbert space valued random variables is to use Rademacher functions and Khintchin inequalities (cf. also Meyer (1983), Watanabe (1984)). We only state the result here, and refer to the Sugita (1985a) for the detailed proof.

Let $\mathcal{K}$ be a separable Hilbert space. By $L^p(\mathcal{K})$, $p\geq 1$, we mean the space of $\mathcal{K}$-valued random variables φ on $(\mathcal{S}'(\mathbb{R}),\mathcal{B},\mu)$, which are finite in the norm

$$\| \varphi \|_{L^p(\mathcal{K})} := \Big(\int | \varphi(x) |_{\mathcal{K}}^p \, d\mu(x)\Big)^{1/p},$$

where $|\cdot|_{\mathcal{K}}$ is the norm of $\mathcal{K}$. Let us denote $\mathcal{H} = L^2(\mathbb{R})$. Then we can write (7.60) as follows

$$\| \nabla\varphi \|_{L^p(\mathcal{H})} \lesssim \| N^{1/2}\varphi \|_p \lesssim \| \nabla\varphi \|_{L^p(\mathcal{H})}. \qquad (7.60)'$$

Let $\mathcal{P}(\mathcal{K})$ be the space of $\mathcal{K}$-valued polynomials, and set $\mathcal{P}_k(\mathcal{K}) = (1-\sum_{m=0}^{k-1} J_m)\mathcal{P}(\mathcal{K})$. Based on (7.60) Sugita (1985a) shows the following result.

Theorem 7.32. *For all* $k\in\mathbb{N}$, $p\in(1,+\infty)$ *the following holds*:

(i) $$\| \nabla^k\varphi \|_{L^p(\mathcal{K}\otimes\mathcal{H}^k)} \lesssim \| N^{k/2}\varphi \|_{L^p(\mathcal{K})}, \quad \varphi\in\mathcal{P}(\mathcal{K}); \qquad (7.62)$$

(ii) $$\| N^{k/2}\varphi \|_{L^p(\mathcal{K})} \lesssim \| \nabla^k\varphi \|_{L^p(\mathcal{K}\otimes\mathcal{H}^k)}, \quad \varphi\in\mathcal{P}_k(\mathcal{K}), \qquad (7.63)$$

where $\mathcal{H}^k = \mathcal{H}^{\otimes k}$.

From Theorem 7.32, Theorem 7.28, the product rule (5.80) for N, and induction one derives now one of the two main results of this chapter (cf. Meyer (1983) and Sugita (1985a)).

Theorem 7.33. *Let* $\varphi, \psi \in \mathfrak{D}$, $p \in (1,+\infty)$, $k \in \mathbb{N}$. *If* $q^{-1}+r^{-1} = p^{-1}$, $1<q, r<+\infty$, *then*

$$|||\, \varphi\psi \,|||_{p,k} \lesssim |||\, \varphi \,|||_{q,k} \, |||\, \psi \,|||_{r,k}. \tag{7.64}$$

In particular, $\mathfrak{D}$ *is an algebra with continuous pointwise multiplication.*

Remark. If φ, ψ in Theorem 7.33 take values in $\mathcal{H}$, then the statement holds when pointwise products are replaced by tensor products of elements in $\mathcal{H}$.

Finally we turn to a sketch of the theorem of Watanabe (1983) on the composition of tempered distributions with elements in $\mathfrak{D}$. We follow the proof in Watanabe (1983) quite closely.

Suppose that $T \in \mathcal{S}'(\mathbb{R})$. Let H be the self-adjoint operator that we used in Chapters 3 and 4 (cf. (3.28)). As before, we consider $H^m T$, $m \in \mathbb{Z}$, as being defined by duality $\langle H^m T, \xi \rangle = \langle T, H^m \xi \rangle$ (which makes sense since H^m maps $\mathcal{S}(\mathbb{R})$ continuously 2into itself). Because the family of Schwartz space seminorms is equivalent to the family of norms given by $\{|\, H^m \cdot\, |_2, m \in \mathbb{N}_0\}$ (cf. Appendix A.5), it follows that for every $p \in \mathbb{N}$ there exists $n \in \mathbb{N}$, with $H^{-n} T \in \mathcal{S}_p(\mathbb{R})$. (For a definition of the spaces $\mathcal{S}_p(\mathbb{R})$ we refer to Appendix A.5.) Thus if $(T_k, k \in \mathbb{N})$ is a sequence in $\mathcal{S}(\mathbb{R})$ converging strongly to T in $\mathcal{S}'(\mathbb{R})$, then $H^{-n} T_k$ converges to $H^{-n} T$ in $\mathcal{S}_p(\mathbb{R})$ (cf. Proposition A.5.1). In particular, for some $n \in \mathbb{N}$, $H^{-n} T_k$ converges pointwise uniformly to $H^{-n} T$:

$$\sup_{u \in \mathbb{R}} | H^{-n}T_k(u) - H^{-n}T(u) | \to 0, \quad k \to +\infty.$$

Let $\varphi \in \mathfrak{D}$, and assume for simplicity that φ is real. We want to prove that $T \circ \varphi \in \mathfrak{D}^*$, by which we mean that the sequence of well-defined random variables $(T_k \circ \varphi, k \in \mathbb{N})$ converges weakly to some element $T \circ \varphi$ in $\mathfrak{D}^*$. Since $\mathfrak{D}^*$ is weakly sequentially complete, it is sufficient to show that the sequence $(T_k \circ \varphi, k \in \mathbb{N})$ is weakly Cauchy in $\mathfrak{D}^*$. Write

$$T_k \circ \varphi = (H^n H^{-n} T_k) \circ \varphi \equiv (H^n \zeta_k) \circ \varphi,$$

with $\zeta_k \equiv H^{-n}T_k \in \mathcal{S}(\mathbb{R})$. Working out the action of H^n, we obtain

$$(H^n \zeta_k) \circ \varphi = \sum_{l,m} a_{l,m} \, \varphi^m \zeta_k^{(l)} \circ \varphi,$$

where the sum is over $l,m = 1,2, \ldots ,2n$, the $a_{l,m}$ are some numerical coefficients, and $\zeta_k^{(l)}$ is the l-th derivative of the function ζ_k. Let $\psi \in \mathfrak{D}$. Because $\mathfrak{D}$ is an algebra we have now

$$<T_j \circ \varphi - T_k \circ \varphi, \psi> = \sum_{l,m} a_{l,m} <\zeta_j^{(l)} \circ \varphi - \zeta_k^{(l)} \circ \varphi, \varphi^m \psi>.$$

Thus it is sufficient to show that for every $l = 1,2, \ldots ,2n$, $(\zeta_k^{(l)} \circ \varphi, k \in \mathbb{N})$ is weakly Cauchy in $\mathfrak{D}^*$. We shall only prove this for $l = 1$; — the general case will follow by the induction.

Consider $\zeta \in C_b^1(\mathbb{R})$, and $\varphi \in \mathfrak{D}$. By the chain rule for the gradient, we may write

$$\nabla \zeta \circ \varphi = \zeta' \circ \varphi \cdot \nabla \varphi.$$

If $| \nabla \varphi |_2 > 0$, μ–a.e., then we have, the following formula:

$$\zeta'\circ\varphi = |\nabla\varphi|_2^{-2}\,(\nabla\varphi, \nabla\zeta\circ\varphi)_{L^2(R)}, \quad \mu\text{-a.e.}$$

Now we use again the product rule (5.80) for the number operator N, to write this as

$$\zeta'\circ\varphi = \tfrac{1}{2}|\nabla\varphi|_2^{-2}\{(N\varphi)\zeta\circ\varphi + \varphi N\zeta\circ\varphi - N(\varphi\cdot\zeta\circ\varphi)\}. \qquad (7.65)$$

Apply formula (7.64) to our situation above, and assume in addition that $|\nabla\varphi|_2^{-2}$ belongs to $\mathfrak{D}$. Then we have for every $\psi\in\mathfrak{D}$,

$$<\zeta_j'\circ\varphi-\zeta_k'\circ\varphi, \psi> = \tfrac{1}{2}<\zeta_j\circ\varphi-\zeta_k\circ\varphi, \{(N\varphi) + N\varphi - \varphi N\}|\nabla\varphi|_2^{-2}\psi>.$$

Now we can estimate

$$|<\zeta_j'\circ\varphi-\zeta_k\circ\varphi, \psi>| \le \tfrac{1}{2}\sup_{u\in\mathbb{R}}|\zeta_j(u)-\zeta_k(u)|\,\|\{(N\varphi) + N\varphi - \varphi N\}|\nabla\varphi|_2^{-2}\psi\|_1.$$

But from above we know that the right hand side of the last inequality vanishes as j, k tend to infinity.

We remark that inequality (7.63) and the chain rule imply that if $\varphi\in\mathfrak{D}$, and $|\nabla\varphi|_2^{-1} > 0$ belongs to all (L^p), $p>1$, then also $|\nabla\varphi|_2^{-1}\in\mathfrak{D}$. Moreover, the above computation is readily generalized to the multidimensional case: Namely for a d-tuple $(\varphi_1, \ldots, \varphi_d)$ of random variables φ_j, $j = 1,2, \ldots ,d$, in $\mathfrak{D}$, such that the so-called *Malliavin covariance matrix* $(\sigma_{ij}$, $i,j = 1,2, \ldots ,d)$ given by

$$\sigma_{ij} = (\nabla\varphi_i, \nabla\varphi_j)_{L^2(R)}, \quad i,j = 1,2, \ldots ,d,$$

is μ-a.e. strictly positive, with an inverse whose entries belong to all (L^p), $p>1$. Then the theorem of Watanabe (1983) can be stated as follows.

Theorem 7.34. *Assume that* $\varphi_1, \ldots, \varphi_d \in \mathfrak{D}$, *so that the matrix* (σ_{ij}) *is* μ–a.e. *strictly positive, with an inverse whose entries belong to all* (L^p), $p>1$. *Denote* $\varphi = (\varphi_1, \cdots, \varphi_d)$. *If* $T \in \mathcal{S}'(\mathbb{R}^d)$, *then* $T \circ \varphi \in \mathfrak{D}^*$.

Remark. With a little more care in the above computation (cf. Sugita (1985a) or Watanabe (1983, 1984)), one can prove that actually the sequence $(T_k \circ \varphi, k \in \mathbb{N})$ converges strongly in $\mathfrak{D}^*$, i.e., in one of the Banach spaces $(\mathfrak{D}^{p,k})^*$, for some $p>1$, $k \in \mathbb{N}_0$.

We conclude this chapter with the remark that Watanabe's Theorem 7.34 gives an alternative proof of the theorem of Malliavin (1978), which states that under the assumptions on φ in Theorem 7.34, the law of φ on $\mathbb{R}^d$ has a smooth density with respect to Lebesgue measure: one only has to choose T as the Dirac distribution δ_u at $u \in \mathbb{R}^d$, to obtain $<\delta_u \circ \varphi, 1>$ as the density. (For more details we refer the reader to Sugita (1985a) and Watanabe (1983, 1984).)

8. Stochastic Integration

8.A Introduction

In the early to mid-seventies Hitsuda (1972), King (1975) and Skorokhod (1975) have formulated new approaches to construct and extend Itô's stochastic integral with respect to the paths of Brownian motion (for excellent accounts on the "traditional" approach we refer for example to McKean (1969), and Ikeda and Watanabe (1981)). The methods of these new approaches were based on the Wiener–Itô decomposition, and on the use of a "creation operator". In the early eighties it was observed independently by Gaveau and Trauber (1982), and Kubo and Takenaka (1981) that this "creation operator" is given by ∂_t^* – i.e., the adjoint of the gradient ∇ (cf. Chapter 5). Actually, $\int_0^T dt\ \partial_t^* \cdot$ is a divergence operator in the sense of Krée and Krée (1983) (cf. also Shigekawa (1980) and Sugita (1985a)). This observation became the basis of a mathematical investigation of this formulation of Itô's integral. It has been exploited in a large numbers of recent publications, especially for the stochastic analysis of anticipating processes. Without aiming at a complete list of references, we want to mention here the following articles: Asch and Potthoff (1991), Buckdahn (1989), Kuo and Russek (1988), Nualart and Pardoux (1988), Nualart and Zakai (1986), Ocone (1984), Ustunel (1986).

Let us start with a simple computation based on the rules of Chapter 5. Let $B(t) = \langle \cdot\ , 1_{[0,t]} \rangle$ be our standard version of Brownian motion. As usual, $: \langle \cdot, 1_{[0,t]} \rangle^n : \ = \ : B(t)^n :$ is the n-th Hermite polynomial of B(t) with parameter $t>0$, i.e.

$$: B(t)^n : = \frac{d^n}{d\lambda^n} \exp(\lambda B(t) - \lambda^2/2\ t) \Big|_{\lambda=0}, \quad n\in\mathbb{R},\ t>0,$$

(cf. Appendix A.1). Let us compute

$$\int_0^1 \partial_t^* : B(t)^n : dt,$$

where — for the moment — we refrain from giving the integral a rigorous sense.

The $\mathcal{S}$-transform of this expression at $\xi \in \mathcal{S}(\mathbb{R})$ (interchanging the integrals without justification here) is given by

$$\int_0^1 \xi(t)\,(\xi, 1_{[0,t)})^n dt = \int_0^1 \xi(t)\,(\int_0^t \xi(u)\,du)^n dt$$

$$= \frac{1}{n+1}\,(\int_0^1 \xi(t)\,dt)^{n+1},$$

which is the $\mathcal{S}$-transform of $\frac{1}{n+1}$: $B(1)^{n+1}$: at ξ. By the Wiener–Itô theorem we therefore have found the formula

$$\int_0^1 \partial_t^* : B(t)^n : dt = \frac{1}{n+1} : B(1)^{n+1} :, \quad n \in \mathbb{N}.$$

For example, if n = 1, we obtain

$$\int_0^1 \partial_t^* B(t)\,dt = \tfrac{1}{2} : B(1)^2 : = \tfrac{1}{2}(B(1)^2 - 1).$$

It is well-known (e.g., McKean (1969)) that

$$\int_0^1 : B(t)^n : dB(t) = \frac{1}{n+1} : B(1)^{n+1} :,$$

where the dB(t)–integral is Itô's stochastic integral. Thus the Itô integral of $: B(t)^n :$ and its $\int_0^1 dt\, \partial_t^*$ — integral coincide.

Let us redo the above computation in a slightly informal way without using the $\mathcal{S}$ - transform:

$$:B(t)^n:(x) = :<x\,,\,1_{[0,t)}>^n:$$

$$= <:x^{\otimes n}:\,,\,1_{[0,t)}{}^{\otimes n}>.$$

Thus

$$\partial_t^* :B(t)^n:(x) = <:x^{\otimes(n+1)}:\,,\,\delta_t\hat{\otimes}\,(1_{[0,t]})^{\otimes n}>,$$

and therefore

$$\int_a^b \partial_t^*:B(t)^n:(x)\,dt = \int_a^b \Big(\int_a^t \cdots \int_a^t :x(t)\,x(u_1)\cdots x(u_n):du_n\cdots du_1 dt$$

$$= \frac{1}{n+1}\int_a^b \cdots \int_a^b :x(u_1)\cdots x(u_{n+1}):du_{n+1}\cdots du_1$$

$$= \frac{1}{n+1} <:x^{\otimes(n+1)}:\,,\,1_{[a,b]}{}^{\otimes(n+1)}>$$

$$= \frac{1}{n+1}(:B(b)^{n+1}: - :B(a)^{n+1}:),$$

where we treated x informally as if it were a continuous function. In the sequel we shall use often the last way of computing, when its rigorous interpretation (typically through the $\mathcal{S}$ - transform) is obvious.

After this heuristic motivation, let us give a precise meaning to the integral used above.

Let $(X, \mathfrak{B}, m)$ be a σ–finite measure space. Assume that that Ψ: X $\rightarrow$ $(\mathcal{S})^*$ is weakly in $L^1(X, \mathfrak{B}, dm)$. I.e., we assume that for every $\varphi\in(\mathcal{S})$, the complex valued function $x \mapsto <\Psi(x), \varphi>$ belongs to $L^1(X, \mathfrak{B}, dm)$. Then we have the following result.

Proposition 8.1. *The mapping* $F: (\mathcal{S}) \to \mathbb{R}$, *given by*

$$<F, \varphi> = \int_X <\Psi(x), \varphi> dm(x), \ \varphi \in (\mathcal{S}),$$

belongs to $(\mathcal{S})^*$.

Proof. We follow closely the argument given in Hille and Phillipps (1957) for Banach space valued functions on X. The mapping F is obviously everywhere defined and linear. Define a linear mapping W from $(\mathcal{S})$ into $L^1(X, \mathfrak{B}, dm)$ as follows:

$$W : \varphi \mapsto <\Psi(\cdot), \varphi>.$$

We prove now that W is closed. Assume that $\varphi_n \to \varphi$ in $(\mathcal{S})$, and that $W(\varphi_n) \to V$ in $L^1(X, \mathfrak{B}, dm)$, as n tends to infinity. Then we have for m–a.e. $x \in X$ that $<\Psi(x), \varphi_n - \varphi> \to 0$. Moreover, $W(\varphi_n) \to V$ implies that a subsequence converges pointwise m–a.e. to V. Therefore this subsequence converges a.e. to $W(\varphi)$. Thus $V = W(\varphi)$ m–a.e. and W is closed. Since $(\mathcal{S})$ is a Fréchet space, closedness of W is equivalent to continuity. Thus there is a constant $C>0$ and $p \in \mathbb{N}$ so that

$$\| W(\varphi) \|_1 \leq C \| \varphi \|_{2,p}.$$

Therefore

$$| F(\varphi) | \leq \| W(\varphi) \|_1 \leq C \| \varphi \|_{2,p},$$

which shows that F is continuous on $(\mathcal{S})$. □

From now on we mean by $\int_X \Psi(x)\, dm(x)$ the *Pettis–integral* of Ψ, i.e., the unique element $F \in (\mathcal{S})^*$ constructed above:

$$<\int_X \Psi(x)\, dm(x), \varphi> = \int_X <\Psi(x), \varphi>\, dm(x),$$

for all $\varphi \in (\mathcal{S})$. Since the $\mathcal{S}$–transform of $\Psi(t)$ is given by pairing with $:e^{<\cdot\,,\,\xi>}:$, $\xi \in \mathcal{S}(\mathbb{R})$, we find immediately that $\mathcal{S}$–transform and Pettis integral of Ψ commute:

$$\mathcal{S}\Big(\int_X \Psi(x)\, dm(x)\Big)(\xi) = \int_X \mathcal{S}(\Psi(x))(\xi)\, dm(x). \tag{8.1}$$

From now on we shall assume for convenience that X is a bounded interval T, equipped with the Borel sets and Lebesgue measure dt. Moreover, we suppose that all random variables and processes are real valued unless mentioned otherwise.

Consider an a.e. defined mapping $\Phi : T \to (\mathcal{S})^*$. Because ∂_t^* maps $(\mathcal{S})^*$ into itself (cf. Chapter 5), we know that for a.e. $t \in T$, $t \mapsto \partial_t^* \Phi(t)$ is a well defined mapping from T into $(\mathcal{S})^*$. If this mapping is weakly in $L^1(T, dt)$, we have a well defined Pettis–integral of $\partial_t^* \Phi(t)$.

Definition 8.2. *Assume that* $\Phi : T \to (\mathcal{S})^*$ *is such that* $t \mapsto \partial_t^* \Phi(t)$ *is weakly in* $L^1(T, dt)$. *Then we call* $\int_T \partial_t^* \Phi(t)\, dt \in (\mathcal{S})^*$ *the* Hitsuda–Skorokhod integral of Φ.

A convenient criterion for the existence of the Hitsuda–Skorokhod integral is given in the following result.

Proposition 8.3. *Assume that there exists* $p \in \mathbb{N}$ *so that for* a.e. $t \in T$, $\Phi(t) \in (\mathcal{S})_{-p}$. *Suppose furthermore that* $t \mapsto \Phi(t)$ *is weakly continuous as a mapping from* T *into* $(\mathcal{S})_{-p}$ *and that* $t \mapsto \|\Phi(t)\|_{2,-p}$ *is integrable. Then the Hitsuda–Skorokhod integral exists in* $(\mathcal{S})^*$.

Proof. Since for a.e. $t \in T$, $<\partial_t^* \Phi(t), \varphi> = <\Phi(t), \partial_t \varphi>$, and $t \mapsto \partial_t \varphi$ belongs to $\mathcal{S}(\mathbb{R}) \otimes (\mathcal{S})$ (cf. Theorem 5.13) for all $\varphi \in (\mathcal{S})$, it is clear that weak continuity of Φ implies that $t \mapsto <\Phi(t), \partial_t \varphi>$ is continuous from T into $\mathbb{R}$. Thus this map is measurable and therefore $t \mapsto \partial_t^* \Phi(t)$ is weakly measurable. By Pettis' theorem (cf. e.g., Hille and Phillips (1957)) this implies that this mapping is strongly measurable. Furthermore the following estimate shows that it is weakly in $L^1(T, dt)$:

$$| <\partial_t^* \Phi(t), \varphi> | \leq \|\partial_t \varphi\|_{2,p} \|\Phi(t)\|_{2,-p} \leq \text{const.} \|\varphi\|_{2,p} \|\Phi(t)\|_{2,-p}$$

(cf. Corollary 5.12). □

Remark. Actually, the hypothesis of Proposition 8.3 implies that $t \mapsto \partial_t^* \Phi(t)$ is Bochner integrable, and its Bochner and Pettis integrals coincide (cf. Hille and Phillips (1957)).

The following is a direct consequence of (8.1) and the rules of Chapter 5.

Proposition 8.4. *If* $\Phi : T \to (\mathcal{S})^*$ *is such that* $t \mapsto \partial_t^* \Phi(t)$ *is weakly in* $L^1(T, dt)$ *then*

$$\mathcal{S}\Big(\int_T \partial_t^* \Phi(t)\, dt\Big)(\xi) = \int_T \xi(t)\, \mathcal{S}(\Phi(t))(\xi)\, dt, \tag{8.2}$$

holds for all $\xi \in \mathcal{S}(\mathbb{R})$.

We leave it as an exercise to the interested reader to reconsider the computations at the beginning of this section, and to show their validity within the framework given above.

The remainder of this chapter is organized as follows. In Section 8.B we shall consider integrals of Itô type and some of their properties in the above described approach. In Section 8.C we give an account of the theory of stochastic integrals with anticipating integrands. Properties of such integrals will be given in Section 8.D. In particular, we prove there Itô's Lemma for the forward integrals of Section 8.C. Finally, in Section 8.E we treat a class of stochastic differential equations with anticipating coefficients.

8.B Integrals of Itô Type

In this section we shall develop the motivating example at the beginning of Section 8.A to the statement, that for non-anticipating processes the Hitsuda-Skorokhod integral and the Itô integral coincide. Furthermore, we shall compute the variance of such integrals, and prove Itô's formula for the composition of a tempered distribution and Brownian motion. Throughout, we shall choose without loss of generality $T=[0,1]$.

For $t\in[0,1]$, let $\mathcal{F}_t$ denote the σ-algebra generated by the random variables $\{B(s);\ 0\leq s\leq t\}$, and set $\mathcal{F}_1=\mathcal{F}$. As usual, a process $X(t)$, $0\leq t\leq 1$, is called *non-anticipating* (or *adapted to* $\mathcal{F}_t$) if and only if $X(t)$ is measurable with respect to $\mathcal{F}_t$. If $t\mapsto X(t)$ is (L^2)–valued with chaos decomposition given by $(f^{(n)}(t);\ n\in\mathbb{N}_0)$, $f^{(n)}(t)\in L^2(\hat{\mathbb{R}^n})$, then it is easy to see that the condition of being non-anticipating implies that for all n and a.e. $u\in([0,t]^n)^c$, $f^{(n)}(t;\ u)=0$.

Recall that $\mathcal{D}^{p,k}$, $p\geq 1$, $k\in\mathbb{N}_0$, denotes the (completed) (L^p)–domain of $(1+N)^{k/2}$, where N is the number operator (cf. Chapter 3 and Chapter 7).

The next result can for example be found in Kuo and Russek (1988), Nualart and Zakai (1986), or Ocone (1984).

Theorem 8.5. *Assume that* $X\in L^2([0,1];\ (L^2))$ *is non–anticipating. Then its Hitsuda–Skorokhod integral belongs to* (L^2), *and*

$$\left\| \int_0^1 \partial_t^* X(t)\ dt \right\|_2^2 = \int_0^1 \| X(t) \|_2^2\ dt. \tag{8.3}$$

The proof of Theorem 8.5 relies on the following result (cf. Kuo and Russek (1988), Nualart and Zakai (1986), Ocone (1984)).

Proposition 8.6. *Assume that* $X \in L^2([0,1]; \mathfrak{D}^{2,1})$. *Then the Hitsuda–Skorokhod integral belongs to* (L^2) *and*

$$\| \int_0^1 \partial_t^* X(t)\, dt \|_2^2 = \int_0^1 \| X(t) \|_2^2\, dt + \int_0^1\int_0^1 (\partial_t X(s), \partial_s X(t))\, ds\, dt. \tag{8.4}$$

Proof. Let $t \mapsto X(t)$ be a mapping from $[0,1]$ into $\mathfrak{P}$, which is continuous with respect to the $\mathfrak{D}^{2,1}$ - norm. Then we can compute as follows:

$$\| \int_0^1 \partial_t^* X(t)\, dt \|_2^2 = \int_0^1\int_0^1 (\partial_t^* X(t), \partial_s^* X(s))\, dt\, ds$$

$$= \int_0^1\int_0^1 (\partial_s \partial_t^* X(t), X(s))\, dt\, ds$$

$$= \int_0^1\int_0^1 (\partial_t^* \partial_s X(t), X(s))\, dt\, ds + \int_0^1 (X(t), X(t))\, dt$$

$$= \int_0^1 \| X(t) \|_2^2\, dt + \int_0^1\int_0^1 (\partial_t X(s), \partial_s X(t))\, dt\, ds,$$

where we used the CCR (5.52). Since

$$| \int_0^1\int_0^1 (\partial_t X(s), \partial_s X(t))\, dt\, ds \,| \le \int_0^1\int_0^1 \| \partial_s X(t) \|_2^2\, ds\, dt$$

$$\le \int_0^1 \| N^{1/2} X(t) \|_2^2,$$

it is clear that by approximating a general element in $L^2([0,1]; \mathfrak{D}^{2,1})$ by $\mathfrak{D}^{2,1}$ -

continuous mappings from [0,1] into $\mathcal{P}$ we have proved the statement of the proposition. □

Proof of Theorem 8.5. The fact that $t \mapsto X(t)$ is non-anticipating implies that for a.e. $(s,t)\in[0,1]^2$ with $s>t$, $\partial_s X(t) = 0$ (μ-a.e.). Thus the second term in (8.4) vanishes for non-anticipating X in $L^2([0,1]; \mathcal{D}^{2,1})$. Since a general element in $L^2([0,1]; (L^2))$ can be approximated by elements in $L^2([0,1]; \mathcal{D}^{2,1})$, we see that this is also true for $X\in L^2([0,1]; (L^2))$. □

The next result can be found in Gaveau and Trauber (1982), Kubo and Takenaka (1981).

Theorem 8.7. *If* $X\in L^2([0,1]; (L^2))$ *is non-anticipating, then its Hitsuda-Skorokhod integral and its Itô integral coincide:*

$$\int_0^1 X(t)\, dB(t) = \int_0^1 \partial_t^* X(t)\, dt.$$

Proof. It suffices to show the statement for a process X with only one homogeneous component, say $X^{(n)}(t)$: the general statement is then implied by summing over $n\in\mathbb{N}_0$. Assume that $X^{(n)}(t)$ is the multiple Wiener integral of $f^{(n)}\in L^2([0,1]; \hat{L}^2(\mathbb{R}^n))$. We have by (8.2)

$$\mathcal{S}\Big(\int_0^1 \partial_t^* X^{(n)}(t)\, dt\Big)(\xi) = \int_0^1 \xi(t)\Big(\int_{\mathbb{R}^n} f^{(n)}(t;u)\, \xi^{\otimes n}(u)\, du\Big)\, dt$$

$$= \int_0^1 \xi(t)\Big(\int_{[0,t]^n} f^{(n)}(t;u)\, \xi^{\otimes n}(u)\, du\Big)\, dt,$$

where we used the fact that X is non-anticipating. Let $\hat{f}^{(n)}$ be the symmetrization of $f^{(n)}$ considered as a function of n+1 variables:

$$\hat{f^{(n)}}(u_1, \dots, u_{n+1}) = \frac{1}{n+1} \sum_{k=1}^{n+1} f^{(n)}(u_k; u_1, \dots, \not{u_k}, \dots, u_{n+1}).$$

Then

$$\mathcal{S}\Big(\int_0^1 \partial_t^* X^{(n)}(t)\, dt\Big)(\xi) =$$

$$= \int_0^1 \xi(t)\, n! \int_{0 \le u_1 \le \dots \le u_n \le t} f^{(n)}(t; u_1, \dots, u_n)\, \xi^{\otimes n}(u)\, du\, dt$$

$$= \frac{(n+1)!}{n+1} \int_{0 \le u_1 \le \dots \le u_{n+1} \le 1} \hat{f^{(n)}}(u)\, \xi^{\otimes(n+1)}(u)\, du$$

$$= \frac{1}{n+1} \int_{[0,1]^{n+1}} \hat{f^{(n)}}(u)\, \xi^{\otimes(n+1)}(u)\, du.$$

Therefore

$$\Big(\int_0^1 \partial_t^* X^{(n)}(t)\, dt\Big)(x) = \frac{1}{n+1} \langle :x^{\otimes(n+1)}:, \hat{f^{(n)}} \rangle,$$

and according to the results in Chapter 2 (cf. eq. (2.46)), the last expression is equal to the n-fold Wiener integral of $\hat{f^{(n)}}$ divided by (n+1) (at $x \in \mathcal{S}'(\mathbb{R})$):

$$\int_0^1 \partial_t^* X^{(n)}(t)\, dt = \frac{1}{n+1} \int_{[0,1]^{n+1}} \hat{f^{(n)}}(u_1, \dots, u_{n+1})\, dB(u_1) \cdots dB(u_{n+1}).$$

On the other hand, by similar reasoning we have

$$\int_0^1 X(t)\, dB(t) = \int_0^1 \Big(\int_{[0,t]^n} f^{(n)}(t; u)\, dB(u_1) \cdots dB(u_n)\Big)\, dB(t)$$

$$= \quad n! \int\limits_{0 \le u_1 \le \cdots \le u_{n+1} \le 1} \hat{f^{(n)}}(u_1, \ldots, u_{n+1})\, dB(u_1) \cdots dB(u_{n+1})$$

$$= \frac{1}{n+1} \int\limits_{[0,1]^{n+1}} \hat{f^{(n)}}(u_1, \ldots, u_{n+1})\, dB(u_1) \cdots dB(u_{n+1}).$$

□

We conclude this section with a proof of Itô's formula. Due to our general framework (as described in Section 8.A) we may consider compositions of tempered distributions with Brownian motion. In this context this has already been done by Kubo (1983). Subsequently, such compositions (actually of tempered distributions in $\mathcal{S}'(\mathbb{R}^n)$ with certain elements in $\mathcal{D}$, cf. Theorem 7.34) have been discussed by Watanabe (1983, 1984), Sugita (1985), Potthoff (1988a,b) and others. Itô's formula in this case is due to Kubo (1983). We shall give a different proof here.

Note that for $t>0$, $B(t)$ satisfies the hypothesis of Theorem 7.34. In particular, we have that $\sigma = |\nabla B(t)|_2 = |1_{[0,t)}|_2 = \sqrt{t}>0$. Thus σ is invertible and belongs to $\mathcal{D}$ (trivially). Thus Watanabe's theorem applies, and we have that for all $t>0$, $T \circ B(t) \in \mathcal{D}^* \subset (\mathcal{S})^*$. Another way to see this is to prove the convergence of the series expansion of $T \circ B(t)$ found in Example (3.4) in $(\mathcal{S})^*$, which is an easy exercise. Still another proof can be based on Theorem 4.38. It is also not very hard to see, that from any of the above proofs that $T \circ B(t) \in (\mathcal{S})^*$, $t>0$, it follows that $T \circ B(t)$ satisfies the hypothesis of Proposition 7.3, for intervals $T = [a,b]$ with $a>0$. Thus the Hitsuda–Skorokhod integral of $T \circ B(t)$ over such T belongs to $(\mathcal{S})^*$, too.

Theorem 8.8. *Let* $T \in \mathcal{S}'(\mathbb{R})$. *Then for all* s,t *so that* $0<s\le t<+\infty$,

$$T \circ B(t) - T \circ B(s) = \int_s^t \partial_u^*\, T' \circ B(u)\, du + \tfrac{1}{2} \int_s^t T'' \circ B(u)\, du, \qquad (8.5)$$

where $T^{(k)}$ *denotes the* k-*th distributional derivative of* T.

Proof. First we show Itô's formula for the function $T(y) = e^{iky}$, $y,k \in \mathbb{R}$. The general case follows then by Fourier transformation and a limiting argument. For this choice of T we have

$$\int_s^t \partial_u^* T' \circ B(u)\, du = ik \int_s^t \partial_u^*\, e^{ikB(u)}\, du$$

$$= ik \int_s^t \partial_u^* : e^{ikB(u)} : e^{-\frac{k^2}{2} u}\, du.$$

By (8.2) and integration by parts we find

$$\mathcal{S}\Big(\int_s^t \partial_u^* T' \circ B(u)\, du\Big)(\xi) =$$

$$= ik \int_s^t \xi(u) \exp(ik \int_0^u \xi(v)\, dv)\, e^{-\frac{k^2}{2} u}\, du$$

$$= \exp(ik \int_0^t \xi(v)\, dv - \frac{k^2}{2} t) - \exp(ik \int_0^s \xi(v)\, dv - \frac{k^2}{2} s) +$$

$$+ \tfrac{1}{2} k^2 \int_s^t \exp(ik \int_0^u \xi(v)\, dv - \frac{k^2}{2} u)\, du.$$

But the last expression is just the $\mathcal{S}$-transform of

$$e^{ikB(t)} - e^{ikB(s)} + \tfrac{1}{2} k^2 \int_s^t e^{ikB(u)}\, du$$

at ξ. Thus we have shown eq. (8.5) in this special case.

Next let $T \in \mathcal{S}(\mathbb{R})$. Write $T(y) = (2\pi)^{-1/2} \int e^{iky}\, \tilde{T}(k)\, dk$, where $\tilde{T}$ is the Fourier transform of T. Because we have formula (8.2) for the $\mathcal{S}$-transform of the

Hitsuda-Skorokhod integral, and since on the right hand side of (8.2) we have a standard Lebesgue integral, we may interchange the orders of integration and use the preceding formula with the result that (8.5) remains true in this case. Finally, approximating $T \in \mathcal{S}'(\mathbb{R})$ by functions in $\mathcal{S}(\mathbb{R})$ and using, e.g. the arguments in the proof of Theorem 7.3, the formula (8.5) remains true in the limit. □

8.C Stochastic Integrals with Anticipating Integrands

During the recent years there has been an increasing interest to define and study stochastic integrals with integrands which anticipate the future. We shall consider such integrals with respect to Brownian motion in this section. Applications of such integrals arise in the theory of filtering, in the investigation of stochastic flows and in other fields of stochastic analysis. Basically there have been two approaches: enlargement of filtration (cf. for example Itô (1978b), Jeulin (1981) and the literature quoted there), and various ways to use the chaos decomposition and "creation operators". The latter approach goes back to Hitsuda (1972), King (1975) and Skorokhod (1975). It is based on the observations in Section 8.A.

In this section we shall introduce a kind of stochastic integral which is different from the Hitsuda–Skorokhod integral. This stochastic integral is motivated directly by our white noise approach. We shall see in the next section that this integral has a number of appealing properties.

Throughout this section, we shall confine ourselves to discuss stochastic integrals of ordinary stochastic processes (i.e., with values in (L^2), in contrast to the generalized processes considered in Section 8.A and 8.B). We begin by recalling the Hitsuda–Skorokhod integral mentioned in the previous section. For an account of the theory of the Hitsuda–Skorokhod integral for (L^2)–valued processes, we refer the interested reader also to the article by Nualart and Pardoux (1988). Without loss of generality we shall consider processes X over the interval [0,1].

From Proposition 8.6 we know that for $X \in L^2([0,1]; \mathfrak{D}^{2,1})$, i.e., such that

$$\int_0^1 \| (1+N)^{1/2} X(t) \|_2^2 \, dt < \infty,$$

the Hitsuda-Skorokhod integral $\int_0^1 \partial_t^* X(t)\, dt$ belongs to (L^2), and that

$$\| \int_0^1 \partial_t^* X(t)\, dt \|_2 \leq \text{const.} \Big(\int_0^1 \| (1+N)^{1/2} X(t) \|_2^2 \, dt \Big)^{1/2}.$$

Later it will be helpful to have a criterion for whether the Hitsuda–Skorokhod integral of X belongs to (L^p), $p \geq 2$. This can be achieved by the application of Meyer's inequalities (cf. Meyer (1983), Potthoff (1988a,b), Sugita (1985), Watanabe (1984)). The following result can be found in Asch and Potthoff (1991), and in a slightly sharper form in Nualart and Pardoux (1988). We shall give its proof for a special case ("k=1") only; – the general case is left as an exercise to the interested reader. We recall from Chapter 3 and Chapter 7, that $||| \cdot |||_{p,k}$ denotes the norm given by $\| (1+N)^{k/2} \cdot \|_p$, where the latter is the norm of (L^p). $\mathfrak{D}^{p,k}$ is the completion of $\mathfrak{P}$ under the norm $||| \cdot |||_{p,k}$.

Proposition 8.9. *Assume that* $X \in L^2([0,1]; \mathfrak{D}^{p,k})$, $p>1$, $k \geq 1$. *Then*

$$\int_0^1 \partial_t^* X(t)\, dt \in \mathfrak{D}^{p,k-1},$$

and

$$\| \int_0^1 \partial_t^* X(t)\, dt \|_{p,k-1} \leq C_{p,k} \Big(\int_0^1 \| X(t) \|_{p,k}^2 \, dt \Big)^{1/2},$$

where $C_{p,k} > 0$ *is a constant depending on* p *and* k *only.*

To prove Proposition 8.9 we first show the following result.

Lemma 8.10. *Assume that* $X \in L^q([0,1]; (L^p))$, $1 \le p \le q$. *Then* $|X|_q \in (L^p)$ *and*

$$\| \, |X|_q \, \|_p \le \Big(\int_0^1 \| X(s) \|_p^q \, ds \Big)^{1/q}.$$

Proof. Consider the mapping

$$\varphi \mapsto \| \, |\varphi|^{1/q} \, \|_p^q$$

from (L^∞) into $\mathbb{R}$. Since $\| \, |\varphi|^{1/q} \, \|_p^q = \| \varphi \|_{p/q}$, this mapping is convex. Therefore Jensen's inequality implies that for X: $[0,1] \to (L^\infty)$, which is strongly measurable and bounded, we have

$$\| \, | \int_0^1 | X(t) |^q dt \, |^{1/q} \, \|_p^q \le \int_0^1 \| X(t) \|_p^q \, ds.$$

The general case $X \in L^q([0,1]; (L^p))$ follows by taking limits. □

Proof of Proposition 8.9 for k=1. Let $\mathcal{E}$ denote the algebra of functions on $\mathcal{S}'(\mathbb{R})$ generated by the functions $\sin \circ X_\xi$, and $\cos \circ X_\xi$, for $\xi \in \mathcal{S}(\mathbb{R})$, and where $X_\xi(x) = \langle x, \xi \rangle$, $x \in \mathcal{S}'(\mathbb{R})$. Also, choose $r \ge 2$ so that $p^{-1} + r^{-1} = \frac{1}{2}$, and denote $Q_0 = 1 - P_0$, where $P_0 \varphi = \mathbb{E}(\varphi)$. We have

$$\| \int_0^1 \partial_t^* X(t) \, dt \, \|_p = \sup_\varphi | (\int_0^1 \partial_t^* \, X(t) \, dt, \varphi) |,$$

where the supremum is taken over all $\varphi \in \mathcal{E}$ with $\| \varphi \|_r = 1$. Thus

$$\| \int_0^1 \partial_t^* X(t) \, dt \, \|_p =$$

$$= \sup_\varphi | \int_0^1 (X(t), \partial_t \varphi) \, dt \, |$$

$$= \sup_\varphi | \int_0^1 ((1+N)^{1/2} X(t), (1+N)^{-1/2} \partial_t \varphi) \, dt \, |$$

$$= \sup_{\varphi} \left| \int_0^1 ((1+N)^{1/2} X(t), \partial_t N^{-1/2} Q_0 \varphi)\, dt \right|$$

$$= \sup_{\varphi} \left| \int \int_0^1 ((1+N)^{1/2} X(t))(x)\, (\partial_t N^{-1/2} Q_0 \varphi)(x)\, dt\, d\mu(x) \right|$$

$$\leq \sup_{\varphi} \int | (1+N)^{1/2} X |_2(x)\, | \nabla N^{-1/2} Q_0 \varphi |_2(x)\, d\mu(x)$$

$$\leq \sup_{\varphi} \| \, | (1+N)^{1/2} X |_2 \, \|_p \, \| \, | \nabla N^{-1/2} Q_0 \varphi |_2 \, \|_r .$$

By Meyer's inequalities (7.60) we have the bound

$$\| \, | \nabla N^{-1/2} Q_0 \varphi |_2 \, \|_r \leq C_p' \| Q_0 \varphi \|_r$$

$$\leq C_p \| \varphi \|_r ,$$

Thus by Lemma 8.11 we obtain the estimation

$$\| \int_0^1 \partial_t^* X(t)\, dt \|_p \leq C_p \| \, | (1+N)^{1/2} X(t) |_2 \, \|_p$$

$$\leq C_p \left(\int_0^1 \| (1+N)^{1/2} X(t) \|_p^2\, dt \right)^{1/2}$$

$$= C_p \left(\int_0^1 \| X(t) \|_{p,1}^2\, dt \right)^{1/2} .$$

□

The Hitsuda-Skorokhod integral of anticipating processes has some disadvantages as compared to the Itô integral of a non-anticipating process: it is not approximated by Riemann (forward) sums (however, cf. Nualart and Pardoux

(1988) for a modified Riemann approximation), and the Itô Lemma does not hold in its usual form.

On the other hand from the rules in Chapter 5 we know that for a random variable $\varphi \in \mathfrak{D}^{2,1}$ multiplication by Brownian motion B(t) is given by

$$B(t)\cdot\varphi = (D^*_{1_{[0,t]}} + D_{1_{[0,t]}})\,\varphi$$
$$= \int_0^1 (\partial_t^* + \partial_t)\,\varphi\,dt.$$

Thus from this point of view it is natural, to attempt to construct a stochastic integral with respect to Brownian motion as

$$\int_0^1 (\partial_t^* + \partial_t)\,X(t)\,dt,$$

adding the second term to the Hitsuda-Skorokhod integral. Integrals of this type have been first discussed by Berger and Mizel (1982), and by Kuo and Russek (1988).

Clearly, the additional term above is in general not well-defined. Let us look at the simplest example: X(t) = B(t). Note that both, $B_1(t) = \langle\cdot\,,\,1_{[0,t)}\rangle$ and $B_2 = \langle\cdot\,,\,1_{[0,t]}\rangle$ are versions of B(t). But according to the rules of Chapter 5 we would have

$$\partial_t B_1(t) = 1_{[0,t)}(t) = 0,$$

and

$$\partial_t B_2(t) = 1_{[0,t]}(t) = 1.$$

It was proposed in Berger and Mizel (1982) and in Kuo and Russek (1988) to use appropriate limits of $\partial_s X(t)$, as s approaches t, to give a well-defined meaning to this additional term. Actually, for both versions of Brownian motion we have

$$\lim_{s\downarrow t} \partial_s B_i(t) = 0, \ \ i{=}1{,}2,$$

and comparing with the results in Sections 8.A and 8.B, this is what we should have in order to recover the Itô integral of B(t).

For the treatment of these limits we shall follow here the formulation of Asch and Potthoff (1991).

Consider a process X in $L^2([0,1]; \mathfrak{D}^{p,1})$, $p\geq 2$. Note that this means that for a.e. $t\in[0,1]$, $X(t)\in\mathfrak{D}^{p,1}$. By Meyer's inequalities (7.60) we have (a.e.) $|\nabla X(t)|_2\in(L^p)$, and also

$$\| \int_0^1 | \nabla X(t) |_2^2 \, dt \|_p = \| | \nabla X |_{L^2(\mathbb{R})\otimes L^2([0,1])} \|_p$$

$$\leq \| | N^{1/2} X |_{L^2([0,1])} \|_p,$$

by Sugita's generalization of Meyer's inequalities to Hilbert space valued random variables, cf. Theorem 7.32 with $\mathcal{H} = L^2(\mathbb{R})$, $\mathcal{K} = L^2([0,1])$.

Consider ∇X as a mapping from $[0,1]^2$ into (L^p):

$$\nabla X : (s,t) \mapsto \nabla X(s,t) = \partial_s X(t).$$

Definition 8.11. X *is said to admit a* right derivative in $L^{q;p}$, $q\geq 1$, $p\geq 1$, *if and only if the following conditions hold:* X *has a version* $\tilde{X}$ *so that:*

(i) *for all sequences* $(\epsilon_n, n\in\mathbb{N})$ *with* $\epsilon_n\to 0$, $\epsilon_n>0$, *and for a.e.* $t\in[0,1]$, $\nabla\tilde{X}(t, t+\epsilon_n)$ *is Cauchy in* (L^p) *with a limit denoted by* $\partial_{t+}X(t)$;

(ii) *the mapping* $t \mapsto \partial_{t+}X(t)$ *from* $[0,1]$ *into* (L^p) *is well-defined: if* $\tilde{Y}$ *is another version yielding the limit* $\partial_{t+}Y(t)$, *then* $\partial_{t+}X(t)(x) = \partial_{t+}Y(t)(x)$, $dt\otimes d\mu$ *a.e.*;

(iii) *the mapping* $t \mapsto \partial_{t+}X(t)$ *is such that*

$$\| \left(\int_0^1 | \partial_{t+}X(t) |^q dt \right)^{1/q} \|_p < \infty.$$

Similarly, by taking sequences $(\epsilon_n, n\in\mathbb{N})$ with $\epsilon_n \to 0$, $\epsilon_n < 0$, we define $\partial_{t-}X(t)$, and say that X admits a *left derivative in* $L^{q;p}$ if the corresponding conditions hold.

In the sequel, we always assume that X is a process in $L^2([0,1]; \mathfrak{D}^{p,1})$, $p\geq 2$, admitting a right (or left or both) derivative in $L^{1;p}$.

Definition 8.12.

(i) $\int_0^1 (\partial_t^* + \partial_{t+})\, X(t)\, dt$ *is called the* forward integral of X, *and it will also be denoted by* $\int_0^1 X(t)\, dB(t+)$;

(ii) $\int_0^1 (\partial_t^* + \partial_{t-})\, X(t)\, dt$ *is called the* backward integral of X, *and it will also be denoted by* $\int_0^1 X(t)\, dB(t-)$;

(iii) $\frac{1}{2}\left(\int_0^1 X(t)\, dB(t+) + \int_0^1 X(t)\, dB(t-)\right)$ *is called the* (generalized) Stratonovich integral of X, *and it will also be denoted by* $\int_0^1 X(t) \circ dB(t)$.

Note that by our hypotheses, all the above integrals are random variables in (L^p).

Proposition 8.13. *If* X *is non–anticipating then*

$$\partial_{t+}X(t) = 0, \qquad dt\otimes d\mu \text{ - a.e.}$$

Proof. X being non-anticipating implies that for a.e. (t,s), $s\geq t$, and μ-a.e. $x\in\mathcal{S}'(\mathbb{R})$, $\partial_s X(t)(x) = 0$. In particular, X has a version $\tilde{X}$ so that for all t,s, with $s\geq t$, $\partial_s\tilde{X}(t) = 0$. Thus the statement is evident. □

From the last proposition and Theorem 8.7 it follows that for non-anticipating X, the forward, Hitsuda–Skorokhod, and Itô integrals coincide. Analogously one proves that the backward integral coincides with Itô's backward integral for a process which is backward measurable. Also, upon comparison with the results in Nualart and Pardoux (1988), one sees that our generalized Stratonovich integral coincides with theirs, and that for non-anticipating processes both are equal to the standard Stratonovich integral.

Next we shall construct a class of processes admitting right derivatives. An analogous construction is readily carried out to produce processes with a left derivative. The idea is due to Kuo and Russek (1988), however we shall follow the formulation of Asch and Potthoff (1991).

Consider the chaos expansion of X(t), $t\in[0,1]$: for a.e. $t\in[0,1]$, X(t) corresponds to a sequence $(f^{(n)}(t),\, n\in\mathbb{N}_0)$, $f^{(n)}(t)\in L^2(\mathbb{R}^n)\hat{}$. Moreover, we have

$$\int_0^1 \| X(t) \|_{2,1}^2 \, dt = \sum_{n=0}^{\infty} n!\,(1+n) \int_0^1 | f^{(n)}(t) |^2_{L^2(\mathbb{R}^n)} \, dt.$$

For $s\in\mathbb{R}^n$, we denote $f^{(n)}(t)(s)$ as $f^{(n)}(t;\, s)$, and

$$\Big(\int_0^1 | f^{(n)}(t;\,\cdot) |^2_{L^2(\mathbb{R}^n)} \, dt\Big)^{1/2} \equiv | f^{(n)} |_2 .$$

The basic observation in Kuo and Russek (1988) is the following.

In order to have $t \mapsto \partial_{t+}X(t)\in L^{2;2}$, we have to have for every $n\in\mathbb{N}$ a trace $t \mapsto f^{(n)}(t,\, t+,\cdot)$ so that

$$\int_0^1 | f^{(n)}(t;\, t+,\cdot) |^2_{L^2(\mathbb{R}^{n-1})}\, dt < \infty.$$

The existence of such a trace (from the right) can be provided by the Sobolev trace theorem (e.g., Adams (1975)). These remarks lead to the following construction (cf. also Asch and Potthoff (1991)).

Fix some $\epsilon > 0$. Let $n \in \mathbb{N}$, and denote the Cartesian coordinates of $s \in \mathbb{R}^n$ as $s = (s_1, \ldots, s_n)$. Define the open region

$$\Delta^n_{+\epsilon} = \{ (t;\, s) \in (0,1) \times \mathbb{R}^n;\ t < s_1 < t+\epsilon \}.$$

Note that $D^n = \{(t;\, s) \in [0,1] \times \mathbb{R}^n;\ t = s_1\}$ is a subset of the boundary of $\Delta^n_{+\epsilon}$. A point in D^n is denoted by $(t;\, s)$, $t \in [0,1]$, $s \in \mathbb{R}^{n-1}$. For $\alpha \in \mathbb{R}$, let $H_\alpha(\Delta^n_{+\epsilon})$ be the Sobolev space of order α over the region $\Delta^n_{+\epsilon}$, and let $| \cdot |_{2,\alpha}$ denote the norm given by

$$| f^{(n)} |^2_{2,\alpha} = | f^{(n)} |^2_2 + | 1_{\Delta^n_{+\epsilon}} f^{(n)} |^2_{H_\alpha(\Delta^n_{+\epsilon})}.$$

Denote by $H^{(n)}_\alpha$ the completed subspace of $L^2([0,1];\, L^2(\hat{\mathbb{R}}^n))$, consisting those elements so that $| f^{(n)} |_{2,\alpha}$ is finite.

For $f^{(n)} \in C^\infty(\overline{\Delta^n_{+\epsilon}})$ define

$$\begin{aligned} \gamma_+ f^{(n)}:\ D^n &\to \mathbb{R} \\ (s;\, t) &\mapsto f^{(n)}(t;\, t,\, s). \end{aligned}$$

Consider γ_+ as a mapping from $C^\infty(\overline{\Delta^n_{+\epsilon}})$ into $L^2(D^n)$, and note that $C^\infty(\overline{\Delta^n_{+\epsilon}})$ is dense in $H_\alpha(\Delta^n_{+\epsilon})$.

Lemma 8.14. *Let* $\alpha > \frac{1}{2}$. *Then* γ_+ *extends continuously to a mapping from* $H_\alpha(\Delta^n_{+\epsilon})$ *into* $L^2(D^n)$. *I.e., if* $f^{(n)} \in H^{(n)}_\alpha$ *then* $f^{(n)}$ *has a version (denoted by the* *same symbol) admitting a trace* $\gamma_+ f^{(n)}$ *on* D^n *with*

$$|\gamma_+ f^{(n)}|_{L^2(D^n)} \leq C\,|f^{(n)}|_{2,\alpha},$$

where the constant $C>0$ *is independent of* n. *Moreover,* $\gamma_+ f^{(n)}$ *is approximated from the interior of* $\Delta^n_{+\epsilon}$. *Namely, there exists a constant* $C'>0$, *independent of* n, *so that for all* ϵ', $0<\epsilon'<\epsilon$, *and a.e.* $t\in[0,1]$,

$$|\gamma_+ f^{(n)}(t;\cdot) - f^{(n)}(t;t+\epsilon',\cdot)|_{L^2(R^{n-1})} \leq C'\,\zeta_\alpha(\epsilon')^{1/2}|f^{(n)}|_{2,\alpha},$$

where

$$\zeta_\alpha(\epsilon') = \int_{\mathbb{R}} (1+u^2)^{-\alpha}\,|1-e^{i\epsilon' u}|^2\,du \to 0, \text{ as } \epsilon' \to 0.$$

The statement of this lemma follows almost directly from the results in Adams (1975), and therefore we omit a detailed proof. However, the following remarks are in order: below it will be crucial that the constants C and C' above are independent of $n\in\mathbb{N}$ (a fact which is not quite obvious in Adams (1975)). This can be shown by proving that for every $n\in\mathbb{N}$ the region $\Delta^n_{+\epsilon}$ admits a total extension operator in the sense of Adams (1975, p. 83) with a bound which is independent of n. In order to prove the last statement one deforms $\Delta^n_{+\epsilon}$ into a rectangular region, and then applies Seeley's reflection method as described in Adams (1975). Next one applies standard techniques from Fourier transformation to obtain the result of Lemma 8.17. The details are left to the interested reader.

Now we construct a space of processes of Fock type, having in each homogeneous component a space with the above described smoothness property. Note that under the $\mathcal{S}$-transform $L^2([0,1];\,(L^2))$ is isomorphic to $L^2([0,1];\,\hat{\mathfrak{F}})$, where $\hat{\mathfrak{F}}$ is the symmetric Fock space over $L^2(\mathbb{R})$. For $\alpha\geq 0$, define a subspace $L^2([0,1];\,\hat{\mathfrak{F}}_\alpha)$ of $L^2([0,1];\,\hat{\mathfrak{F}})$, consisting of those sequences $f = (f^{(n)},\ n\in\mathbb{N}_0)$ of mappings $f^{(n)}\colon [0,1] \to L^2(\hat{\mathbb{R}}^n)$ so that

$$\sum_{n=0}^{\infty} n!\,|f^{(n)}|^2_{2,\alpha} < \infty. \tag{8.6}$$

Denote $(H_\alpha) = \mathcal{S}^{-1}L^2([0,1]; \mathcal{F}_\alpha)$ and equip this space with the norm whose square is given by (8.6). Clearly, (H_α) is a Hilbert subspace of $L^2([0,1]; (L^2))$ and we denote its norm by $| X |_{2,\alpha}$: if $u\in(H_\alpha)$ is in one-to-one correspondence with the sequence $(f^{(n)}, n\in\mathbb{N}_0)$, then

$$| X |^2_{2,\alpha} = \sum_{n=0}^{\infty} n! \, | f^{(n)} |^2_{2,\alpha}.$$

The following result is due to Kuo and Russek (1988) (cf. also Asch and Potthoff (1991)).

Proposition 8.15. *Assume that* X *is a process on* [0,1] *so that* $(1+N)^{1/2}X\in(H_\alpha)$, $\alpha>\frac{1}{2}$. *Then* X *has a right derivative in* $L^{2;2}$. *Moreover, the forward integral of* X *exists and belongs to* (L^2).

Proof. By hypothesis, for a.e. $t\in[0,1]$, $X(t)\in\mathcal{D}^{2,1}$, so that a priori $\nabla X(t)\in L^2(\mathbb{R}; (L^2))$, with $\nabla X(t)(s) = \partial_s X(t)$. Moreover, $\partial_s X(t)$ has chaos expansion given by $(n\, f^{(n)}(t;\, s,\, \cdot);\, n\in\mathbb{N})$ if the chaos expansion of X(t) is given by $(f^{(n)}(t);\, n\in\mathbb{N}_0)$. Fix $t\in[0,1]$, and let s decrease to t. By assumption, for every $n\in\mathbb{N}$ we have $f^{(n)}\in H^{(n)}_\alpha$. In particular, we know that $1_{\Delta^n_{+\epsilon}} f^{(n)}$ belongs to the Sobolev space $H_\alpha(\Delta^n_{+\epsilon})$, and so by Lemma 8.14 it admits a trace in $L^2(D^n)$, which is approached from the interior of $\Delta^n_{+\epsilon}$. We denote this trace by $\gamma_+ f^{(n)}$. Define $\partial_{t+}X(t)$ as being given by $(n\, \gamma_+ f^{(n)};\, n\in\mathbb{N})$. Then it follows from the estimate in Lemma 8.14 that $\partial_{t+}X(t)\in L^{2;2}$. Moreover, the second statement of Lemma 8.14 entails that for a.e. $t\in[0,1]$, $\partial_s X(t) \to \partial_{t+}X(t)$ in (L^2) as $s\downarrow t$. Thus X has a right derivative in $L^{2;2}$.

On the other hand, our assumption on X implies that $X\in L^2([0,1]; \mathcal{D}^{2,1})$. Thus the Hitsuda–Skorokhod integral of X exists in (L^2), and therefore its forward integral is in (L^2). □

Remark. It is clear that one can do analogous constructions, and find corresponding results, for the backward integral by using above the region

$$\Delta^n_{-\epsilon} = \{ (t, s)\in[0,1]\times\mathbb{R}^n;\ t-\epsilon<s_1<t \}.$$

Also, using both $\Delta^n_{+\epsilon}$ and $\Delta^n_{-\epsilon}$, we can construct a class of processes admitting both: right and left derivatives, and therefore have a generalized Stratonovich integral in (L^2).

8.D Properties of Anticipating Stochastic Integrals

In this section we present some of the properties of the forward, backward and Hitsuda–Skorokhod integrals. Essentially we shall review here the articles by Asch and Potthoff (1991), and by Kuo and Russek (1988), although some of the results concerning the Hitsuda–Skorokhod integral can also be found in the articles mentioned at the beginning of Section 8.C, notably in the paper by Nualart and Pardoux (1988).

We start with the proof that the forward integral is linear with respect to "constant processes", i.e., random variables (satisfying a certain smoothness condition).

Proposition 8.16. *Assume that* $\varphi\in\mathfrak{D}^{2,1}$ *and that* $\nabla\varphi$: $[0,1] \to (L^2)$ *has a right continuous version. Then for all* $a,b\in[0,1]$, $a\le b$,

$$\int_a^b \varphi\ dB(t+) = \varphi \cdot (B(b) - B(a)).$$

Proof. Without loss of generality we may set $a=0$, $b=1$.

Note that we have by assumption that for a.e. $t\in[0,1]$, $\partial_t\varphi = \nabla\varphi(t) = \partial_{t+}\varphi$ (μ-a.e.). Let $\psi\in\mathcal{S}$, and observe that $\nabla\psi\in C([0,1];\ \mathcal{S})$ (continuity with respect to the (L^2)-topology). Compute as follows.

$$(\psi, \int_0^1 \partial_t^*\varphi\ dt\ ) = \int_0^1 (\partial_t\psi, \varphi)\ dt$$

$$= \int_0^1 \left\{(\partial_t \psi\varphi, 1) - (\psi, \partial_t \varphi)\right\} dt$$

$$= (\psi, \varphi \int_0^1 \partial_t^* 1 \, dt) - \int_0^1 (\psi, \partial_{t+} \varphi) \, dt$$

$$= (\psi, \varphi \, B(1)) - (\psi, \int_0^1 \partial_{t+} \varphi \, dt),$$

where we used the product rule for ∂_t. Thus we have

$$(\psi, \int_0^1 \varphi \, dB(t+)) = (\psi, \varphi \, B(1)),$$

for all $\psi \in \mathcal{E}$. Since $\mathcal{E}$ is dense in (L^2), the proof is finished. □

Assume from now on that X is a fixed process on [0,1], so that $X \in L^\alpha([0,1]; \mathfrak{D}^{p,1})$, α, $p \geq 2$, with a right derivative in $L^{2;p}$. Set

$$Y(t) := \int_0^t X(s) \, dB(s+),$$

and for suitable processes Z define

$$\int_0^1 Z(t) \, dY(t+) := \int_0^1 Z(t) \, X(t) \, dB(t+).$$

Here is a result, stating sufficient conditions for the existence of the last integral. It follows essentially from the product rule for ∇ and Meyer's inequalities (7.60), for a detailed proof we refer to Asch and Potthoff (1991).

Proposition 8.17. *Assume that* $\beta \geq 2$, *and* q, $r \geq 2$ *are such that* $\alpha^{-1} + \beta^{-1} = \frac{1}{2}$, $p^{-1} + q^{-1} = r^{-1}$. *Suppose that* $Z \in L^\beta([0,1]; \mathfrak{D}^{q,1})$ *and that* Z *has a right derivative in* $L^{2;q}$. *Then*

$$\int_0^1 Z(t)\, dY(t+) = \int_0^1 Z(t)\, X(t)\, dB(t+)$$

exists and belongs to (L^r).

It is not hard to generalize the proof of Proposition 8.16 for the dY(t+)–integral.

Proposition 8.18. *Let* $q \geq 2$ *be such that* $q^{-1}+p^{-1} = \frac{1}{2}$. *Assume that* $\varphi \in \mathfrak{D}^{q,1}$, *and that* $\nabla\varphi$: $[0,1] \to (L^q)$ *has a right continuous version. Then for all* a, b$\in$[0,1], $a \leq b$,

$$\int_a^b \varphi\, dY(t+) = \varphi \cdot (Y(b) - Y(a)), \quad \mu\text{-a.e.}$$

The proof can be found in Asch and Potthoff (1991). We also remark that an analogous property holds for the backward integral, whenever φ has a version so that $\nabla\varphi$ is a.e. left continuous.

Now we can prove that the forward integral admits approximations by forward Riemann sums.

Proposition 8.19. *Assume that* $Z \in C([0,1]; \mathfrak{D}^{2,1})$ *and that* ∇Z *has a version (denoted by the same symbol) which is continuous from the set*

$$D_{+\epsilon} := \{(s, t) \in [0,1] \times \mathbb{R};\ s \in [0,1],\ s \leq t \leq s+\epsilon\},$$

into (L^2). *Let* $\{\Delta_k,\ k=0, 1, \ldots, n-1\}$, $\Delta_k = [t_k, t_{k+1})$, *be a partition of* [0,1). *Then*

$$\sum_{k=0}^{n-1} Z(t_k)(B(t_{k+1}) - B(t_k)) \to \int_0^1 Z(t)\, dB(t+)$$

in (L^2) *as the mesh of the partition tends to zero.*

Proof. Without loss of generality we may assume that $\Delta = \max_k |\Delta_k| < \epsilon$. Then the mapping

$$t \mapsto \nabla Z(t_k, t) = \partial_t Z(t_k)$$

is right continuous from Δ_k into (L^2). Thus Proposition 8.17 implies that

$$Z(t_k)(B(t_{k+1}) - B(t_k)) = \int_{t_k}^{t_{k+1}} Z(t_k)\, dB(t+).$$

Define

$$Z_\Delta(t) := \sum_{k=0}^{n-1} Z(t_k)\, 1_{\Delta_k}(t).$$

Then

$$\sum_{k=0}^{n-1} Z(t_k)(B(t_{k+1}) - B(t_k)) - \int_0^1 Z(t)\, dB(t+) = \int_0^1 (Z_\Delta(t) - Z(t))\, dB(t+).$$

Thus we only need to prove that

$$\left\| \int_0^1 (Z_\Delta(t) - Z(t))\, dB(t+) \right\|_2$$

converges to zero as $\Delta \to 0$. The last norm is easily estimated by Proposition 8.9:

$$\left\| \int_0^1 (Z_\Delta(t) - Z(t))\, dB(t+) \right\|_2^2 \le$$
$$\le \text{const.} \int_0^1 \| Z_\Delta(t) - Z(t) \|_{2,1}^2 dt + \left(\int_0^1 \| \partial_{t+} Z_\Delta(t) - \partial_{t+} Z(t) \|_2\, dt \right)^2.$$

The assumption that Z is continuous from [0,1] into $\mathfrak{D}^{2,1}$ shows that the first of the last two terms vanishes as $\Delta \to 0$. The second term converges to zero because we assumed that ∇Z is continuous (and therefore uniformly continuous) from $D_{+\epsilon}$ into (L^2). □

By obvious modifications one finds the following result (for details we refer to Asch and Potthoff (1991)).

Proposition 8.20. *Assume that* $q \geq 2$ *is such that* $p^{-1}+q^{-1}=\frac{1}{2}$. *Suppose furthermore that* $Z \in ([0,1]; \mathfrak{D}^{q,1})$ *and that Z has a version (denoted by the same symbol) with* ∇Z *belonging to* $C(D_{+\epsilon}; (L^q))$. *If* $\{\Delta_k, k = 0, 1, \dots, n\text{-}1\}$ *is a partition of* [0,1) *as above, with mesh* Δ, *then*

$$\sum_{k=0}^{n-1} Z(t_k)(Y(t_{k+1}) - Y(t_k)) \to \int_0^1 Z(t)\, dY(t+)$$

in (L^2), *as the mesh* Δ *of the partition converges to zero.*

Next we shall compute the quadratic variation of the forward integral. The essential step will be to do this for the Hitsuda–Skorokhod integral. This has also been done in Nualart and Pardoux (1988). We give here a result of weak convergence which has been proved in the article by Asch and Potthoff (1991), and which will be sufficient for our purposes.

In the following we shall assume that Z is continuous from [0,1] into $\mathfrak{D}^{2p,1}$, $p \geq 1$. Let $\varphi \in \mathcal{S}$. Then we may compute as follows.

$$((\int_a^b \partial_t^* Z(t)\, dt)^2, \varphi) = \int_{[a,b]^2} (Z(s), \partial_s(\partial_t^* Z(t)\, \varphi))\, ds\, dt$$

$$= \int_a^b (Z(t)^2, \varphi)\, dt + \int_{[a,b]^2} (Z(s), (\partial_t^* \partial_s\, Z(t))\, \varphi)\, ds\, dt +$$

$$+ \int_{[a,b]^2} (Z(s), (\partial_t^* Z(t))\, \partial_s \varphi)\, ds\, dt$$

$$= \int_a^b (Z(t)^2, \varphi)\, ds\, dt + \int_{[a,b]^2} (Z(s)\, \partial_s Z(t), \partial_t \varphi)\, ds\, dt +$$

$$+ \int_{[a,b]^2} (Z(s)\, \partial_t^* Z(t), \partial_s \varphi)\, ds\, dt + \tag{8.7}$$

$$+ \int_{[a,b]^2} ((\partial_t Z(s))(\partial_s Z(t)), \varphi)\, ds\, dt,$$

where we used the product rule for ∂_t and the CCR (5.51), (5.52).

We shall study the behavior of

$$\frac{1}{b-a} ((\int_a^b \partial_t^* Z(t)\, dt)^2, \varphi),$$

as $| b-a | \to 0$, in the case that φ is of the form $\sin \circ <\cdot\,, \xi>$ or $\cos \circ <\cdot\,, \xi>$. This will be sufficient due to the following lemma.

Lemma 8.21. *Assume that* $\epsilon_n \downarrow 0$, $t \in [0,1)$. *Then the sequence*

$$\{ \epsilon_n^{-1} (\int_t^{t+\epsilon_n} \partial_t^* Z(t)\, dt)^2, n \in \mathbb{N} \}$$

is bounded in (L^p).

Proof. By Proposition 8.14 we have

$$\| \epsilon_n^{-1} (\int_t^{t+\epsilon_n} \partial_t^* Z(t)\, dt)^2 \|_p = \epsilon_n^{-1} \| \int_t^{t+\epsilon_n} \partial_t^* Z(t)\, dt \|_{2p}^2$$

$$\leq \text{const.}\ \epsilon_n^{-1} \int_t^{t+\epsilon_n} \| Z(t) \|_{2p,1}^2\, dt,$$

and the continuity of $t \mapsto \| Z(t) \|_{2p,1}$ implies the statement of the lemma. □

Proposition 8.22. *For all* $t \in [0,1]$, $\epsilon^{-1}\left(\int_t^{t+\epsilon} \partial_s^* X(s)\, ds \right)^2$ *converges weakly in* (L^p) *to* $X(t)^2$ *as* ϵ *tends to zero, uniformly in* $t \in [0,1)$.

Proof. By Lemma 8.21 it is sufficient to show the claimed convergence on a total set. Here we choose the set consisting of the functions $\sin \circ \langle \cdot\,, \xi \rangle$, $\cos \circ \langle \cdot\,, \xi \rangle$, $\xi \in \mathcal{S}(\mathbb{R})$. The proof will only be carried out for $\sin \circ \langle \cdot\,, \xi \rangle$, for $\cos \circ \langle \cdot\,, \xi \rangle$ the arguments are analogous.

Consider formula (8.7) which reads

$$\epsilon^{-1}\Big(\Big(\int_t^{t+\epsilon} \partial_s^* X(s)\, ds\Big)^2, \sin \circ \langle \cdot\,, \xi \rangle\Big) =$$

$$= \epsilon^{-1} \int_t^{t+\epsilon} (X(s)^2, \sin \circ \langle \cdot\,, \xi \rangle)\, ds +$$

$$+ \epsilon^{-1} \int_t^{t+\epsilon}\int_t^{t+\epsilon} \xi(s_2)\, (X(s_1)\, \partial_{s_2} X(s_2), \cos \circ \langle \cdot\,, \xi \rangle)\, ds_1 ds_2$$

$$+ \epsilon^{-1} \int_t^{t+\epsilon}\int_t^{t+\epsilon} \xi(s_1)\, (X(s_1)\, \partial_{s_2}^* X(s_2), \cos \circ \langle \cdot\,, \xi \rangle)\, ds_1 ds_2 \tag{8.8}$$

$$+ \epsilon^{-1} \int_t^{t+\epsilon}\int_t^{t+\epsilon} (\partial_{s_1} X(s_2))\, (\partial_{s_2} X(s_1)), \sin \circ \langle \cdot\,, \xi \rangle)\, ds_1 ds_2.$$

It is obvious that the first term on the right hand side of (8.8) converges to $(X(t)^2, \sin \circ \langle \cdot\,, \xi \rangle)$ as ϵ tends to zero. Thus it remains to show that the other three terms on the right hand side of (8.8) vanish in the limit. The proof of this fact is for the second and third term very similar, and therefore we only show it for the third expression:

$$\epsilon^{-1} \Big| \int_t^{t+\epsilon} \int_t^{t+\epsilon} \xi(s_1) \, (X(s_1) \, \partial^*_{s_2} X(s_2), \cos \circ <\cdot \, , \xi>) \, ds_1 ds_2 \Big| \le$$

$$\le \epsilon^{-1} \int_t^{t+\epsilon} (| \xi(s_1) | \, \| X(s_1) \|_2) ds_1 \, \| \int_t^{t+\epsilon} \partial^*_{s_2} X(s_2) \, ds_2 \|_2$$

$$\le C \, \epsilon^{-1} \int_t^{t+\epsilon} (| \xi(s_1) | \, \| X(s_1) \|_2) ds_1 \, (\int_t^{t+\epsilon} \| X(s_2) \|^2_{2,1} \, ds_2)^{1/2},$$

where $C>0$ is some constant, and where we used the estimate of Proposition 8.9. The continuity of X: $s \mapsto X(s)$, from [0,1] into $\mathfrak{D}^{2p,1}$, $p \ge 1$, (entailing uniform continuity of X from [0,1] into (L^2) and into $\mathfrak{D}^{2,1}$) implies that this term vanishes as $\epsilon \to 0$, uniformly in t.

Now consider the last term on the right hand side of (8.8):

$$\epsilon^{-1} \Big| \int_t^{t+\epsilon} \int_t^{t+\epsilon} (\partial_{s_1} X(s_2)) \, (\partial_{s_2} X(s_1)), \sin \circ <\cdot \, , \xi>) \, ds_1 ds_2 \Big| \le$$

$$\le \epsilon^{-1} \Big\| \int_t^{t+\epsilon} \int_t^{t+\epsilon} | \partial_{s_1} X(s_2)) \, (\partial_{s_2} X(s_1)) | \, ds_1 ds_2 \Big\|_1$$

$$\le \epsilon^{-1} \Big\| \int_t^{t+\epsilon} \int_t^{t+\epsilon} (\partial_{s_1} X(s_2))^2 \, ds_1 \, ds_2 \Big\|_1,$$

where we used Hölder's and Schwarz' inequalities. For $s,t \in [0,1]$ define

$$\phi(t,s) := \int_0^s (\partial_\sigma X(t))^2 \, d\sigma.$$

It follows from Meyer's inequalities (7.60) that $\sigma \mapsto \partial_\sigma X(t)(x)$ is for μ-a.e. $x \in \mathcal{S}'(\mathbb{R})$ in $L^2(\mathbb{R})$, because for every $t \in [0,1]$, $X(t) \in \mathcal{D}^{2p,1}$, $p \geq 1$. Thus for all $t \in [0,1]$ and a.e. $x \in \mathcal{S}'(\mathbb{R})$, we have that

$$\phi(t,s')(x) - \phi(t,s)(x) \to 0,$$

as $| s' - s | \to 0$. Moreover, for all $s,t \in [0,1]$, $\phi(t,s) \leq \phi(t,1) \in (L^1)$. Thus the dominated convergence theorem implies that for every $t \in [0,1]$, $\phi(t,\cdot\,)$ is continuous from $[0,1]$ into (L^1). Therefore $\phi(t,\cdot\,)$ is for every $t \in [0,1]$ uniformly continuous from $[0,1]$ into (L^1). In addition, a compactness argument shows that the family $\{\phi(t,\cdot\,);\ t \in [0,1]\}$ is uniformly equicontinuous (see Asch and Potthoff (1991)) for details). Thus

$$\epsilon^{-1} \left\| \int_t^{t+\epsilon} \int_t^{t+\epsilon} (\partial_{s_1} X(s_2))^2 \, ds_1 \, ds_2 \right\|_1 = \epsilon^{-1} \left\| \int_t^{t+\epsilon} (\phi(s,t+\epsilon) - \phi(s,t)) \, ds \right\|_1$$

$$\leq \epsilon^{-1} \int_t^{t+\epsilon} \| \phi(s,t+\epsilon) - \phi(s,t) \|_1 \, ds$$

$$\leq \sup_{s \in [0,1]} \| \phi(s,t+\epsilon) - \phi(s,t) \|_1,$$

which vanishes (uniformly in t) as ϵ tends to zero. □

An obvious direct consequence is the following result.

Theorem 8.23. *Assume that* $X \in C([0,1]; \mathcal{D}^{2p,1})$, $p \geq 1$, *and that* X *has a right derivative which is continuous from* $[0,1]$ *into* (L^{2p}). *Then*

$$\lim_{\epsilon \downarrow 0} \epsilon^{-1} \Big(\int_t^{t+\epsilon} X(s) \, dB(s+) \Big)^2 = X(t)^2,$$

weakly in (L^p), *uniformly in* $t \in [0,1)$.

Next we shall state some continuity properties of the forward integrals. These will be of importance for the proof of Itô's Lemma at the end of this section.

Proposition 8.24. *Assume that* $C \in L^2([0,1]; \mathfrak{D}^{p,k})$, $p \geq 2$, $k \in \mathbb{N}$, *and that* X *has a right derivative in* $L^1([0,1]; \mathfrak{D}^{p,k-1})$. *Then*

$$t \mapsto Y(t) = \int_0^t X(s)\, dB(s+)$$

is continuous from $[0,1]$ *into* $\mathfrak{D}^{p,k-1}$.

Proof. Consider first the Hitsuda-Skorokhod integral of X. Let $t > t'$:

$$\left\| \int_0^t \partial_s^* X(s)\, ds - \int_0^{t'} \partial_s^* X(s)\, ds \right\|_{p,k-1} = \left\| \int_0^1 \partial_s^* 1_{[t',t]}(s)\, X(s)\, ds \right\|_{p,k-1}$$

$$\leq C_{p,k} \left(\int_0^1 1_{[t',t]}(s)\, \| X(s) \|_{p,k}^2\, ds \right)^{1/2},$$

by Proposition 8.9, and the claimed continuity holds for this part of the forward integral. The obvious estimate

$$\left\| \int_{t'}^t \partial_{s+} X(s)\, ds \right\|_{p,k-1} \leq \int_{t'}^t \| \partial_{s+} X(s) \|_{p,k-1}\, ds$$

concludes the proof. □

Define the set

$$B_{+\epsilon} := \{ (s;v) \in [0,1] \times \mathbb{R}^2;\ s \in [0,1],\ v_1 \in [s, s+\epsilon],\ v_2 \in [s, 1+\epsilon] \}.$$

Then one can show the following result (for a proof we refer to the article of Asch and Potthoff (1991)).

Proposition 8.25. *Assume that* $X \in L^2([0,1]; \mathfrak{D}^{p,k})$, $p \geq 2$, $k \geq 2$, *has a version (denoted by the same symbol) such that* ∇X *is continuous from* $D_{+\epsilon}$ *into* $\mathfrak{D}^{p,k-2}$, *and* $\nabla^2 X$ *is continuous from* $B_{+\epsilon}$ *into* $\mathfrak{D}^{p,k-2}$. *Then* Y *has a version (denoted also by* Y) *so that* ∇Y *is continuous from* $D_{+\epsilon}$ *into* $\mathfrak{D}^{p,k-2}$.

Finally we give a path–continuity result for Y. As in the proof of Theorem 8.23, the essential part is the treatment of the Hitsuda–Skorokhod integral of X. This has been done in Nualart and Pardoux (1988).

Proposition 8.26. *Assume that* $X \in L^\infty([0,1]; \mathfrak{D}^{p,1})$, $p>2$, *and that* X *has a right derivative in* $L^\infty([0,1]; (L^q))$, $q>1$. *Then* $Y(t) = \int_0^t X(s)\, dB(s+)$, $t \in [0,1]$, *has a* μ-a.e. *continuous modification.*

Proof. Proposition 8.9 gives the estimate $(t,t' \in [0,1],\ t'<t)$

$$\left\| \int_0^t \partial_s^* X(s)\, ds - \int_0^{t'} \partial_s^* X(s)\, ds \right\|_p \leq C_p \left(\int_{t'}^t \| X(s) \|_{p,1}^2\, ds \right)^{1/2}$$

$$\leq C_p \operatorname*{ess\,sup}_{s \in [0,1]} \| X(s) \|_{p,1} \, | t-t' |^{1/2}.$$

Therefore an application of Kolmogoroff's Lemma implies the existence of a continuous version of the Hitsuda-Skorokhod integral Y of X. Similarly, the estimate

$$\left\| \int_0^t \partial_{s+} X(s)\, ds - \int_0^{t'} \partial_{s+} X(s)\, ds \right\|_q \leq \operatorname*{ess\,sup}_{s \in [0,1]} \| \partial_{s+} X(s) \|_q \, | t-t' |,$$

and another application of Kolomgoroff's Lemma conclude the proof. □

In the remainder of this section we shall always choose the continuous modification of Y.

Now we are going to discuss Itô's Lemma for the forward integral of X. Similarly as in the non–anticipating case, its proof has two essential ingredients: the Riemann approximation of the forward integral and the result (Theorem 8.23) about the quadratic variation. Therefore we shall first establish the Riemann approximation for integrands of the form $g\circ Y(t)$, $t\in[0,1]$, where $g\in C_b^1(\mathbb{R})$.

For the remainder of this section we shall make the following assumptions on $X: [0,1] \to \mathfrak{D}^{2,2}$ (except if mentioned otherwise):

(A.1) $X\in C([0,1]; \mathfrak{D}^{p,2})$, $p>4$

(A.2) $\nabla X\in C(D_{+\epsilon}; \mathfrak{D}^{p,1})$, $\epsilon>0$, $p>4$

(A.3) $\nabla^2 X\in C(B_{+\epsilon}; (L^p))$, $\epsilon>0$, $p>4$.

Proposition 8.27. *Let* $s,t\in[0,1]$, $s\leq t$. *If* $\{\Delta_k;\ k=0,1 \ldots ,n-1\}$ *is a partition of* $[s,t)$ *into intervals* $\Delta_k = [s_k,s_{k+1})$, *then*

$$\sum_{k=0}^{n-1} g\circ Y(s_k)(Y(s_{k+1})-Y(s_k)) \to \int_s^t g\circ Y(s)\, dY(s+)$$

in (L^2) *as the mesh of the partition tends to zero.*

Proof. In view of Proposition 8.20 it suffices to show that for q with $q^{-1}+p^{-1}=\frac{1}{2}$, we have $g\circ Y\in C([0,1]; \mathfrak{D}^{q,1})$, and $\nabla g\circ Y\in C(D_{+\epsilon}; (L^q))$.

Let $q\geq 2$ be such that $q^{-1}+p^{-1}=\frac{1}{2}$, and $s,t\in[0,1]$. Then

$$\begin{aligned}\| g\circ Y(t) - g\circ Y(s) \|_{q,1} &= \| N^{1/2}(g\circ Y(t) - g\circ Y(s)) \|_q \\ &\leq C_p \| \,| \nabla(g\circ Y(t) - g\circ Y(s)) |_2 \|_q \\ &= C_p \| \,| g'\circ Y(t)\nabla Y(t) - g'\circ Y(s)\nabla Y(s) |_2 \|_q\end{aligned}$$

$$\le C_p\{|g'|_\infty \| |\nabla(Y(t)-Y(s))|_2 \|_q +$$

$$+ \| g'\circ Y(t) - g'\circ Y(s) \|_r \| |\nabla Y(s)|_2 \|_p\}$$

$$\le C_p\{|g'|_\infty \| Y(t)-Y(s) \|_{q,1} +$$

$$+ \| g'\circ Y(t) - g'\circ Y(s) \| \| Y(s) \|_{p,1}\},$$

where we used Meyer's inequalities (7.60) repeatedly, and Hölder's inequality with $r>2$ so that $r^{-1} + p^{-1} = q^{-1}$. (Also, the constant C_p may be changing from step to step above.) Note that assumption (A.2) implies that X has a right derivative in $C([0,1]; \mathfrak{D}^{p,1})$ and therefore in $L^\infty([0,1]; \mathfrak{D}^{p,1})$. By Proposition 8.24 we conclude that $t \mapsto Y(t)$ is continuous from [0,1] into $\mathfrak{D}^{p,1}$. Since $p>q$, it follows that Y belongs to $C([0,1]; \mathfrak{D}^{q,1})$. Thus the first terms in the last expression above vanishes in the limit $|s-t| \to 0$.

Now consider the second term of the last expression above. The continuity of $s \to Y(s)$ from [0,1] into $\mathfrak{D}^{p,1}$ implies that $\| Y(s) \|_{p,1}$ is bounded. On the other hand, since g' is continuous, and Y is a.s. continuous (Proposition 8.26), $g'\circ Y$ is a.s. continuous. Boundedness of g' and the dominated convergence theorem imply that the second term vanishes in the limit $|s-t| \to 0$, too. Therefore we have shown that $g\circ Y$ belongs to $C([0,1]; \mathfrak{D}^{q,1})$.

Note that $\nabla g\circ Y = g'\circ Y \cdot \nabla Y$, by the chain rule. Using now assumption (A.3) and Proposition 8.25, one can show in a similar way that $\nabla g\circ Y \in C(D_{+\epsilon}; (L^q))$. The details are left as an exercise to the interested reader. □

Now we can prove Itô's Lemma for the forward integral.

Theorem 8.28. *Suppose that* X *is a process satisfying assumptions* (A.1), (A.2) *and* (A.3), *and set* $Y(t) = \int_0^t X(s)\, dB(s+)$. *Assume furthermore that* $f \in C_b^2(\mathbb{R})$. *Then* *for all* s,t$\in$[0,1], s$\leq$t,

$$f(Y(t))-f(Y(s)) = \int_s^t f'(Y(u))\, dY(u+) + \tfrac{1}{2} \int_s^t f''(Y(u))\, X(u)^2\, du, \qquad (8.9)$$

μ-a.s.

Proof. Note that every term in (8.9) is in (L^1). Thus it suffices to show that for all $\phi\in(L^\infty)$ we have the following:

$$\Big(\phi,\ f(Y(t)) - f(Y(s)) - \int_s^t f'(Y(u))\, dY(u+) - \tfrac{1}{2} \int_s^t f''(Y(u))\, X(u)^2\, du\Big) = 0.$$

Let $\{\Delta_k,\ k=0, 1, \dots, n\text{-}1\}$ be a partition of [s,t) into intervals $\Delta_k=[s_k, s_{k+1})$. Use Taylor's theorem to write

$$f(Y(t)) - f(Y(s)) = \sum_{k=0}^{n-1} (f(Y(s_{k+1})) - f(Y(s_k))$$

$$= \sum_{k=0}^{n-1} f'(Y(s_k))\, (Y(s_{k+1}) - Y(s_k)) +$$

$$+ \sum_{k=0}^{n-1} \Big(\int_0^1 (1-\theta)\, f''(Y_\theta(s_k))\, d\theta\Big)\, (Y(s_{k+1})-Y(s_k))^2,$$

where we have set

$$Y_\theta(s_k) = \theta\, Y(s_{k+1}) + (1-\theta)\, Y(s_k).$$

As the mesh of the partition tends to zero, we find by Proposition 8.29 that

$$\sum_{k=0}^{n-1} f'(Y(s_k))\, (Y(s_{k+1}) - Y(s_k)) - \int_s^t f'(Y(u))\, dY(u+)$$

converges to zero in (L^1). Thus it remains to show that the following three expressions vanish as the mesh of the partition converges to zero:

$$\text{(i)}\quad \Big(\phi, \sum_{k=0}^{n-1} \Big[\int_0^1 (1-\theta)\, f''(Y_\theta(s_k))\, d\theta - \tfrac{1}{2} f''(Y(s_k))\Big] \times$$

$$\times \mid \Delta_k \mid (\mid \Delta_k \mid^{-1} (Y(s_{k+1})) - Y(s_k)^2) \Big),$$

$$\text{(ii)}\quad \Big(\phi, \sum_{k=0}^{n-1} f''(Y(s_k))\, [\, X(s_k)^2 - \mid \Delta_k \mid^{-1} (Y(s_{k+1}) - Y(s_k))^2 \,] \mid \Delta_k \mid \Big),$$

$$\text{(iii)}\quad \Big(\phi, \sum_{k=0}^{n-1} f''(Y(s_k))\, X(s_k)^2 \mid \Delta_k \mid - \int_s^t f''(Y(u))\, X(u)^2\, du \Big).$$

The first of these terms vanishes in the limit because of the following two facts: boundedness of f'' and a.e. continuity of Y imply that

$$\sup_{u,\theta} \mid f''(Y(u)) - f''(\theta Y(u+\epsilon) + (1-\theta) Y(u)) \mid$$

tends to zero as $\epsilon \to 0$ in (L^∞), where the supremum is taken over $u \in [s,t)$, $\theta \in [0,1]$. On the other hand Lemma 8.21 shows that $\sup_u \epsilon^{-1}(Y(u+\epsilon) - Y(u))^2$ is bounded in (L^1) as $\epsilon \to 0$.

Consider the second expression (ii). Since $v \mapsto f''(Y(v))$ is continuous from [s,t] into (L^∞), the set $\{\phi \cdot f''(Y(v)),\ v \in [s,t]\}$ is compact in (L^∞). Thus by Theorem 8.23,

$$\sup_{u,v} \Big(\phi \cdot f''(Y(v)),\ X(u)^2 - \epsilon^{-1}(Y(u+\epsilon) - Y(u))^2\Big) \to 0, \text{ as } \epsilon \to 0,$$

where the supremum is taken over $u,v \in [s,t)$, since weak convergence is uniform on compact sets. But this implies that (ii) vanishes as the mesh of the partition tends to zero.

The third term (iii) converges to zero as a direct consequence of the continuity assumptions on X. □

Of course, the assumption that f belongs to $C_b^2(\mathbb{R})$ in Theorem 8.28 is not very satisfying. One can relax this condition considerably by using the localization principle of Nualart and Pardoux (1988). Moreover, it is straightforward to generalize the Itô Lemma to the case of a Brownian motion in $\mathbb{R}^d$. We merely state the result which is established in Asch and Potthoff (1991) and refer the interested reader to this article for details.

Take d independent copies of $(\mathcal{S}'(\mathbb{R}),\mathcal{B},\mu)$, and denote by $x=(x_1, \ldots ,x_d)$ the typical element in $\mathcal{S}'(\mathbb{R})^{\times d}$. Then $B(t) = (B_1(t), \ldots ,B_d(t))$, $t\geq 0$, with

$$B_k(t)(x) = \langle x_k, 1_{[0,t)}\rangle, \quad x\in\mathcal{S}'(\mathbb{R})^{\times d},$$

is a d–dimensional Brownian motion starting at zero.

Consider a matrix–valued process $X = (X_{jk};\ j, k=1,2, \ldots ,d)$, such that for every j, $k\in\{1,2, \ldots ,d\}$, X_{jk}: $[0,1] \to \mathcal{D}^{p,2}$, $p>4$, satisfies the assumptions (A.1), (A.2) and (A.3). Denote

$$Y_j(t) = \sum_{k=1}^{d} \int_0^t X_{jk}(s)\, dB_k(s+),$$

and

$$\sigma_{jk}(t) = \sum_{l=1}^{d} X_{jl}(t)\, X_{kl}(t), \quad j,k=1,2, \ldots ,d.$$

Let f be in $C^2(\mathbb{R}_+\times\mathbb{R}^d)$, with D_0f denoting the partial derivative of f with respect to the variable in $\mathbb{R}_+$, D_kf denoting the partial derivative of f in the k-th direction of $\mathbb{R}^d$.

Then on can prove the following theorem (cf. Asch and Potthoff (1991)).

Theorem 8.29. *For all* s,t∈[0,1], s≤t, *the following formula holds* μ-a.s.

$$f(t,Y(t)) - f(s,Y(s)) = \int_s^t D_0 f(u,Y(u))\,du + \sum_{k=1}^{d} \int_s^t D_k f(u,Y(u))\,dY_k(u+) + \tag{8.10}$$

$$+ \tfrac{1}{2} \sum_{k,l=1}^{d} \int_s^t D_k D_l f(u,Y(u))\,\sigma_{kl}(u)\,du.$$

9. Fourier and Fourier–Mehler Transforms

Recall that in Section 4.A we have the following $\mathcal{T}$– and $\mathcal{S}$–transformations defined on the space $(\mathcal{S})^*$ of generalized white noise functionals: for $\Phi \in (\mathcal{S})^*$ and $\xi \in \mathcal{S}(\mathbb{R})$,

$$(\mathcal{T}\Phi)(\xi) = \langle \overline{\Phi}, e^{i\langle \cdot, \xi \rangle} \rangle,$$

$$(\mathcal{S}\Phi)(\xi) = \langle \overline{\Phi}, :e^{\langle \cdot, \xi \rangle}: \rangle = e^{-\frac{1}{2}|\xi|_2^2} \langle \overline{\Phi}, e^{\langle \cdot, \xi \rangle} \rangle.$$

These two transformations play important roles in the previous chapters. Note that for $\varphi \in (L^2)$, we have

$$(\mathcal{T}\varphi)(\xi) = \int_{\mathcal{S}'(\mathbb{R})} \exp(i\langle x, \xi \rangle)\, \varphi(x)\, d\mu(x), \quad \xi \in \mathcal{S}(\mathbb{R}), \tag{9.1}$$

$$(\mathcal{S}\varphi)(\xi) = \int_{\mathcal{S}'(\mathbb{R})} \varphi(x+\xi)\, d\mu(x), \quad \xi \in \mathcal{S}(\mathbb{R}), \tag{9.2}$$

where μ is the standard Gaussian measure on $\mathcal{S}'(\mathbb{R})$. Thus both $\mathcal{T}$– and $\mathcal{S}$–transformations are generalizations of the finite dimensional Fourier transform. But unlike the finite dimensional case, they take generalized white noise functionals on the space $\mathcal{S}'(\mathbb{R})$ to functions defined on another space, i.e., $\mathcal{S}(\mathbb{R})$. Of course, for φ in (L^2), $\mathcal{T}\varphi$ in (9.1) can be defined on $\mathcal{S}'(\mathbb{R})$ μ-a.e. and belongs to (L^2). On the other hand, a slight modification in the definition of $\mathcal{S}\varphi$ in (9.2) by constants gives the Fourier–Wiener transform which is a unitary operator on (L^2), see Cameron–Martin (1945) and Segal (1956). However, the Fourier–Wiener transform does not possess similar properties as in the finite dimensional case with respect to the differentiation operators D_h, see Kuo (1974).

Thus it is natural to ask whether it is possible to define a Fourier transform taking "functions" defined on $\mathcal{S}'(\mathbb{R})$ into "functions" also defined on $\mathcal{S}'(\mathbb{R})$ with similar properties as in the finite dimensional case, e.g., taking the coordinate differentiation ∂_t to multiplication by ix(t). In Kuo (1982), such a Fourier transform was defined and some properties were obtained. But, as pointed out by B. Driver and L. Gross, there was a fundamental difficulty in the definition of the Fourier transform given in Kuo (1982). Later, Kuo (1989) used the previous definition as a motivation to define the Fourier transform in terms of the $\mathcal{S}$-transformation. However, it was unknown whether or not this Fourier transform was defined on the whole space $(\mathcal{S})^*$ of generalized white noise functionals. But in view of the characterization theorem in Potthoff and Streit (1991), i.e., Theorem 4.38, it is clear that the Fourier transform is indeed defined on the whole space $(\mathcal{S})^*$. See Kuo (1990a, 1990b). Independently, Yan (1989) also used the $\mathcal{S}$-transformation to define this Fourier transform. In addition he used the Wick product to indicate that this Fourier transform was defined on the whole space $(\mathcal{S})^*$. We remark that the Fourier transform has also been studied by Dôku (1990), Hida, Kuo and Obata (1990), Ito, Kubo and Takenaka (1990), J. S. Lee (1992), and Y. J. Lee (1991), among others.

9.A A Heuristic Motivation

The Fourier transform $\hat{f}$ of an integrable function on $\mathbb{R}^k$ is defined by

$$\hat{f}(y) = (2\pi)^{-k/2} \int_{\mathbb{R}^k} e^{-i\langle x, y\rangle} f(x)\, dx.$$

There are several difficulties to generalize the Fourier transform to an infinite dimensional space, i.e., 1) the factor $(2\pi)^{-k/2}$ tends to 0 as $k \to +\infty$, 2) the Lebesgue measure does not exist, and 3) the inner product $\langle\cdot,\cdot\rangle$ may not be well-defined. However, we can overcome these difficulties by reformulating the above expression of the Fourier transform.

Suppose y is an $\mathbb{R}^k$-valued Gaussian random variable with mean 0 and covariance operator Id. Then for fixed $x \in \mathbb{R}^k$, the expectation of $e^{-i\langle x, y\rangle}$ with respect to y is given by

$$E_y\, e^{-i\langle x,\, y\rangle} = e^{-\frac{1}{2}|x|^2}.$$

Thus if we define the renormalization $:e^{-i\langle x,\, y\rangle}:_y$ of $e^{-i\langle x,\, y\rangle}$ with respect to the y-variable by

$$:e^{-i\langle x,\, y\rangle}:_y = \frac{e^{-i\langle x,\, y\rangle}}{E_y\, e^{-i\langle x,\, y\rangle}}.$$

Then we have

$$:e^{-i\langle x,\, y\rangle}:_y = e^{-i\langle x,\, y\rangle + \frac{1}{2}|x|^2}.$$

Therefore, we can rewrite the Fourier transform $\hat{f}$ as follows:

$$\hat{f}(y) = (2\pi)^{-k/2}\int_{\mathbb{R}^k} :e^{-i\langle x,\, y\rangle}:_y\, f(x) e^{-\frac{1}{2}|x|^2}\, dx.$$

Let μ_k be the standard Gaussian measure on $\mathbb{R}^k$. Then $\hat{f}(y)$ can be expressed by

$$\hat{f}(y) = \int_{\mathbb{R}^k} :e^{-i\langle x,\, y\rangle}:_y\, f(x)\, d\mu_k(x).$$

Now, observe that for the infinite dimensional space $\mathcal{S}'(\mathbb{R})$, we have the Gaussian measure μ and for fixed $x \in \mathcal{S}'(\mathbb{R})$, $:e^{-i\langle x,\, y\rangle}:_y$ is a generalized white noise functional (see Examples 4.29 and 4.42). Therefore, it looks promising that we can generalize the finite dimensional Fourier transform to the space $(\mathcal{S})^*$ of generalized white noise functionals.

9.B Definition and Examples of the Fourier Transform

From the discussion in the previous section it appears that we can simply define the Fourier transform $\hat{\Phi}$ of a generalized white noise functional Φ by

$$\hat{\Phi}(y) = \int_{\mathcal{S}'(\mathbb{R})} :e^{-i<x,\, y>}:_y \, \Phi(x) \, d\mu(x). \tag{9.3}$$

This was taken as the definition of the Fourier transform in Kuo (1982). However, note that

$$:e^{-i<x,\, y>}:_y = \sum_{n=0}^{\infty} \frac{(-i)^n}{n!} <:y^{\otimes n}:, x^{\otimes n}>.$$

Therefore, for fixed y in $\mathcal{S}'(\mathbb{R})$, the renormalization $:e^{-i<\cdot,\, y>}:_y$ belongs to $(\mathcal{S})^*$. Thus if we replace the integral in (9.3) with the pairing of $(\mathcal{S})^*$ and $(\mathcal{S})$ then (9.3) makes sense only for Φ in $(\mathcal{S})$. To overcome this difficulty, regard the Fourier transform as informally defined by (9.3), and take the $\mathcal{S}$–transformation to get

$$(\mathcal{S}\hat{\Phi})(\xi) = \int_{\mathcal{S}'(\mathbb{R})} e^{-i<x,\, \xi>} \Phi(x) \, d\mu(x), \; \xi \in \mathcal{S}(\mathbb{R}). \tag{9.4}$$

Observe that for any $\xi \in \mathcal{S}(\mathbb{R})$, the function $e^{-i<\cdot,\, \xi>}$ is in $(\mathcal{S})$. Therefore, we can use the pairing of $(\mathcal{S})^*$ and $(\mathcal{S})$ to rewrite the right hand side of (9.4) as follows:

$$(\mathcal{S}\hat{\Phi})(\xi) = <\overline{\Phi}, e^{-i<\cdot,\, \xi>}>, \; \xi \in \mathcal{S}(\mathbb{R}).$$

On the other hand, it is natural to express the $\mathcal{S}$–transformation of $\hat{\Phi}$ in terms of the $\mathcal{S}$–transformation of Φ. To do this, recall that for each $\xi \in \mathcal{S}(\mathbb{R})$ the function $(\mathcal{S}\Phi)(\lambda\xi)$, $\lambda \in \mathbb{R}$, has an entire extension $(\mathcal{S}\Phi)(z\xi)$, $z \in \mathbb{C}$. In particular, when $z=-i$, we have

$$(\mathcal{S}\Phi)(-i\xi) = e^{\frac{1}{2}|\xi|_2^2} <\overline{\Phi}, e^{-i<\cdot,\, \xi>}>, \; \xi \in \mathcal{S}(\mathbb{R}).$$

Therefore, we can rewrite the above $(\mathcal{S}\hat{\Phi})(\xi)$ as follows:

$$(\mathcal{S}\hat{\Phi})(\xi) = (\mathcal{S}\Phi)(-i\xi) e^{-\frac{1}{2}|\xi|_2^2}, \; \xi \in \mathcal{S}(\mathbb{R}).$$

Now, given a generalized white noise functional Φ in $(\mathcal{S})^*$, the right hand sides of the last equation is a U–functionals in view of Theorem 4.38. And so, by Theorem 4.38, it is the $\mathcal{S}$–transform of a unique generalized white noise functional.

The above discussion is the motivation for the following definition of the Fourier transform.

Definition 9.1. *The* Fourier transform *of a generalized white noise functional* Φ *is the generalized white noise functional, denoted by* $\hat{\Phi}$, *whose* $\mathcal{S}$*–transformation is given by*

$$(\mathcal{S}\hat{\Phi})(\xi) = <\bar{\Phi}, e^{-i<\cdot,\ \xi>}>,\ \xi\in\mathcal{S}(\mathbb{R}), \tag{9.5}$$

or equivalently,

$$(\mathcal{S}\hat{\Phi})(\xi) = (\mathcal{S}\Phi)(-i\xi)e^{-\frac{1}{2}|\xi|_2^2},\ \xi\in\mathcal{S}(\mathbb{R}). \tag{9.6}$$

We will use $\mathcal{F}$ to denote the Fourier transform mapping, i.e., $\mathcal{F}\Phi=\hat{\Phi}$. Let ν be a measure on $\mathcal{S}'(\mathbb{R})$ satisfying the conditions in Theorem 4.57. In this chapter we will use $\frac{d\nu}{d\mu}$ (as in Chapter 4) and $\tilde{\nu}$ (for the sake of convenience) to denote the generalized white noise functional associated with ν, i.e.,

$$<\frac{d\nu}{d\mu}, \varphi> = <\tilde{\nu}, \varphi> = \int_{\mathcal{S}'(\mathbb{R})} \tilde{\varphi}(x)\ d\nu(x),\ \ \varphi\in(\mathcal{S}),$$

where $\tilde{\varphi}$ is the continuous version of φ. In particular, for y in $\mathcal{S}'(\mathbb{R})$, we have the delta white noise functional $\tilde{\delta}_y$ at y (see Kubo and Yokoi (1989)). Note that the $\mathcal{S}$–transformation of $\tilde{\delta}_y$ is given by $(\mathcal{S}\tilde{\delta}_y)(\xi)=\exp(-\frac{1}{2}|\xi|_2^2 + <y, \xi>)$.

Example 9.2. $\hat{1}=\tilde{\delta}_0$ and $(\tilde{\delta}_0)^\wedge=1$. To check this, just note that $(\mathcal{S}1)(\xi)=1$ and $(\mathcal{S}\tilde{\delta}_0)(\xi)=\exp(-\frac{1}{2}|\xi|_2^2)$. Incidentally, if we use the formal expression (9.3), then the fact that $\hat{1}=\tilde{\delta}_0$ can be stated as

$$\tilde{\delta}_0(y) = \int_{\mathcal{S}'(\mathbb{R})} :e^{-i<x,\, y>}:_y \, d\mu(x).$$

Example 9.3. Let $y \in \mathcal{S}'(\mathbb{R})$. Then $(: \exp(i<\cdot,\, y>) :)^\wedge = \tilde{\delta}_y$. To check this, simply note that $(\mathcal{S}: \exp(i<\cdot,\, y>) :)(\xi) = \exp(i<y,\, \xi>)$, and $(\mathcal{S}\tilde{\delta}_y)(\xi) = \exp(-\frac{1}{2}|\xi|_2^2 + <y,\, \xi>)$.

Example 9.4. For a, $c \in \mathbb{C}$, $c \neq 0, -1$, and $w \in \mathcal{S}'(\mathbb{R})$, let $G_{aw,\, c}$ denote the Gaussian white noise functional $\mathcal{N}\exp(a<\cdot,\, w> - \frac{1}{2c}|\cdot|_2^2)$ whose $\mathcal{S}$-transformation is given by

$$(\mathcal{S}G_{aw,\, c})(\xi) = \exp\Big(\frac{ac}{1+c}<w,\, \xi> - \frac{1}{2(1+c)}|\xi|_2^2\Big).$$

With a little bit of computation it can be checked easily that the Fourier transform of $G_{aw,\, c}$ is also a Gaussian functional,

$$(G_{aw,\, c})^\wedge = G_{-aciw,\, c^{-1}}.$$

We will use g_c to denote $G_{0,\, c}$. Then $\hat{g}_c = g_{c^{-1}}$. This is a well-known property of Gaussian functions in finite dimensional spaces. Note that $\lim_{c\to\infty} g_c = 1$ and $\lim_{c\to 0} g_c = \tilde{\delta}_0$. Thus we will use the convention $g_\infty = 1$ and $g_0 = \tilde{\delta}_0$.

Example 9.5. The white noise analog of the finite dimensional Hermite function of order n is $\mathcal{H}_n(t) = (\partial_t^*)^n \mathcal{N}\exp(-\frac{1}{2}\,|\cdot|_2^2)$. Its $\mathcal{S}$-transformation is given by

$$(\mathcal{S}\mathcal{H}_n(t))(\xi) = \xi(t)^n \exp(-\tfrac{1}{4}|\xi|_2^2\,).$$

Then it is easy to check that the Fourier transform of $\mathcal{H}_n$ is given by

$$(\mathcal{H}_n(t))^\wedge = (-i)^n \mathcal{H}_n(t).$$

Again, this is a well-known property of Hermite functions in finite dimensional spaces.

Example 9.6. $(:\dot{B}(t)^n:)^\wedge=(-i)^n(\partial_t^*)^n\tilde{\delta}_0$. To check this, note that $\mathcal{S}(:\dot{B}(t)^n:)(\xi)=\xi(t)^n$ and so, by (9.6), we have $\mathcal{S}((:\dot{B}(t)^n:)^\wedge)(\xi)=(-i)^n\xi(t)^n\exp(-\frac{1}{2}|\xi|_2^2)$. On the other hand, since $\mathcal{S}(\partial_t^*\Phi)(\xi)=\xi(t)\mathcal{S}\Phi(\xi)$ for any Φ in $(\mathcal{S})^*$,

$$(\mathcal{S}(\partial_t^*)^n\tilde{\delta}_0)(\xi) = \xi(t)^n\,(\mathcal{S}\tilde{\delta}_0)(\xi)$$

$$= \xi(t)^n\exp(-\tfrac{1}{2}|\xi|_2^2).$$

By comparing the $\mathcal{S}$-transformations, we see that $(:\dot{B}(t)^n:)^\wedge=(-i)^n(\partial_t^*)^n\tilde{\delta}_0$.

Example 9.7. To find the Fourier transform of $:\ddot{B}(t)^n:$, note that

$$:\ddot{B}(t)^n: = (-1)^n:\langle\cdot,\delta_t'\rangle^n:,$$

where δ_t' is the distribution derivative of δ_t. Thus $(\mathcal{S}:\ddot{B}(t)^n:)(\xi)=\xi'(t)^n$ and so

$$(\mathcal{S}(:\ddot{B}(t)^n:)^\wedge)(\xi) = (-i)^n\,\xi'(t)^n e^{-\frac{1}{2}|\xi|_2^2}. \tag{9.7}$$

On the other hand, recall that for any $x\in\mathcal{S}'(\mathbb{R})$, the adjoint D_x^* of the differential operator D_x is a continuous linear operator from $(\mathcal{S})^*$ into itself and

$$(\mathcal{S}D_x^*\Phi)(\xi) = \langle x,\xi\rangle\mathcal{S}\Phi(\xi).$$

When $x=\delta_t'$, we have $(\mathcal{S}D_{\delta_t'}^*\Phi)(\xi)=-\xi'(t)\,\mathcal{S}\Phi(\xi)$. This yields immediately that

$$\Big(\mathcal{S}(D_{\delta_t'}^*)^n\,\Phi\Big)(\xi) = (-1)^n\xi'(t)^n\mathcal{S}\Phi(\xi).$$

In particular, when $\Phi=\tilde{\delta}_0$, we get

$$\left(\mathcal{S}(D^*_{\delta'_t})^n\,\tilde{\delta}_0\right)(\xi) = (-1)^n \xi'(t)^n e^{-\frac{1}{2}|\xi|_2^2}. \tag{9.8}$$

By comparing (9.7) and (9.8), we conclude that

$$(:\ddot{B}(t)^n:)^\wedge = i^n (D^*_{\delta'_t})^n \tilde{\delta}_0.$$

Example 9.8. For the higher derivatives $B^{(k)}(t)$, $k \geq 3$, of $B(t)$, we can use the same argument as in Example 9.7 to find its Fourier transform. Note that

$$:B^{(k)}(t)^n: = (-1)^{n(k-1)} \langle :x^{\otimes n}:, (\delta_t^{(k-1)})^{\otimes n}\rangle.$$

Therefore, the $\mathcal{S}$–transformation of $:B^{(k)}(t)^n:$ is given by

$$(\mathcal{S}:B^{(k)}(t)^n:)(\xi) = (\xi^{(k-1)}(t))^n,$$

and its Fourier transform is given by

$$(:B^{(k)}(t)^n:)^\wedge = i^n\,(-1)^{nk}\left(D^*_{\delta_t^{(k-1)}}\right)^n \tilde{\delta}_0.$$

9.C Elementary Properties

In this section we will prove some elementary properties of the Fourier transform. These properties have been proved, in one way or another, e.g., in Ito, Kubo and Takenaka (1990), Kuo (1982, 1989, 1990a), Y. J. Lee (1991) and Yan (1989).

The finite dimensional Fourier transform on $\mathbb{R}^k$ has the following relationships with partial differentiation, multiplication, and translation. For $1 \leq j \leq k$ and $a \in \mathbb{R}^k$,

$$(\partial_{x_j} f)^\wedge (y) = i y_j \hat{f}(y), \tag{9.9}$$

$$(i x_j f)^\wedge (y) = -\partial_{y_j} \hat{f}(y), \tag{9.10}$$

$$(f(\cdot + a))^\wedge (y) = e^{i<y,\, a>} \hat{f}(y).$$

We will show that the white noise Fourier transform has the same relationships with white noise differentiation, multiplication, and translation. For the white noise differentiation, recall from Chapters 5 and 6 that for any $\Phi \in (\mathcal{S})^*$, and $\xi, \eta \in \mathcal{S}(\mathbb{R})$,

$$<D_\eta \Phi, e^{<\cdot,\, \xi>}> = -<\xi, \eta><\Phi, e^{<\cdot,\, \xi>}> + <\Phi, <\cdot, \eta> e^{<\cdot,\, \xi>}>. \tag{9.11}$$

As for the translation recall that in Theorem 4.15 we showed that for any $y \in \mathcal{S}'(\mathbb{R})$, the translation operator τ_y in the direction y, i.e., $(\tau_y \varphi)^\sim(x) = \tilde{\varphi}(x+y)$, is continuous from $(\mathcal{S})$ into itself. Suppose $\eta \in \mathcal{S}(\mathbb{R})$ and $\varphi \in (\mathcal{S})$. Then it is easy to see that $(\mathcal{S}\tau_\eta \varphi)(\xi) = (\mathcal{S}\varphi)(\xi + \eta)$, $\xi \in \mathcal{S}(\mathbb{R})$. This suggests that for $\eta \in \mathcal{S}(\mathbb{R})$, we can define the *translation* τ_η on $(\mathcal{S})^*$ by

$$\tau_\eta \Phi = \mathcal{S}^{-1}\Big((\mathcal{S}\Phi)(\cdot + \eta)\Big). \tag{9.12}$$

Theorem 9.9. *The Fourier transform has the following properties:*
(i) *For any* $\Phi \in (\mathcal{S})^*$ *and* $\eta \in \mathcal{S}(\mathbb{R})$,

$$(D_\eta \Phi)^\wedge = i<\cdot, \eta>\hat{\Phi},$$

$$(i<\cdot, \eta>\Phi)^\wedge = -D_\eta \hat{\Phi},$$

$$(\tau_\eta \Phi)^\wedge = e^{i<\cdot,\, \eta>} \hat{\Phi},$$

$$(e^{i\langle\cdot,\,\eta\rangle}\Phi)^\wedge = \tau_{-\eta}\hat{\Phi}.$$

(ii) *For any* $\Phi\in(\mathcal{S})^*$ *and* $y\in\mathcal{S}'(\mathbb{R})$, $(D_y^*\Phi)^\wedge = -iD_y^*\hat{\Phi}$; *in particular, for any* $t\in\mathbb{R}$,

$$(\partial_t^*\Phi)^\wedge = -i\partial_t^*\hat{\Phi}.$$

Remark. If we put $\eta=\delta_t$ informally in the first two equalities in (i), then we get the symbolic expressions

$$(\partial_t\Phi)^\wedge\,(x) = ix(t)\,\hat{\Phi}(x).$$

$$(ix(t)\Phi)^\wedge = -\partial_t\hat{\Phi}.$$

These equalities can be interpreted as the white noise analogs of (9.9) and (9.10). Recall that $x(t)=\partial_t+\partial_t^*$.

Proof. From (9.5) and (9.11), we have

$$\mathcal{S}(D_\eta\Phi)^\wedge\,(\xi) = \langle D_\eta\overline{\Phi}, e^{-i\langle\cdot,\,\xi\rangle}\rangle$$

$$= i\langle\xi,\eta\rangle\langle\overline{\Phi}, e^{-i\langle\cdot,\,\xi\rangle}\rangle + \langle\overline{\Phi}, \langle\cdot,\eta\rangle e^{-i\langle\cdot,\,\xi\rangle}\rangle. \tag{9.13}$$

On the other hand, by Theorem 5.33, $\langle\cdot,\eta\rangle\hat{\Phi} = (D_\eta + D_\eta^*)\hat{\Phi}$. Therefore,

$$\mathcal{S}(i\langle\cdot,\eta\rangle\hat{\Phi})(\xi) = e^{-\frac{1}{2}|\xi|_2^2}\langle\overline{i\langle\cdot,\eta\rangle\hat{\Phi}}, e^{\langle\cdot,\,\xi\rangle}\rangle$$

$$= ie^{-\frac{1}{2}|\xi|_2^2}\langle(D_\eta + D_\eta^*)\overline{\hat{\Phi}}, e^{\langle\cdot,\,\xi\rangle}\rangle$$

$$= ie^{-\frac{1}{2}|\xi|_2^2}\langle D_\eta\overline{\hat{\Phi}}, e^{\langle\cdot,\,\xi\rangle}\rangle + ie^{-\frac{1}{2}|\xi|_2^2}\langle\overline{\hat{\Phi}}, D_\eta e^{\langle\cdot,\,\xi\rangle}\rangle$$

$$= i\mathcal{S}(D_\eta \hat{\Phi})(\xi) + ie^{-\frac{1}{2}|\xi|_2^2} \langle \overline{\hat{\Phi}}, \langle \xi, \eta \rangle e^{\langle \cdot, \xi \rangle} \rangle$$

$$= i\langle (\mathcal{S}\hat{\Phi})'(\xi), \eta \rangle + i\langle \xi, \eta \rangle (\mathcal{S}\hat{\Phi})(\xi)$$

$$= \langle \overline{\Phi}, \langle \cdot, \eta \rangle e^{-i\langle \cdot, \xi \rangle} \rangle + i\langle \xi, \eta \rangle \langle \overline{\Phi}, e^{-i\langle \cdot, \xi \rangle} \rangle. \quad (9.14)$$

It follows from (9.13) and (9.14) that $(D_\eta \Phi)^\wedge = i\langle \cdot, \eta \rangle \hat{\Phi}$. On the other hand, from (9.5), we have

$$\mathcal{S}(D_y^* \Phi)^\wedge (\xi) = \langle D_y^* \overline{\Phi}, e^{-i\langle \cdot, \xi \rangle} \rangle$$

$$= \langle \overline{\Phi}, D_y e^{-i\langle \cdot, \xi \rangle} \rangle$$

$$= \langle \overline{\Phi}, -i\langle y, \xi \rangle e^{-i\langle \cdot, \xi \rangle} \rangle$$

$$= -i\langle y, \xi \rangle \langle \overline{\Phi}, e^{-i\langle \cdot, \xi \rangle} \rangle$$

$$= -i\langle y, \xi \rangle \mathcal{S}\hat{\Phi}(\xi)$$

$$= \mathcal{S}(-iD_y^* \hat{\Phi})(\xi).$$

In the last step above, we have used the fact that $\mathcal{S}(D_y^* \Psi)(\xi) = \langle y, \xi \rangle \mathcal{S}\Psi(\xi)$ for any y in $\mathcal{S}'(\mathbb{R})$ and Ψ in $(\mathcal{S})^*$. Therefore, $(D_y^* \Phi)^\wedge = -iD_y^* \hat{\Phi}$. In particular, when $y = \delta_t$, we get $(\partial_t^* \Phi)^\wedge = -i\partial_t^* \hat{\Phi}$. Moreover, note that $\langle \cdot, \eta \rangle \Psi = (D_\eta + D_\eta^*)\Psi$ for any η in $\mathcal{S}(\mathbb{R})$ and Ψ in $(\mathcal{S})^*$. Therefore,

$$(i\langle \cdot, \eta \rangle \Phi)^\wedge = i(D_\eta \Phi)^\wedge + i(D_\eta^* \Phi)^\wedge$$

$$= -\langle \cdot, \eta \rangle \hat{\Phi} + D_\eta^* \hat{\Phi}$$

$$= -D_\eta \hat{\Phi}.$$

From (9.6) and (9.12), we obtain that

$$(\mathcal{I}(\tau_\eta \Phi)^\wedge)(\xi) = e^{-\frac{1}{2}|\xi|_2^2} (\mathcal{I}\tau_\eta \Phi)(-i\xi)$$

$$= e^{-\frac{1}{2}|\xi|_2^2} (\mathcal{I}\Phi)(-i\xi+\eta). \tag{9.15}$$

On the other hand, note that by (9.6) we have $<\overline{\hat{\Phi}}, e^{<\cdot,\ \xi>}> = (\mathcal{I}\Phi)(-i\xi)$. Therefore,

$$\mathcal{I}(e^{i<\cdot,\ \eta>} \hat{\Phi})(\xi) = e^{-\frac{1}{2}|\xi|_2^2} <\overline{e^{i<\cdot,\ \eta>} \hat{\Phi}}, e^{<\cdot,\ \xi>}>$$

$$= e^{-\frac{1}{2}|\xi|_2^2} <\overline{\hat{\Phi}}, e^{<\cdot,\ \xi+i\eta>}>$$

$$= e^{-\frac{1}{2}|\xi|_2^2} (\mathcal{I}\Phi)(-i\xi+\eta). \tag{9.16}$$

Obviously, (9.15) and (9.16) imply that $(\tau_\eta \Phi)^\wedge = e^{i<\cdot,\ \eta>} \hat{\Phi}$. Now, by (9.5), we have

$$\mathcal{I}((e^{i<\cdot,\ \eta>} \Phi)^\wedge)(\xi) = <\overline{e^{i<\cdot,\ \eta>} \Phi}, e^{-i<\cdot,\ \xi>}>$$

$$= <\overline{\Phi}, e^{-i<\cdot,\ \xi-\eta>}>$$

$$= \mathcal{I}\hat{\Phi}(\xi-\eta)$$

$$= \mathcal{I}(\tau_{-\eta}\hat{\Phi})(\xi).$$

Hence $(e^{i<\cdot,\ \eta>} \Phi)^\wedge = \tau_{-\eta}\hat{\Phi}$ and the theorem has been proved. □

In the next theorem we will show that the Fourier transform $\mathfrak{F}$ can be expressed in terms of the $\mathcal{T}$– and $\mathcal{S}$–transformations. We define the *reflection map* ρ from $(\mathcal{S})^*$ into itself by

$$\mathcal{S}(\rho\Phi)(\xi) = \mathcal{S}\Phi(-\xi),\ \xi \in \mathcal{S}(\mathbb{R}).$$

Theorem 9.10. $\mathcal{F} = \mathcal{T}^{-1}\,\mathcal{S} = \rho\mathcal{S}^{-1}\,\mathcal{T}$.

Proof. Note that the $\mathcal{T}$– and $\mathcal{S}$–transformations are related by

$$(\mathcal{T}\Psi)(\xi) = e^{-\frac{1}{2}|\xi|_2^2}\,(\mathcal{S}\Psi)(i\xi).$$

From this relation and (9.5), we obtain that for any $\xi \in \mathcal{S}(\mathbb{R})$

$$\begin{aligned}(\mathcal{T}\hat{\Phi})(\xi) &= e^{-\frac{1}{2}|\xi|_2^2}\,(\mathcal{S}\hat{\Phi})(i\xi)\\ &= e^{-\frac{1}{2}|\xi|_2^2}\,<\overline{\Phi},\, e^{<\cdot,\,\xi>}>\\ &= (\mathcal{S}\Phi)(\xi).\end{aligned}$$

This implies that $\mathcal{T}\hat{\Phi} = \mathcal{S}\Phi$ for any Φ in $(\mathcal{S})^*$ and so $\mathcal{T}\mathcal{F} = \mathcal{S}$, i.e., $\mathcal{F} = \mathcal{T}^{-1}\mathcal{S}$. On the other hand, from the definition of the reflection map ρ, we have

$$\begin{aligned}\mathcal{S}(\rho\mathcal{S}^{-1}\,\mathcal{T}\Phi)(\xi) &= (\mathcal{T}\Phi)(-\xi)\\ &= e^{-\frac{1}{2}|\xi|_2^2}\,(\mathcal{S}\Phi)(-i\xi).\end{aligned}$$

Thus, in view of (9.6), we have $\mathcal{F} = \rho\mathcal{S}^{-1}\,\mathcal{T}$. □

Theorem 9.11. $\mathcal{F}^2 = \rho$ *and* $\mathcal{F}^4 = \mathrm{Id}$. *Here* ρ *is the reflection map on* $(\mathcal{S})^*$.

Proof. By (9.5), the $\mathcal{S}$–transformation of $\mathcal{F}\Phi$ is given by

$$\mathcal{S}(\mathfrak{F}\Phi)(\xi) = <\overline{\Phi}, e^{-i<\cdot,\ \xi>}>,\ \xi\in\mathcal{S}(\mathbb{R}).$$

Hence, by applying (9.6), we get the following

$$\mathcal{S}(\mathfrak{F}^2\Phi)(\xi) = <\overline{\Phi}, e^{-<\cdot,\ \xi>}>e^{-\frac{1}{2}|\xi|_2^2},\ \xi\in\mathcal{S}(\mathbb{R}).$$

On the other hand, the $\mathcal{S}$–transformation of $\rho\Phi$ is given by

$$\mathcal{S}(\rho\Phi)(\xi) = <\overline{\Phi}, e^{-<\cdot,\ \xi>}>e^{-\frac{1}{2}|\xi|_2^2},\ \xi\in\mathcal{S}(\mathbb{R}).$$

Thus we get immediately $\mathfrak{F}^2=\rho$. But $\rho^2=\mathrm{Id}$ and, so, we have $\mathfrak{F}^4=\mathrm{Id}$. □

Theorem 9.12. *Let* p *be a real number. Then the following equality holds for all* Φ *in* $(\mathcal{S})_p$:

$$|\hat{\Phi}:g_{-2}|_{2,p} = |\Phi|_{2,p}, \tag{9.17}$$

or equivalently,

$$|\hat{\Phi}|_{2,p} = |\Phi:g_{-2}|_{2,p}, \tag{9.18}$$

where $g_{-2} = \mathcal{N}\exp(\frac{1}{4}|\cdot|_2^2)$.

Remark. This theorem is the white noise analog of the Plancherel theorem. In fact, it can be checked easily that (9.17) with p=0 reduces to $|\hat{f}|_2 = |f|_2$ for all f in $L^2(\mathbb{R}^n)$ with respect to the Lebesgue measure.

Proof. From (9.6) we have

$$(\mathcal{S}\hat{\Phi})(\xi)e^{\frac{1}{2}|\xi|_2^2} = (\mathcal{S}\Phi)(-i\xi).$$

But the $\mathcal{S}$–transformation of g_{-2} is given by $\exp(\frac{1}{2}|\xi|_2^2)$. Therefore,

$$\mathcal{S}(\hat{\Phi}:g_{-2})(\xi) = (\mathcal{S}\Phi)(-i\xi).$$

This means that if Φ has the following chaos decomposition

$$\Phi(x) = \sum_{n=0}^{\infty} <: x^{\otimes n} :, f_n >, \tag{9.19}$$

then $\hat{\Phi}:g_{-2}$ is represented by

$$\hat{\Phi}:g_{-2}\,(x) = \sum_{n=0}^{\infty} <: x^{\otimes n} :, (-i)^n f_n>. \tag{9.20}$$

It is obvious from (9.19) and (9.20) that $|\hat{\Phi}:g_{-2}|_{2,p} = |\Phi|_{2,p}$ for any $p \in \mathbb{R}$. To see that (9.17) and (9.18) are equivalent, note that the reflection map ρ preserves all $|\cdot|_{2,p}$ norms and $(\rho\Phi):g_{-2}=\rho(\Phi:g_{-2})$. Therefore, by replacing Φ in (9.17) and using the fact $\mathfrak{F}^2=\rho$ from Theorem 9.11, we get

$$\begin{aligned} |\hat{\Phi}|_{2,p} &= |(\rho\Phi):g_{-2}|_{2,p} \\ &= |\rho(\Phi:g_{-2})|_{2,p} \\ &= |\Phi:g_{-2}|_{2,p}. \end{aligned}$$

The same argument as above shows also that (9.18) implies (9.17). □

We can use Theorem 9.12 to find the image of $(\mathcal{S})$ under the Fourier transform. By (9.17), we have $\varphi \in (\mathcal{S})$ if and only if $\hat{\varphi}:g_{-2} \in (\mathcal{S})$. But $g_{-2}:\tilde{\delta}_0=1$. Hence $\hat{\varphi}:g_{-2} \in (\mathcal{S})$ if and only if $\hat{\varphi} \in (\mathcal{S}):\tilde{\delta}_0$. Thus we have proved the following result.

Corollary 9.13. *The image of* $(\mathcal{S})$ *under the Fourier transform* $\mathfrak{F}$ *is given by*

$$(\mathcal{S}):\tilde{\delta}_0 \equiv \{\varphi:\tilde{\delta}_0;\ \varphi\in(\mathcal{S})\},$$

where $\tilde{\delta}_0$ *is the delta function at* 0.

9.D The Fourier Transform as an Adjoint Operator

For the finite dimensional Fourier transform, we have $<\mathcal{F}f, g>=<f, \mathcal{F}g>$ for all f, $g\in\mathcal{S}(\mathbb{R}^k)$. Thus the Fourier transform has a continuous extension to $\mathcal{S}'(\mathbb{R}^k)$ given by $<\mathcal{F}F, g>=<F, \mathcal{F}g>$, $g\in\mathcal{S}(\mathbb{R}^k)$. The situation is different in white noise calculus. We have defined the Fourier transform directly on the space $(\mathcal{S})^*$ of generalized white noise functionals. We will show that the Fourier transform is the adjoint of a continuous linear operator $\mathcal{G}$ from $(\mathcal{S})$ into itself. This result is due independently to Hida, Kuo and Obata (1991), Y. J. Lee (1989), J. S. Lee (1992), and Kondratiev (private communication 1991). The following presentation is taken from Hida, Kuo and Obata (1991).

Lemma 9.14. *For any* $p>\frac{1}{4}$ *and* $f_{n+2m}\in\mathcal{S}(\mathbb{R}^{n+2m})$, *we have*

$$|<\mathrm{Tr}^{\otimes m}, f_{n+2m}>|_{2,p} \leq a_p^m |f_{n+2m}|_{2,p}, \tag{9.21}$$

where $a_p=(\sum_{j=0}^{\infty} (2j+2)^{-4p})^{1/2}$.

Proof. Let e_j be the Hermite function of order j. Then $Ae_j=(2j+2)e_j$, $j\geq 0$. Therefore, for any $p>\frac{1}{4}$,

$$|\mathrm{Tr}|_{2,-p}^2 = \sum_{j=0}^{\infty} |A^{-p}e_j|_2^4$$
$$= \sum_{j=0}^{\infty} (2j+2)^{-4p}.$$

Hence $|\mathrm{Tr}|_{2,-p}\leq a_p$ with a_p given as in the lemma. This implies immediately that for any $F_n\in\mathcal{S}_p'(\mathbb{R}^n)$, we have

$$|\langle \mathrm{Tr}^{\otimes m} \otimes F_n, f_{n+2m} \rangle| \leq |\mathrm{Tr}|_{2,-p}^m |F_n|_{2,-p} |f_{n+2m}|_{2,p}$$

$$\leq a_p^m |F_n|_{2,-p} |f_{n+2m}|_{2,p}.$$

By taking the supremum over the set $\{F_n \in \mathcal{S}_p'(\mathbb{R}^n);\ |F_n|_{2,-p} \leq 1\}$, we get (9.21). □

Recall that the $\mathcal{S}$-transformation of $\Phi \in (\mathcal{S})^*$ is defined by

$$(\mathcal{S}\Phi)(\xi) = \langle \overline{\Phi}, :e^{\langle \cdot, \xi \rangle}: \rangle, \ \xi \in \mathcal{S}(\mathbb{R}).$$

Observe that for any fixed $x \in \mathcal{S}'(\mathbb{R})$, the renormalization $:e^{\langle \cdot, x \rangle}:$ is in $(\mathcal{S})^*$. Hence if $\varphi \in (\mathcal{S})$, then the function $\mathcal{S}\varphi(\xi)$, $\xi \in \mathcal{S}(\mathbb{R})$, extends to $\mathcal{S}\varphi(x)$, $x \in \mathcal{S}'(\mathbb{R})$. This is the motivation for defining the following map Υ on $(\mathcal{S})$, i.e., for $\varphi \in (\mathcal{S})$,

$$(\Upsilon\varphi)(x) = \langle :e^{\langle \cdot, x \rangle}:, \varphi \rangle, \ x \in \mathcal{S}'(\mathbb{R}). \tag{9.22}$$

We remark that this map has been also considered by Yan (1990).

Theorem 9.15. *Υ is a continuous linear operator from $(\mathcal{S})$ into itself.*

Proof. This proof is based on the same argument as in the proof of Proposition 4.3 in Hida, Kuo and Obata (1991). Note that for $x \in \mathcal{S}'(\mathbb{R})$,

$$:e^{\langle \cdot, x \rangle}:(y) = \sum_{n=0}^{\infty} \frac{1}{n!} \langle :y^{\otimes n}:, x^{\otimes n} \rangle.$$

Hence if φ is represented by

$$\varphi(y) = \sum_{n=0}^{\infty} \langle :y^{\otimes n}:, f_n \rangle, \ f_n \in \mathcal{S}(\mathbb{R}^n), \tag{9.23}$$

then $\Upsilon\varphi$ is given by

$$(\Upsilon\varphi)(x) = \sum_{n=0}^{\infty} \langle x^{\otimes n}, f_n \rangle. \tag{9.24}$$

But $x^{\otimes n}$ has the following decomposition

$$x^{\otimes n} = \sum_{m=0}^{[n/2]} \binom{n}{2m}(2m-1)!! : x^{\otimes(n-2m)} : \hat{\otimes} \mathrm{Tr}^{\otimes m}.$$

With this decomposition, we can check that $\Upsilon\varphi$ is represented by

$$\Upsilon\varphi = \sum_{n=0}^{\infty} \langle : x^{\otimes n} :, g_n \rangle, \quad g_n = \sum_{m=0}^{\infty} \frac{(n+2m)!}{2^m m! n!} \langle \mathrm{Tr}^{\otimes m}, f_{n+2m} \rangle. \tag{9.25}$$

Now, use (9.21) and the fact that $|f_{n+2m}|_{2,p} \leq 2^{-q(n+2m)} |f_{n+2m}|_{2,p+q}$ to get

$$|g_n|_{2,p} \leq \sum_{m=0}^{\infty} \frac{(n+2m)!}{2^m m! n!} a_p^m \, 2^{-q(n+2m)} |f_{n+2m}|_{2,p+q}.$$

Then by the Schwarz inequality and the fact that $4^m (m!)^2 \geq (2m)!$, we have

$$|g_n|_{2,p}^2 \leq \Big(\sum_{m=0}^{\infty} (n+2m)! |f_{n+2m}|_{2,p+q}^2 \Big) \Big(\sum_{m=0}^{\infty} \frac{(n+2m)!}{(n!)^2 (2m)!} a_p^{2m} 2^{-2q(n+2m)} \Big)$$

$$\leq |\varphi|_{2,p+q}^2 \sum_{m=0}^{\infty} \frac{(n+2m)!}{(n!)^2 (2m)!} a_p^{2m} 2^{-2q(n+2m)}.$$

Therefore, from the representation in (9.25), we get

$$|\Upsilon\varphi|_{2,p}^2 \leq \sum_{n=0}^{\infty} n! |g_n|_{2,p}^2$$

$$\leq |\varphi|^2_{2,p+q} \sum_{n=0}^{\infty} \sum_{m=0}^{\infty} \binom{n+2m}{n} a_p^{2m} 2^{-2q(n+2m)}$$

$$= |\varphi|^2_{2,p+q} \sum_{m=0}^{\infty} \sum_{j=2m}^{\infty} \binom{j}{j-2m} a_p^{2m} 2^{-2qj}$$

$$= |\varphi|^2_{2,p+q} \sum_{j=0}^{\infty} \sum_{m=0}^{[j/2]} \binom{j}{j-2m} (a_p 2^{-2q})^{2m} (2^{-2q})^{j-2m}$$

$$\leq |\varphi|^2_{2,p+q} \sum_{j=0}^{\infty} (a_p 2^{-2q} + 2^{-2q})^j.$$

By choosing q large enough the last series is convergent and this shows that Υ is a continuous linear operator from $(\mathcal{S})$ into itself. □

Now, we are ready to define the transformation $\mathcal{G}$ mentioned before Lemma 9.14. For $\varphi \in (\mathcal{S})$, define $\mathcal{G}\varphi$ by

$$\mathcal{G}\varphi(x) = \Upsilon\varphi(ix), \ x \in \mathcal{S}'(\mathbb{R}). \tag{9.26}$$

Note that if φ is represented as in (9.23), then in view of (9.24) $\mathcal{G}\varphi$ is given by

$$(\mathcal{G}\varphi)(x) = \sum_{n=0}^{\infty} \langle x^{\otimes n}, (-i)^n f_n \rangle. \tag{9.27}$$

Theorem 9.16. *The transformation $\mathcal{G}$ is a continuous linear operator from $(\mathcal{S})$ into itself. Moreover, the Fourier transform $\mathcal{F}$ is the adjoint of $\mathcal{G}$.*

Proof. By comparing the representations (9.24) for Υ and (9.27) for $\mathcal{G}$, we see that the proof of Theorem 9.15 shows that $\mathcal{G}$ is a continuous linear operator from $(\mathcal{S})$ into itself. The continuity of $\mathcal{G}$ can also be proved in another way by observing that $\mathcal{G} = \sigma_i \circ \Upsilon$. The map σ_i is continuous from $(\mathcal{S})$ into itself (cf. Theorem 4.15). Therefore, the continuity of $\mathcal{G}$ follows from that of Υ in Theorem 9.15. To show that the Fourier transform $\mathcal{F}$ is the adjoint of $\mathcal{G}$, note that for any $\xi\in\mathcal{S}(\mathbb{R})$ we have

$$(\Upsilon\!: e^{<\cdot,\, \xi>}:)(x) = \langle\!: e^{<\cdot,\, x>}:, : e^{<\cdot,\, \xi>}:\!\rangle = e^{<x,\, \xi>}, \quad x\in\mathcal{S}'(\mathbb{R}).$$

Therefore, by (9.26),

$$(\mathcal{G}\!: e^{<\cdot,\, \xi>}:)(x) = e^{-i<x,\, \xi>}, \quad x\in\mathcal{S}'(\mathbb{R}).$$

Hence for any $\Phi\in(\mathcal{S})^*$,

$$\langle\!\langle \mathcal{G}^*\overline{\Phi}, : e^{<\cdot,\, \xi>}:\rangle\!\rangle = \langle\!\langle \overline{\Phi}, \mathcal{G}\!: e^{<\cdot,\, \xi>}:\rangle\!\rangle = \langle\!\langle \overline{\Phi}, e^{-i<\cdot,\, \xi>}\rangle\!\rangle, \ \xi\in\mathcal{S}(\mathbb{R}). \qquad (9.28)$$

On the other hand, from (9.5) we have

$$\langle\!\langle \mathcal{F}\overline{\Phi}, : e^{<\cdot,\, \xi>}:\rangle\!\rangle = (\mathcal{S}\hat{\overline{\Phi}})(\xi) = \langle\!\langle \overline{\Phi}, e^{-i<\cdot,\, \xi>}\rangle\!\rangle, \ \xi\in\mathcal{S}(\mathbb{R}). \qquad (9.29)$$

Since the linear span of the set $\{: e^{<\cdot,\, \xi>}:;\ \xi\in\mathcal{S}(\mathbb{R})\}$ is dense in $(\mathcal{S})$, we conclude from (9.28) and (9.29) that $\mathcal{F}\Phi=\mathcal{G}^*\Phi$ for any $\Phi\in(\mathcal{S})^*$. Thus we have $\mathcal{F}=\mathcal{G}^*$. □

Theorem 9.17. *The Fourier transform $\mathcal{F}$ is an injective, strongly continuous linear operator from $(\mathcal{S})^*$ onto itself.*

Proof. The fact $\mathcal{F}^4=\mathrm{Id}$ from Theorem 9.11 implies that $\mathcal{F}$ is injective and onto. The linearity is obvious. The strong continuity follows from Theorem 9.16. □

9.E A Characterization of the Fourier Transform

As in the finite dimensional case, the white noise Fourier transform can be characterized by its relationships with differentiation and multiplication. This result (Theorem 9.20 below) is due to Hida, Kuo and Obata (1991). In fact, without much additional effort, we can also characterize the operator $\mathcal{G}$ defined in

(9.26).

For each $\xi\in\mathcal{S}(\mathbb{R})$, consider the differentiation operator D_ξ in the direction ξ and the multiplication operator q_ξ given by $q_\xi\varphi=i<\cdot, \xi>\varphi$. We know that both D_ξ and q_ξ are continuous linear operators from $(\mathcal{S})$ into itself. Moreover, they can be extended continuously to continuous linear operators from $(\mathcal{S})^*$ into itself. We will use the same notation D_ξ and q_ξ for the extensions. Note that $q_\xi^*=q_\xi$ and $q_\xi=i(D_\xi+D_\xi^*)$ as operators from $(\mathcal{S})$ into itself and as operators from $(\mathcal{S})^*$ into itself.

Lemma 9.18. *Let* L *be a continuous linear operator from* $(\mathcal{S})$ *into itself such that* (*i*) $LD_\xi = D_\xi L$ *and* (*ii*) $Lq_\xi = q_\xi L$ *for all* ξ *in* $\mathcal{S}(\mathbb{R})$. *Then there exists a constant* c *such that* $L = c\ \mathrm{Id}$.

Proof. Let $\lambda(x)=(L1)(x)$ and $\psi_\xi(x)=(i<x, \xi>)^n$. Then we have $\psi_\xi=q_\xi^n 1$. It follows from the assumption (ii) that

$$L\psi_\xi = Lq_\xi^n 1 = q_\xi^n L1 = q_\xi^n\ \lambda = \lambda\psi_\xi. \tag{9.30}$$

But the span of the set $\{\psi_\xi;\ \xi\in\mathcal{S}(\mathbb{R})\}$ is dense in $(\mathcal{S})$ and the multiplication $(\varphi, \psi)\mapsto\varphi\psi$ is continuous from $(\mathcal{S})\times(\mathcal{S})$ to $(\mathcal{S})$. Thus (9.30) implies that $L\varphi = \lambda\varphi$ for all $\varphi\in(\mathcal{S})$. On the other hand, let τ_ξ be the translation in the direction ξ, i.e., $\tau_\xi\varphi(x) = \varphi(x+\xi)$. By Corollary 5.16, for any $\varphi\in(\mathcal{S})$ and $\xi\in\mathcal{S}(\mathbb{R})$,

$$\tau_\xi\varphi = \sum_{n=0}^{\infty}\frac{1}{n!}D_\xi^n\ \varphi. \tag{9.31}$$

It follows from (9.31) and the assumption (i) that $L\tau_\xi\varphi = \tau_\xi L\varphi$ for all φ in $(\mathcal{S})$. In particular, with $\varphi=1$, we have $\lambda=\tau_\xi\lambda$ for all $\xi\in\mathcal{S}(\mathbb{R})$. That is, $\lambda(\cdot)=\lambda(\cdot+\xi)$ μ-a.e. for all $\xi\in\mathcal{S}(\mathbb{R})$. But, the translations τ_ξ, $\xi\in\mathcal{S}(\mathbb{R})$, acts on $\mathcal{S}'(\mathbb{R})$ ergodically with respect to μ by Theorem 5.8 in Hida (1980a). Therefore, λ is a constant function and the theorem is proved. □

By taking the adjoint operators in Lemma 9.18 we obtain easily the following result for an operator acting on the space $(\mathcal{S})^*$ of generalized white noise functionals. The operator q_ξ is regarded as from $(\mathcal{S})^*$ into itself.

Lemma 9.19. *Let* Λ *be a strongly continuous linear operator from* $(\mathcal{S})^*$ *into itself such that* (i) $\Lambda D_\xi^* = D_\xi^* \Lambda$ *and* (ii) $\Lambda q_\xi = q_\xi \Lambda$ *for all* ξ *in* $\mathcal{S}(\mathbb{R})$. *Then there exists a constant* c *such that* $\Lambda = c\ \mathrm{Id}$.

Now, we are ready to prove a theorem characterizing the Fourier transform due to Hida, Kuo and Obata (1991).

Theorem 9.20. *The Fourier transform* $\mathfrak{F}$ *has the following properties:*

(1) $\mathfrak{F}$ *is a continuous linear operator from* $(\mathcal{S})^*$ *into itself;*
(2) $\mathfrak{F} D_\xi = q_\xi \mathfrak{F}$ *for any* $\xi \in \mathcal{S}(\mathbb{R})$;
(3) $\mathfrak{F} q_\xi = -D_\xi \mathfrak{F}$ *for any* $\xi \in \mathcal{S}(\mathbb{R})$.

Conversely, suppose $\mathcal{A}$ *is a linear operator from* $(\mathcal{S})^*$ *into itself satisfying the conditions* (1)–(3). *Then there exists a constant* c *such that* $\mathcal{A} = c\mathfrak{F}$.

Proof. By Theorems 9.9 and 9.17 the Fourier transform $\mathfrak{F}$ has the properties (1)–(3). Conversely, suppose $\mathcal{A}$ is a linear operator from $(\mathcal{S})^*$ into itself satisfying conditions (1)–(3). By Theorem 9.11 $\mathfrak{F}$ is invertible and, in fact, $\mathfrak{F}^{-1} = \mathfrak{F}^3$. Let $\Lambda = \mathfrak{F}^{-1}\mathcal{A}$. Then Λ is a strongly continuous linear operator from $(\mathcal{S})^*$ into itself. Moreover, it is easy to check that $\Lambda D_\xi = D_\xi \Lambda$ and $\Lambda q_\xi = q_\xi \Lambda$ for any $\xi \in \mathcal{S}(\mathbb{R})$. But $q_\xi = i(D_\xi + D_\xi^*)$. Hence we have $\Lambda D_\xi^* = D_\xi^* \Lambda$ for any $\xi \in \mathcal{S}(\mathbb{R})$. Thus Λ satisfies the conditions in Lemma 9.19. Therefore, there exists a constant c such that $\Lambda = c\ \mathrm{Id}$, i.e., $\mathcal{A} = c\mathfrak{F}$. □

We can also characterize the operator $\mathfrak{G}$ as follows:

Theorem 9.21. *The operator* $\mathcal{G}$ *has the following properties:*

(1) $\mathcal{G}q_\xi = D_\xi^*\mathcal{G}$ *and* $\mathcal{G}D_\xi^* = -q_\xi\mathcal{G}$ *for any* $\xi \in \mathcal{S}(\mathbb{R})$;

(2) $\mathcal{G}D_y = iD_y\mathcal{G}$ *for any* $y \in \mathcal{S}'(\mathbb{R})$;

(3) $\mathcal{G}^2 = \rho$ *and* $\mathcal{G}^4 = I$, *where* $(\rho\varphi)(x) = \varphi(-x)$.

Moreover, $\mathcal{G}$ *is the unique continuous linear operator (up to some constant) from* $(\mathcal{S})$ *into itself satisfying condition* (1).

Proof. To show (1), simply take the adjoint of Theorem 9.9 (i) or Theorem 9.20 (2) and (3). To get (2), take the adjoint in both sides of $(D_y^*\Phi)^\wedge = -iD_y^*\hat{\Phi}$ from Theorem 9.9 (ii). (3) follows from Theorem 9.11 by taking the adjoint in view of the fact that $\langle\rho\Phi, \varphi\rangle = \langle\Phi, \rho\varphi\rangle$ for any $\Phi \in (\mathcal{S})^*$ and $\varphi \in (\mathcal{S})$. The last assertion is a consequence of Theorem 9.20. On the other hand, we can also prove this assertion directly by the same idea as in the proof of Theorem 9.20 except that we use Lemma 9.18 instead of Lemma 9.19. □

9.F Relations with Laplacians, Wick Product, and Convolution

In this section we will study the relationships between the Fourier transform and various Laplacian operators. The results are taken from the following papers: Hida and Saitô (1988), Kuo (1989, 1990a), and Kuo, Obata and Saitô (1990). We will also study the relationships of the Fourier transform with the Wick product and convolution due to Kuo (1990a, 1991).

Recall from Chapter 6 that we have the Gross Laplacian Δ_G, the number operator N, the Lévy Laplacian Δ_L and the Volterra Laplacian Δ_V.

Theorem 9.22. $(\Delta_G^*\Phi)^\wedge = -\Delta_G^*\hat{\Phi}$ *for all* Φ *in* $(\mathcal{S})^*$.

Proof. It is easy to check that for any $z \in \mathbb{C}$ and $\xi \in \mathcal{S}(\mathbb{R})$,

$$\Delta_G e^{z<\cdot,\ \xi>} = z^2|\xi|_2^2 e^{z<\cdot,\ \xi>}. \tag{9.32}$$

Therefore, by (9.5) and (9.32), we have

$$\begin{aligned}\mathcal{F}(\Delta_G^*\Phi)^\wedge(\xi) &= <\Delta_G^*\overline{\Phi},\ e^{-i<\cdot,\ \xi>}> \\ &= <\overline{\Phi},\ \Delta_G e^{-i<\cdot,\ \xi>}> \\ &= -|\xi|_2^2<\overline{\Phi},\ e^{-i<\cdot,\ \xi>}> \\ &= -|\xi|_2^2(\mathcal{F}\hat{\Phi})(\xi).\end{aligned} \tag{9.33}$$

On the other hand, by the definition of $\mathcal{F}$–transformation and (9.32),

$$\begin{aligned}\mathcal{F}(\Delta_G^*\ \hat{\Phi})(\xi) &= <\Delta_G^*\overline{\hat{\Phi}},\ e^{<\cdot,\ \xi>}>e^{-\frac{1}{2}|\xi|_2^2} \\ &= <\overline{\hat{\Phi}},\ \Delta_G e^{<\cdot,\ \xi>}>e^{-\frac{1}{2}|\xi|_2^2} \\ &= <\overline{\hat{\Phi}},\ |\xi|_2^2 e^{<\cdot,\ \xi>}>e^{-\frac{1}{2}|\xi|_2^2} \\ &= |\xi|_2^2(\mathcal{F}\hat{\Phi})(\xi).\end{aligned} \tag{9.34}$$

It follows from (9.33) and (9.34) that $\mathcal{F}(\Delta_G^*\Phi)^\wedge(\xi)=-\mathcal{F}(\Delta_G^*\hat{\Phi})(\xi)$ for all $\xi\in\mathcal{S}(\mathbb{R})$ and so $(\Delta_G^*\Phi)^\wedge=-\Delta_G^*\hat{\Phi}$. □

Theorem 9.23. $(N\Phi)^\wedge = N\hat{\Phi} + \Delta_G^*\ \hat{\Phi}$ *for all* Φ *in* $(\mathcal{S})^*$.

Proof. First note that for any $\xi\in\mathcal{S}(\mathbb{R})$,

$$Ne^{-i<\cdot,\ \xi>} = (|\xi|_2^2 - i<\cdot,\ \xi>)e^{-i<\cdot,\ \xi>}, \tag{9.35}$$

$$N{:}\, e^{<\cdot,\, \xi>}{:} = (<\cdot, \xi> - |\xi|_2^2){:}\, e^{<\cdot,\, \xi>}{:}. \tag{9.36}$$

Let $\Phi \in (\mathcal{S})^*$. Then by (9.5), (9.34) and (9.35), we have

$$\begin{aligned}
\mathcal{S}(N\Phi)^\wedge(\xi) &= <N\overline{\Phi}, e^{-i<\cdot,\, \xi>}> \\
&= <\overline{\Phi}, Ne^{-i<\cdot,\, \xi>}> \\
&= |\xi|_2^2<\overline{\Phi}, e^{-i<\cdot,\, \xi>}> - i<\overline{\Phi}, <\cdot, \xi>e^{-i<\cdot,\, \xi>}> \\
&= |\xi|_2^2\mathcal{S}\hat{\Phi}(\xi) - i<\overline{\Phi}, <\cdot, \xi>e^{-i<\cdot,\, \xi>}> \\
&= \mathcal{S}(\Delta_G^* \hat{\Phi})(\xi) - i<\overline{\Phi}, <\cdot, \xi>e^{-i<\cdot,\, \xi>}>.
\end{aligned} \tag{9.37}$$

Now, by (9.5), the following holds for any $t \in \mathbb{R}$,

$$(\mathcal{S}\hat{\Phi})(t\xi) = <\overline{\Phi}, e^{-it<\cdot,\, \xi>}>.$$

On the other hand, by the definition of the $\mathcal{S}$-transformation, we have

$$(\mathcal{S}\hat{\Phi})(t\xi) = e^{-\frac{1}{2}t^2|\xi|_2^2}<\overline{\hat{\Phi}}, e^{t<\cdot,\, \xi>}>.$$

Therefore, for any $t \in \mathbb{R}$,

$$<\overline{\Phi}, e^{-it<\cdot,\, \xi>}> = e^{-\frac{1}{2}t^2|\xi|_2^2}<\overline{\hat{\Phi}}, e^{t<\cdot,\, \xi>}>.$$

We can differentiate both sides with respect to t and then put t=1 to get

$$-i<\overline{\Phi}, <\cdot, \xi>e^{-i<\cdot,\, \xi>}> = <\overline{\hat{\Phi}}, (<\cdot, \xi> - |\xi|_2^2){:}\, e^{<\cdot,\, \xi>}{:}>.$$

Hence, by (9.36), we have

$$-i\langle\overline{\Phi}, \langle\cdot, \xi\rangle e^{-i\langle\cdot, \xi\rangle}\rangle = \langle\overline{\hat{\Phi}}, N{:}\, e^{\langle\cdot, \xi\rangle}{:}\rangle$$

$$= \langle N\overline{\hat{\Phi}}, {:}\, e^{\langle\cdot, \xi\rangle}{:}\rangle$$

$$= \mathcal{S}(N\hat{\Phi})(\xi). \tag{9.38}$$

It follows from (9.37) and (9.38) that $\mathcal{S}(N\Phi)^\wedge(\xi)=\mathcal{S}(\Delta_G^*\,\hat{\Phi})(\xi)+\mathcal{S}(N\hat{\Phi})(\xi)$ for all $\xi\in\mathcal{S}(\mathbb{R})$. This implies that $(N\Phi)^\wedge=N\hat{\Phi}+\Delta_G^*\,\hat{\Phi}$. □

In order to find out how the Fourier transform is related to the Lévy and Volterra Laplacians, we need to consider functional derivatives. Let $\Phi\in(\mathcal{S})^*$, $F=\mathcal{S}\Phi$ and $G=\mathcal{S}\hat{\Phi}$. Then, by (9.6), $G(\xi)=F(-i\xi)\exp(-\frac{1}{2}|\xi|_2^2)$. It is easy to check that

$$G'(\xi;\, t) = -\, iF'(-i\xi;\, t)e^{-\frac{1}{2}|\xi|_2^2} - \xi(t)F(-i\xi)e^{-\frac{1}{2}|\xi|_2^2}, \tag{9.39}$$

$$\begin{aligned} G''(\xi;\, s,\, t) = &-\, F''(-i\xi;\, s,\, t)e^{-\frac{1}{2}|\xi|_2^2} + i\xi(s)F'(-i\xi;\, t)e^{-\frac{1}{2}|\xi|_2^2} - \\ &- \mathrm{Tr}(s,\, t)F(-i\xi)e^{-\frac{1}{2}|\xi|_2^2} + i\xi(t)F'(-i\xi;\, s)e^{-\frac{1}{2}|\xi|_2^2} + \\ &+ \xi(s)\xi(t)e^{-\frac{1}{2}|\xi|_2^2}. \end{aligned} \tag{9.40}$$

Recall that in Sections 6.C and 6.D we use $\mathcal{D}(\Delta_V)$ and $\mathcal{D}(\Delta_L)$ to denote the sets of V–functionals and L–functionals, respectively. Moreover, recall that if $\Phi\in\mathcal{D}(\Delta_L)$ then $\Delta_L^T\Phi$ is defined for any finite interval T.

Theorem 9.24. *If* $\Phi\in\mathcal{D}(\Delta_L)$, *then* $\hat{\Phi}\in\mathcal{D}(\Delta_L)$ *and* $\Delta_L^T\hat{\Phi} = -(\Delta_L^T\Phi)^\wedge - \hat{\Phi}$ *for any* *finite interval* T.

Proof. It is obvious from (9.39) and (9.40) that if F is an L–functional then G is also an L–functional. Hence $\Phi\in\mathfrak{D}(\Delta_L)$ implies that $\hat{\Phi}\in\mathfrak{D}(\Delta_L)$. Moreover, it follows from (9.40) that the Lévy part G_L'' of G is given by

$$G_L''(\xi;\, t) = -F_L''(-i\xi;\, t)e^{-\frac{1}{2}|\xi|_2^2} - F(-i\xi)e^{-\frac{1}{2}|\xi|_2^2},$$

where F_L'' is the Lévy part of F. For simplicity we denote Δ_L^T by Δ_L. Then

$$\mathcal{S}(\Delta_L\hat{\Phi})(\xi) = -e^{-\frac{1}{2}|\xi|_2^2}\frac{1}{|T|}\int_T F_L''(-i\xi;\, t)\, dt - F(-i\xi)e^{-\frac{1}{2}|\xi|_2^2}. \tag{9.41}$$

Now, note that by (9.6) and the definition of Δ_L we have

$$\begin{aligned}\mathcal{S}(\Delta_L\Phi)^\wedge\,(\xi) &= \mathcal{S}(\Delta_L\Phi)\,(-i\xi)e^{-\frac{1}{2}|\xi|_2^2}\\ &= e^{-\frac{1}{2}|\xi|_2^2}\frac{1}{|T|}\int_T F_L''(-i\xi;\, t)\, dt.\end{aligned} \tag{9.42}$$

Note also that by (9.6)

$$(\mathcal{S}\hat{\Phi})(\xi) = F(-i\xi)e^{-\frac{1}{2}|\xi|_2^2}. \tag{9.43}$$

It follows from (9.41), (9.42) and (9.43) that $\Delta_L\hat{\Phi}=-(\Delta_L\Phi)^\wedge-\hat{\Phi}$. □

Theorem 9.25. *If* $\Phi\in\mathfrak{D}(\Delta_V)$, *then* $\hat{\Phi}\in\mathfrak{D}(\Delta_V)$ *and* $\Delta_V\hat{\Phi} = -(\Delta_V\Phi)^\wedge - 2N\hat{\Phi} - \Delta_G^*\hat{\Phi}$.

Proof. It is obvious from (9.39) and (9.40) that if F is a V–functional then G is also a V–functional. Hence $\Phi\in\mathfrak{D}(\Delta_V)$ implies that $\hat{\Phi}\in\mathfrak{D}(\Delta_V)$. Moreover, it follows from (9.40) that the Volterra part G_V'' of G is given by

$$G''_V(\xi; s, t) = -F''_V(-i\xi; s, t)e^{-\frac{1}{2}|\xi|_2^2} + i\xi(s)F'(-i\xi; t)e^{-\frac{1}{2}|\xi|_2^2}$$

$$+ i\xi(t)F'(-i\xi; s)e^{-\frac{1}{2}|\xi|_2^2} + \xi(s)\xi(t)e^{-\frac{1}{2}|\xi|_2^2},$$

where F''_V is the Volterra part of F. By the assumption $F''_V(\xi; \cdot, \cdot)$ is the kernel function of a trace class operator. Therefore, $G''_G(\xi; \cdot,\cdot)$ is also the kernel function of a trace class operator and

$$\mathcal{F}(\Delta_V \hat{\Phi})(\xi) = \operatorname{trace}_{L^2(\mathbb{R})} G''_V(\xi)$$

$$= -\operatorname{trace}_{L^2(\mathbb{R})} F''_V(-i\xi)e^{-\frac{1}{2}|\xi|_2^2} - 2\langle F'(-i\xi), -i\xi\rangle e^{-\frac{1}{2}|\xi|_2^2}$$

$$+ |\xi|_2^2 F(-i\xi)e^{-\frac{1}{2}|\xi|_2^2}. \tag{9.44}$$

Now, note that from (9.6) and the definition of $\Delta_V\Phi$

$$\mathcal{F}(\Delta_V\Phi)^\wedge(\xi) = \operatorname{trace}_{L^2(\mathbb{R})} F''_V(-i\xi)e^{-\frac{1}{2}|\xi|_2^2}. \tag{9.45}$$

From (9.6) and (9.34), we get

$$\mathcal{F}(\Delta_G^* \hat{\Phi})(\xi) = |\xi|_2^2 F(-i\xi)e^{-\frac{1}{2}|\xi|_2^2}. \tag{9.46}$$

On the other hand, from Theorem 6.11, we have

$$\mathcal{F}(N\Phi)(\xi) = \langle F'(\xi), \xi\rangle.$$

Hence, from (9.6)

$$\mathcal{F}(N\Phi)^\wedge(\xi) = \langle F'(-i\xi), -i\xi\rangle e^{-\frac{1}{2}|\xi|_2^2}. \tag{9.47}$$

It follows from (9.44)–(9.47) that

$$\Delta_V \hat{\Phi} = -(\Delta_V \Phi)^\wedge - 2(N\Phi)^\wedge + \Delta_G^* \hat{\Phi}. \tag{9.48}$$

From (9.48) and Theorem 9.23 we obtain immediately that $\Delta_V \hat{\Phi} = -(\Delta_V \Phi)^\wedge - 2N\hat{\Phi} - \Delta_G^* \hat{\Phi}$. □

It is worthwhile to point out that the above equations relating the Fourier transform $\mathcal{F}$ to the various Laplacians can be rewritten as:

$$\mathcal{F}\Delta_G^* = -\Delta_G^* \mathcal{F},$$

$$\mathcal{F}(N + \tfrac{1}{2}\Delta_G^*) = (N + \tfrac{1}{2}\Delta_G^*)\mathcal{F},$$

$$\mathcal{F}(\Delta_L + \tfrac{1}{2}\mathrm{Id}) = -(\Delta_L + \tfrac{1}{2}\mathrm{Id})\mathcal{F},$$

$$\mathcal{F}(\Delta_V + N) = -(\Delta_V + N)\mathcal{F}.$$

We now study the relationships of the Fourier transform with the Wick product and convolution. In the finite dimensional case, it is well-known that the Fourier transform carries convolution to multiplication. However, in the white noise calculus, the relationship between the Fourier transform and convolution appears to be different since the Gaussian measure μ is used. But we will see below that the white noise Fourier transform carries the convolution of Φ and Ψ to the Wick product of $\hat{\Phi}$ and $\hat{\Psi}$. This indicates that convolution and the Wick product are related by the Fourier transform in the same way as in the finite dimensional spaces. The Gaussian white noise functional $g_{-2} = \mathcal{N}\exp(\frac{1}{4}|\cdot|_2^2)$ will play an important role in relating the Fourier transform to the Wick product and convolution. Recall that $\mathcal{S}g_{-2}(\xi) = \exp(\frac{1}{2}|\xi|_2^2)$, $\xi \in \mathcal{S}(\mathbb{R})$.

Theorem 9.26. *For any Φ and Ψ in $(\mathcal{S})^*$, the following holds*

$$(\Phi{:}\Psi)^{\wedge} = \hat{\Phi}{:}\hat{\Psi}{:}g_{-2}, \tag{9.49}$$

where $g_{-2} = \mathcal{N}\exp(\frac{1}{4}|\cdot|_2^2)$.

Proof. Simply use (9.6) from the definition of the Fourier transform and compare the $\mathcal{S}$–transformations of both sides in (9.49). □

Now, we consider the convolution $\nu{*}\sigma$ of two measures ν and σ on $\mathcal{S}'(\mathbb{R})$ defined by

$$\nu{*}\sigma(dx) = \int_{\mathcal{S}'(\mathbb{R})} \nu(dx-y)\,\sigma(dy).$$

Theorem 9.27. *Suppose* $\frac{d\nu}{d\mu}$ *and* $\frac{d\sigma}{d\mu}$ *are in* $(\mathcal{S})^*$. *Then* $\frac{d(\nu*\sigma)}{d\mu}$ *is also in* $(\mathcal{S})^*$. *Moreover, we have*

$$\frac{d(\nu*\sigma)}{d\mu} = \frac{d\nu}{d\mu}{:}\frac{d\sigma}{d\mu}{:}g_{-2}. \tag{9.50}$$

Proof. The first assertion can be proved easily by checking the conditions in Theorem 4.57. To show (9.50), we take the $\mathcal{S}$–transformation

$$\begin{aligned}
\mathcal{S}\Big(\frac{d(\nu*\sigma)}{d\mu}\Big)(\xi) &= e^{-\frac{1}{2}|\xi|_2^2}\Big\langle\frac{d(\nu*\sigma)}{d\mu},\, e^{\langle\cdot,\,\xi\rangle}\Big\rangle \\
&= e^{-\frac{1}{2}|\xi|_2^2}\int_{\mathcal{S}'(\mathbb{R})}\int_{\mathcal{S}'(\mathbb{R})} e^{\langle x+y,\,\xi\rangle}\, d\nu(x)\, d\sigma(y) \\
&= e^{-\frac{1}{2}|\xi|_2^2}\Big\langle\frac{d\nu}{d\mu},\, e^{\langle\cdot,\,\xi\rangle}\Big\rangle\Big\langle\frac{d\sigma}{d\mu},\, e^{\langle\cdot,\,\xi\rangle}\Big\rangle \\
&= \mathcal{S}\Big(\frac{d\nu}{d\mu}\Big)(\xi)\ \mathcal{S}\Big(\frac{d\sigma}{d\mu}\Big)(\xi)e^{\frac{1}{2}|\xi|_2^2} \\
&= \mathcal{S}\Big(\frac{d\nu}{d\mu}\Big)(\xi)\ \mathcal{S}\Big(\frac{d\sigma}{d\mu}\Big)(\xi)\ \mathcal{S}g_{-2}(\xi)
\end{aligned}$$

$$= \mathcal{S}(\frac{d\nu}{d\mu}:\frac{d\sigma}{d\mu}:g_{-2})(\xi). \qquad \square$$

Theorem 9.28. *For any two measures ν and σ on $\mathcal{S}'(\mathbb{R})$ such that $\frac{d\nu}{d\mu}, \frac{d\sigma}{d\mu} \in (\mathcal{S})^*$, the following equality holds*

$$(\frac{d(\nu*\sigma)}{d\mu})^{\wedge} = (\frac{d\nu}{d\mu})^{\wedge}:(\frac{d\sigma}{d\mu})^{\wedge}.$$

Proof. It follows from Theorems 9.26 and 9.27 that

$$(\frac{d(\nu*\sigma)}{d\mu})^{\wedge} = (\frac{d\nu}{d\mu})^{\wedge}:(\frac{d\sigma}{d\mu})^{\wedge}:(g_{-2})^{\wedge}:g_{-2}:g_{-2}. \tag{9.51}$$

Note that for any a, b$\in\mathbb{R}$, a$\neq -1$, b$\neq -1$

$$g_a:g_b = g_c, \quad c = \frac{ab-1}{2+a+b}.$$

On the other hand, from Example 9.4 we have $(g_{-2})^{\wedge} = g_{-1/2}$. Hence

$$(g_{-2})^{\wedge}:g_{-2}:g_{-2} = 1. \tag{9.52}$$

Obviously, (9.51) and (9.52) yield the conclusion. $\square$

In view of Theorem 9.27 we can define the convolution of two generalized white noise functionals Φ and Ψ as follows:

$$\Phi*\Psi = \Phi:\Psi:g_{-2}. \tag{9.53}$$

Note that for any ξ in $\mathcal{S}(\mathbb{R})$, we have

$$\mathcal{T}(\Phi*\Psi)(\xi) = \exp(-\tfrac{1}{2}|\xi|_2^2)\mathcal{S}(\Phi*\Psi)(i\xi)$$

$$= \exp(-\tfrac{1}{2}|\xi|_2^2)\ \mathcal{S}\Phi(i\xi)\ \mathcal{S}\Psi(i\xi)\ \exp(-\tfrac{1}{2}|\xi|_2^2)$$

$$= \mathcal{T}\Phi(\xi)\ \mathcal{T}\Psi(\xi).$$

Therefore, the convolution $\Phi*\Psi$ given in (9.53) is the same as defined in Corollary 4.39. The following theorem states the relationships of the Fourier transform with the Wick product and convolution.

Theorem 9.29. *For any* Φ *and* Ψ *in* $(\mathcal{S})^*$, *the following equalities hold:*

$$(\Phi*\Psi)^\wedge = \hat{\Phi}{:}\hat{\Psi}, \quad (\Phi{:}\Psi)^\wedge = \hat{\Phi}*\hat{\Psi}.$$

Proof. The first equality follows from the same argument as in the proof of Theorem 9.28, while the second equality is a restatement of Theorem 9.26 by using convolution. □

As is well-known in finite dimensional spaces, the Fourier transform takes convolution into multiplication and vice versa. On the other hand, in the white noise calculus, the above theorem indicates that the Fourier transform takes convolution into the Wick product and vice versa. Thus the Wick product plays the role of multiplication in the study of generalized white noise functionals. This is not surprising because the Gaussian white noise measure μ is used.

9.G Relation with the Finite Dimensional Fourier Transform

It is a natural question to ask how the white noise Fourier transform is related to the finite dimensional Fourier transform. We will present some results from Kubo and Kuo (1991), and Kuo (1990a).

Suppose ζ_1, ζ_2, $\cdots$, and ζ_n are orthonormal in $L^2(\mathbb{R})$ and $F \in \mathcal{S}'(\mathbb{R}^n)$. Let

$$\Phi = F\circ(\langle\cdot, \zeta_1\rangle, \langle\cdot, \zeta_2\rangle, \cdots, \langle\cdot, \zeta_n\rangle).$$

Then Φ is a generalized white noise functional (e.g., by Theorem 7.34 and the fact that $\mathcal{D}^* \subseteq (\mathcal{S})^*$) with $\mathcal{S}$–transformation given by

$$(\mathcal{S}\Phi)(\xi) = <\overline{\hat{F}}, \theta_\xi>, \ \xi \in \mathcal{S}(\mathbb{R}), \tag{9.54}$$

where $\theta_\xi \in \mathcal{S}(\mathbb{R}^n)$ is the following function

$$\theta_\xi(y_1, y_2, \ldots, y_n) =$$

$$= (2\pi)^{-n/2} \exp(i[<\zeta_1, \xi>y_1 + <\zeta_2, \xi>y_2 + \cdots + <\zeta_n, \xi>y_n] - \tfrac{1}{2}|y|^2).$$

The question is to find a relationship between the Fourier transform $\hat{\Phi}$ and the finite dimensional Fourier transform $\hat{F}$.

To motivate what kind of results we can expect to get, let us look at a function $\phi(x, y) = f(x)$ on $\mathbb{R}^2$ depending only on x. Its Fourier transform is given by

$$\hat{\phi}(u, v) = \frac{1}{2\pi} \int_{\mathbb{R}^2} e^{-i(ux+vy)} f(x)\, dxdy$$

$$= \hat{f}(u) \frac{1}{\sqrt{2\pi}} \int_{\mathbb{R}} e^{-ivy}\, dy.$$

Note that informally the Dirac delta function δ_0 at 0 can be expressed as follows:

$$\delta_0(v) = \frac{1}{2\pi} \int_{\mathbb{R}} e^{ivy}\, dy.$$

Therefore, we have

$$\hat{\phi}(u, v) = \hat{f}(u)\sqrt{2\pi}\, \delta_0(-v). \tag{9.55}$$

On the other hand, we can express $\hat{\phi}$ in a different form. Let $(\cdot, \cdot)$ be the pairing of distributions and test functions with respect to the standard Gaussian measure μ_2 on $\mathbb{R}^2$. Then for any test function η on $\mathbb{R}^2$,

$$<\hat{\phi}, \eta>= \int_{\mathbb{R}^2} \hat{f}(u)\sqrt{2\pi}\ \delta_0(-v)\ \eta(u, v)\ d\mu_2(u, v)$$

$$= \int_{\mathbb{R}} (\tilde{\delta}_{(u,0)}, \eta)\ \hat{f}(u)\ d\mu_1(u), \tag{9.56}$$

where μ_1 is the standard Gaussian measure on $\mathbb{R}$ and $\tilde{\delta}_x$ is the distribution defined by $(\tilde{\delta}_x, \eta)=\eta(x)$. It is easy to see that

$$\delta_x = \frac{1}{2\pi}\, e^{-\frac{1}{2}|x|^2}\, \tilde{\delta}_x,$$

where δ_x is the usual Dirac delta function at $x\in\mathbb{R}^2$, i.e., $<\delta_x, \eta>=\eta(x)$. In (9.56) we have chosen $\tilde{\delta}_x$ instead of δ_x because $\tilde{\delta}_x$ has a counterpart (i.e., the delta white noise functional) in $(\mathcal{S})^*$, while δ_x has none.

We will generalize (9.55) and (9.56) to cylindrical generalized white noise functionals in Theorem 9.30 (due to Kuo (1990a)) and Theorem 9.32 (due to Kubo and Kuo (1991)).

Theorem 9.30. *Suppose* $F\in\mathcal{S}'(\mathbb{R}^n)$, *and* $\zeta_1, \zeta_2, \cdots, \zeta_n\in L^2(\mathbb{R})$ *are orthonormal. Then the Fourier transform of* $\Phi = F\circ(<\cdot, \zeta_1>, <\cdot, \zeta_2>, \cdots, <\cdot, \zeta_n>)$ *is given by*

$$\hat{\Phi} = (\hat{F}\circ(<\cdot, \zeta_1>, <\cdot, \zeta_2>, \cdots, <\cdot, \zeta_n>))\otimes(\tilde{\delta}_0\circ Q),$$

where $Q=Id-P$, *and* P *is the orthogonal projection onto the span of* $\zeta_1, \zeta_2, \cdots, \zeta_n$.

Proof. For simplicity, let $X_\zeta=(<\cdot, \zeta_1>, <\cdot, \zeta_2>, \cdots, <\cdot, \zeta_n>)$. By (9.6) and (9.54),

$$(\mathcal{S}\hat{\Phi})(\xi) = <\bar{\hat{F}}, \theta_{-i\xi}>e^{-\frac{1}{2}|\xi|_2^2},\ \xi\in\mathcal{S}(\mathbb{R}). \tag{9.57}$$

On the other hand, note that

$$\mathcal{S}(\tilde{\delta}_0 \circ Q)(\xi) = e^{-\frac{1}{2}|Q\xi|_2^2}, \ \xi \in \mathcal{S}(\mathbb{R}).$$

Hence by (9.54) we get

$$\mathcal{S}((\hat{F} \circ X_\zeta) \otimes (\tilde{\delta}_0 \circ Q))(\xi) = <\overline{(\hat{F})^\wedge}, \theta_\xi > e^{-\frac{1}{2}|Q\xi|_2^2}$$

$$= <\overline{\hat{F}}, \hat{\theta}_\xi > e^{-\frac{1}{2}|Q\xi|_2^2}.$$

But it is easy to check that $\hat{\theta}_\xi = \theta_{-i\xi} \exp(-\frac{1}{2}|P\xi|_2^2)$. Therefore, from the last equality, we obtain that

$$\mathcal{S}((\hat{F} \circ X_\zeta) \otimes (\tilde{\delta}_0 \circ Q))(\xi) = <\overline{\hat{F}}, \theta_{-i\xi} > e^{-\frac{1}{2}|\xi|_2^2}, \ \xi \in \mathcal{S}(\mathbb{R}). \tag{9.58}$$

The conclusion of the theorem follows obviously from (9.57) and (9.58). □

We need the following lemma for generalizing (9.56) to cylindrical generalized white noise functionals.

Lemma 9.31. *Let* $\zeta_1, \zeta_2, \cdots, \zeta_n \in L^2(\mathbb{R})$ *be orthonormal. Then for any* $\varphi \in (\mathcal{S})$, *the function* Θ_φ *defined by*

$$\Theta_\varphi(y_1, y_2, \cdots, y_n) = <\tilde{\delta}_{y_1\zeta_1 + y_2\zeta_2 + \cdots + y_n\zeta_n}, \varphi> (2\pi)^{-n/2} e^{-\frac{1}{2}|y|^2}$$

is in $\mathcal{S}(\mathbb{R}^n)$. *Here* $y = (y_1, y_2, \cdots, y_n) \in \mathbb{R}^n$ and $|y|^2 = y_1^2 + y_2^2 + \cdots y_n^2$.

Proof. It has been shown in Kubo and Yokoi (1989) that for any $x \in \mathcal{S}_p'(\mathbb{R})$ and $p \geq 1$,

$$|\tilde{\delta}_x|_{2,-p} \leq \exp[\tfrac{1}{2}(b_p + |x|_{2,-p}^2)], \tag{9.59}$$

where $b_p=2^{-2p}(1-2^{-4p})^{-1}\sum_{j=1}^{\infty}(2j)^{-2p}$. As pointed out in Kubo and Kuo (1991) that the inequality in Theorem 4.1 (c) in Kubo and Yokoi (1989) should be

$$\|\delta_y\|_{\mathcal{H}(-p)} \le \exp[\tfrac{1}{2}(\frac{\rho^{2p}}{1-\rho^{4p}}\|\iota_{0,p}\|_{H.S.}^2 + |y|_{-p}^2)].$$

Note also that for any $p \ge 0$,

$$|y_1\zeta_1+y_2\zeta_2+\cdots+y_n\zeta_n|_{2,-p} \le 2^{-p}|y|(|\zeta_1|_2^2+|\zeta_2|_2^2+\cdots+|\zeta_n|_2^2). \qquad (9.60)$$

It follows easily from (9.59) and (9.60) with large enough p that for any positive integer k,

$$\lim_{|y|\to\infty} |y|^k|\Theta_\varphi(y_1, y_2, \cdots, y_n)| = 0. \qquad (9.61)$$

Moreover, the function Θ_φ is differentiable with partial derivatives given by

$$\frac{\partial}{\partial y_j}\Theta_\varphi = <\tilde{\delta}_{y_1\zeta_1+y_2\zeta_2+\cdots+y_n\zeta_n}, D_{\zeta_j}\varphi>(2\pi)^{-n/2}e^{-\frac{1}{2}|y|^2} -$$

$$- <\tilde{\delta}_{y_1\zeta_1+y_2\zeta_2+\cdots+y_n\zeta_n}, \varphi>(2\pi)^{-n/2}y_je^{-\frac{1}{2}|y|^2}.$$

Note that D_{ζ_j} is continuous from $(\mathcal{S})$ into itself. Therefore, by the same reason as above, the partial derivatives of Θ_φ satisfy the growth condition (9.61). We can show inductively that Θ_φ has higher order derivatives which satisfy the growth condition (9.61). Hence Θ_φ is in $\mathcal{S}(\mathbb{R}^n)$. □

Theorem 9.32. *Suppose* $F\in\mathcal{S}'(\mathbb{R}^n)$ *and* $\zeta_1, \zeta_2, \cdots, \zeta_n\in L^2(\mathbb{R})$ *are orthonormal. Then the Fourier transform of* $\Phi = F\circ(<\cdot, \zeta_1>, <\cdot, \zeta_2>, \cdots, <\cdot, \zeta_n>)$ *is given by*

$$<\hat{\Phi}, \varphi> = <\hat{F}, \Theta_\varphi>, \ \varphi\in(\mathcal{S}), \qquad (9.62)$$

where Θ_φ *is defined by*

$$\Theta_\varphi(y_1, y_2, \cdots, y_n) = <\tilde{\delta}_{y_1\zeta_1+y_2\zeta_2+\cdots+y_n\zeta_n}, \varphi> (2\pi)^{-n/2} e^{-\frac{1}{2}|y|^2}.$$

Remark. We point out that the pairing $<\cdot, \cdot>$ in the right hand side of (9.62) is with respect to the Lebesgue measure. Symbolically, $\hat{\Phi}$ can be expressed as follows:

$$\hat{\Phi} = \int_{\mathbb{R}^n} \tilde{\delta}_{y_1\zeta_1+y_2\zeta_2+\cdots+y_n\zeta_n} \hat{F}(y)\, d\mu_n(y),$$

where μ_n is the standard Gaussian measure on $\mathbb{R}^n$.

Proof. Let $\varphi =: e^{<\cdot, \xi>}:$. Then from (9.57), we have

$$\begin{aligned} <\overline{\hat{\Phi}}, \varphi> &= (\mathcal{S}\hat{\Phi})(\xi) \\ &= <\overline{\hat{F}}, \theta_{-i\xi}> e^{-\frac{1}{2}|\xi|_2^2}, \end{aligned} \tag{9.63}$$

where θ_ξ is given as in (9.54), i.e.,

$$\theta_\xi(y_1, y_2, \ldots, y_n) =$$

$$= (2\pi)^{-n/2} \exp(i[<\zeta_1, \xi>y_1+<\zeta_2, \xi>y_2+\cdots+<\zeta_n, \xi>y_n]-\tfrac{1}{2}|y|^2).$$

On the other hand, note that

$$(\mathcal{S}\tilde{\delta}_{y_1\zeta_1+y_2\zeta_2+\cdots+y_n\zeta_n})(\xi) =$$

$$= e^{-\frac{1}{2}|\xi|_2^2+<\zeta_1, \xi>y_1+<\zeta_2, \xi>y_2+\cdots+<\zeta_n, \xi>y_n}.$$

Hence, with $\varphi =: e^{<\cdot,\, \xi>}:$, we have

$$\Theta_\varphi(y_1, y_2, \cdots, y_n) =$$

$$= (\mathcal{S}\tilde{\delta}_{y_1\zeta_1+y_2\zeta_2+\cdots+y_n\zeta_n})(\xi)(2\pi)^{-n/2} e^{-\frac{1}{2}|y|^2}$$

$$= e^{-\frac{1}{2}|\xi|_2^2 + <\zeta_1,\, \xi>y_1 + <\zeta_2,\, \xi>y_2 + \cdots + <\zeta_n,\, \xi>y_n} (2\pi)^{-n/2} e^{-\frac{1}{2}|y|^2}$$

$$= e^{-\frac{1}{2}|\xi|_2^2}\, \theta_{-i\xi}(y_1, y_2, \cdots, y_n).$$

This shows that with $\varphi =: e^{<\cdot,\, \xi>}:$,

$$<\overline{\hat{F}}, \Theta_\varphi> = <\overline{\hat{F}}, \theta_{-i\xi}> e^{-\frac{1}{2}\,|\xi|_2^2}. \tag{9.64}$$

By taking complex conjugate in both sides of (9.63) and (9.64), we see that the conclusion (9.62) holds for all $\varphi =: e^{<\cdot,\, \xi>}:$, $\xi \in \mathcal{S}(\mathbb{R})$. But the linear span of $\{: e^{<\cdot,\, \xi>} :;\ \xi \in \mathcal{S}(\mathbb{R})\}$ is dense in $(\mathcal{S})$. Moreover, observe that both sides of (9.62) are continuous linear functionals on $(\mathcal{S})$. Therefore, the equality in (9.62) holds for all $\varphi \in (\mathcal{S})$. □

The next theorem shows how the delta white noise functional $\tilde{\delta}_x$ at $x \in \mathcal{S}'(\mathbb{R})$ is related to the finite dimensional Dirac delta function δ_x at $x \in \mathbb{R}^n$. Recall that they are defined by

$$<\delta_x, f> = f(x), \quad f \in \mathcal{S}(\mathbb{R}^n),$$

$$<\tilde{\delta}_x, \varphi> = \tilde{\varphi}(x), \quad \varphi \in (\mathcal{S}),$$

where $\tilde{\varphi}$ denotes the continuous version of φ. Here the finite dimensional pairing is with respect to the Lebesgue measure.

Theorem 9.33. *Let* $\{\xi_n;\ n\geq 1\}\subset\mathcal{S}(\mathbb{R})$ *be an orthonormal basis for* $L^2(\mathbb{R})$. *Then for any* $x\in\mathcal{S}'(\mathbb{R})$ *the following converges in* $(\mathcal{S})^*$

$$\lim_{n\to\infty}(2\pi)^{n/2}\, e^{\frac{1}{2}|\theta_n(x)|^2}\,\delta_{\theta_n(x)}\circ(<\cdot,\ \xi_1>,\ <\cdot,\ \xi_2>,\ \cdots,\ <\cdot,\ \xi_n>) = \tilde{\delta}_x,$$

where $\theta_n(x) = (<x,\ \xi_1>,\ <x,\ \xi_2>,\ \cdots,\ <x,\ \xi_n>)\in\mathbb{R}^n$.

Proof. For the clarity of the proof, let

$$F_n = (2\pi)^{n/2} e^{\frac{1}{2}|\theta_n(x)|^2}\,\delta_{\theta_n(x)},$$

$$\Phi_n = F_n\circ(<\cdot,\ \xi_1>,\ <\cdot,\ \xi_2>,\ \cdots,\ <\cdot,\ \xi_n>).$$

Then

$$\hat{F}_n(y_1,\ y_2,\ \ldots,\ y_n) =$$

$$= e^{-i(<x,\ \xi_1>y_1+<x,\ \xi_2>y_2+\cdots+<x,\ \xi_n>y_n) + \frac{1}{2}|\theta_n(x)|^2}.$$

By using (9.54), we can check easily that

$$(\mathcal{S}\Phi_n)(\xi) = \exp\Big\{-\tfrac{1}{2}\sum_{k=1}^{n}<\xi,\ \xi_k>^2 + \sum_{k=1}^{n}<\xi,\ \xi_k><x,\ \xi_k>\Big\}.$$

Clearly, the bound in (4.31) holds uniformly in n. Moreover, for any $\xi\in\mathcal{S}(\mathbb{R})$,

$$\lim_{n\to\infty}(\mathcal{S}\Phi_n)(\xi) = e^{-\frac{1}{2}|\xi|_2^2+<x,\ \xi>}$$

$$= (\mathcal{S}\tilde{\delta}_x)(\xi).$$

Therefore, by Theorem 4.41, Φ_n converges to $\tilde{\delta}_x$ in $(\mathcal{S})^*$. □

9.H Fourier–Mehler Transforms

In this section we will study the Fourier–Mehler transforms of generalized white noise functionals. The results are from Hida, Kuo and Obata (1991), and Kuo (1983b, 1990a, 1990b, 1991b). The k–dimensional Fourier–Mehler transform with parameter $\theta \in \mathbb{R}$ is defined as follows:

$$(\mathcal{F}_\theta f)(y) = \int_{\mathbb{R}^k} L(x, y) f(x)\, dx, \tag{9.65}$$

where the kernel function L(x, y) is given by

$$L(x, y) = (\pi(1-e^{2i\theta}))^{-k/2} \exp\Big(\frac{i}{\sin\theta}\langle x, y\rangle - \frac{i}{2\tan\theta}(|x|^2 + |y|^2)\Big).$$

For the case k=1, see Hida (1980a). Just as in the case of the Fourier transform, there are several difficulties to generalize $\mathcal{F}_\theta$ to infinite dimensional spaces. However, the same trick in defining the Fourier transform on $(\mathcal{S})^*$ can be used to generalize $\mathcal{F}_\theta$ to $(\mathcal{S})^*$. Consider the function

$$g(x, y) = \exp\Big(\frac{i}{\sin\theta}\langle x, y\rangle - \frac{i}{2\tan\theta}(|x|^2 + |y|^2)\Big).$$

Suppose y is an $\mathbb{R}^k$–valued Gaussian random variable with mean 0 and covariance operator Id. Let $E_y\, g(x, y)$ denote the expectation of g with respect to the variable y and let $\mathcal{N}_y\, g(x, y)$ be defined by

$$\mathcal{N}_y g(x, y) = \frac{g(x, y)}{E_y\, g(x, y)}.$$

Then with direct calculation we get

$$\mathcal{N}_y g(x, y) = (\pi(1-e^{2i\theta}))^{-\frac{k}{2}} (2\pi)^{\frac{k}{2}} e^{\frac{1}{2}|x|^2} \exp\Big(\frac{i}{\sin\theta}\langle x, y\rangle - \frac{i}{2\tan\theta}(|x|^2 + |y|^2)\Big).$$

Therefore, we can rewrite $\mathcal{F}_\theta f$ in (9.65) as follows:

$$(\mathcal{F}_\theta f)(y) = \int_{\mathbb{R}^k} f(x)\ \mathcal{N}_y \exp\Big(\frac{i}{\sin\theta}<x, y> - \frac{i}{2\tan\theta}(|x|^2 + |y|^2)\Big)\ d\mu_k(x),$$

where μ_k is the standard Gaussian measure on $\mathbb{R}^k$. Thus the same consideration as in Section 9.B leads to the following definition of Fourier–Mehler transform.

Definition 9.34. *The* Fourier–Mehler Transform $\mathcal{F}_\theta$, $\theta\in\mathbb{R}$, *of a generalized white noise functional* Φ *in* $(\mathcal{S})^*$ *is the generalized white noise functional, denoted by* $\mathcal{F}_\theta\Phi$, *whose* $\mathcal{S}$*–transformation is given by*

$$\mathcal{S}(\mathcal{F}_\theta\Phi)(\xi) = <\overline{\Phi}, \exp[e^{i\theta}<\cdot, \xi> - \tfrac{1}{2}e^{i\theta}\cos\theta\ |\xi|_2^2]>, \quad \xi\in\mathcal{S}(\mathbb{R}), \tag{9.66}$$

or equivalently,

$$\mathcal{S}(\mathcal{F}_\theta\Phi)(\xi) = (\mathcal{S}\Phi)(e^{i\theta}\xi)\ \exp\Big(\tfrac{i}{2}e^{i\theta}\sin\theta\ |\xi|_2^2\Big), \quad \xi\in\mathcal{S}(\mathbb{R}). \tag{9.67}$$

By Theorem 4.38, the right hand sides in (9.66) and (9.67) are U–functionals. Therefore, $\mathcal{F}_\theta\Phi$ exists for each Φ in $(\mathcal{S})^*$. Note that $\mathcal{F}_0=\mathrm{Id}$, and $\mathcal{F}_{-\pi/2}$ is the Fourier transform $\mathcal{F}$. It is easy to see that $\mathcal{F}_{\pi/2}$ is the inverse Fourier transform $\mathcal{F}^{-1}$. Hence by (9.66) and (9.67) we have

$$(\mathcal{S}\mathcal{F}^{-1}\Phi)(\xi) = <\overline{\Phi}, e^{i<\cdot, \xi>}> = (\mathcal{S}\Phi)(i\xi)e^{-\frac{1}{2}|\xi|_2^2}, \quad \xi\in\mathcal{S}(\mathbb{R}).$$

We give some examples of the Fourier–Mehler transform below. Recall the notations $G_{ax,\,c}$ and g_c that we used in Example 9.4, i.e.,

$$G_{ax,\,c} = \mathcal{N}\exp(a<\cdot, x>-\tfrac{1}{2c}|\cdot|_2^2), \quad g_c = \mathcal{N}\exp(-\tfrac{1}{2c}|\cdot|_2^2).$$

Example 9.35. $\mathcal{F}_\theta 1 = g_{i\ \cot\theta}$.

Example 9.36. $\mathcal{F}_\theta \tilde{\delta}_x = G_{i\ \csc\theta\ x,\ -i\ \tan\theta}$.

Example 9.37. $\mathcal{F}_\theta$: $e^{a<\cdot,\ y>} := G_{a\ \sec\theta,\ i\ \cot\theta}$.

Example 9.38. $\mathcal{F}_\theta \mathcal{H}_n(t) = e^{in\theta}\mathcal{H}_n(t)$. Here $\mathcal{H}_n$ is the Hermite white noise functional given in Example 9.5.

Example 9.39. $\mathcal{F}_\theta G_{ax,\ c} = G_{a(\theta)x,\ c(\theta)}$. Here $a(\theta)$ and $c(\theta)$ are given by

$$a(\theta) = \frac{ac}{c\ \cos\theta - i\ \sin\theta},\quad c(\theta) = \frac{c - i\ \tan\theta}{1 - ic\ \tan\theta}.$$

The Fourier–Mehler transform $\mathcal{F}_\theta$ has similar properties as the Fourier transform $\mathcal{F}$. To verify these properties we need to modify the proofs for the Fourier transform case by taking the parameter θ into account. We will omit the proofs if they require only obvious modifications and straightforward calculations. For instance, we state the next two theorems without proofs.

Theorem 9.40. *The Fourier–Mehler transform $\mathcal{F}_\theta$ has the following properties:*

(i) *For any* $\Phi \in (\mathcal{S})^*$ *and* $\eta \in \mathcal{S}(\mathbb{R})$,

$$\mathcal{F}_\theta(D_\eta\Phi) = (\cos\theta\ D_\eta - \sin\theta\ q_\eta)\mathcal{F}_\theta\Phi,$$

$$\mathcal{F}_\theta(q_\eta\Phi) = (\sin\theta\ D_\eta + \cos\theta\ q_\eta)\mathcal{F}_\theta\Phi,$$

$$\mathcal{F}_\theta(\tau_\eta\Phi) = \exp\Big(i\ \sin\theta <\cdot,\ \eta> - \tfrac{i}{2}\sin\theta\ \cos\theta\ |\eta|_2^2\Big)\ \tau_{\cos\theta\ \eta}(\mathcal{F}_\theta\Phi).$$

(ii) *For any* $\Phi \in (\mathcal{S})^*$ *and* $y \in \mathcal{S}'(\mathbb{R})$,

$$\mathcal{F}_\theta(D_y^*\Phi) = e^{i\theta} D_y^*(\mathcal{F}_\theta\Phi).$$

In particular, for any $t\in\mathbb{R}$,

$$\mathcal{F}_\theta(\partial_t^*\Phi) = e^{i\theta} \partial_t^*(\mathcal{F}_\theta\Phi).$$

Theorem 9.41. *For any* $\Phi\in(\mathcal{S})^*$, *the following equalities hold for any* ξ *in* $\mathcal{S}(\mathbb{R})$:

$$\mathcal{S}(\mathcal{F}_\theta\Phi)(\xi) = \exp\Big(-\tfrac{1}{2}e^{i\theta}\cos\theta\ |\xi|_2^2\ \Big)(\mathcal{T}\Phi)(-ie^{i\theta}\xi),$$

$$\mathcal{T}(\mathcal{F}_\theta\Phi)(\xi) = \exp\Big(-\tfrac{1}{2}e^{i\theta}\cos\theta\ |\xi|_2^2\ \Big)(\mathcal{S}\Phi)(ie^{i\theta}\xi).$$

To generalize Theorem 9.12, we use the same idea in its proof. Let Φ be in $(\mathcal{S})^*$. Then for any $\xi\in\mathcal{S}(\mathbb{R})$, by (9.67), we have

$$\mathcal{S}(\mathcal{F}_\theta\Phi)(\xi)\ \exp(-\tfrac{i}{2}\ e^{i\theta}\sin\theta\ |\xi|_2^2) = \mathcal{S}\Phi(e^{i\theta}\xi). \tag{9.68}$$

On the other hand, let $c(\theta) = -i\cot\theta-2$. It is easy to check that

$$\mathcal{S}g_{c(\theta)}(\xi) = \exp(-\tfrac{i}{2}\ e^{i\theta}\sin\theta\ |\xi|_2^2). \tag{9.69}$$

From (9.68) and (9.69) we obtain that

$$\mathcal{S}((\mathcal{F}_\theta\Phi){:}g_{c(\theta)})(\xi) = \mathcal{S}\Phi(e^{i\theta}\xi),\ \ \xi\in\mathcal{S}(\mathbb{R}).$$

This implies that if $\Phi(x) = \sum_{n=0}^\infty \langle :x^{\otimes n}:, f_n\rangle$, then $(\mathcal{F}_\theta\Phi){:}g_{c(\theta)}$ is given by

$$((\mathcal{F}_\theta\Phi){:}g_{c(\theta)})(x) = \sum_{n=0}^\infty e^{in\theta} \langle :x^{\otimes n}:, f_n\rangle.$$

The last equality yields immediately the following theorem.

Theorem 9.42. *Let* $p \in \mathbb{R}$. *Then the following equality holds for all* Φ *in* $(\mathcal{S})_p$:

$$|(\mathcal{F}_\theta \Phi):g_{-i\cot\theta-2}|_{2,p} = |\Phi|_{2,p}. \tag{9.70}$$

or equivalently,

$$|\mathcal{F}_\theta \Phi|_{2,p} = |\Phi:g_{i\cot\theta-2}|_{2,p}.$$

where $g_c = \mathcal{N}\exp(-\frac{1}{2c}|\cdot|_2^2)$.

To find the image of $(\mathcal{S})$ under the Fourier–Mehler transform, we need the following fact which can be checked easily

$$g_a:g_b = g_c, \quad c = \frac{ab-1}{a+b+2}.$$

In particular, $g_a:g_{-2-a}=1$. Therefore, we have

$$g_{-i\cot\theta-2}:g_{i\cot\theta}=1. \tag{9.71}$$

It is easy to see from (9.70) and (9.71) that $\varphi \in (\mathcal{S})$ if and only if $\mathcal{F}_\theta\varphi \in (\mathcal{S}):g_{i\cot\theta}$. Hence we have proved the following generalization of Corollary 9.13.

Corollary 9.43. *The image of* $(\mathcal{S})$ *under* $\mathcal{F}_\theta$ *is given by*

$$(\mathcal{S}):g_{i\cot\theta} \equiv \{\varphi:g_{i\cot\theta};\ \varphi \in (\mathcal{S})\},$$

where $g_c = \mathcal{N}\exp(-\frac{1}{2c}|\cdot|_2^2)$.

We now show that the Fourier–Mehler transform $\mathcal{F}_\theta$ is the adjoint of a continuous linear operator $\mathcal{G}_\theta$ from $(\mathcal{S})$ into itself. The idea to obtain $\mathcal{G}_\theta$ is the same as in the case of the Fourier transform in Theorem 9.16. Recall that the map Υ as defined in (9.22) and the dilation σ_λ, $\lambda\in\mathbb{C}$, are continuous linear operators from $(\mathcal{S})$ into itself. Define

$$\mathcal{G}_\theta = \sigma_{\beta(\theta)}\circ\Upsilon\circ\sigma_{\alpha(\theta)},$$

where $\alpha(\theta)$ and $\beta(\theta)$ are solutions of the equations $\alpha\beta=e^{-i\theta}$ and $1-\alpha^2=e^{-i\theta}\cos\theta$. Then $\mathcal{G}_\theta$ is a continuous linear operator from $(\mathcal{S})$ into itself.

Theorem 9.44. *$\mathcal{F}_\theta$ is the adjoint operator of $\mathcal{G}_\theta$ for any $\theta\in\mathbb{R}$.*

Proof. First note that $\Upsilon\colon e^{c\langle\cdot,\,\xi\rangle}\colon = e^{c\langle\cdot,\,\xi\rangle}$ for any $c\in\mathbb{C}$ and $\xi\in\mathcal{S}(\mathbb{R})$. Hence if we let $\mathcal{G}=\sigma_\beta\circ\Upsilon\circ\sigma_\alpha$, α, $\beta\in\mathbb{C}$, then for any $\xi\in\mathcal{S}(\mathbb{R})$

$$\mathcal{G}\colon e^{\langle\cdot,\,\xi\rangle}\colon = \exp[\overline{\alpha\beta}\langle\cdot,\,\xi\rangle - \tfrac{1}{2}(1-\overline{\alpha}^2)|\xi|_2^2].$$

Therefore, for any $\xi\in\mathcal{S}(\mathbb{R})$,

$$\begin{aligned}\langle\mathcal{G}^*\overline{\Phi},\, \colon e^{\langle\cdot,\,\xi\rangle}\colon\rangle &= \langle\overline{\Phi},\, \mathcal{G}\colon e^{\langle\cdot,\,\xi\rangle}\colon\rangle \\ &= \langle\overline{\Phi},\, \exp[\overline{\alpha\beta}\langle\cdot,\,\xi\rangle - \tfrac{1}{2}(1-\overline{\alpha}^2)|\xi|_2^2]\rangle. \qquad (9.72)\end{aligned}$$

On the other hand, from (9.66) in the definition of $\mathcal{F}_\theta$,

$$\begin{aligned}\langle\mathcal{F}_\theta\overline{\Phi},\, \colon e^{\langle\cdot,\,\xi\rangle}\colon\rangle &= \mathcal{S}(\mathcal{F}_\theta\Phi)(\xi) \\ &= \langle\overline{\Phi},\, \exp[e^{i\theta}\langle\cdot,\,\xi\rangle - \tfrac{1}{2}e^{i\theta}\cos\theta\, |\xi|_2^2]\rangle. \qquad (9.73)\end{aligned}$$

Note that the linear span of the set $\{\colon e^{\langle\cdot,\,\xi\rangle}\colon;\ \xi\in\mathcal{S}(\mathbb{R})\}$ is dense in $(\mathcal{S})$. Thus upon comparing (9.72) and (9.73), we can conclude immediately that $\mathcal{F}_\theta=\mathcal{G}^*$ if α and β satisfy the equations $\alpha\beta=e^{-i\theta}$ and $1-\alpha^2=e^{-i\theta}\cos\theta$, i.e., $\mathcal{G}=\mathcal{G}_\theta$. Hence

$\mathcal{F}_\theta = \mathcal{G}_\theta^*$. □

It is easy to check from (9.67) that $\mathcal{F}_\theta^{-1} = \mathcal{F}_{-\theta}$. Hence $\mathcal{F}_\theta$ is injective and onto. Moreover, $\mathcal{F}_\theta$ is strongly continuous from $(\mathcal{S})^*$ into itself since it is the adjoint of a continuous linear operator, i.e., $\mathcal{G}_\theta$, from $(\mathcal{S})$ into itself. Thus we have the following theorem.

Theorem 9.45. *The Fourier–Mehler transform* $\mathcal{F}_\theta$ *is an injective strongly continuous linear operator from* $(\mathcal{S})^*$ *onto itself.*

For the next theorem, recall that the operators N and Δ_G^* are strongly continuous from $(\mathcal{S})^*$ into itself.

Theorem 9.46. *The set* $\{\mathcal{F}_\theta;\ \theta \in \mathbb{R}\}$ *forms a one parameter group of strongly continuous linear operators acting on* $(\mathcal{S})^*$ *and its infinitesimal generator is given by* $\mathrm{iN} + \frac{\mathrm{i}}{2}\Delta_G^*$.

Proof. By using (9.67), we can check easily that for any $\Phi \in (\mathcal{S})^*$, and $\theta_1, \theta_2 \in \mathbb{R}$,

$$\mathcal{S}(\mathcal{F}_{\theta_1+\theta_2}\Phi)(\xi) = \mathcal{S}(\mathcal{F}_{\theta_1}\mathcal{F}_{\theta_2}\Phi)(\xi), \quad \xi \in \mathcal{S}(\mathbb{R}).$$

Therefore, $\mathcal{F}_{\theta_1+\theta_2} = \mathcal{F}_{\theta_1}\mathcal{F}_{\theta_2}$, and so $\{\mathcal{F}_\theta;\ \theta \in \mathbb{R}\}$ forms a one parameter group acting on $(\mathcal{S})^*$. To find its infinitesimal generator, let $F = \mathcal{S}\Phi$ and $F_\theta = \mathcal{S}(\mathcal{F}_\theta\Phi)$. Then by (9.67)

$$F_\theta(\xi) = F(e^{i\theta}\xi)\exp\Big(\frac{i}{2}e^{i\theta}\sin\theta\ |\xi|_2^2\Big).$$

Note that F is Fréchet differentiable. Hence the function $F_\theta(\xi)$ is differentiable in θ. It is easy to check that

$$\lim_{\theta\to 0}\frac{F_\theta(\xi)-F(\xi)}{\theta} = i\langle F'(\xi), \xi\rangle + \frac{i}{2}|\xi|_2^2\, F(\xi), \quad \xi \in \mathcal{S}(\mathbb{R}). \tag{9.74}$$

Moreover, it is easy to see that the U–functionals $\theta^{-1}(F_\theta - F)$, $\theta \in \mathbb{R}$, satisfy the bound (4.29) uniformly. Hence, by Theorem 4.41, $\theta^{-1}(\mathcal{F}_\theta \Phi - \Phi)$ converges in $(\mathcal{S})^*$ as $\theta \to 0$. But by Theorems 6.11 and 6.20 we have

$$\mathcal{S}(N\Phi + \tfrac{1}{2}\Delta_G^*\Phi)(\xi) = \langle F'(\xi), \xi \rangle + \tfrac{1}{2}|\xi|_2^2\, F(\xi), \quad \xi \in \mathcal{S}(\mathbb{R}). \tag{9.75}$$

Therefore, by (9.74) and (9.75), the infinitesimal generator of $\{\mathcal{F}_\theta;\ \theta \in \mathbb{R}\}$ is given by $iN + \frac{i}{2}\Delta_G^*$. □

The next theorem follows from Theorems 9.44 and 9.46.

Theorem 9.47. *The set* $\{\mathcal{G}_\theta;\ \theta \in \mathbb{R}\}$ *forms a one parameter group of continuous linear operators acting on* $(\mathcal{S})$ *with infinitesimal generator* $iN + \frac{i}{2}\Delta_G$.

The proofs of Theorems 9.20 and 9.21 can be easily modified to characterize the Fourier–Mehler transform $\mathcal{F}_\theta$ and the transformation $\mathcal{G}_\theta$.

Theorem 9.48. $\mathcal{F}_\theta$ *has the following properties:*

(1) $\mathcal{F}_\theta$ *is a strongly continuous linear operator from* $(\mathcal{S})^*$ *into itself;*

(2) $\mathcal{F}_\theta D_\xi = (\cos\theta\, D_\xi - \sin\theta\, q_\xi)\mathcal{F}_\theta$ *for any* $\xi \in \mathcal{S}(\mathbb{R})$;

(3) $\mathcal{F}_\theta q_\xi = (\sin\theta\, D_\xi + \cos\theta\, q_\xi)\mathcal{F}_\theta$ *for any* $\xi \in \mathcal{S}(\mathbb{R})$.

Conversely, suppose $\mathcal{A}$ *is a linear operator from* $(\mathcal{S})^*$ *into itself satisfying the conditions* (1)–(3). *Then there exists a constant* c *depending on* θ *such that* $\mathcal{A} = c\mathcal{F}_\theta$.

Theorem 9.49. $\mathcal{G}_\theta$ *has the following properties:*

(1) $\mathcal{G}_\theta q_\xi = \cos\theta\, q_\xi \mathcal{G}_\theta - \sin\theta\, D_\xi^* \mathcal{G}_\theta$ *for any* $\xi \in \mathcal{S}(\mathbb{R})$;

(2) $\mathcal{G}_\theta D_\xi^* = \sin\theta\, q_\xi \mathcal{G}_\theta + \cos\theta\, D_\xi^* \mathcal{G}_\theta$ *for any* $\xi \in \mathcal{S}(\mathbb{R})$;

(3) $\mathcal{G}_\theta D_y = e^{-i\theta} D_y \mathcal{G}_\theta$ *for any* $y \in \mathcal{S}'(\mathbb{R})$.

Moreover, $\mathcal{G}_\theta$ *is the unique continuous linear operator (up to some constant depending on* θ*) from* $(\mathcal{S})$ *into itself satisfying conditions* (1) *and* (2).

Finally, we state the relationships of the Fourier-Mehler transform with the Laplacian operators, the Wick product, and the convolution. We omit the proofs since they involve tedious and straightforward calculations. The relationships of the operator $\mathcal{G}_\theta$ with the Gross Laplacian Δ_G and the number operator N can also be easily obtained.

Theorem 9.50. (1) $\mathcal{F}_\theta \Delta_G^* = e^{2i\theta} \Delta_G^* \mathcal{F}_\theta$; (2) $\mathcal{G}_\theta \Delta_G = e^{-2i\theta} \Delta_G \mathcal{G}_\theta$.

Theorem 9.51. (1) $\mathcal{F}_\theta N = N\mathcal{F}_\theta - ie^{i\theta}\sin\theta\ \Delta_G^* \mathcal{F}_\theta$; (2) $\mathcal{G}_\theta N = N\mathcal{G}_\theta + ie^{i\theta}\sin\theta\ \Delta_G \mathcal{G}_\theta$.

Theorem 9.52. *For any* $\Phi \in \mathcal{D}(\Delta_L)$ *and any finite interval* T, *we have*

$$\Delta_L^T(\mathcal{F}_\theta \Phi) = e^{2i\theta} \mathcal{F}_\theta(\Delta_L^T \Phi) + ie^{i\theta}\sin\theta\ \mathcal{F}_\theta \Phi.$$

Theorem 9.53. *For any* $\Phi \in \mathcal{D}(\Delta_V)$, *we have*

$$\Delta_V(\mathcal{F}_\theta \Phi) = e^{2i\theta} \mathcal{F}_\theta(\Delta_V \Phi) + 2ie^{i\theta}\sin\theta\ N(\mathcal{F}_\theta \Phi) + e^{2i\theta}\sin^2\theta\ \Delta_G^*(\mathcal{F}_\theta \Phi).$$

Theorem 9.54. *For any* Φ *and* Ψ *in* $(\mathcal{S})^*$, *the following holds*

$$\mathcal{F}_\theta(\Phi{:}\Psi) = (\mathcal{F}_\theta \Phi){:}(\mathcal{F}_\theta \Psi){:}g_{-i\cot\theta - 2},$$

where $g_c = \mathcal{N}\exp(-\frac{1}{2c}|\cdot|_2^2)$.

Theorem 9.55. *For any two measures* ν *and* σ *on* $\mathcal{S}'(\mathbb{R})$ *such that* $\frac{d\nu}{d\mu}, \frac{d\sigma}{d\mu} \in (\mathcal{S})^*$, *the following equality holds*

$$\mathfrak{F}_\theta \frac{d(\nu*\sigma)}{d\mu} = (\mathfrak{F}_\theta \frac{d\nu}{d\mu}):(\mathfrak{F}_\theta \frac{d\sigma}{d\mu}):_{g_{i\tan\theta - 1}}.$$

Theorem 9.56. *For any Φ and Ψ in $(\mathcal{S})^*$, the following equality holds*

$$\mathfrak{F}_\theta(\Phi*\Psi) = (\mathfrak{F}_\theta\Phi):(\mathfrak{F}_\theta\Psi):_{g_{i\tan\theta - 1}}.$$

10. Dirichlet Forms

Dirichlet forms are various generalizations of the classical Dirichlet integral

$$\int_{\Omega} |\operatorname{grad} f|^2(x)\, dx,$$

where Ω is a domain in $\mathbb{R}^n$, $f \in C_0^\infty(\mathbb{R}^n)$. The relation between the above Dirichlet integral, the solution of the Dirichlet problem for the Laplacian on Ω, "stopped" Brownian motion, and the associated heat equation (and its semigroup) has long been known. We refer to Fukushima (1980) and Silverstein (1974) for systematic, self-contained expositions, and for comprehensive references.

The investigation of forms of the particular type

$$\mathcal{E}(f) := \int_{\mathbb{R}^n} |\operatorname{grad} f|^2(x)\, d\nu(x), \quad f \in C_0^\infty(\mathbb{R}^n), \tag{10.1}$$

for appropriate (Borel) measures ν on $\mathcal{B}(\mathbb{R}^n)$ has important applications in quantum mechanics (c.f., e.g., Albeverio et al. (1977)) since it allows to implement very singular forces acting on quantum mechanical particles. In the same spirit one can use Dirichlet forms like (10.1) to construct diffusions with singular drifts. Moreover, the formulation of quantum mechanical Hamiltonians via (10.1) provides the possibility of applying probabilistic techniques in quantum mechanics, such as large deviation estimates for a mathematical treatment of the tunneling problem (cf. Albeverio et al. (1981)).

In this chapter we shall consider generalizations of the Dirichlet integral to infinite dimensions. Motivated by problems in quantum field theory, this has been done first by Albeverio and Høegh-Krohn (1976) (cf. also Chapter 11 and literature quoted there). Here we shall essentially review the work of Albeverio et al. (1989, 1990), Hida, Potthoff, and Streit (1988), Potthoff and Röckner (1990). The essential ingredients of our formulation are the theory of Gâteaux derivatives on $(\mathcal{S})$ (Chapter 5), and positive generalized functionals in $(\mathcal{S})^*$, which provide the analogue of the measure in (10.1). For a more general framework (which does not make use of the last ingredient, and which allows the underlying space to be a general Suslin space), we refer the interested reader to the work of Albeverio and Röckner (1989, 1990) and the literature quoted there. We also want to mention in this context the work of Feyel and de la Pradelle (1991).

Because of the above mentioned relation with the Hamilton operator of quantum mechanical particles, forms of the type (10.1) are also called *energy forms.* In this chapter we shall only consider this special class of Dirichlet forms.

First we shall construct infinite dimensional v1ersions of the form (10.1) in Section 10.A. $\mathbb{R}^n$ will be replaced by $\mathcal{S}'(\mathbb{R})$, as an a-priori domain of the form we shall use $(\mathcal{S})$ (instead of $C_0^\infty(\mathbb{R}^n)$), ν will be given by certain positive elements in $(\mathcal{S})^*$, and as gradient we shall take the operator ∇ discussed in Chapter 5. Also, we shall compare the forms constructed in Section 10.A with those of other approaches. Finally, under the assumption of closability, we shall prove that the closures of our forms admit the (Markovian) contraction property, i.e., that they are Dirichlet forms in the sense of Fukushima (1980).

In Section 10.B we provide some results concerning the closability of the forms constructed in Section 10.A. Also, the associated diffusion processes will be discussed in Section 10.B.

10.A Construction of Energy Forms over $\mathcal{S}'(\mathbb{R})$

In the whole chapter we shall consider for simplicity (and without loss of generality) all vector spaces (in particular $\mathcal{P}$, $\mathcal{A}$, $(\mathcal{S})$, $(\mathcal{S})^*$, (L^2), $L^2(\mathcal{S}'(\mathbb{R}),\mathcal{B},\nu)$ and so on) as over the field of real numbers. Moreover, we shall consider $\mathcal{S}'(\mathbb{R})$ as being equipped with the strong topology, unless mentioned otherwise.

Let $\Phi\in(\mathcal{S})^*$ be positive in the sense of Definition 4.25. Furthermore, let ν be the measure on $\mathcal{B}$ associated with Φ by Theorem 4.26. We begin this section by introducing the following notion:

Definition 10.1. Φ *is called* strictly positive, *if and only if for all open sets* O *in* $\mathcal{S}'(\mathbb{R})$, $\nu(O)>0$.

Remarks. In other words, Φ is strictly positive, if and only if the (topological) support of ν is full: $\operatorname{supp}\nu = \mathcal{S}'(\mathbb{R})$. In the literature on Dirichlet forms in infinite dimensions, one has a notion of *strictly positive measures* (e.g., Kusuoka (1982)), which is different from the definition above. In particular, if Φ is strictly positive in the sense of Definition 10.1, the associated measure ν does *not* need to be strictly positive in the sense of Kusuoka (1982).

For example, if we consider 1 as an element in $(\mathcal{S})^*$, then the associated measure is the white noise measure μ, having this property: 1 is strictly positive in this sense. On the other hand, consider the evaluation mapping $\delta_x : \varphi \mapsto \tilde{\varphi}(x)$, $x \in \mathcal{S}'(\mathbb{R})$, $\varphi \in (\mathcal{S})$, of Corollary 4.8. Then the associated (Dirac–) measure on $\mathfrak{B}$ is equal to 1 on all sets in $\mathfrak{B}$ containing x, and otherwise zero. Therefore the support of this measure is not full, and δ_x is not strictly positive. The notion of strict positivity will be needed in order to make forms well-defined on ν–classes; see below.

Let us now quickly recall the gradient operator ∇ from Chapter 5 (cf. also Chapter 7 for additional information). For $t \in \mathbb{R}$, ∂_t denoted the (extension of) the Gâteaux derivative in direction of the Dirac distribution $\delta_t \in \mathcal{S}'(\mathbb{R})$ on $(\mathcal{S})$. Then we have set for $\varphi \in (\mathcal{S})$, $\nabla\varphi(t,x) = \partial_t\varphi(x)$, $t \in \mathbb{R}$, $x \in \mathcal{S}'(\mathbb{R})$. It turned out (Theorems 5.13 and 5.14) that $\nabla\varphi$ is the Fréchet derivative of $\varphi \in (\mathcal{S})$, and that $\nabla\varphi \in \mathcal{S}(\mathbb{R}) \otimes (\mathcal{S})$. It will be convenient to consider $\nabla\varphi$ as an element in $L^2(\mathbb{R}) \otimes (\mathcal{S})$. If $(e_k, k \in \mathbb{N})$ is a complete orthonormal system of $L^2(\mathbb{R})$, we expand $\nabla\varphi$ as follows:

$$\nabla\varphi = \sum_{k=1}^{\infty} e_k \otimes (e_k, \nabla\varphi) = \sum_{k=1}^{\infty} e_k \otimes \partial_k\varphi, \tag{10.2}$$

where the scalar product is the one in $L^2(\mathbb{R})$, and ∂_k denotes the Gâteaux derivative D_{e_k} in direction of e_k (cf. also Theorem 5.13). By $|\nabla\varphi|_2^2$ we mean the $L^2(\mathbb{R})$–norm of $\nabla\varphi$, i.e.

$$|\nabla\varphi|_2^2(x) = \int_{\mathbb{R}} (\partial_t\varphi(x))^2\, dt = \sum_{k=1}^{\infty} (\partial_k\varphi(x))^2, \quad x\in\mathcal{S}'(\mathbb{R}). \tag{10.3}$$

The following result will be useful.

Lemma 10.2. *If* $\varphi\in(\mathcal{S})$, *then* $|\nabla\varphi|_2^2\in(\mathcal{S})$. *Furthermore, there exists a constant* $C>0$, *such that for all* $\varphi, \psi\in(\mathcal{S})$, $p\in\mathbb{N}_0$, *the following estimate holds:*

$$\|\,|\nabla\varphi|_2^2 - |\nabla\psi|_2^2\,\|_{2,p} \le C\,\|\varphi-\psi\|_{2,p+2}\,(\|\varphi\|_{2,p+2} + \|\psi\|_{2,p+2}).$$

In particular, the mapping $\varphi \mapsto |\nabla\varphi|_2^2$ *is continuous from* $(\mathcal{S})$ *into itself.*

Proof. Let $\varphi\in(\mathcal{S})$. Theorem 5.7 implies that for every $k\in\mathbb{N}$, $\partial_k\varphi\in(\mathcal{S})$, and by Theorem 4.9 we therefore have that also $(\partial_k\varphi)^2$ belongs to $(\mathcal{S})$. Therefore, we only need to prove that the sum on the right hand side of (10.3) converges in $(\mathcal{S})$. To this end we estimate as follows. By Proposition 4.10, we have for some constant $K>0$ (independent of p, as shown in the proof of this proposition),

$$\|(\partial_k\varphi)^2\|_{2,p} \le K\,\|\partial_k\varphi\|_{2,p+1}^2.$$

The last norm can be estimated by help of Proposition 5.8 (replacing p and q there by p+1), and we obtain

$$\|(\partial_k\varphi)^2\|_{2,p} \le 4K\,|e_k|_{2,-p-1}^2\,\|\varphi\|_{2,p+2}^2$$

$$\le 4K\,|e_k|_{2,-1}^2\,\|\varphi\|_{2,p+2}^2.$$

Therefore

$$\| \sum_{k=1}^{\infty} (\partial_k \varphi)^2 \|_{2,p} \leq \sum_{k=1}^{\infty} \| (\partial_k \varphi)^2 \|_{2,p}$$

$$\leq 4K \| \varphi \|_{2,p+2}^2 \sum_{k=1}^{\infty} | e_k |_{2,-1}^2$$

$$= C \| \varphi \|_{2,p+2}^2,$$

where we have set $C = 4K \| H^{-1} \|_{H.S.}^2$, and used the fact that $\sum_k | e_k |_{2,-1}^2 = \| H^{-1} \|_{H.S.}^2 < +\infty$. Therefore the first statement of the lemma is proved.

To prove the second statement, write

$$\| | \nabla \varphi |_2^2 - | \nabla \psi |_2^2 \|_{2,p} = \| \sum_{k=1}^{\infty} (\partial_k(\varphi - \psi)) (\partial_k(\varphi + \psi)) \|_{2,p},$$

and estimate the term on the right hand side by

$$\sum_{k=1}^{\infty} \| (\partial_k(\varphi - \psi))(\partial_k(\varphi + \psi)) \|_{2,p} \leq K \sum_{k=1}^{\infty} \| \partial_k(\varphi - \psi) \|_{2,p+1} \| \partial_k(\varphi + \psi) \|_{2,p+1},$$

where we used again Proposition 4.10. The norms under the last sum are estimated as before, and one obtains the claimed bound. □

As a consequence of Lemma 10.2 we know that for strictly positive $\Phi \in (\mathcal{S})^*$, the following quadratic form is well-defined on $(\mathcal{S})$:

$$\mathcal{E}(\varphi,\varphi) := \langle \Phi, | \nabla \varphi |_2^2 \rangle = \int_{\mathcal{S}'(\mathbb{R})} | \nabla \tilde{\varphi} |_2^2(x) \, d\nu(x), \quad \varphi \in (\mathcal{S}). \qquad (10.4)$$

(Here we used the convention from Chapter 4 to denote the continuous version of $\varphi \in (\mathcal{S})$ by $\tilde{\varphi}$. Note that $(| \nabla \varphi |_2^2)^{\sim} = | \nabla \tilde{\varphi} |_2^2$, as one easily verifies.) By polarization, one obtains from (10.4) a symmetric bilinear form which will be denoted as follows:

$$\mathcal{E}(\varphi,\psi) = <\Phi, (\nabla\varphi,\nabla\psi)> = \int_{\mathcal{S}'(\mathbb{R})} (\nabla\tilde{\varphi},\nabla\tilde{\psi})(x)\, d\nu(x), \quad \varphi,\ \psi\in(\mathcal{S}).$$

We remark that forms on $\mathcal{S}'(\mathbb{R})$, more general than (10.4) and its polarization, have been discussed in Razafimanantena (1991).

It will be important to consider the form $\mathcal{E}$ as a form on the L^2–space defined by the measure ν, which we denote as $L^2(\nu) \equiv L^2(\mathcal{S}'(\mathbb{R}), \mathcal{B}, \nu)$.

Proposition 10.3. *The form* $\mathcal{E}$ (10.4) *is a positive symmetric densely defined form on* $L^2(\nu)$.

Proof. Let us first show that $\mathcal{E}$ is well-defined on $(\mathcal{S})$ as a subspace of $L^2(\nu)$. Let $\varphi\in(\mathcal{S})$, with everywhere defined continuous version $\tilde{\varphi}$. Denote by $\hat{\varphi}$ its ν–class. Then $\tilde{\varphi}$ is the unique continuous representative of $\hat{\varphi}$, because the support of ν is full. Let φ_1 and φ_2 be two representatives of $\hat{\varphi}$. Then φ_i, i=1,2, is ν–a.e. Fréchet differentiable with unique continuous version given by $\nabla\tilde{\varphi} = (\nabla\varphi)^{\sim}$ (cf. Chapter 5.B). Therefore $| \nabla\varphi_1 |_2^2 = | \nabla\varphi_2 |_2^2$, ν–a.e., and consequently $\mathcal{E}$ is well-defined as a form on $L^2(\nu)$.

It remains to show that $\mathcal{E}$ is densely defined. To this end it suffices to note that $\mathcal{P}\subset(\mathcal{S})$ is dense in $L^2(\nu)$, which can be proved as indicated in the sketch of the proof of Theorem 1.9, because ν is finite and hence a Radon measure on $(\mathcal{S}'(\mathbb{R}),\mathcal{B})$. □

Let us now recall some terminology about forms, related to their closability. We refer the reader to Fukushima (1980) and Kato (1976) for more background material and further results.

Consider a densely defined, positive symmetric form $(F,\mathcal{D}(F))$ on a separable real Hilbert space $\mathcal{H}$, i.e., F is a bilinear symmetric mapping from

$\mathfrak{D}(F) \times \mathfrak{D}(F)$ into $\mathbb{R}_+$, where $\mathfrak{D}(F)$ is a dense linear subspace of $\mathcal{K}$. We shall also simply call $(F, \mathfrak{D}(F))$ a form on $\mathcal{K}$. The form $(F, \mathfrak{D}(F))$ is called *closed on* $\mathcal{K}$, if and only if the following holds: if $(u_n, n \in \mathbb{N})$ is a sequence in $\mathfrak{D}(F)$, converging to u in $\mathcal{K}$, and if $(u_n, n \in \mathbb{N})$ is Cauchy in F, i.e., $F(u_n - u_m, u_n - u_m) \to 0$, as $n, m \to +\infty$, then $u \in \mathfrak{D}(F)$ and $F(u_n - u, u_n - u) \to 0$, as $n \to +\infty$. This can rephrased as follows. Introduce the following form $(F_1, \mathfrak{D}(F))$ on $\mathcal{K}$:

$$F_1(u, u) = F(u, u) + (u, u)_{\mathcal{K}}, \quad u \in \mathfrak{D}(F).$$

It is clear that F_1 is a scalar product on $\mathfrak{D}(F)$, and therefore $(\mathfrak{D}(F), F_1)$ is a pre-Hilbert space. $(F, \mathfrak{D}(F))$ is closed if and only if $(\mathfrak{D}(F), F_1)$ is complete, i.e., if and only if it is a Hilbert space. We quote the following result (e.g., Kato (1976)):

Theorem 10.4. *Let* $(F, \mathfrak{D}(F))$ *be a densely defined positive symmetric closed form on* $\mathcal{K}$. *Then there exists a unique positive self-adjoint operator* T *with* $\mathfrak{D}(\sqrt{T}) = \mathfrak{D}(F)$, *so that for all* $u, v \in \mathfrak{D}(F)$, $F(u, v) = (\sqrt{T}u, \sqrt{T}v)_{\mathcal{K}}$.

A form $(F, \mathfrak{D}(F))$ on $\mathcal{K}$ is called *closable*, if and only if it has a closed extension, i.e., if and only if there exists a closed form $(\hat{F}, \mathfrak{D}(\hat{F}))$ on $\mathcal{K}$ so that $\mathfrak{D}(\hat{F}) \supset \mathfrak{D}(F)$, and for all $u, v \in \mathfrak{D}(F)$, $F(u, v) = \hat{F}(u, v)$. If $(F, \mathfrak{D}(F))$ is closable then the smallest of all closed extension is called its *closure*, and we shall denote it by $(\bar{F}, \mathfrak{D}(\bar{F}))$. Closability of the form $(F, \mathfrak{D}(F))$ can be described equivalently as follows. Consider the completion $\overline{\mathfrak{D}(F)}$ of $\mathfrak{D}(F)$ with respect to the norm defined by F_1. Then the injection $\iota\colon \mathfrak{D}(F) \to \mathcal{K}$, extends, as a mapping from the pre-Hilbert space $(\mathfrak{D}(F), F_1)$ into $\mathcal{K}$, continuously to a mapping from $(\overline{\mathfrak{D}(F)}, F_1)$ into $\mathcal{K}$, because it it densely defined and bounded. $(F, \mathfrak{D}(F))$ is closable if and only if this extension is injective. In this case, the closure $(\bar{F}, \mathfrak{D}(\bar{F}))$ of $(F, \mathfrak{D}(F))$ coincides with the form $(\tilde{F}, \overline{\mathfrak{D}(F)})$, where $\tilde{F}$ is the continuous extension of F to the completion $\overline{\mathfrak{D}(F)}$ of $\mathfrak{D}(F)$ with respect to F_1. A more concrete characterization of the closability of $(F, \mathfrak{D}(F))$ on $\mathcal{H}$ is the following (cf., e.g., Kato (1976)): Assume that $\{u_n, n \in \mathbb{N}\}$ is a sequence in $\mathfrak{D}(F)$. $(F, \mathfrak{D}(F))$ is closable on $\mathcal{H}$ if and only if

$F(u_n, u_n) \to 0$ is implied by $u_n \to 0$ in $\mathcal{H}$, and $F(u_n-u_m, u_n-u_m) \to 0$, as m,n $\to +\infty$. In fact, an inspection of the proof in Kato (1976) shows that it is sufficient to consider subsequences: $(F,\mathcal{D}(F))$ is closable if and only if every sequence $\{u_n, n\in\mathbb{N}\}$ in $\mathcal{D}(F)$ with $u_n \to 0$ in $\mathcal{H}$, and $F(u_n-u_m, u_n-u_m) \to 0$, has a subsequence $\{u_{n_k}, k\in\mathbb{N}\}$ so that $F(u_{n_k}, u_{n_k}) \to 0$, as $k \to +\infty$. If S is an operator from $\mathcal{H}$ to another (or the same) Hilbert space then the form (Su, Su) is closable if S is, i.e., whenever its adjoint is densely defined.

We shall discuss a number of criteria for the closability of the energy form (10.4) on $L^2(\nu)$ in the next section. Here we shall derive the result that, under the hypothesis of being closable, the closure of the form (10.4) is a Dirichlet form. Thus we recall first the definition of a Dirichlet form from Fukushima (1980).

Assume from now on that the separable real Hilbert space $\mathcal{H}$ above is equal to $L^2(X,\mathcal{B},m)$, where $(X,\mathcal{B},m)$ is a σ-finite measure space. Assume that $(F,\mathcal{D}(F))$ is a form on $L^2(X,\mathcal{B},m)$. We say that $(F,\mathcal{D}(F))$ has the (*Markovian*) *contraction property*, if and only if one (and therefore all) of the following equivalent properties holds:

(C.10.A) if C is a function on $\mathbb{R}$, with $C(0) = 0$, and $|C(t) - C(s)| \leq |t-s|$ for all $t,s\in\mathbb{R}$, then $C\circ u\in\mathcal{D}(F)$ for all $u\in\mathcal{D}(F)$, and $F(C\circ u,C\circ u) \leq F(u,u)$;

(C.10.B) if $u\in\mathcal{D}(F)$, then $u^\# \equiv (u\vee 0)\wedge 1\in\mathcal{D}(F)$, and $F(u^\#,u^\#)\leq F(u,u)$;

(C.10.C) for all $\epsilon>0$, there exists C_ϵ: $\mathbb{R} \to [-\epsilon, 1+\epsilon]$ so that: (a) $C_\epsilon(t) = t$ for $t\in[0,1]$, $0\leq C_\epsilon(t)-C_\epsilon(s)\leq t-s$, for all $t,s\in\mathbb{R}$, $s\leq t$; (b) for all $u\in\mathcal{D}(F)$, $C_\epsilon\circ u\in\mathcal{D}(F)$ and $F(C_\epsilon\circ u,C_\epsilon\circ u)\leq F(u,u)$.

The fact that these three properties of $(F,\mathcal{D}(F))$ are equivalent is proved in Fukushima (1980) for locally compact separable metric spaces X. For a proof in the general case we refer to Albeverio and Röckner (1990).

Definition 10.5. *A densely defined closed symmetric positive form* $(F,\mathfrak{D}(F))$ *on a separable Hilbert space* $\mathcal{K}$ *which has the contraction property is called a* Dirichlet form. *If* $(F,\mathfrak{D}(F))$ *is a densely defined closable symmetric positive form on* $\mathcal{K}$ *whose closure is a Dirichlet form then it is called a* pre–Dirichlet form.

Let us now return to energy forms like (10.4) over $\mathcal{S}'(\mathbb{R})$, with strictly positive $\Phi\in(\mathcal{S})^*$. We make another definition.

Definition 10.6. $\Phi\in(\mathcal{S})^*$ *is called* admissible, *if and only if it is strictly positive and if* $(\mathcal{E},(\mathcal{S}))$ *is closable on* $L^2(\nu)$, *where* ν *is the measure determined by* Φ.

For the remainder of this section we shall assume that $\Phi\in(\mathcal{S})^*$ is admissible. We shall consider besides $(\mathcal{S})$ also the following domains for $\mathcal{E}$: the (real) algebra $\mathcal{P}$ of polynomials (cf. Chapter 1), and the algebra $\mathcal{C}$ finitely generated by $\sin\circ X_\xi$, $\cos\circ X_\xi$, where $\xi\in\mathcal{S}(\mathbb{R})$ and $X_\xi(x) = \langle x, \xi\rangle$, $x\in\mathcal{S}'(\mathbb{R})$. Later we shall also take the space $\mathcal{F}C_b^\infty$ of finitely based smooth cylinder functions (see below). As before, we shall consider $\mathcal{P}$ and $\mathcal{C}$ as naturally embedded into $(\mathcal{S})$, by identifying $\varphi\in\mathcal{P}$ or $\mathcal{C}$ with its μ–class. On the other hand, in order to keep our notation simple, we shall also usually identify an element in $(\mathcal{S})$ with its ν–class. Although in general μ and ν will be singular to each other, this will not lead to confusion since $\varphi\in(\mathcal{S})$ has a unique continuous representative $\tilde{\varphi}$ which is also the unique continuous representative of its ν–class because $\operatorname{supp}\nu$ is full. Note that if $\varphi\in(\mathcal{S})$, then it belongs to $L^p(\nu)$ for every $p\geq1$ (the above interpretation about classes being understood). In fact, for even $p\geq2$ we have

$$\|\varphi\|^p_{L^p(\nu)} = \int (\varphi(x))^p \, d\nu(x) = \langle\Phi, \varphi^p\rangle,$$

which is finite because $(\mathcal{S})$ is an algebra, and therefore $\varphi^p\in(\mathcal{S})$. For general $p\geq1$, we now only have to apply Hölder's inequality using the fact that ν is a finite measure (since $\nu(\mathcal{S}'(\mathbb{R})) = \langle\Phi, 1\rangle < +\infty$).

We shall make use of the following result.

Lemma 10.7. *For* $\varphi \in (\mathcal{S})$ *set*

$$\mathcal{E}_1(\varphi,\varphi) := \mathcal{E}(\varphi,\varphi) + \| \varphi \|^2_{L^2(\nu)}. \tag{10.5}$$

Denote by $\overline{\mathcal{C}}$, $\overline{\mathcal{P}}$, $\overline{(\mathcal{S})}$ *the completions of* $\mathcal{C}$, $\mathcal{P}$ *and* $(\mathcal{S})$ *with respect to the norm defined by* $\mathcal{E}_1$. *Then*

$$\overline{\mathcal{C}} = \overline{\mathcal{P}} = \overline{(\mathcal{S})}. \tag{10.6}$$

Proof. We only need to show that $\mathcal{C}$ and $\mathcal{P}$ are dense in $(\mathcal{S})$ with respect to the norm defined by (10.5). On the other hand, $\mathcal{P}$ is dense in $(\mathcal{S})$ with respect to its usual topology by construction. Moreover, $\mathcal{C}$ is the real form of the algebra $\mathcal{E}$ (cf. Chapter 1) which has been proved to be dense in $(\mathcal{S})$ (over $\mathbb{C}$) in Proposition 4.14. Hence $\mathcal{C}$ is dense in $(\mathcal{S})$ (over $\mathbb{R}$), too. Now we apply Lemma 10.2 to conclude the proof, since

$$\mathcal{E}_1(\varphi,\varphi) = <\Phi, | \nabla\varphi |_2^2 + \varphi^2>,$$

for $\varphi \in (\mathcal{S})$, $\mathcal{P}$, $\mathcal{C}$ and $\Phi \in (\mathcal{S})^*$. □

Corollary 10.8. *Assume that* $\Phi \in (\mathcal{S})^*$ *is admissible. Then* $(\mathcal{E}, \mathcal{C})$ *and* $(\mathcal{E}, \mathcal{P})$ *are closable, and their closures coincide with the closure of* $(\mathcal{E}, (\mathcal{S}))$.

Proof. Since $\overline{\mathcal{C}} = \overline{\mathcal{P}} = \overline{(\mathcal{S})}$, it follows directly from our second description of closability that $(\mathcal{E}, \mathcal{C})$ and $(\mathcal{E}, \mathcal{P})$ are closable. Moreover, their closures are given by the continuous extension of $\mathcal{E}$ to $\overline{\mathcal{C}} = \overline{\mathcal{P}} = \overline{(\mathcal{S})}$. □

To simplify our notation, we shall from now on denote the closures of $(\mathcal{E}, \mathcal{C})$, $(\mathcal{E}, \mathcal{P})$ and $(\mathcal{E}, (\mathcal{S}))$ by $(\mathcal{E},\overline{\mathcal{C}})$, $(\mathcal{E}, \overline{\mathcal{P}})$ and $(\mathcal{E}, \overline{(\mathcal{S})})$, respectively. Now we are ready to prove the following result.

Theorem 10.9. *If* Φ *is admissible then* $(\mathcal{E}, (\mathcal{S}))$ *is a pre–Dirichlet form.*

Proof. We have to show the contraction property for the closure $(\mathcal{E}, \overline{(\mathcal{F})})$ of $(\mathcal{E}, (\mathcal{F}))$, and by Corollary 10.8 it is sufficient to prove the contraction property of $(\mathcal{E},\overline{\mathcal{C}})$.

Let $\varphi\in\mathcal{C}$, so that φ is bounded, say $\sup_x |\varphi(x)| \leq R$. Let $g\in C_b^1(\mathbb{R})$. We are going to show that $g\circ\varphi\in\overline{\mathcal{C}}$. Weierstrass' theorem implies that there exists a sequence $(g_n', n\in\mathbb{N})$ of polynomials g_n' converging uniformly to g' on $[-R, R]$. Moreover,

$$g_n(t) := \int_{-R}^{t} g_n'(s)\, ds + g(-R)$$

defines a sequence of polynomials g_n converging uniformly to g on $[-R, R]$. Since $\mathcal{C}$ is an algebra, we have $g_n\circ\varphi\in\mathcal{C}$. Moreover, $g_n\circ\varphi \to g\circ\varphi$ in $L^2(\nu)$ because ν is finite. Next we show that $\mathcal{E}(g_n\circ\varphi-g_m\circ\varphi, g_n\circ\varphi-g_m\circ\varphi) \to 0$, as $m,n \to +\infty$. Namely, we can estimate in the following way:

$$\mathcal{E}(g_n\circ\varphi-g_m\circ\varphi, g_n\circ\varphi-g_m\circ\varphi) = \int | \nabla(g_n\circ\varphi-g_m\circ\varphi) |_2^2\, d\nu$$

$$= \int | g_n'\circ\varphi-g_m'\circ\varphi || \nabla\varphi |_2^2\, d\nu$$

$$\leq \sup_{t\in[-R,R]} | g_n'(t)-g_m'(t) | \,\mathcal{E}(\varphi, \varphi).$$

Clearly, the right hand side of the last inequality vanishes as $m,n \to +\infty$. Now the closability of $(\mathcal{E}, \mathcal{C})$ implies that $g\circ\varphi\in\overline{\mathcal{C}}$.

Now we prove that for all $\varphi\in\overline{\mathcal{C}}$, we have $g\circ\varphi\in\overline{\mathcal{C}}$. Since $\overline{\mathcal{C}}$ is the completion of $\mathcal{C}$ with respect to $\mathcal{E}_1$, there exists a sequence $(\varphi_n, n\in\mathbb{N})$ in $\mathcal{C}$ so that $\mathcal{E}_1(\varphi-\varphi_n, \varphi-\varphi_n) \to 0$, as $n \to +\infty$. In particular, $\varphi_n \to \varphi$ in $L^2(\nu)$, and we may choose a subsequence (denoted in the same way) which converges pointwise ν-a.e. to φ. This implies that $g\circ\varphi_n$ converges to $g\circ\varphi$ in $L^2(\nu)$, since g is continuous and bounded, and ν is finite. Moreover, we can estimate in the following way:

$$\mathcal{E}(g\circ\varphi_n - g\circ\varphi_m,\, g\circ\varphi_n - g\circ\varphi_m) =$$

$$= \int |\, g'\circ\varphi_n\, \nabla\varphi_n - g'\circ\varphi_m\, \nabla\varphi_m\, |_2^2\, d\nu$$

$$\leq 2\int |\, g'\circ\varphi_n - g'\circ\varphi_m\, |^2\, |\, \nabla\varphi_n\, |_2^2\, d\nu + 2\int |\, g'\circ\varphi_m\, |^2\, |\, \nabla(\varphi_n - \varphi_m)\, |_2^2\, d\nu$$

$$\leq 2\int |\, g'\circ\varphi_n - g'\circ\varphi_m\, |^2 |\, \nabla\varphi_n\, |_2^2\, d\nu + 2 \sup_{t\in\mathbb{R}} |\, g'(t)\, |^2 \int |\, \nabla(\varphi_n - \varphi_m)\, |_2^2\, d\nu.$$

Now the first of the last two expressions vanishes as m,n $\to +\infty$ because $\varphi_n \to \varphi$ ν–a.e., g is continuous and bounded, and because $\varphi_n \to \varphi$ with respect to $\mathcal{E}_1$. The second term converges to zero since g' is bounded, and because $\varphi_n \to \varphi$ in the norm defined by $\mathcal{E}_1$. Since $(\mathcal{E}, \bar{\mathcal{C}})$ is closed we have $g\circ\varphi\in\bar{\mathcal{C}}$, and $g\circ\varphi_n \to g\circ\varphi$ with respect to $\mathcal{E}_1$.

Finally we show the contraction property of $(\mathcal{E}, \bar{\mathcal{C}})$, by verifying property C.10.C. Let $\epsilon>0$, and let C_ϵ be a function on $\mathbb{R}$ satisfying the conditions stated in C.10.C. (An example of such a function is given in Albeverio et al. (1977).) Note that $|\, C'_\epsilon\, | \leq 1$. Let $\varphi\in\bar{\mathcal{C}}$. By the chain rule for ∇ we have now

$$\begin{aligned}\mathcal{E}(C_\epsilon\circ\varphi,\, C_\epsilon\circ\varphi) &= \int |\, \nabla C_\epsilon\circ\varphi\, |_2^2\, d\nu\\ &= \int (C'_\epsilon\circ\varphi)^2\, |\, \nabla\varphi\, |_2^2\, d\nu\\ &\leq \int |\, \nabla\varphi\, |_2^2\, d\nu\\ &= \mathcal{E}(\varphi,\, \varphi).\end{aligned}$$

□

Corollary 10.10. *If* $\Phi\in(\mathcal{S})^*$ *is admissible, then* $(\mathcal{E}, \mathcal{C})$ *and* $(\mathcal{E}, \mathcal{P})$ *are pre–Dirichlet forms.*

Remark. We want to point out that we obtained the Markovian contraction property for the closure of forms with polynomial domain. Without using the space $(\mathcal{S})$, this seems to be rather difficult, since the contraction mappings do not operate naturally on polynomials. On the other hand, since the polynomials give the moments of the measure ν, they form a quite natural domain. The problem of proving the contraction property of energy forms with polynomial a-priori domains was first addressed in Potthoff and Röckner (1990), the solution being given above. (However, we refer also to Schmuland (1990).)

Assume that $\Phi\in(\mathcal{S})^*$ is admissible, and consider the closed from $(\bar{\mathcal{E}},\overline{(\mathcal{S})})$. Theorem 10.4 guarantees the existence of a unique self–adjoint, positive operator T on $L^2(\nu)$, where ν is the measure on $(\mathcal{S}'(\mathbb{R}),\mathcal{B})$ associated with Φ, such that $\bar{\mathcal{E}}(\varphi,\varphi) = (\sqrt{T}\varphi,\sqrt{T}\varphi)_{L^2(\nu)}$, for all $\varphi\in\overline{(\mathcal{S})}$. General theory (e.g., Fukushima (1980)), and Theorem 10.9 imply now the following result.

Theorem. 10.11. *The semigroup* $\{P_t,\ t\geq 0\}$ *on* $L^2(\nu)$ *given by* $P_t = \exp(-tT)$, $t\geq 0$, *is Markovian, i.e., for all* $t\geq 0$, $0\leq P_t u\leq 1$, ν–a.e., *for every* u *with* $0\leq u\leq 1$, ν–a.e.

There are a number of different formulations of energy forms like (10.4) in infinite dimensions in the literature. For a unified presentation we refer the reader to the article by Albeverio and Röckner (1990). There forms of the type (10.4) are called *classical Dirichlet forms.* The a-priori domain is chosen to be the space of all smooth finitely based cylinder functions $\mathcal{F}C_b^\infty$. In our context an element in this space is of the following form: $\psi\in\mathcal{F}C_b^\infty$, if and only if there there exists $n\in\mathbb{N}$, $\xi_1, \ldots, \xi_n\in\mathcal{S}(\mathbb{R})$, and a function f in $C_b^\infty(\mathbb{R}^n)$, so that for all $x\in\mathcal{S}'(\mathbb{R})$,

$$\psi(x) = f(\langle x,\xi_1\rangle, \ldots, \langle x,\xi_n\rangle). \tag{10.7}$$

Note that $\psi\in\mathcal{F}C_b^\infty$ is infinitely often Gâteaux differentiable in all directions of $\mathcal{S}'(\mathbb{R})$, and its Gâteaux derivative D_y in direction $y\in\mathcal{S}'(\mathbb{R})$ is given by

$$D_y\psi(x) = \sum_{k=1}^{n} D_k f(<x,\xi_1>, \dots ,<x,\xi_n>) <y,\xi_k>.$$

Here, we chose ψ of the form (10.7), and $D_k f$ denotes the k–th partial derivative of f.

As before, let $(e_k,\ k\in\mathbb{N})$ be a complete orthonormal system of $L^2(\mathbb{R})$, and set

$$\nabla\psi = \sum_{k=1}^{\infty} e_k \otimes \partial_k\psi,$$

$$|\nabla\psi|_2^2 = \sum_{k=1}^{\infty} (\partial_k\psi)^2,$$

where $\partial_k \equiv D_{e_k}$. Clearly, these operations coincide with the previously defined ones on $(\mathcal{S})\cap\mathcal{F}C_b^\infty$. It is not hard to see, that for every $x\in\mathcal{S}'(\mathbb{R})$, $(\nabla\psi)(x)$ belongs to $L^2(\mathbb{R})$, and therefore $|\nabla\psi|_2^2$ exists. In fact, we have

$$|\nabla\psi(x)|_2^2 = \sum_{k,l=1}^{n} (D_k f)(<x,\xi_1>, \dots ,<x,\xi_n>)\,(D_l f)(<x,\xi_1>, \dots ,<x,\xi_n>)\,(\xi_k,\xi_l), \tag{10.8}$$

and $x \mapsto |\nabla\psi(x)|_2^2$ is a bounded continuous function on $\mathcal{S}'(\mathbb{R})$. In particular, we can define for a finite measure ν on $(\mathcal{S}'(\mathbb{R}),\mathcal{B})$ with full support the quadratic form

$$\mathcal{E}(\psi,\psi) := \int |\nabla\psi|_2^2(x)\, d\nu(x), \quad \psi\in\mathcal{F}C_b^\infty.$$

As before, we shall always assume that ν is determined by a strictly positive element in $(\mathcal{S})^*$. Moreover, we shall identify $\mathcal{E}$ with the symmetric bilinear form on $\mathcal{F}C_b^\infty$ defined on it. Then $(\mathcal{E},\mathcal{F}C_b^\infty)$ is a densely defined positive symmetric form on $L^2(\nu)$. (The fact that it is densely defined follows for example from $\mathcal{C}\subset\mathcal{F}C_b^\infty$.) In case of the closability of this form, the question arises how its closure is related to the closure of $(\mathcal{E},(\mathcal{S}))$. Since $\mathcal{C}\subset\mathcal{F}C_b^\infty$, we have immediately that also $(\mathcal{E},\mathcal{C})$, and therefore also $(\mathcal{E},(\mathcal{S}))$, $(\mathcal{E},\mathcal{P})$ must be closable (i.e., we find that necessarily Φ must be admissible). Moreover, the closure of $(\mathcal{E},\mathcal{F}C_b^\infty)$ must contain the closure of $(\mathcal{E},(\mathcal{S}))$. It has been proved in Potthoff and Röckner (1990) that the closures are actually identical. We conclude this section by giving the proof of this fact.

We shall show that that $\mathcal{F}C_b^\infty \cap (\mathcal{S})$ is dense in $\mathcal{F}C_b^\infty$ with respect to $\mathcal{E}_1$, given in (10.5). To this end, we introduce another space $\mathcal{F}C_{b,0}^\infty$ of functions on $\mathcal{S}'(\mathbb{R})$ as follows. $\psi \in \mathcal{F}C_b^\infty$ of the form (10.7) is in $\mathcal{F}C_{b,0}^\infty$, if and only if $f \in C_b^\infty(\mathbb{R}^n)$ is the Fourier transform of a function $\tilde{f} \in L^1(\mathbb{R}^n)$ with compact support. From Proposition 4.12 we then have the following result.

Lemma 10.12. $\mathcal{F}C_{b,0}^\infty \subset (\mathcal{S})$.

Our next step is

Proposition 10.13. $\mathcal{F}C_{b,0}^\infty$ *is dense in* $\mathcal{F}C_b^\infty$ *with respect to* $\mathcal{E}_1$. *In particular,* $\overline{\mathcal{F}C_b^\infty} = \overline{\mathcal{F}C_{b,0}^\infty}$, *where* $\overline{\cdot}$ *denotes completion with respect to the norm defined by* $\mathcal{E}_1$.

Proof. Let $\psi \in \mathcal{F}C_b^\infty$ be of the form (10.7), where we assume without loss of generality that $\xi_1, \xi_2, \dots, \xi_n$ are linearly independent. Note that we have the following estimate from (10.8).

$$\mathcal{E}(\psi, \psi) \le \left(\sum_{k=1}^n |\xi_k|_2^2\right) \left(\sum_{k=1}^n \int D_k f(\langle\cdot, \xi_1\rangle, \dots, \langle\cdot, \xi_n\rangle)^2 \, d\nu\right).$$

First we argue that it is sufficient to consider $f \in C_b^\infty(\mathbb{R}^n)$ with compact support: $f \in C_0^\infty(\mathbb{R}^n)$. Namely, let χ_m, $m \in \mathbb{N}$, be a smooth cut-off function with $\operatorname{supp}\chi_m \subset (-2m,2m)$, $\chi_m = 1$ on $[-m,m]$, and with $|D_k \chi_m| \le 2$, for all $k = 1,2, \dots$ Then $\chi_m f \to f$, and $D_k(\chi_m f) \to D_k f$ as $m \to +\infty$, uniformly on every compact subset of $\mathbb{R}^n$. Moreover, $|D_k \chi_m f| \le 2|f| + |D_k f|$ shows that $(D_k \chi_m f, m \in \mathbb{N})$ is uniformly bounded for every $k = 1,2, \dots, n$. Thus we may apply the dominated convergence theorem to conclude that for every finite Borel measure σ on $(\mathbb{R}^n, \mathcal{B}(\mathbb{R}^n))$ we have

$$\int (D_k f - D_k \chi_m f)(u)^2 \, d\sigma(u) \to 0,$$

as m tends to infinity. In particular we may choose for σ the measure which is the projection of ν onto $\mathbb{R}^n$.

According to the preceding argument, we assume from now on that $f \in C_0^\infty(\mathbb{R}^n)$. Choose $\eta \in C_{b,0}^\infty(\mathbb{R}^n) \cap L^1(\mathbb{R}^n)$ such that $\eta \geq 0$, $\int \eta(u)\, du = 1$. For example, if we take

$$\zeta(t) = \pi^{-1}\left(\frac{\sin t}{t}\right)^2, \ t \in \mathbb{R},$$

then $\eta = \zeta^{\otimes n}$ has these properties. Set $\eta_m(u) = m^n \eta(mu)$, $u \in \mathbb{R}^n$, and define $f_m := \eta_m * f$, where $*$ denotes convolution. Note that f_m belongs to $C_{b,0}^\infty$. Also, $f_m \to f$, and $D_k f_m \to D_k f$ for every $k = 1,2, \ldots ,n$, uniformly on every compact subset of $\mathbb{R}^n$, as m tends to infinity. Moreover, we have $|\, f_m \,| \leq |\, f \,|$, and for every k, $|\, D_k f_m \,| \leq |\, D_k f \,|$. Therefore, if ψ_m denotes the function in $\mathcal{F}C_{b,0}^\infty$ defined by f_m, then ψ_m converges to ψ with respect to the norm defined by $\mathcal{E}_1$. □

Corollary 10.14. $(\mathcal{S}) \cap \mathcal{F}C_b^\infty$ *is dense in* $\mathcal{F}C_b^\infty$ *with respect to* $\mathcal{E}_1$, *and*

$$\overline{\mathcal{F}C_b^\infty} = \overline{(\mathcal{S})} = \overline{\mathcal{P}} = \overline{\mathcal{C}}. \tag{10.9}$$

In particular, we have derived the following result.

Theorem 10.15. *Let* $\Phi \in (\mathcal{S})^*$ *be strictly positive. Then the following are equivalent:*

(i) $(\mathcal{E}, \mathcal{F}C_b^\infty)$ *is closable;*

(ii) $(\mathcal{E}, (\mathcal{S}))$ *is closable;*

(iii) $(\mathcal{E}, \mathcal{P})$ *is closable;*

(iv) $(\mathcal{E}, \mathcal{C})$ *is closable.*

Moreover, if one and hence all of the above forms are closable, then their closures coincide.

10.B Closability and the Associated Diffusion Processes

In this section we present some sufficient conditions for the admissibility of $\Phi \in (\mathcal{S})^*$. We begin with two criteria which have been proved in Hida, Potthoff, and Streit (1988). Let us recall that $\mathfrak{D}^{p,k}$ denotes the (completed) (L^p)–domain of $(1+N)^{k/2}$, where N is the number operator (cf. also Chapter 3, and Chapter 7). $\mathfrak{D}^{p,k}$, $p>1$, $k \in \mathbb{R}$, is a Banach space with norm denoted by $|||\cdot|||_{p,k}$.

Theorem 10.16. *Assume that* Φ *is a* $\mathfrak{B}$*–measurable function on* $\mathcal{S}'(\mathbb{R})$, *so that* $\Phi>0$, μ–a.e., *and* $\Phi^{1/2} \in \mathfrak{D}^{2p,1}$, *for* $p>1$. *Then* Φ *is admissible.*

Proof. Our strategy of proof will be to show that ∇ has a densely defined adjoint from $L^2(\mathbb{R}) \otimes L^2(\nu)$ into $L^2(\nu)$, where $\nu = \Phi \cdot \mu$.

By Theorem 7.33 we have the following inequality

$$|||\,\Phi\,|||_{p,1} = |||\,\Phi^{1/2}\,\Phi^{1/2}\,|||_{p,1}$$

$$\leq |||\,\Phi^{1/2}\,|||^2_{2p,1}.$$

Therefore our assumptions imply that $\Phi \in \mathfrak{D}^{p,1}$, and therefore Φ is in the (L^p)–domain, $p>1$, of ∇ by Theorem 7.31. An application of the chain rule (or of the product rule, both being valid because ∇ is closed on $\mathfrak{D}^{p,k}$, $p>1$, $k \geq 0$) yields $\nabla \Phi^{1/2} = \frac{1}{2}\Phi^{-1/2}\nabla\Phi$.

Let $(e_k,\, k \in \mathbb{N})$ be a complete orthonormal system of $L^2(\mathbb{R})$. Consider the subspace $\mathcal{L}_0^2(\nu)$ of $\mathcal{L}^2(\nu) \equiv L^2(\mathbb{R}) \otimes L^2(\nu)$, consisting of elements F of the form

$$F = \sum_k e_k \otimes F_k,$$

where $F_k \in \mathcal{C}$, and only a finite number of the F_k's are non-zero. Observe that $\mathcal{L}_0^2(\nu)$ is dense in $\mathcal{L}^2(\nu)$. On $\mathcal{L}_0^2(\nu)$ we define the following linear operator

$$F \mapsto \Phi^{-1}(\nabla\Phi)F = \Phi^{-1} \sum_k (\partial_k \Phi) F_k,$$

where as before ∂_k denotes the Gâteaux derivative in direction e_k. Now we show that for all $F \in \mathcal{L}_0^2(\nu)$, $\Phi^{-1}(\nabla\Phi)F$ belongs to $L^2(\nu)$. It is clearly sufficient to show that for every $k \in \mathbb{N}$, and all $\varphi \in \mathcal{C}$, $\Phi^{-1}(\partial_k\Phi)\varphi \in L^2(\nu)$. But

$$\begin{aligned}\int (\Phi^{-1}(\partial_k\Phi)\varphi)^2 \, d\nu &= \int \Phi^{-1} (\partial_k\Phi)^2 \varphi^2 \, d\mu \\ &\leq \| \varphi \|_\infty^2 \int (\Phi^{-1/2} \partial_k \Phi)^2 \, d\mu \\ &= 4 \| \varphi \|_\infty^2 \int (\partial_k \Phi^{1/2})^2 \, d\mu \\ &\leq 4 \| \varphi \|_\infty^2 \, ||| \Phi^{1/2} |||_{2,1}^2,\end{aligned}$$

where the last inequality follows from Theorem 5.24 (or from Theorem 7.31). Thus the operator $\Phi^{-1}(\nabla\Phi)$ is densely defined from $\mathcal{L}^2(\nu)$ into $L^2(\nu)$.

Let ∇^* be the (L^2)–adjoint of ∇ (cf. Chapter 5). We prove now that ∇^* maps $\mathcal{L}_0^2(\nu)$ into $L^2(\nu)$. It is sufficient to prove that for all $k \in \mathbb{N}$, $\varphi \in \mathcal{C}$, $\partial_k^* \varphi \in L^2(\nu)$. Since $\mathcal{C} \subset (\mathcal{S}) \subset \mathcal{D}$ (cf. Chapter 4), we know that $\varphi \in \mathcal{C}$ belongs to $\mathcal{D}^{p,k}$ for all $p>1$, $k \geq 0$. Also, we know from Chapter 5 that $\partial_k^* \varphi = \langle \cdot, e_k \rangle \varphi - \partial_k \varphi$. Thus we estimate in the following way.

$$\begin{aligned}\int (\partial_k^*\varphi)^2(x) \, d\nu(x) &= \int (\langle x, e_k \rangle \varphi(x) - \partial_k \varphi(x))^2 \, \Phi(x) \, d\mu(x) \\ &\leq 2 \| \varphi \|_\infty^2 \, \| \langle \cdot, e_k \rangle \|_{2q}^2 \, \| \Phi \|_p + 2 \| \partial_k \varphi \|_{2q}^2 \, \| \Phi \|_p \\ &\leq C_p(\| \varphi \|_\infty^2 + ||| \varphi |||_{2q,1}^2) \, \| \Phi \|_p,\end{aligned}$$

where we made use of the fact that $<\cdot, e_k>$ belongs to all (L^p), $p \geq 1$, and we have chosen $q>1$, such that $q^{-1}+p^{-1} = 1$. Since $\Phi \in (L^p)$, the last expression is finite, and we have shown that ∇^* maps $\mathcal{L}_0^2(\nu)$ into $L^2(\nu)$. Therefore

$$\nabla^{\dagger} := \nabla^* - \Phi^{-1}(\nabla\Phi),$$

is a densely defined operator from $\mathcal{L}^2(\nu)$ into $L^2(\nu)$. We finally show that $\nabla^{\dagger}$ is the adjoint of ∇ defined on $\mathcal{C}$ as an operator from $L^2(\nu)$ into $\mathcal{L}^2(\nu)$. Let $\varphi \in \mathcal{C}$, and $F \in \mathcal{L}_0^2(\nu)$. Then we have

$$\begin{aligned}
(\varphi, \nabla^{\dagger} F)_{L^2(\nu)} &= \int \varphi\, (\nabla^* F - \Phi^{-1}(\nabla\Phi)F)\, \Phi\, d\mu \\
&= \int ((\nabla\varphi\Phi), F)_{L^2(\mathbb{R})}\, d\mu - \int \varphi\, ((\nabla\Phi)F)\, d\mu \\
&= \int ((\nabla\varphi), F)_{L^2(\mathbb{R})}\, \Phi d\mu \\
&= (\nabla\varphi, F)_{L^2(\mathbb{R})\otimes L^2(\nu)}
\end{aligned}$$

where we used the product rule for ∇.

Altogether we have shown that $\nabla : L^2(\nu) \to \mathcal{L}_0^2(\nu)$ has a densely defined adjoint $\nabla^{\dagger}$. Now we can apply the general theory found in, e.g., Kato (1976) to conclude that $\mathcal{E}(\varphi, \varphi) = \int |\nabla\varphi|_2^2\, d\nu$ defines a closable form on $L^2(\nu)$. □

Remark. A similar result has also been proven in Takeda (1985) under slightly stronger conditions.

Next we establish a closability criterion for genuinely generalized functionals $\Phi \in (\mathcal{S})^*$.

Theorem 10.17. *Assume that* $\Phi \in (\mathcal{S})^*$ *is strictly positive. If* $\nabla\Phi$ *is of the form* $B\cdot\Phi$, *where* $B \in \mathcal{S}'(\mathbb{R}) \otimes (\mathcal{S})$, *then* Φ *is admissible.*

Remark. We have shown in Theorem 5.34 that $\Phi\in(\mathcal{S})^*$ implies that $\nabla\Phi\in\mathcal{S}'(\mathbb{R})\otimes(\mathcal{S})^*$. Thus our assumption reads that for all $\eta\in\mathcal{S}(\mathbb{R})$, $\varphi\in(\mathcal{S})$ we must have

$$<\nabla\Phi,\ \eta\otimes\varphi> \;=\; <\Phi,\ <B,\ \eta>\cdot\varphi>,$$

where the inner pairing on the right hand side is the one between $\mathcal{S}'(\mathbb{R})$ and $\mathcal{S}(\mathbb{R})$. Since $<B,\ \eta>\in(\mathcal{S})$, and $(\mathcal{S})$ is an algebra, we have $<B,\ \eta>\cdot\varphi\in(\mathcal{S})$, and the equality makes sense.

Proof. We use the same notations as in the proof of Theorem 10.16, and show again that ∇ has a densely defined adjoint from $\mathcal{L}^2(\nu)$ into $L^2(\nu)$ (ν denotes the measure defined by Φ).

Let $(e_k,\ k\in\mathbb{N})$ be a complete orthonormal system of $L^2(\mathbb{R})$ in $\mathcal{S}(\mathbb{R})$ (e.g., we can choose the set of Hermite functions, cf. Appendix A.1). We choose again as a dense domain for the adjoint $\nabla^\dagger$ of ∇ the space $\mathcal{L}_0^2(\nu)$. For $F\in\mathcal{L}_0^2(\nu)$, we set

$$\nabla^\dagger F := \nabla^* F \;-\; <B,\ F>,$$

where

$$<B,\ F> \equiv \sum_k <B,\ e_k>\cdot\ F_k.$$

Note that $<B,\ F>$ belongs to $(\mathcal{S})$ for all F of the above form, and therefore $<B,\ F>$ is in $L^2(\nu)$ (cf. also the previous section). Consider $\nabla^*F = \sum_k (<\cdot,\ e_k> - \partial_k)F_k$. Since $<\cdot,\ e_k>\in(\mathcal{S})$, $F_k\in(\mathcal{S})$, and $\partial_k F_k\in(\mathcal{S})$ (cf. Chapter 5), we see immediately that ∇^*F belongs to $(\mathcal{S})$ and therefore to $L^2(\nu)$. Note that we have shown that $\nabla^\dagger F\in(\mathcal{S})$.

It remains to show that $\nabla^\dagger$ as defined above is the adjoint of ∇ in $L^2(\nu)$. Let F be as before, and let $\varphi\in\mathcal{C}$. Then

$$
\begin{aligned}
(\varphi, \nabla^\dagger F)_{L^2(\nu)} &= <\Phi, \varphi\cdot\nabla^\dagger F> \\
&= <\Phi, \varphi\cdot(\nabla^* F - <B, F>)> \\
&= <\nabla(\varphi\cdot\Phi), F> - <\Phi, <B, \varphi\cdot F>> \\
&= <\Phi, ((\nabla\varphi), F)_{L^2(\mathbb{R})}> + <\nabla\Phi, \varphi\cdot F> - <\Phi, <B, \varphi\cdot F>> \\
&= <\Phi, ((\nabla\varphi), F)_{L^2(\mathbb{R})}> \\
&= (\nabla\varphi, F)_{\mathcal{L}^2(\nu)},
\end{aligned}
$$

where for $\Psi\in(\mathcal{S})^*$, $G\in\mathcal{L}_0^2(\nu)$, we have used the notation $<\nabla\Psi, G> \equiv \sum_k <\partial_k\Psi, G_k>$. Now we can apply again the results in, e.g., Kato (1976) to conclude the proof. □

Example 10.17. Let us choose $\Phi\in(\mathcal{S})^*$ as given in Example 4.23, Theorem 4.24. Informally,

$$\Phi(x) = \mathcal{N}\exp(-\tfrac{1}{2}<x, Kx>), \quad x\in\mathcal{S}'(\mathbb{R}),$$

and K is an operator on $L^2(\mathbb{R})$ satisfying the conditions of Theorem 4.24. It is clear that Φ is positive — note that $\mathcal{T}\Phi$ is positive definite (cf. (4.18)) and use Theorem 4.28. The Gaussian measure ν on $(\mathcal{S}'(\mathbb{R}),\mathcal{B})$ associated with Φ has the characteristic functional

$$
\begin{aligned}
\int e^{i<x, \xi>}\, d\nu(x) &= <\Phi, e^{i<\cdot, \xi>}> \\
&= \mathcal{T}\Phi(\xi)
\end{aligned}
$$

$$= \exp(-\tfrac{1}{2}(\xi, (K+1)^{-1}\xi)),$$

for $\xi \in \mathcal{S}(\mathbb{R})$. It follows that Φ is strictly positive in the sense of Definition 10.1. Let us compute $\nabla\Phi \in \mathcal{S}'(\mathbb{R}) \otimes (\mathcal{S})^*$. For $\eta \in \mathcal{S}(\mathbb{R})$ we use Theorem 5.35 and Theorem 4.24 to calculate as follows:

$$(\mathcal{S}<\cdot, K\eta>\Phi)(\xi) = D_{K\eta}\mathcal{S}\Phi(\xi) + (K\eta, \xi)\mathcal{S}\Phi(\xi)$$

$$= [-(K\eta, \frac{K}{K+1}\xi) + (K\eta, \xi)]\ \mathcal{S}\Phi(\xi)$$

$$= -(K\eta, (K+1)^{-1}\xi)\mathcal{S}\Phi(\xi)$$

$$= (\eta, K(K+1)^{-1}\xi)\mathcal{S}\Phi(\xi).$$

On the other hand, it follows from the mentioned theorems that

$$(\mathcal{S}D_\eta\Phi)(\xi) = (D_\eta\mathcal{S}\Phi)(\xi)$$

$$= -(\eta, K(K+1)^{-1}\xi)\mathcal{S}\Phi(\xi).$$

Since the $\mathcal{S}$-transform is injective on $(\mathcal{S})^*$ (Theorem 4.38), we have for all $\varphi \in (\mathcal{S})$, $\eta \in \mathcal{S}(\mathbb{R})$ the relation

$$<D_\eta\Phi, \varphi> = <\Phi, <\cdot, K\eta>\varphi>,$$

or

$$\nabla\Phi = B_K\Phi$$

with

$$<B_K, \eta> = <\cdot, K\eta> \in (\mathcal{S}), \quad \eta \in \mathcal{S}(\mathbb{R}).$$

We have shown that such Φ is admissible. Moreover, we found that for F of the form

$$F(t,x) = \sum_m e_m(t)\, F_m(x),\ t\in\mathbb{R},\ x\in\mathcal{S}'(\mathbb{R}),$$

and only finitely many non-zero $F_m \in \mathcal{C}$, we have

$$\nabla^\dagger F(x) = (\nabla^* F)(x) - \sum_m <x, Ke_m> F_m(x)$$

$$= \sum_m \{<x, (1-K)e_m> F_m(x) - \partial_m F_m(x)\}.$$

Therefore the self-adjoint operator H_K determined by the closure of the pre-Dirichlet form associated with Φ acts on $\varphi \in \mathcal{C}$ as follows:

$$H_K\varphi(x) = \nabla^\dagger \nabla\varphi(x)$$

$$= \sum_m \{<x, (1-K)e_m> \partial_m\varphi(x) - \partial_m^2\varphi(x)\}.$$

If K has an integral kernel K(t,s), t,s$\in\mathbb{R}$, this can also be written informally as

$$H_K\varphi(x) = (N\varphi)(x) - \int_{\mathbb{R}^2} K(t,s)\, x(t)\partial_s\varphi(x)\, dtds.$$

It is clear that we can readily generalize this example to white noise with d–dimensional time parameter, $d\in\mathbb{N}$. Let us put $A = 1-K$, and write H_A instead of H_K, i.e.

$$H_A\varphi(x) = \sum_m \{<x, Ae_m> \partial_m\varphi(x) - \partial_m^2\varphi(x)\},$$

where $\{e_m,\ m\in\mathbb{N}\}$ is now a CONS of $L^2(\mathbb{R}^d)$ in $\mathcal{S}(\mathbb{R}^d)$.

H_A can be identified with another operator as we shall show now.

Consider the dense subspace $\mathcal{E}_r$ generated by (real) exponential functions in $L^2(\nu)$. Define for $\lambda \in \mathbb{R}$, $\xi \in \mathcal{S}(\mathbb{R}^d)$,

$$:e^{\lambda<\cdot,\,\xi>}: \equiv e^{\lambda<\cdot,\,\xi> - \frac{\lambda^2}{2}|A^{-1}\xi|_2^2},$$

and note that $:e^{\lambda<\cdot,\,\xi>}:$ is the generating functional of orthogonal polynomials in $L^2(\nu)$. Moreover, we set for a suitable operator B on $L^2(\mathbb{R}^d)$

$$\gamma(B):e^{\lambda<\cdot,\,\xi>}: = :e^{\lambda<\cdot,\,B\xi>}:.$$

Let us compute the generator $d\gamma(A)$ of the strongly continuous semigroup defined by $\gamma(\exp(-tA))$, $t \geq 0$ (cf., e.g., Simon (1974)). We have

$$\begin{aligned}(d\gamma(A):e^{<\cdot,\,\xi>}:)(x) &= -\frac{d}{dt}\gamma(\exp(-tA):e^{<\cdot,\,\xi>}:\Big|_{t=0} \\ &= -\frac{d}{dt}:e^{<\cdot,\,e^{-tA}\xi>}:\Big|_{t=0} \\ &= (<x, A\xi> - |\xi|_2^2):e^{<x,\,\xi>}:.\end{aligned}$$

On the other hand,

$$\begin{aligned}H_A:e^{<\cdot,\,\xi>}:(x) &= e^{-\frac{1}{2}|A^{-1}\xi|_2^2}\,. \\ &\quad\cdot \sum_m \{<x, Ae_m><e_m, \xi> - (e_m, \xi)^2\}\, e^{<x,\,\xi>} \\ &= (<x, A\xi> - |\xi|_2^2):e^{<x,\,\xi>}:.\end{aligned}$$

Therefore, (the closure of) H_A can be identified with the (closure of the) generator of $\gamma(\exp(-tA))$: $H_A = d\gamma(A)$.

The particular choice $A = (-\Delta+m^2)^{\frac{1}{2}}$, where Δ is the Laplacian on $\mathbb{R}^d$ and $m^2>0$, leads to the Dirichlet form and Hamiltonian of the free Boson field in d space dimensions (cf. Example 1.4, and also Chapter 11).

The following result will be useful for several purposes. It is taken from the article by Albeverio et al. (1990a). A similar, slightly more general theorem can be found in the paper by Albeverio and Röckner (1990).

Let $(S,\mathcal{F},m)$ be a measure space with a σ–finite measure m. Assume that we are given a mapping $\sigma \mapsto (\mathcal{H}_\sigma, (\cdot\,,\,\cdot)_\sigma)$, $\sigma \in S$, where $(\mathcal{H}_\sigma, (\cdot\,,\,\cdot)_\sigma)$ is a Hilbert space. Suppose furthermore that there exists a linear set $\mathcal{D} \subset \bigcap_\sigma \mathcal{H}_\sigma$, which is dense in every $\mathcal{H}_\sigma$, $\sigma \in S$, and that for every $\sigma \in S$ there is a positive quadratic form F_σ defined on $\mathcal{D}$. Moreover, we require that for every $u \in \mathcal{D}$ the following holds:

(i) $$(u, u) := \int (u, u)_\sigma \, dm(\sigma) < +\infty,$$

(ii) $$F(u, u) := \int F_\sigma(u, u) \, dm(\sigma) < +\infty.$$

Define a Hilbert space $\mathcal{H}$ as the completion of $\mathcal{D}$ under the norm determined by $(\cdot\,,\,\cdot)$.

Theorem 10.19. *Suppose that for* m–a.e. $\sigma \in S$, $(F_\sigma, \mathcal{D})$ *is closable on* $\mathcal{H}_\sigma$. *Then* $(F, \mathcal{D})$ *is closable on* $\mathcal{H}$.

Proof. Let $\{u_n, n \in \mathbb{N}\}$ be a sequence in $\mathcal{D}$ which converges to 0 in $\mathcal{H}$, and which is Cauchy w.r.t. F, i.e., $F(u_n - u_m, u_n - u_m) \to 0$, as $m,n \to +\infty$. We have to show that there exists a subsequence (denoted again by $\{u_n, n \in \mathbb{N}\}$, for convenience) so that $F(u_n, u_n) \to 0$, as $n \to +\infty$ (cf. Section 10.A).

By assumption, we have

$$\int (u_n, u_n)_\sigma \, dm(\sigma) \to 0, \; n \to +\infty,$$

and

$$\int F_\sigma(u_n - u_m, u_n - u_m) \, dm(\sigma) \to 0, \; m,n \to +\infty,$$

We may select a subsequence (denoted again by $\{u_n, n\in\mathbb{N}\}$), so that for m–a.e. $\sigma\in S$, $(u_n, u_n)_\sigma \to 0$, and $F_\sigma(u_n-u_m, u_n-u_m) \to 0$. The assumption that for m–a.e. $\sigma\in S$, $(F_\sigma,\mathfrak{D})$ is closable on $\mathcal{H}_\sigma$ implies that for m–a.e. $\sigma\in S$, $F_\sigma(u_n, u_n) \to 0$, $n\to +\infty$.

Note that for every $v\in\mathfrak{D}$, and m–a.e. $\sigma\in S$ we then have

$$F_\sigma(v, v) = \lim_{m\to\infty} F_\sigma(v-u_m, v-u_m).$$

To see this, we only have to observe that $F_\sigma(v-u_m, v-u_m) = F_\sigma(v, v)-2F_\sigma(v, u_m) + F_\sigma(u_m, u_m)$. Since for m–a.e. $\sigma\in S$, $F_\sigma(u_m, u_m) \to 0$, the third vanishes in the limit, and the second term vanishes too after an application of Schwarz' inequality for the positive bilinear form F_σ.

Now we choose v as u_n, and apply Fatou's lemma in the following way.

$$\begin{aligned}
\lim_{n\to\infty} F(u_n, u_n) &= \lim_{n\to\infty} \int F_\sigma(u_n, u_n)\, dm(\sigma) \\
&= \lim_{n\to\infty} \int \lim_{m\to\infty} F_\sigma(u_n-u_m, u_n-u_m)\, dm(\sigma) \\
&\leq \lim_{n\to\infty} \liminf_{m\to\infty} \int F_\sigma(u_n-u_m, u_n-u_m)\, dm(\sigma) \\
&= \lim_{n\to\infty} \liminf_{m\to\infty} F(u_n-u_m, u_n-u_m).
\end{aligned}$$

The assumption that $\{u_n\}$ is Cauchy w.r.t. F implies that the last term can be made arbitrarily small, and consequently we have $F(u_n, u_n) \to 0$ as desired. □

In the following we consider certain special cases of Theorem 10.17 which are worth being pointed out.

Choose $S = \mathbb{N}$, and m as the counting measure. Then we obtain the following result.

Corollary 10.20. *Let* $\{(\mathcal{H}_n, (\cdot\,,\,\cdot)_n), n\in\mathbb{N}\}$ *be a sequence of Hilbert spaces. Assume that* $\mathfrak{D}$ *is a dense linear subspace of every* $\mathcal{H}_n$, $n\in\mathbb{N}$. *Suppose furthermore that for every* $n\in\mathbb{N}$, $(F_n, \mathfrak{D})$ *is a positive, quadratic, closable form on* $\mathcal{H}_n$. *If for every* $u\in\mathfrak{D}$, $\sum_n (u, u)_n < +\infty$, *and* $\sum_n F_n(u, u) < +\infty$, *then the positive, quadratic form* $(F,\mathfrak{D})$ *defined by*

$$F(u, u) = \sum_{n=1}^{\infty} F_n(u, u), \quad u\in\mathfrak{D},$$

is closable on the Hilbert space $\mathcal{H}$ *which is given by the completion of* $\mathfrak{D}$ *with respect to the norm determined by the inner product* $(u, u) = \sum_n (u, u)_n$.

Let us now consider some consequences for the case of energy forms determined by admissible elements in $(\mathcal{S})^*$.

Assume that we are given a sequence $\{\Phi_k, k\in\mathbb{N}\}$ in $(\mathcal{S})^*$, such that

$$\Phi := \sum_{k=1}^{\infty} \Phi_k$$

converges weakly (and therefore strongly) in $(\mathcal{S})^*$.

Corollary 10.21. *If for every* $k\in\mathbb{N}$, Φ_k *is admissible, then* Φ *is admissible.*

As before let $(S,\mathcal{F},m)$ be a σ-finite measure space. Suppose that we are given a mapping $\Phi(\cdot)$ from S into $(\mathcal{S})^*_+$ which is weakly in $L^1(S,m)$. We have proved in Chapter 8.A then the Pettis integral $\Phi := \int \Phi(\sigma)\, dm(\sigma)$ belongs to $(\mathcal{S})^*$, and we have for every $\varphi\in(\mathcal{S})$,

$$<\int \Phi(\sigma)\, dm(\sigma), \varphi> = \int <\Phi(\sigma), \varphi>\, dm(\sigma).$$

Let $(\mathcal{E}_\sigma,(\mathcal{S}))$ be the energy form defined by Φ_σ, while $(\mathcal{E},(\mathcal{S}))$ is the form given by

Φ. The Hilbert spaces $\mathcal{H}_\sigma$ are the spaces $L^2(\mathcal{S}'(\mathbb{R}),\ \nu_\sigma)$, where ν_σ is the measure associated with $\Phi(\sigma)$, while $\mathcal{H}$ is the Hilbert space $L^2(\mathcal{S}'(\mathbb{R}),\nu)$, and ν is determined by Φ. Note that $(\mathcal{S})$ is densely included in all $\mathcal{H}_\sigma$, $\sigma \in S$, and in $\mathcal{H}$. For $\varphi \in (\mathcal{S})$, we have

$$\begin{aligned} \mathcal{E}(\varphi, \varphi) &= <\Phi, |\nabla\varphi|_2^2> \\ &= <\int \Phi(\sigma)\, dm(\sigma), |\nabla\varphi|_2^2> \\ &= \int <\Phi(\sigma), |\nabla\varphi|_2^2>\, dm(\sigma) \\ &= \int \mathcal{E}_\sigma(\varphi, \varphi)\, dm(\sigma). \end{aligned}$$

Thus we obtain from Theorem 10.19 the following result.

Corollary 10.22. *Let* $\Phi(\sigma)$, $\sigma \in S$, *be as above. If for* m–a.e. $\sigma \in S$, $\Phi(\sigma)$ *is admissible, then* $\Phi = \int \Phi(\sigma)\, dm(\sigma)$ *is admissible.*

Example 10.23. Consider the generalized Gaussian functional of Example 10.18 with the operator K depending on a parameter $\sigma \in S$, where $(S,\mathcal{F},m)$ is some σ–finite measure space. Assume that the family of operators $\{K(\sigma),\ \sigma \in S\}$ satisfies the conditions of Theorem 4.24, uniformly in $\sigma \in S$. Denote the associated admissible elements in $(\mathcal{S})^*$ by $\Phi(\sigma)$, $\sigma \in S$. Suppose furthermore that $\sigma \mapsto \Phi(\sigma)$ is weakly in $L^1(S,m)$. Then $\int \Phi(\sigma)\, dm(\sigma) \in (\mathcal{S})^*$ is admissible by Corollary 10.22. We write informally

$$\int \mathcal{N} \exp(-\tfrac{1}{2}<x, K(\sigma)x>)\, dm(\sigma), \quad x \in \mathcal{S}'(\mathbb{R})$$

for this generalized functional. Observe that this element of $(\mathcal{S})^*$ is no longer Gaussian.

As a special choice, let us put

$$K(\sigma) = (-\Delta+\sigma)^{1/2} - 1,$$

where $\sigma \in S=(\sigma_0,\sigma_1)$, $\sigma_0>0$, $\sigma_1<+\infty$, and we choose $\mathcal{F}$ as the Borel sets of (σ_0,σ_1), m as Lebesgue measure $d\sigma$. It is obvious, that we may let the time parameter be d dimensional, $d\in\mathbb{N}$. In this case, $\Phi(\sigma)$ is associated with the measure of a free Boson quantum field of mass σ (cf. Example 1.4 and also Example 10.18). Using the arguments leading to Theorem 4.24, it is not hard to to see that $\sigma \mapsto \Phi(\sigma)$ is weakly continuous, and therefore it is weakly integrable over any closed interval I contained in (σ_0, σ_1). Thus,

$$\int_I \mathcal{N}\exp(-\tfrac{1}{2}<x, [(-\Delta+\sigma)^{1/2}-1]x>\, d\sigma$$

is an admissible functional in $(\mathcal{S})^*$. We remark that the covariance of the associated measure on $(\mathcal{S}'(\mathbb{R}^d), \mathcal{B})$ is the one of a generalized free field (cf., e.g., Simon (1974)), but that this measure being non-Gaussian is not the one of a generalized free field.

Before we conclude our discussion of closability criteria by describing a very general result due to Albeverio and Röckner (1990), we want to mention that in the article of Albeverio et al. (1990a) further results are proved about how to construct admissible elements in $(\mathcal{S})^*$ from given ones. In particular one finds there the result that weak limits of admissible elements are admissible.

Let $\Phi\in(\mathcal{S})^*_+$, and choose a CONS $\{e_k, k\in\mathbb{N}\}$ of $L^2(\mathbb{R}^d)$. As before, denote the Gâteaux derivative in direction of e_k, $k\in\mathbb{N}$, by ∂_k. For $k\in\mathbb{N}$, consider the following form $\mathcal{E}_k$ on $(\mathcal{S})$:

$$\mathcal{E}_k(\varphi, \varphi) := <\Phi, (\partial_k\varphi)^2>.$$

By (10.3) (continuity of Φ, and Lemma 10.2) we have for every $\varphi \in (\mathcal{S})$,

$$\mathcal{E}(\varphi, \varphi) = \sum_{k=1}^{\infty} \mathcal{E}_k(\varphi, \varphi),$$

where $\mathcal{E}(\varphi, \varphi) := <\Phi, | \nabla\varphi |_2^2>$. Due to Corollary 10.20, it is sufficient to prove the closability of $\mathcal{E}_k$, for every $k \in \mathbb{N}$, in order to conclude closability of $(\mathcal{E},(\mathcal{S}))$, or equivalently of $(\mathcal{E},\mathcal{F}C_b^\infty)$, on $L^2(\nu)$, where ν is associated with Φ. Since we have in $\mathcal{E}_k$ only a derivative acting in one direction, the idea in Albeverio and Röckner (1990) is to disintegrate the measure ν according to a decomposition of $\mathcal{S}'(\mathbb{R}^d)$ into the subspace spanned by e_k and its complement. Then the resulting problem is essentially a one-dimensional one, and this can be solved completely (cf. Fukushima (1980), Hamza (1975) and Rullkötter and Spönemann (1983)).

Consider the measure ν on $(\mathcal{S}'(\mathbb{R}^d),\mathcal{B})$ given by $\Phi \in (\mathcal{S})_+^*$, and assume without loss of generality that ν is a probability measure (normalizing ν if necessary). Since $\mathcal{S}'(\mathbb{R}^d)$ is a Suslin space (actually a Lusin space), ν is a Radon measure on $(\mathcal{S}'(\mathbb{R}^d),\mathcal{B})$ (cf. Schwartz (1973)). Consider e_k as an element in $\mathcal{S}'(\mathbb{R}^d)$, and decompose $\mathcal{S}'(\mathbb{R}^d)$ as $E_k' \oplus \mathbb{R}\cdot e_k$, where E_k' is some closed linear subspace of $\mathcal{S}'(\mathbb{R}^d)$. For example, choose $\eta_k \in \mathcal{S}(\mathbb{R}^d)$ so that $<e_k, \eta_k> = 1$, and define the projections

$$\pi_k(x) = <x, \eta_k>e_k, \; \pi_k'(x) = x - \pi_k(x), \quad x \in \mathcal{S}'(\mathbb{R}^d).$$

Set $E_k' = \pi_k'\mathcal{S}'(\mathbb{R}^d)$. From now let η_k be fixed; – it will be easy to see that our considerations are independent of the concrete choice of η_k. Thus from now on every $x \in \mathcal{S}'(\mathbb{R}^d)$ has a unique decomposition $x = x_k + s\cdot e_k$, $x_k = \pi_k'(x) \in E_k'$, $s = <x, \eta_k> \in \mathbb{R}$.

Since $(\mathcal{S}'(\mathbb{R}^d),\mathcal{B},\nu)$ is a Radon space, we may disintegrate ν according to the decomposition $\mathcal{S}'(\mathbb{R}^d) = E_k' \oplus \mathbb{R}\cdot e_k$, cf., e.g., Dellacherie and Meyer (1978, III, no. 70–73). Namely, there exists a kernel ρ_k: $E_k' \times \mathcal{B}(\mathbb{R}) \to [0,1]$, so that for all bounded, measurable functions u on $\mathcal{S}'(\mathbb{R}^d)$ we have

$$\int u(x)\, d\nu(x) = \int_{E_k'} \Big(\int_{\mathbb{R}} u(x_k+s\cdot e_k)\, \rho_k(x_k, ds)\Big)\, d\nu_k(x_k),$$

where ν_k is the projection of ν onto E_k': $\nu_k = \nu\circ\pi_k'^{-1}$, and ds denotes Lebesgue measure on $(\mathbb{R},\mathcal{B}(\mathbb{R}))$.

Assume that u is (everywhere) Gâteaux differentiable in direction e_k. If we denote

$$\hat{u}(x_k, s) := u(x_k+s\cdot e_k), \quad x_k\in E_k',\ s\in\mathbb{R},$$

then it is clear that

$$\partial_k u(x) = (\tfrac{\partial}{\partial s}\hat{u})(x_k,s), \quad x = x_k + s\cdot e_k.$$

Therefore we are interested in the problem of closability of the form

$$\mathcal{E}_k(u, u) = \int_{E_k'} \Big(\int_{\mathbb{R}} (\tfrac{\partial}{\partial s}\hat{u})(x_k,s)^2\, \rho(x_k,ds)\Big)\, d\nu_k(x_k),$$

for $u\in\mathcal{F}C_b^\infty$, say. Theorem 10.19 reduces this problem to proving the closability of the form

$$\int_{\mathbb{R}} (\tfrac{\partial}{\partial s}\hat{u}(x_k,s))^2\, \rho(x_k,ds), \quad u\in\mathcal{F}C_b^\infty,$$

on $L^2(\mathbb{R}, \rho(x_k,ds))$ for ν_k-a.e. $x_k\in E_k'$. But this is now for fixed x_k a one-dimensional problem, and the crucial property of the measure $\rho(x_k, \cdot)$ which is necessary and sufficient for the closability of the above form has been isolated by Hamza (1975). Let ω be a positive, $\mathcal{B}(\mathbb{R})$–measurable function on $\mathbb{R}$, and define the following set:

$$R(\omega) := \{t\in\mathbb{R},\ \int_{t-\epsilon}^{t+\epsilon} \omega(s)^{-1}ds < +\infty, \text{ for some } \epsilon>0\}.$$

Hamza's condition is the following:

$$\omega = 0, \quad \text{ds–a.e. on } \mathbb{R}\backslash R(\omega). \tag{H}$$

The above mentioned result of Albeverio and Röckner (1990) can now be stated as follows.

Theorem 10.24.

(i) *Assume that for* ν_k-a.e. $x_k \in E_k{}'$, $\rho(x_k,ds)$ *has a density* $\omega(x_k,s)$ *with respect to Lebesgue measure which satisfies* (H), *then the form* $(\mathcal{E}_k,\mathcal{F}C_b^\infty)$ *is closable on* $L^2(\nu)$.

(ii) *If* $(\mathcal{E}_k,\mathcal{F}C_b^\infty)$ *is closable on* $L^2(\nu)$, *then for* ν_k-a.e. $x_k \in E_k{}'$, $\rho(x_k,ds) = \omega(x_k,s)ds$ *for some measurable function* $\omega(x_k,\cdot)$: $\mathbb{R} \to \mathbb{R}_+$, *which satisfies* (H).

In particular, Φ is admissible if for every $k \in \mathbb{N}$ and ν_k-a.e. $x_k \in E_k{}'$, $\rho(x_k,ds)$ has a density satisfying (H).

We conclude this chapter with a result concerning the stochastic process associated with a closed (or closable) energy form over $\mathcal{S}'(\mathbb{R}^d)$. This result can be found in Albeverio et al. (1990b), and is a special case of a theorem in Albeverio and Röckner (1989) which is formulated there for a general Suslin space equipped with a Radon measure.

Let $\Phi \in (\mathcal{S})^*$ be admissible, and let us simply denote its closure on $L^2(\nu)$ by $(\mathcal{E},\mathcal{D}(\mathcal{E}))$, ν being the measure on $(\mathcal{S}'(\mathbb{R}^d),\mathcal{B})$ determined by Φ. T denotes the self-adjoint, positive operator associated with $(\mathcal{E},\mathcal{D}(\mathcal{E}))$, P_t, $t \geq 0$, is the Markovian semigroup generated by T (cf. Theorem 10.11).

Theorem 10.25. *There exists a diffusion process* $\mathfrak{X}$ *associated with* $(\mathcal{E},\mathcal{D}(\mathcal{E}))$ (*i.e., with* Φ). *Namely,* $\mathfrak{X} = (\Omega, \mathcal{F}, (X_t, t \geq 0), (P_x, x \in \mathcal{S}'(\mathbb{R})))$ *is a Markov process with state space* $\mathcal{S}'(\mathbb{R}^d)$, *and a.s. (strongly) continuous sample paths. Moreover, for any* *bounded, measurable function* u *on* $\mathcal{S}'(\mathbb{R}^d)$, *all* $t \geq 0$, *and* ν–a.e. $x \in \mathcal{S}'(\mathbb{R}^d)$, *the following equality holds:*

$$(P_t u)(x) = \int_\Omega u(X_t)\, dP_x. \tag{10.10}$$

Let us give a very rough description of the proof of this result. For details we refer the reader to Albeverio and Röckner (1989), Fukushima (1980), and Kusuoka (1982). In Fukushima (1980) the analogue of the above theorem for the case of a locally compact state space has been proved, by constructing from the Markovian transition functions of P_t a stochastic process by means of the Kolmogorov extension theorem. Then, the capacity of the energy form is used to prove that this process is indeed a diffusion. In Albeverio and Röckner (1989) and in Kusuoka (1982) this has been extended to state spaces E which are not locally compact, by first compactifying the state space in an appropriate way, and then proving that the (extended) capacity associated with the (extended) energy form does not charge the difference between E and its compactification. Thereby, one obtains a diffusion process on the original state space.

11. Applications to Quantum Field Theory

11.A Introduction

In this chapter we shall discuss various aspects of quantum field theory which can be formulated in the framework developed in Chapters 1 to 5. We begin with this section containing an introduction to quantum field theory, which is intended to give the reader an idea about the basic questions and notions in this area. Here we make a number of heuristic and informal considerations, referring to the pertaining literature for mathematically more rigorous treatments (when available). For a general background on quantum field theory which is more oriented towards physics we refer to the books of Bjørken and Drell (1965), Bogoliubov and Shirkov (1959), Itzykson and Zuber (1980), Schweber (1961). Mathematically rigorous treatments of the foundations of quantum field theory are given in the monographs by Bogoliubov et al. (1975), Jost (1965) and Streater and Wightman (1964). Modern aspects of mathematical quantum field theory, in particular Euclidean quantum field theory, are treated in the books by Simon (1974), and Glimm and Jaffe (1981).

We warn the reader that, due to our ignorance and to the lack of space, we will provide only a very rough picture of quantum field theory in this introduction. Moreover, throughout this chapter we will only consider models of massive Boson fields. In order to conform with the traditional notations of quantum field theory, we shall sometimes deviate slightly from the notations used in the previous chapters.

Quantum field theory is concerned with the construction and study of representations of the so-called *canonical commutation relations* (*CCR*): let φ and π be two mappings from $\mathbb{R}^s$, $s \in \mathbb{N}$, into an algebra with a unit element 1, and let $[\cdot\ ,\ \cdot]$ denote the commutator. Then, informally the CCR read:

$$[\varphi(x),\ \varphi(y)] = [\pi(x),\ \pi(y)] = 0,$$

$$[\varphi(x),\ \pi(x)] = i\delta(x-y)1, \tag{11.1}$$

where $x,y \in \mathbb{R}^s$, and δ is the Dirac distribution. In the sequel we will use the same symbols for the abstract fields $\varphi(x)$, $\pi(x)$, and their various representations as self-adjoint operators on Hilbert spaces. In addition, one assumes the existence of an element H in the algebra, called *Hamiltonian*, which is a function of φ and π such that

$$[H,\ \varphi(x)] = -i\pi(x). \tag{11.2}$$

In physics, $\varphi(x)$ is interpreted as the (operator valued) field at the point $x \in \mathbb{R}^s$, a dynamical variable associated with this point in space. $\pi(x)$ is its canonical counterpart. The Hamiltonian H represents the energy operator, and it generates the unitary group of time translations $\exp(-itH)$, $t \in \mathbb{R}$, of the system, i.e., it implements the dynamics. Different models of quantum field theory are obtained by specific choices of H. In addition to the CCR, one requires in quantum field theory the relativistic invariance of models: the representation of time translations and of Euclidean transformations of $\mathbb{R}^s$ are part of a unitary representation of the Poincaré group in $d = s+1$ dimensions. Moreover, stability of the system is formulated by the requirement that H is a positive operator on a Hilbert space $\mathcal{H}$ which has a unique ground state $\Omega \in \mathcal{H}$, $H\Omega = 0$, called the *vacuum*.

Instead of the CCR (11.1), (11.2) one sometimes considers the *Weyl relations* for the unitary groups $U(\xi) = \exp(i\varphi\xi))$, $V(\xi) = \exp(i(\pi(\xi))$, $\xi \in \mathcal{S}(\mathbb{R}^s)$, generated by "smeared" fields $\varphi(\xi) = \int \varphi(x)\xi(x)\,dx$, $\pi(\xi) = \int \pi(x)\ \xi(x)\,dx$. They read as follows:

$$U(\xi)\ U(\eta) = U(\xi+\eta),\quad V(\xi)\ V(\eta) = V(\xi+\eta), \tag{11.3}$$

$$U(\xi)\ V(\eta) = \exp(-i(\xi,\eta))\ V(\eta)\ U(\xi), \tag{11.4}$$

where $\xi,\eta\in\mathcal{S}(\mathbb{R}^s)$, and $(\cdot\ ,\ \cdot)$ is the inner product of $L^2(\mathbb{R}^s)$. The advantage of this point of view is that it does not involve any difficulties concerning domains of unbounded operators. Relation (11.2) is then replaced by (cf. Araki (1960))

$$[H,\ U(\xi)] = U(\xi)\ \pi(\xi)\ +\tfrac{1}{2}(\xi,\ \xi)\ U(\xi). \tag{11.5}$$

In order to avoid the operator $\pi(\xi)$ on the right hand side of (11.5), Araki (1960) proposed the following relation

$$[U(\xi),\ [H,\ U(\eta)]] = -(\xi,\ \eta)\ U(\xi+\eta), \tag{11.6}$$

which turns out to be equivalent to (the smeared form of) (11.2) and to (11.5) under the additional (but natural) assumption that the Hamiltonian H is invariant under time reversal. In any case, either from (11.6) or from (11.5) and the CCR (or the Weyl relations) Araki (1960) deduces the following important formula

$$(U(\xi)\Omega,\ H\ U(\eta)\Omega) = \tfrac{1}{2}(\xi,\ \eta)\ (U(\xi)\Omega,\ U(\eta)\Omega), \tag{11.7}$$

which is also called the *Araki relation.* The Araki relation (11.7) will be at the basis of our arguments in Section 11.C.

It is common to assume that all informations about a model of quantum field theory are contained in the vacuum expectation values of the time translated fields: for $t\in\mathbb{R}$ denote $U(t) = \exp(-itH)$, and $\varphi(t,x) := U(t)^{-1}\varphi(x)U(t)$. Let $n\in\mathbb{N}$, $t_1, \ldots ,t_n\in\mathbb{R}$, $x_1, \ldots ,x_n\in\mathbb{R}^s$, and consider the so-called *n–point* or *Wightman functions*

$$W_n(t_1,x_1,\ t_2,x_2,\ \ldots ,t_n,x_n) = (\Omega,\ \varphi(t_1,x_1)\ \varphi(t_2,x_2)\ \cdots\ \varphi(t_n,x_n)\ \Omega), \tag{11.8}$$

where $(\cdot\ ,\ \cdot)$ denotes the inner product of $\mathcal{H}$. (The axiom of the above mentioned assumption goes under the name "cyclicity of the vacuum".)

The construction of representations of the CCR with the properties sketched above turned out to be very difficult from a mathematical point of view. In particular, the computations done by physicists with the field operators and the Hamiltonian involved many informal manipulations of infinities. In order to avoid such manipulations, Wightman (1956) proposed to consider, instead of the field operators, the vacuum expectation values W_n as the basic objects of quantum field theory. He formulated a set of axioms – known as the Wightman axioms – for the distributions W_n, and these axioms reflect all the physically necessary properties mentioned above of a representation of the CCR (cf. also Streater and Wightman (1964)). The investigation of the distributions W_n led to the following observation (e.g., Jost (1965), Streater and Wightman (1964)): the W_n are boundary values of analytic functions on certain domains in $(\mathbb{C}^{s+1})^n$. In particular the points of $(\mathbb{C}^{s+1})^n$ which are of the form $(it_1,x_1, it_2,x_2, \dots , it_n,x_n)$, with $x_i \neq x_j$, $i \neq j$, which are called *Euclidean points*, belong to this domain. Assume that $t_1 \geq t_2 \geq \cdots \geq t_n$. Then under $t_k \to it_k$, the right hand side of (11.8) becomes

$$S_n(t_1,x_1, t_2,x_2, \dots ,t_n,x_n) =$$
$$= (\Omega, \varphi(x_1) \exp(-(t_1-t_2)H) \varphi(x_2) \cdots \exp(-(t_{n-1}-t_n)H) \varphi(x_n) \Omega). \tag{11.9}$$

The S_n are called *Schwinger functions.* Symanzik (1969) initiated a program to study the Schwinger functions systematically, and he pointed out that they can be obtained from *Euclidean fields* – which essentially are fields in the sense of probability theory. He furthermore advocated the use of tools from statistical mechanics and probability theory in the construction and investigation of the Schwinger functions. On one hand, these ideas led to two new sets of axioms for Euclidean quantum field theories: one set of axioms for the Schwinger functions, which is equivalent to the Wightman axioms, was formulated by Osterwalder and Schrader (1973). Another set of axioms was given by Nelson (1973a) for the Euclidean fields. These axioms imply the Osterwalder–Schrader axioms for the Schwinger functions defined as the moments of the Euclidean fields, and thereby the Wightman axioms for the analytically continued Schwinger functions. In particular, Nelson (1973) stressed the Markovian nature of the Euclidean fields.

On the other hand, Symanzik's observations gave the basis for the construction of models of Euclidean fields through the Feynman–Kac type formula

$$S_n(t_1,x_1, t_2,x_2, \dots, t_n,x_n) = \int \phi(t_1,x_1)\, \phi(t_2,x_n) \cdots \phi(t_n,x_n)\, d\nu(\phi), \qquad (11.10)$$

for the Schwinger functions. ν is a probability measure on the configuration space of the Euclidean field ϕ, given (informally) by

$$d\nu(\phi) = Z^{-1} \exp\Big(-\int V(\phi(t,x))\, dt\, dx\Big)\, d\mu_0(\phi), \qquad (11.11)$$

where Z is a normalizing factor, V is an appropriate function of the fields ϕ representing their interaction ("potential"), and μ_0 is a suitable Gaussian measure. Typically, μ_0 is the centered Gaussian measure on $(\mathcal{S}'(\mathbb{R}^{s+1}),\mathcal{B})$ with covariance given by the resolvent kernel of the Laplacian in s+1 dimensions. The rigorous construction of measures of the type (11.11) and the study of their properties is the aim of the so-called "constructive quantum field theory", and we refer to Albeverio and Høegh–Krohn (1973), Fröhlich (1974, 1977), Glimm and Jaffe (1981), Guerra, Rosen and Simon (1975), Simon (1974), Velo and Wightman (1973) and literature quoted there for further reading.

Next we are going to sketch another approach to the construction of representations of (11.1) and (11.2): the *Dirichlet strategy.* In quantum field theory this approach goes back to Araki (1960), and it has been worked out by Albeverio and Høegh–Krohn (1976, 1977, 1984) and their followers (cf., e.g., Albeverio and Röckner (1989, 1990) and the literature quoted there). Let us first describe in the finite dimensional case, i.e., in quantum mechanics, the so-called ground state transformation.

Consider the question of finding a representation of (11.1) and (11.2) in the case where x,y take values in a finite set, say {1,2, ... ,d}, and let us write φ_i,

π_j, etc. instead. In particular, the CCR then read $[\varphi_i, \varphi_j] = [\pi_i, \pi_j] = 0$, and $[\varphi_i, \pi_j] = i\delta_{ij}1$, δ_{ij} being the Kronecker symbol. The Schrödinger representation of the CCR is given as follows. Choose $L^2(\mathbb{R}^d, dx)$ as the Hilbert space $\mathcal{H}$ (where dx denotes Lebesgue measure). φ_i is determined by $\varphi_i f(x) = x_i f(x)$, $f \in \mathcal{S}(\mathbb{R}^d)$, $x \in \mathbb{R}^d$, while π_j is the self-adjoint extension of $-i\frac{\partial}{\partial x_j} f(x)$, $f \in \mathcal{S}(\mathbb{R}^d)$, $x \in \mathbb{R}^d$. (11.2) holds automatically for operators H which are of the form $Hf(x) = -\frac{1}{2}\Delta f(x) + V(x)f(x)$, for suitable functions V representing the potential of the force field which acts on the quantum mechanical particle. For large classes of potentials V, H has a unique ground state $\Omega \in L^2(\mathbb{R}^d, dx)$, $H\Omega = 0$. For example, whenever $V \in L^2_{loc}(\mathbb{R}^d, dx)$, $V \geq 0$, is such that $V(x) \to +\infty$, as $|x| \to +\infty$, this is true (cf., e.g., Reed and Simon (1978, Theorem XIII.47)). Moreover, Ω can be chosen strictly positive: $\Omega(x) > 0$, for all $x \in \mathbb{R}^d$. Let us assume from now on that this is true. Then we may perform the ground state transformation U_Ω as follows:

$$U_\Omega : L^2(\mathbb{R}^d, dx) \to L^2(\mathbb{R}^d, d\nu_0)$$

$$f(x) \mapsto \Omega^{-1}(x) f(x), \; x \in \mathbb{R}^d,$$

where $d\nu_0(x) = \Omega(x)^2 dx$ is called the *ground state measure.* It is clear that U_Ω is unitary, and that φ_i and π_j from the Schrödinger representation are mapped to the operators given by multiplication by x_i, and $-i\frac{\partial}{\partial x_j} + (\frac{\partial}{\partial x_j} \ln \Omega)(x)$, respectively. On the other hand, the Hamiltonian is mapped into the following operator:

$$H_\Omega g = U_\Omega H U_\Omega^{-1} g = -\tfrac{1}{2}\Delta g - (\nabla \ln \Omega)\cdot\nabla g,$$

$\cdot$ indicating the Euclidean inner product of $\mathbb{R}^d$. Since H was by assumption self-adjoint, so is H_Ω, and thus H_Ω (and consequently also H) is uniquely determined by its associated quadratic form (e.g., Kato (1976)):

$$(g, H_\Omega h)_\Omega = \tfrac{1}{2} \int_{\mathbb{R}^d} \overline{\nabla g(x)}\cdot\nabla h(x)\, d\nu_0(x), \quad g, h \in \mathcal{D}(H_\Omega^{1/2}).$$

Since the right hand side of the above equation has an extremely simple form, it suggests to try to construct Hamiltonians in the infinite dimensional case through Dirichlet forms as in this equation. As we have shown in Chapter 10, it is not hard to generalize such expressions to the infinite dimensional case, but it is usually a difficult problem to prove for a given choice of the measure ν_0 the closability of the Dirichlet form. We refer to the works by Albeverio and Høegh–Krohn (1977, 1984), Kusuoka (1982), Albeverio and Röckner (1990), Albeverio et al. (1990b), and to the discussion in Chapter 10. In the infinite dimensional quantum field theory case, the ground state measure ν_0 of the Dirichlet form has to be chosen as the restriction of the Euclidean measure ν indicated above to the σ–algebra generated by the the time–zero Euclidean fields $\phi(0,x)$ (cf. also Albeverio and Høegh–Krohn (1977) and Fröhlich (1974, 1977)).

In the next section we give a construction of free fields in terms of the differential operators which where investigated in Chapter 5. Moreover, we shall describe also free Euclidean fields, and show how the fields can be formulated through Dirichlet forms of the type which were constructed in Chapter 10. In section 11.C, we prove that the ground state measures which were constructed in various quantum field theories in two space–time dimensions can be realized as generalized white noise functionals in $(\mathcal{S})^*$. Thus, via the Dirichlet strategy, we can construct non-trivial representations of (11.1) and (11.2) within the white noise framework. Here, the key observation will be the relation between so-called φ–bounds and the representability of a ground state measure by an element in $(\mathcal{S})^*$, which has been made in a paper by Potthoff and Streit (1990).

Throughout this chapter we shall work with white noise with s– or d–dimensional time, $s = d+1$, $s \in \mathbb{N}$. We shall continue to use the notations of the previous chapters also for this case.

11.B Free Fields

From now on we consider the following realization of the CCR (11.1), (11.2). We study a Hilbert space $\mathcal{H}$, linear maps φ, π from $\mathcal{S}(\mathbb{R}^s)$ into the set of self-adjoint operators on $\mathcal{H}$, and a self-adjoint operator $H \geq 0$ on $\mathcal{H}$, so that for all $\xi, \eta \in \mathcal{S}(\mathbb{R}^s)$, we have, on appropriate domains,

$$[\varphi(\xi), \varphi(\eta)] = [\pi(\xi), \pi(\eta)] = 0$$
$$[\varphi(\xi), \pi(\eta)] = i(\xi,\eta), \tag{11.12}$$

where the last inner product is the one of $L^2(\mathbb{R}^s, du)$, du denoting Lebesgue measure. Moreover, we ask for

$$[H, \varphi(\xi)] = -i\pi(\xi), \tag{11.13}$$

and require the existence of a unique (up to constant multiples of modulus 1) ground state $\Omega \in \mathcal{H}$ of H: $H\Omega = 0$, $\| \Omega \|_{\mathcal{H}} = 1$.

In this section we construct and study representations of the CCR on the white noise space $(L^2) \equiv L^2(\mathcal{S}'(\mathbb{R}^s), \mathcal{B}, \mu)$. Basically, these are the same as the corresponding and well-known representations on Fock space (e.g., Reed and Simon (1975)). The ideas in this section are very much inspired by the works of Nelson (1973).

The simplest representation of the CCR (11.7) and (11.8) on the Hilbert space $\mathcal{H} = (L^2)$ can be achieved by setting:

$$\varphi(\xi) = D_\xi^* + D_\xi, \quad \pi(\eta) = \tfrac{i}{2}(D_\eta^* - D_\eta),$$

where – as in Chapter 5 – D_ξ denotes the Gâteaux derivative in direction ξ, D_ξ^* its adjoint, and both have (L^2)–domain $\mathcal{D}^{2,1} = \mathcal{D}(N^{1/2})$ (cf., e.g., Corollary 5.26 and Theorem 5.27, and the remarks thereafter). Theorem 5.28 tells us immediately that (11.12) holds on $\mathcal{D}^{2,2}$. It is clear that both, $\varphi(\xi)$ and $\pi(\eta)$, are self-adjoint on $\mathcal{D}^{2,1}$. Moreover, as the estimates in Chapter 5 show, as mappings from $\mathcal{S}(\mathbb{R}^s)$ into self-adjoint operators on (L^2), both have strongly continuous extensions to $L^2(\mathbb{R}^s)$.

Finally, we choose for the Hamiltonian $\frac{1}{2}$ times the number operator N with domain $\mathfrak{D}^{2,2}$. Note that N is self-adjoint on this domain, and that $\Omega(x) = 1$, $x \in \mathcal{S}'(\mathbb{R}^s)$, is the unique strictly positive ground state of N. It is straightforward to verify relation (11.8) on $\mathfrak{D}^{3,1}$, by use of the representation $N = \sum_k \partial_k^* \partial_k$, where ∂_k is equal to D_{e_k}, ∂_k^* is its adjoint, and $\{e_k, k \in \mathbb{N}\}$ is a CONS of $L^2(\mathbb{R}^s)$.

However, from the point of view of physics this representation is trivial in the sense that it does not implement an interesting dynamics for the fields. Note that $\varphi(f)$, $f \in L^2(\mathbb{R}^s)$, acts on (L^2) by multiplication with the random variable $X_f = <\cdot, f>$. In physics, $\varphi(f)\Omega = X_f 1 = <\cdot, f>$ is interpreted as a one particle state, whenever it is normalized, i.e., when $|f|_2 = 1$. The energy of this one particle state would then be given by $(\varphi(f)\Omega, H\,\varphi(f)\Omega) = \frac{1}{2}(<\cdot, f>, N<\cdot, f>) = \frac{1}{2}(<\cdot, f>, <\cdot, f>) = \frac{1}{2}|f|_2 = \frac{1}{2}$. Thus the energy of all one particle states is equal to one half, independent of their "shape" given by f. Moreover, the unitary time translations would act on all one particle states as multiplication by $\exp(-\frac{i}{2}t)$. Clearly, this is a trivial type of dynamical behavior. In physics, the requirement of relativistic invariance of the theory leads to the expression $(f, (-\Delta + m^2)^{1/2} f)$, $m^2 > 0$, for the energy of a one particle state given by $\varphi(f)\Omega$ (cf., e.g., Reed and Simon (1975, IX.7)). Therefore, we construct now a representation which will have one particle states with this energy.

Let ω denote the self-adjoint operator on $L^2(\mathbb{R}^s)$ given by the pseudo--differential operator $(-\Delta + m^2)^{1/2}$, where m^2 is some strictly positive number (the so-called "bare mass" of the particles). Note that for every $p \in \mathbb{R}$, we have $\omega^p \xi \in \mathcal{S}(\mathbb{R}^s)$ whenever $\xi \in \mathcal{S}(\mathbb{R}^s)$. Thus we may define the self-adjoint operators $\varphi(\xi)$, $\pi(\xi)$, $\xi \in \mathcal{S}(\mathbb{R}^s)$, on $\mathfrak{D}^{2,1}$ by

$$\varphi(\xi) := 2^{-1/2}(D^*_{\omega^{-1/2}\xi} + D_{\omega^{-1/2}\xi}), \tag{11.14}$$

$$\pi(\xi) := i\, 2^{-1/2}(D^*_{\omega^{1/2}\xi} - D_{\omega^{1/2}\xi}). \tag{11.15}$$

It is obvious that these operators satisfy the relations (11.7) on $\mathfrak{D}^{2,2}$. Also, $\varphi(\xi)$

acts as multiplication by $2^{-1/2}X_{\omega^{-1/2}\xi} = 2^{-1/2}<\cdot, \omega^{-1/2}\xi>$. The Hamiltonian H we choose as follows:

$$H := \sum_{k=1}^{\infty} D^*_{\omega^{1/2}e_k} D_{\omega^{1/2}e_k}, \tag{11.16}$$

where $\{e_k, k \in \mathbb{N}\}$ is a complete orthonormal system of $L^2(\mathbb{R}^s)$ which lies in $\mathcal{S}(\mathbb{R}^s)$ (we invite the reader to convince himself that H does not depend on the choice of a particular CONS). As an a–priori domain for H we choose $\mathcal{P}$. Cook (1953) has shown (in Fock space language) that H of this form is essentially self-adjoint on (L^2). For the convenience of the reader, we give a proof here which we base on Nelson's analytic vector theorem (cf., e.g., Reed and Simon (1975)).

Theorem 11.1. H *as in* (11.16) *with domain* $\mathcal{P}$ *is essentially self-adjoint on* (L^2).

We will only give a proof in the case of one space dimension: s = 1. The general case follows by obvious modifications. First we prepare the following lemma.

Lemma 11.2. *Let* e_k *denote the* k*-th Hermite functions (cf., e.g.,* Appendix A.1). *There exists a constant* $C_m>0$, *and for every* $k \in \mathbb{N}_0$ *there exists a constant* $C_k>0$, *so that for all* $n \in \mathbb{N}$,

$$|\,\omega^n e_k\,|_2 \le C_k\, C_m^n \sqrt{n!}.$$

Proof. First note that the Fourier transform $\tilde{e}_k$ of the Hermite function e_k is equal to $i^k e_k$, as an easy computation shows. Use the Plancherel theorem to compute in the following way.

$$|\,\omega^n e_k\,|_2^2 = \int e_k(p)^2\, (p^2+m^2)^n\, dp$$

$$= N_k^2 \int H_k(p)^2\, e^{-p^2} (p^2+m^2)^n\, dp$$

$$\leq 2^{n-1}\, N_k^2 \int H_k(p)^2\, e^{-p^2} (p^{2n}+m^{2n})\, dp,$$

where $H_k(p)$ denotes the Hermite polynomial of order k (generated by $p \mapsto \exp(2tp-t^2)$), and N_k is the normalization constant of the Hermite functions. Now we estimate

$$\int H_k(p)^2\, p^{2n}\, e^{-p^2}\, dp \leq C_k' \sup_{p\in\mathbb{R}} |\, p^{2n} \exp(-\tfrac{1}{2}p^2)\, |,$$

where $C_k' = \int H_k(p)^2 \exp(-\frac{1}{2}p^2)\, dp$. The supremum of the last inequality is easily computed, and using Stirling's theorem we obtain that it is bounded by $2^n n!$. Inserting these estimates into the above bound for $|\, \omega^n e_k\, |_2^2$ we get immediately the bound of the lemma. □

Proof of Theorem 11.1. Consider polynomials of the form

$$:\, <\cdot, \xi_1>\, <\cdot, \xi_2> \cdots <\cdot, \xi_m>\, :,$$

where $m\in\mathbb{N}_0$, $\xi_j\in\mathcal{S}(\mathbb{R})$, $j = 1,2, \ldots ,m$, and note that every element in $\mathcal{P}$ can be written as a finite linear combination of such functions. It straightforward to obtain the following result by use of the formulae given in Chapter 5:

$$H :\, <\cdot, \xi_1>\, <\cdot, \xi_2> \cdots <\cdot, \xi_m>\, : =$$

$$= \sum_{j=1}^{m} :\, <\cdot, \xi_1>\, <\cdot, \xi_2> \cdots <\cdot, \omega\xi_j><\cdot, \xi_{j+1}> \cdots <\cdot, \xi_m>\, :.$$

Since $\xi\in\mathcal{S}(\mathbb{R})$ implies that $\omega\xi\in\mathcal{S}(\mathbb{R})$, we see that $\mathcal{P}$ is invariant under H. Moreover, for $n\in\mathbb{N}$, we have

$$H^n :\, <\cdot, \xi_1>\, <\cdot, \xi_2> \cdots <\cdot, \xi_m>\, : =$$

$$= \sum_{n_1,n_2,\ldots,n_m}' :<\cdot,\omega^{n_1}\xi_1><\cdot,\omega^{n_2}\xi_2>\cdots<\cdot,\omega^{n_m}\xi_m>:,$$

where the summation extends over all $n_1,n_2,\ldots n_m \in \mathbb{N}_0$, so that $n_1+n_2+\cdots+n_m = n$ (note that the sum has m^n many terms).

Consider the set of polynomials in $\mathcal{P}$ which are of the form

$$P_{\mathbf{k}} = :<\cdot,e_{k_1}><\cdot,e_{k_2}>\cdots<\cdot,e_{k_m}>:,$$

where $m\in\mathbb{N}_0$, and e_{k_l}, $l=1,2,\ldots,m$, is a Hermite function. The set of all such polynomials is total in (L^2) (actually, we proved in Chapter 2, that they form an orthogonal basis of (L^2)). Now we estimate the (L^2)–norm of $H^nP_{\mathbf{k}}$:

$$\| H^nP_{\mathbf{k}} \|_2 \leq$$

$$\leq \sum_{n_1,n_2,\ldots,n_m}' \| :<\cdot,\omega^{n_1}e_{k_1}><\cdot,\omega^{n_2}e_{k_2}>\cdots<\cdot,\omega^{n_m}e_{k_m}>: \|_2$$

$$\leq \sqrt{(2m)!} \sum_{n_1,n_2,\ldots,n_m}' |\omega^{n_1}e_{k_1}|_2\,|\omega^{n_2}e_{k_2}|_2 \cdots |\omega^{n_m}e_{k_m}|_2.$$

Applying the bound of Lemma 11.2, and denoting $C_{\mathbf{k}} = C_{k_1}\cdot C_{k_2}\cdot \cdots \cdot C_{k_m}$, we get

$$\| H^nP_{\mathbf{k}} \|_2 \leq C_{\mathbf{k}}\, m^n\, C_m^n\, \sqrt{(2m)!}\,\sqrt{n!}.$$

Obviously we have for every $P_{\mathbf{k}}$ of this form that

$$\sum_{n=0}^{\infty} \frac{\| H^n P_{\mathbf{k}} \|_2}{n!} < +\infty.$$

Therefore, we can now apply Nelson's analytic vector theorem (cf., e.g., Reed and Simon (1975)) to conclude that H with domain $\mathcal{P}$ is essentially self-adjoint on

(L^2). □

The self-adjoint extension of H will be denoted by the same symbol. H has a unique strictly positive ground state Ω given by the constant function 1. Also, it is straightforward to verify that the commutation relation (11.13) holds, e.g., on $\mathcal{P}$. We remark that H can be expressed through φ and π as follows:

$$H = \frac{1}{2} \sum_{k=1}^{\infty} (: \pi(e_k)^2 : + : \varphi(\omega e_k)^2 :),$$

where the *Wick ordering* $: \cdot :$ of the product of operators means that in every appearing product the operators $D.$ are moved to the right. For example, $: D_\xi D_\eta^* : = : D_\eta^* D_\xi : = D_\eta^* D_\xi$. The following fact is well-known (cf. also Appendix 2), and it will be useful later.

Lemma 11.3. H *is the generator* $d\Gamma(\omega)$ *of the semigroup* $\{\Gamma(\exp(-t\omega)),\ t\geq 0\}$ *on* (L^2).

Proof. Let $\xi_1, \xi_2, \dots, \xi_m \in \mathcal{S}(\mathbb{R})$, and compute as follows.

$$\frac{d}{dt}\, \Gamma(\exp(-t\omega)) : <\cdot, \xi_1><\cdot, \xi_2> \cdots <\cdot, \xi_m> : \Big|_{t=0} =$$

$$= \frac{d}{dt} : <\cdot, e^{-t\omega}\, \xi_1><\cdot, e^{-t\omega}\, \xi_2> \cdots <\cdot, e^{-t\omega}\, \xi_m> : \Big|_{t=0}.$$

For example, one can compute this derivative as follows. Take the $\mathcal{S}$-transform of the last expression, and interchange it with the t-derivative. (This is readily justified, e.g., by the dominated convergence theorem). Working out the derivative is now just an application of the product rule. Inverting the $\mathcal{S}$-transform one obtains the following result:

$$-\sum_{j=m}^{\infty} :<\cdot,\xi_1><\cdot,\xi_2>\cdots<\cdot,\omega\xi_j><\cdot,\xi_{j+1}>\cdots<\cdot,\xi_m>:.$$

The last expression shows that the generator of $\Gamma(\exp(-t\omega))$ equals $-H$ on the set of Wick monomials. Therefore they coincide on $\mathcal{P}$, and since $\mathcal{P}$ is a core for H, they are equal on the domain of H. □

Let us now compute the Schwinger functions

$$S_n(t_1,\xi_1,\ldots,t_n,\xi_n) :=$$
$$= (\Omega, \varphi(\xi_n)\exp(-(t_n-t_{n-1})H)\,\varphi(\xi_{n-1})\cdots\exp(-(t_2-t_1)H)\,\varphi(\xi_1)\,\Omega), \tag{11.17}$$

for $t_1,t_2,\ldots,t_n\in\mathbb{R}$ with $t_1\le t_2\le\cdots\le t_n$, and $\xi_1,\xi_2,\ldots,\xi_n\in\mathcal{S}(\mathbb{R}^s)$.

To this end we introduce some notation which is similar to the one used in Chapter 2. For $n\in\mathbb{N}$, let $\mathbf{n}$ denote the set $\{1,2,\ldots,n\}$. If $k\in\mathbb{N}$, $k\le n$, then $\mathbf{r}(k)\subset\mathbf{n}$ denotes a subset of $\mathbf{n}$ with k elements. Let A be a set of 2k elements, $k\in\mathbb{N}$. Then P(A) denotes the set of all pairings of the elements of A. Note that P(A) has $(2k-1)!!$ many elements. If $n,k\in\mathbb{N}$, $k\le[\frac{n}{2}]$, and $\mathbf{r}(2k)\subset\mathbf{n}$, then $p\in P(\mathbf{r}(2k))$ is of the form $p=\{(i_1,j_1),(i_2,j_2),\ldots,(i_k,j_k)\}$, where $\{i_1,j_1,i_2,j_2,\ldots,i_k,j_k\}=\mathbf{r}(2k)$, $i_l\le j_l$, $l=1,2,\ldots,k$.

Lemma 11.4. *Let* $n\in\mathbb{N}$, $n\ge2$, $\xi_1,\xi_2,\ldots,\xi_n\in\mathcal{S}(\mathbb{R})$, $t_1,t_2,\ldots,t_n\in\mathbb{R}$, $t_1\le t_2\le\cdots\le t_n$. *Then the following formula holds:*

$$\varphi(\xi_n)\exp(-(t_n-t_{n-1})H)\,\varphi(\xi_{n-1})\cdots\exp(-(t_2-t_1)H)\varphi(\xi_1)\,1 =$$

$$= 2^{-n/2}\sum_{k=0}^{\infty}\sum_{\mathbf{r}(2k)\subset\mathbf{n}}\sum_{p\in P(\mathbf{r}(2k))}\prod_{(i_m,j_m)\in p}(\xi_{i_m},\exp(-(t_{i_m}-t_{j_m})\omega)\,\omega^{-1}\xi_{j_m})\times \tag{11.18}$$

$$\times : \prod_{l \notin r(2k)} <\cdot, \exp(-(t_n - t_m)\omega)\, \omega^{-1/2} \xi_m> :,$$

where $\varphi(\xi_i)$, $i = 1,2, \dots ,n$, *is the operator in* (11.14), *while* H *is given in* (11.16).

Proof. We proceed by induction. First let $n = 2$. Then we use Lemma 11.3 and the properties of D. and D$^*_\cdot$ given in Chapter 5 to calculate as follows:

$$\varphi(\xi_2) \exp(-(t_2 - t_1)H)\, \varphi(\xi_1)\, 1 =$$

$$= 2^{-1/2} \varphi(\xi_2) \exp(-(t_2 - t_1)H) <\cdot, \omega^{-1/2} \xi_1>$$

$$= 2^{-1/2}\, \varphi(\xi_2) <\cdot, \exp(-(t_2 - t_1)\omega)\, \omega^{-1/2} \xi_1>$$

$$= \tfrac{1}{2} : <\cdot, \omega^{-1/2} \xi_2> <\cdot, \exp(-(t_2 - t_1)\omega)\, \omega^{-1/2} \xi_1> : +$$

$$+ \tfrac{1}{2} (\omega^{-1/2} \xi_2, \exp(-(t_2 - t_1)\omega)\, \omega^{-1/2} \xi_1),$$

which is equal to the right hand side of (11.18) for $n = 2$.

Now assume that (11.18) holds $n \in \mathbb{N}$, $n \geq 2$, and make the following computation:

$$\varphi(\xi_{n+1}) \exp(-(t_{n+1} - t_n)H) : \prod_{l \notin r(2k)} <\cdot, \exp(-(t_n - t_m)\omega)\, \omega^{-1/2} \xi_m> : =$$

$$= \varphi(\xi_{n+1}) : \prod_{l \notin r(2k)} <\cdot, \exp(-(t_{n+1} - t_m)\omega)\, \omega^{-1/2} \xi_m> :$$

$$= 2^{-1/2} \sum_{m \notin r(2k)} (\omega^{-1/2} \xi_{n+1}, \exp(-(t_{n+1} - t_n)\omega)\, \omega^{-1/2} \xi_m) \times$$

$$\times : \prod_{l \notin r(2k), l \neq m} <\cdot, \exp(-(t_{n+1}-t_m)\omega)\, \omega^{-1/2}\xi_m> : +$$

$$+ 2^{-1/2} : <\cdot, \omega^{-1/2}\xi_{n+1}> \cdot \prod_{l \notin r(2k)} <\cdot, \exp(-(t_{n+1}-t_m)\omega)\, \omega^{-1/2}\xi_m> :.$$

If we now substitute the last expression into the induction hypothesis we see that (11.18) holds also when n is replaced by n+1. □

Corollary 11.5. *Let* $n \in \mathbb{N}$, $\xi_1, \xi_2, \ldots, \xi_n \in \mathcal{S}(\mathbb{R}^s)$, $t_1, t_2, \ldots, t_n \in \mathbb{R}$, $t_1 \le t_2 \le \cdots \le t_n$. *Then the Schwinger function* $S_n(t_1,\xi_1, \ldots, t_n,\xi_n)$ *equals zero when* n *is odd. If* n $= 2k$, $k \in \mathbb{N}$, *the following formula holds:*

$$S_n(t_1,\xi_1, \ldots, t_n,\xi_n) = 2^{-k} \sum_{p \in P(n)} \prod_{(i_m,j_m) \in p} (\xi_{i_m}, \exp(-(t_{i_m}-t_{j_m})\omega)\, \omega^{-1}\xi_{j_m}) \tag{11.19}$$

It is because of the Gaussian nature of (11.19), and the corresponding formula for real times $t_k \to -it_k$, that the fields that we constructed here are called "free fields". For real times, the analytic continuation of (11.19) says that the overlap of 2k particles, at times t_k and with densities ξ_k, is given by the sum over all possible products of one particle transitions (under the time development determined by $\omega^{1/2}$): i.e., the particles do not interact with each other.

Next we construct Euclidean field operators denoted by $\phi(f)$, $f \in \mathcal{S}(\mathbb{R}^d)$, where $d = s+1$, and show how these Euclidean fields can be used to reproduce the Schwinger functions (11.19).

We also denote the Laplacian in d dimensions by Δ, and the pseudo-differential operator $(-\Delta+m^2)^{1/2}$, $m^2>0$, on $L^2(\mathbb{R}^d)$ by ω. Define Euclidean field operators on the white noise space $(L^2) \equiv L^2(\mathcal{S}'(\mathbb{R}^d),\mathcal{B},\mu)$ as follows. For $f \in \mathcal{S}(\mathbb{R}^d)$ set

$$\phi(f) := (D^*_{\omega^{-1}f} + D_{\omega^{-1}f}), \tag{11.20}$$

on the domain $\mathfrak{D}^{2,1}$. Note that for every $f \in \mathcal{S}(\mathbb{R}^d)$, $\phi(f)$ is self-adjoint on $\mathfrak{D}^{2,1}$ and it acts via multiplication by $\langle \cdot, \omega^{-1} f \rangle$. Moreover, it is clear that the linear mapping $f \mapsto \phi(f)$ extends strongly continuously to f in the Sobolev space $H_{-1}(\mathbb{R}^d)$, since then $\omega^{-1} f \in L^2(\mathbb{R}^d)$. Let us compute the "moments" of the Euclidean field. For $f_1, f_2, \ldots, f_n \in H_{-1}(\mathbb{R}^d)$ they are given by

$$(\Omega_E, \phi(f_1)\phi(f_2) \cdots \phi(f_n)\, \Omega_E) = \mathbb{E}(\langle \cdot, \omega^{-1} f_1 \rangle \langle \cdot, \omega^{-1} f_2 \rangle \cdots \langle \cdot, \omega^{-1} f_n \rangle),$$

where we have set $\Omega_E := 1$. Clearly, this is zero, unless $n = 2k$, $k \in \mathbb{N}$, and in this case equal to

$$\sum_{p \in P(n)} \prod_{(i_m, j_m) \in p} (f_{i_m}, \omega^{-2} f_{j_m}).$$

In other words, we may identify the Euclidean field operator with a centered Gaussian process indexed by $H^{-1}(\mathbb{R}^d)$, with covariance given by $(-\Delta + m^2)^{-1}$. Note that for every $t \in \mathbb{R}$, $\xi \in L^2(\mathbb{R}^s)$, we have $f = \xi \otimes \delta_t \in H_{-1}(\mathbb{R}^d)$, and we may use f of this form in the argument of ϕ. Then the covariance of $\phi(\xi \otimes \delta_s)$ and $\phi(\eta \otimes \delta_t)$, $\xi, \eta \in L^2(\mathbb{R}^s)$, $s, t \in \mathbb{R}$, is computed as follows.

$$(\Omega_E, \phi(\xi \otimes \delta_s)\, \phi(\eta \otimes \delta_t)\, \Omega_E) =$$

$$= (\omega^{-1/2} \xi \otimes \delta_s, \omega^{-1/2} \eta \otimes \delta_t)_{L^2(\mathbb{R}^d)}$$

$$= (2\pi)^{-1} \int_{\mathbb{R}^s} \overline{\tilde{\xi}(p)}\, \tilde{\eta}(p) \left(\int_{\mathbb{R}} (|p|^2 + p_0^2 + m^2)^{-1} e^{-i(t-s)p_0}\, dp_0 \right) dp$$

$$= \frac{1}{2} \int_{\mathbb{R}^s} \overline{\tilde{\xi}(p)}\, \tilde{\eta}(p) \frac{\exp(-|t-s|(|p|^2+m^2)^{1/2})}{(|p|^2+m^2)^{1/2}}\, dp$$

$$= \tfrac{1}{2}\, (\xi, \exp(-|t-s|\omega)\, \omega^{-1}\eta),$$

where we denoted the Fourier transform by $\tilde{\ }$, used the Plancherel theorem, and $p\in\mathbb{R}^s$, $p_0\in\mathbb{R}$. Therefore, the Euclidean fields ϕ reproduce at "sharp times" the canonical time zero fields which have been "translated in imaginary time".

Finally we come to a discussion of the Dirichlet picture for the free fields, in analogy with the case of quantum mechanics (although with a slightly different normalization of the operators).

As we have indicated in the introduction, the Dirichlet strategy consists in constructing the Dirichlet form on $\mathcal{S}'(\mathbb{R}^s)$ corresponding to the ground state measure. The ground state measure of a model is given by the restriction of the Euclidean measure to the σ-algebra generated by the time zero (Euclidean) fields: $\phi(\xi\otimes\delta_0) = \varphi(\xi)$, $\xi\in\mathcal{S}(\mathbb{R}^s)$. In the case of free fields it is clear from the preceding computations that the Euclidean measure ν on $(\mathcal{S}'(\mathbb{R}^d), \mathcal{B})$, $d = s+1$, is given by the characteristic functional $C_E(f) = \exp(-\frac{1}{2}(f, (-\Delta+m^2)^{-1}f)$, while the restriction of ν to the time zero fields is the measure ν_0 on $(\mathcal{S}'(\mathbb{R}^s),\mathcal{B})$ with characteristic functional

$$C_0(\xi) = \exp(-\tfrac{1}{2}(\xi, (-\Delta+m^2)^{-1/2}\xi)),$$

where we use the convention that the time zero field is given by $2^{-1/2}$ times the canonical coordinate process associated with this measure. Thus we obtain the Dirichlet picture of the free field by making the following choices. $\mathcal{H}$ is the Hilbert space $L^2(\mathcal{S}'(\mathbb{R}^s),\mathcal{B},\nu_0)$, $\varphi(\xi)$ is multiplication by $2^{-1/2}<\cdot, \xi>$, and H is the operator which is determined by the Dirichlet form $\mathcal{E}_0$ associated with ν_0, i.e., for $u,v\in\mathcal{D}(\mathcal{E}_0)$, $(u, Hv)_{\mathcal{H}} = \mathcal{E}_0(u,v)$. Since ν_0 is represented by an admissible element

$\Phi_0 \in (\mathcal{S})^*$ (cf. Chapter 10, Example 10.17), we know that the Dirichlet form $\mathcal{E}_0$ is the closure of

$$\mathcal{E}_0(u,v) = <\Phi_0, (\nabla u, \nabla v)_{L^2(\mathbb{R}^s)}>,$$

for $u,v \in (\mathcal{S})$. The Hamiltonian H determined by this form has been calculated in Example 10.18:

$$H = \sum_{k=1}^{\infty} \{<\cdot, \omega e_k> \partial_k - \partial_k{}^2\},$$

where $\{e_k, k \in \mathbb{N}\}$ is a CONS of $L^2(\mathbb{R}^s)$ in $\mathcal{S}(\mathbb{R}^s)$, and where $\partial_k = D_{e_k}$. Now we can compute the momentum field π in this representation by requiring relation (11.13). The result is $\pi(\xi) = 2^{-1/2} i <\cdot, \omega\xi> - 2^{1/2} i D_\xi$, and the reader checks immediately that (11.11) holds. The Schwinger function are given in this representation as follows.

$$S_n(t_1,\xi_1, t_2, \xi_2, \dots ,t_n,\xi_n) =$$

$$= 2^{-n/2} <\Phi_0, <\cdot, \xi_n> e^{-(t_n-t_{n-1})H} <\cdot, \xi_{n-1}> \cdots e^{-(t_2-t_1)H} <\cdot, \xi_1>>.$$

Using the facts that Φ_0 corresponds to the measure ν_0, that the projections $: \cdot :$ defined in Example 10.18 are orthogonal in $L^2(\nu_0)$, and that H is in this representation given as the generator of the semigroup $\gamma(\exp(-t\omega))$, $t \geq 0$, cf. Example 10.18, it is not hard to compute the Schwinger functions S_n explicitly with the same combinatorics as in Lemma 11.4, and – of course – with the same result as in Corollary 11.5.

We conclude this section with the remark that the use of the Euclidean measure for a Dirichlet form of the above type, leads to an $\mathcal{S}'(\mathbb{R}^d)$–valued process (cf. Theorem 10.25) which is known in physics as the "stochastic quantization" of the field. We refer the reader to Röckner (1990) and the literature quoted there.

11.C Interacting Fields and the Role of φ – Bounds

As we have mentioned in the introduction, the construction of non-trivial representations of the CCR (1.1), (1.2) – in the sense that the Wightman (or Schwinger) functions have a non-Gaussian structure – is a very difficult problem. In analogy with quantum mechanics, a natural ansatz to incorporate interactions is the addition of a "potential" term V to the Hamiltonian $H_0 = d\Gamma(\omega)$ which was constructed in Section 11.B. Relativistic invariance of the theory requires that V be of the form $V = \int_{\mathbb{R}^s} F(\varphi(x))\,dx$, for a certain function F. Clearly, in view of the singular character of the fields φ, this can only be considered as a heuristic expression. Most typically, let F be a polynomial of degree greater than 2 which is bounded from below (in order to achieve stability $H \geq 0$). Also, let us from now on assume that $s = 1$. (Otherwise, the operations described below have to be supplemented by so-called renormalizations which take care of singularities which arise from the increasingly singular behavior of ω^{-1}, as s increases.) Then the standard strategy in quantum field theory is to introduce "cut-offs": let $\delta_{x,n}$, $x \in \mathbb{R}$, $n \in \mathbb{N}$, be an approximation to the Dirac distribution at x as $n \to +\infty$, $\delta_{x,n} \in \mathcal{S}(\mathbb{R})$, and let Λ be a bounded Borel set of $\mathbb{R}$. Define $V_{n,\Lambda} := \int_\Lambda :F(\varphi(\delta_{x,n})):\,dx$, which is obviously densely defined in (L^2), say on $\mathcal{P}$. Here $:\cdot:$ is the Wick ordering of powers of the field φ as explained in Section 11.B, and it has to be applied to every term of the polynomial F. Set $\hat{H}_{n,\Lambda} = H_0 + V_{n,\Lambda}$. Then it can be shown that $\hat{H}_{n,\Lambda}$ is essentially self-adjoint on (L^2) (cf., e.g., Guerra et al. (1975) or Simon (1974)), and we denote its closure by the same symbol. Let $E_{n,\Lambda} = \inf \operatorname{spec}(\hat{H}_{n,\Lambda})$, and set $H_{n,\Lambda} = \hat{H}_{n,\Lambda} - E_{n,\Lambda}$. Then (*loc. cit.*), $H_{n,\Lambda}$ has a unique (strictly positive) ground state $\Omega_{n,\Lambda}$. Of course, the interest is to study the limit as the cut-offs are removed: $n \to +\infty$, $\Lambda \to \mathbb{R}$. The problem is that in the limit $\Omega_{n,\Lambda}$ does not converge in (L^2). In fact, Haag's Theorem (cf., e.g., Bogoliubov et al. (1975)) implies that there is no non-trivial representation of the CCR on (L^2). Moreover, it is impossible to describe directly a suitable Hilbert space $\mathcal{H}$ on which the would-be limit H would be defined as an essentially self-adjoint operator.

The by now most successful approach to overcome these problems is to construct the Euclidean (interaction) measure ν through a limiting process for a cut–off Feynman–Kac formula, cf. (11.10), (11.11). This leads to Schwinger functions which satisfy the Osterwalder–Schrader axioms, and one can apply the Osterwalder–Schrader reconstruction theorem to obtain a Hilbert space $\mathcal{H}$, and a self-adjoint operator H on $\mathcal{H}$ with ground state $\Omega \in \mathcal{H}$. In fact, $\mathcal{H}$ is the L^2–space over $(\mathcal{S}'(\mathbb{R}, \mathcal{B}, \nu_0)$ where ν_0 is the restriction of the measure ν to the σ–algebra generated by the Euclidean time zero fields.

On the other hand, in the early sixties Araki (1960), and Coester and Haag (1960) proposed to consider the vacuum Ω as the central object of a quantum field theory. Their idea is that the whole dynamics of a model of quantum fields is encoded in the vacuum. Typically, as we saw in Section 11.B, it is easy to find a first expression for the field $\varphi(\xi)$, $\xi \in \mathcal{S}(\mathbb{R})$, say as the operator on the Hilbert space $\mathcal{H} = L^2(\mathcal{S}'(\mathbb{R}^s), \mathcal{B}, \mu_0)$ which is given by multiplication by the canonical coordinate process X_ξ (up to a factor $2^{-1/2}$). Here, μ_0 is some appropriate reference measure. Assume that Ω is some vector in $\mathcal{H}$. Then we can define informally a Hamiltonian H as in the Dirichlet picture, i.e., as associated with the form $\mathcal{E}(u,v) = (\Omega, \overline{\nabla u} \cdot \nabla v\, \Omega)_{\mathcal{H}} \equiv \mathbb{E}_\Omega(\overline{\nabla u} \cdot \nabla v)$. Note that $\mathcal{E}(u,v)$ can be written as $\int \overline{\nabla u} \cdot \nabla v \,|\, \Omega \,|^2 d\mu_0$, i.e., $|\,\Omega\,|^2$ should be considered as the density of the measure $d\nu_0 = |\,\Omega\,|^2 d\mu_0$. Next we *define* (still informally) the momentum operator $\pi(\xi)$ as given by $i[H, \varphi(\xi)]$. It straightforward, to verify that informally the canonical commutation relations (11.1) hold true.

If we want to implement the Dirichlet strategy just sketched in the white noise framework, it is natural to choose as reference measure μ_0 the white noise measure μ. The first problem is then to make sense of the expectation $\mathbb{E}_\Omega$ defined by the vacuum. Since we have to face the fact (see above) that Ω is in general not in (L^2), we shall consider the possibility that $|\,\Omega\,|^2 \in (\mathcal{S})^*$, and that we may realize its expectations $\mathbb{E}_\Omega$ as dual pairing between $(\mathcal{S})^*$ and $(\mathcal{S})$: $\mathbb{E}_\Omega(\,\cdot\,) = \langle |\,\Omega\,|^2, \cdot\, \rangle$. The next step in the program would be the construction of the corresponding form $\mathcal{E}$, and to prove that it is a Dirichlet or pre–Dirichlet form, yielding a self–adjoint (essentially self-adjoint, respectively) operator H. Finally, $\pi(\xi)$ has to be constructed, and the CCR have to be verified on appropriate domains.

Step two of this program has been worked out in Chapter 10 (cf. also the references quoted there), and concerning the last step we mention the results in Albeverio and Høegh-Krohn (1977a, 1984), Fröhlich (1974, 1977). Here we shall prove a general criterium which implies the representability of $\mathbb{E}_\Omega$ by an element in $(\mathcal{S})^*$, which settles the first of the above steps in many cases (for examples see below). This criterion was developed in Potthoff and Streit (1990), and it makes use of so-called φ–bounds. φ–bounds estimate the field operator (in the sense of quadratic forms) by the Hamiltonian. This might appear to be somewhat circular, since ultimately we want to *construct* the Hamiltonian from the vacuum expectation. However, we show in examples how the result of Potthoff and Streit (1990) can be applied successfully using φ–bounds for the cut-off theories, plus limit arguments.

From now on, we let $\mathcal{H}$ be a separable complex Hilbert space, and assume the existence of a linear mapping φ from $\mathcal{S}(\mathbb{R}^s)$ into the set of self-adjoint operators on $\mathcal{H}$. Furthermore, we assume the existence of a self-adjoint, positive operator H on $\mathcal{H}$ with domain $\mathcal{D}(H)$, and of a normalized vector $\Omega \in \mathcal{D}(H)$ so that $H\Omega = 0$. The Araki relation (11.7) will be central to our discussion in this section, and we assume from now that the vectors $\exp(i\varphi(\xi))\Omega$, $\xi \in \mathcal{S}(\mathbb{R}^s)$, belong to $\mathcal{D}(H^{1/2})$, and that (11.7) holds for the triple (φ, H, Ω). The following result has been derived from the Araki relation in Herbst (1976), and for the convenience of the reader we reproduce (a version of) his proof.

We shall use the following notations: if $\xi_1, \xi_2, \dots, \xi_n \in \mathcal{S}(\mathbb{R}^s)$, we denote by $\varphi(\xi)$ the vector valued operator $(\varphi(\xi_1), \dots, \varphi(\xi_n))$. If F is a continuous function on $\mathbb{R}^n$ so that $F(\varphi(\xi))\Omega \in \mathcal{H}$, then we set $\hat{F} := F(\varphi(\xi))\Omega$.

Lemma 11.6. *Suppose that* $\xi_1, \xi_2, \dots, \xi_n \in \mathcal{S}(\mathbb{R}^s)$, *and that* $F \in C^1(\mathbb{R}^n)$ *so that* $\hat{F} \in \mathcal{H}$, *and* $(D_jF)^\wedge \in \mathcal{H}$, *for all* $j = 1,2, \dots, n$, *where* D_jF *is the* j*-th partial derivative of* F. *Then* $\hat{F} \in \mathcal{D}(H^{1/2})$, *and the following formula holds:*

$$(\hat{F}, H\hat{F})_{\mathcal{H}} = \tfrac{1}{2}\sum_{i,j=1}^{n} (\xi_i, \xi_j)\, ((D_iF)\hat{}, (D_jF)\hat{})_{\mathcal{H}}, \tag{11.21}$$

where $(\cdot\,,\,\cdot)$ *is the inner product of* $L^2(\mathbb{R},dx)$.

Proof. To simplify the notation we consider only the case $n = 1$, $s = 1$ ($\xi \equiv \xi_1$); – the general case follows from obvious modifications. We have to show that $\hat{F} \in \mathfrak{D}(H^{1/2})$, and that

$$(\hat{F}, H\hat{F}) = \tfrac{1}{2}\,|\,\xi\,|_2^2\, ((F')\hat{}, (F')\hat{}). \tag{11.22}$$

Let $P_\xi(\lambda)$, $\lambda\in\mathbb{R}$, be the family of spectral projections of $\varphi(\xi)$. Then $dm_\xi(\lambda) = d(\Omega, P_\xi(\lambda)\,\Omega)$ is a probability measure on $\mathbb{R}$, so that for all $F, G \in C(\mathbb{R})$ with $\hat{F}$, $\hat{G}\in\mathcal{H}$ we have

$$(\hat{F}, \hat{G}) = \int_{\mathbb{R}} \overline{F(\lambda)}\, G(\lambda)\, dm_\xi(\lambda).$$

We show first the statement of the lemma for $F\in\mathcal{S}(\mathbb{R})$, and then obtain the full statement by a sequence of limit arguments.

Let $F\in\mathcal{S}(\mathbb{R})$, so that

$$F(x) = (2\pi)^{-1/2}\int \tilde{F}(p)\, e^{ipx}\, dp,$$

$$F'(x) = (2\pi)^{-1/2}\int p\tilde{F}(p)\, e^{ipx}\, dp,$$

where $\tilde{F}\in\mathcal{S}(\mathbb{R})$ is the Fourier transform of F. Let $\chi_n\in C_0^\infty(\mathbb{R})$, $n\in\mathbb{N}$, be such that $\chi_n \geq 0$, $\chi_n(x) = 1$, if $|\,x\,| \leq n$, $\chi_n(x) = 0$ for x with $|\,x\,| \geq n+1$, $\|\,\chi_n'\,\|_\infty \leq 2$. Set

$$F_n(x) = (2\pi)^{-1/2}\int \chi_n(p)\, \tilde{F}(p)\, e^{ipx}\, dp.$$

Let $\tilde{F}_m$ be a sequence of step functions converging pointwise to $\tilde{F}$. Then for every $n \in \mathbb{N}$, this sequence can be chosen in such a way that $\chi_n \tilde{F}_m$ converges uniformly to $\chi_n \tilde{F}_m$. Moreover, $\tilde{F}_m$ can be arranged so that $\| \tilde{F}_m \|_\infty \leq \| \tilde{F} \|_\infty + 1$. Let

$$F_{n,m}(x) = (2\pi)^{-1/2} \int \chi_n(p)\, \tilde{F}_m(p)\, e^{ipx}\, dx.$$

Then $F_{n,m}(x)$ is of the form

$$F_{n,m}(x) = \sum_{k=1}^{N} a_{n,m,k}\, e^{ip_k x},$$

for certain numbers $a_{n,m,k}$, p_k, and N finite. By our assumptions (in particular (11.7)), $F_{n,m} \in \mathfrak{D}(H^{1/2})$ and formula (11.22) holds. For every $n \in \mathbb{N}$, $F_{n,m}$ and $F'_{n,m}$ converge pointwise to F_n, F'_n respectively, as m tends to infinity. Moreover, we have the bounds

$$| F_{n,m}(x) | \leq (2\pi)^{-1/2}\, 2(n+1)\, (\| \tilde{F} \|_\infty + 1),$$

and

$$| F'_{n,m}(x) | \leq (2\pi)^{-1/2}\, 2(n+1)^2\, (\| \tilde{F} \|_\infty + 1).$$

Since dm_ξ is a probability measure, we may apply the dominated convergence theorem to conclude that $(F_{n,m})\hat{}$ and $(F'_{n,m})\hat{}$ converge to $(F_n)\hat{}$, and $(F'_n)\hat{}$ respectively, in $\mathcal{H}$. Since H is closed, it follows that $(F_n)\hat{} \in \mathfrak{D}(H^{1/2})$ and that the formula (11.22) holds for F_n. Now we consider the limit $n \to +\infty$. It is obvious that F_n, F'_n converge pointwise to F, F' respectively, and that F_n, and F'_n are uniformly bounded because $\tilde{F}$ belongs to $\mathcal{S}(\mathbb{R})$. Thus we can again apply the same combination of the dominated convergence theorem and closedness of H to finish the proof of the statement for the case $F \in \mathcal{S}(\mathbb{R})$.

Next, we consider the case $F \in C_0^1(\mathbb{R})$. Let δ_n be a smooth approximation of the Dirac distribution at zero, as $n \to +\infty$, with compact support. Then $\delta_n * F \in C_0^\infty(\mathbb{R}) \subset \mathcal{S}(\mathbb{R})$. Now one shows in a similar way that $\delta_n * F$ and its derivative converge to F, F′ respectively, in $L^2(dm_\xi)$, and therefore closedness of H implies the statement of the lemma for $F\in C_0^1(\mathbb{R})$. Finally, for $F\in C^1(\mathbb{R})$ with $\hat{F}\in\mathcal{H}$, $(F')^\wedge\in\mathcal{H}$, one considers $\chi_n F\in C_0^1(\mathbb{R})$, and goes through analogous arguments. □

The second main ingredient for the results of this section is a φ–bound of a certain type. We assume that (φ, H, Ω) are as above.

Definition 11.7. *If there exist constants* α, $\beta>0$, $\gamma\geq 0$ *and* $p\in\mathbb{N}_0$, *so that for all* $\xi\in\mathcal{S}(\mathbb{R}^s)$, *the following inequality holds*

$$\pm\,\varphi(\xi)\leq\alpha H+\beta|\,\xi\,|_{2,p}^2+\gamma, \tag{11.23}$$

in the sense of quadratic forms on the domain $\mathcal{D}_\xi=\{F(\varphi(\xi))\Omega,\ F\in\mathcal{S}(\mathbb{R})\}$, *then it is said that* a φ–bound holds for (φ, H, Ω).

The following result has been proved in Potthoff and Streit (1990).

Theorem 11.8. *Assume that* (φ, H, Ω) *admits the Araki relation* (11.7), *and that a* φ*–bound holds. Then there exists a positive element* $\Xi\in(\mathcal{S})^*$ *so that for all* $\xi\in\mathcal{S}(\mathbb{R}^s)$ *the following equation holds:*

$$(\Omega,\exp(i\varphi(\xi))\Omega)=\langle\Xi,\exp(i2^{-1/2}\langle\cdot,\xi\rangle)\rangle. \tag{11.24}$$

Remark. The factor $2^{-1/2}$ was included in order to have a normalization which is consistent with Section 11.B.

By the analyticity properties of the right hand side of (11.7) (cf. Theorem 4.38) and the estimate of Lemma 11.11 (see below), we obtain the following result.

Corollary 11.9. Let $\xi_1,\xi_2,\dots,\xi_n\in\mathcal{S}(\mathbb{R}^s)$, $n\in\mathbb{N}$. *Under the conditions of Theorem* 11.8 *the following holds:*

$$(\Omega, \varphi(\xi_1)\, \varphi(\xi_2) \cdots \varphi(\xi_n)\, \Omega) = \tag{11.25}$$
$$= 2^{-n/2} <\Xi, <\cdot, \xi_1><\cdot, \xi_2> \cdots <\cdot, \xi_n>>.$$

Moreover, there exists a constants $K>0$, *depending only* α, β, γ, *and* $p\in\mathbb{N}_0$, *so that for all* $n\in\mathbb{N}$, $\xi_1, \xi_2, \ldots, \xi_n \in \mathcal{S}(\mathbb{R})$,

$$|(\Omega, \varphi(\xi_1)\, \varphi(\xi_2) \cdots \varphi(\xi_n)\, \Omega)| = K^n \sqrt{n!} \prod_{k=1}^{n} |\xi_k|_{2,p}. \tag{11.26}$$

The proof of Theorem 11.8 relies on a moment inequality which we prove in the next two lemmas. Inequalities of similar type have been proved on the basis of φ–bounds in Herbst (1976). We follow his method very closely.

We first prove that if the Araki relation and a φ–bound hold for (φ, H, Ω), that then Ω has a first moment. Then we show by an iterative procedure that the n-th moment of Ω exists and behaves like $\sqrt{n!}$ in n.

Lemma 11.10. *Assume that for* (φ, H, Ω) *Araki's relation and a* φ–*bound hold. Then for every* $\xi\in\mathcal{S}(\mathbb{R}^s)$, $\Omega\in\mathfrak{D}(|\varphi(\xi)|^{1/2})$, *and the following bound holds*:

$$|(\Omega, \varphi(\xi)\, \Omega)| \le K\, |\xi|_{2,p}, \tag{11.27}$$

where K *is a constant which depending only on* α, β, γ.

Proof. We use the same notation as in the proof of Lemma 11.6.

Let $s\in C^\infty(\mathbb{R})$ be a monotonely increasing function, so that $s = 0$ on $(-\infty, \frac{1}{2})$, $s = 1$ on $[1, +\infty)$, and $\sup_u |s'(u)| \le 4$. Let χ_n be the cut-off function in the proof of Lemma 11.6, and set $s_n = \chi_n \cdot s$. Note that for all $n\in\mathbb{N}$, $\sup_u |s_n'(u)| \le 4$.

Let $\xi\in\mathcal{S}(\mathbb{R})$, and assume that $|\xi|_{2,p} = 1$, and let $dm_\xi(u) = d(\Omega, P_\xi\Omega)$

be the probability measure on $\mathfrak{B}(\mathbb{R})$ defined by the spectral family P_ξ of $\varphi(\xi)$. We know from Lemma 11.6, that $(s_n)\hat{} \in \mathfrak{D}(H^{1/2})$, and it is clear that also $(s_n)\hat{} \in \mathfrak{D}(|\varphi(\xi)|^{1/2})$. The φ–bound (11.23) gives the bound

$$((s_n)\hat{},\ \varphi(\xi)\ (s_n)\hat{}) \le ((s_n)\hat{},\ (\alpha H + c)\ (s_n)\hat{}),$$

where $c = \beta+\gamma$. Relation (11.21) gives the following estimation:

$$\begin{aligned} ((s_n)\hat{},\ \varphi(\xi)\ (s_n)\hat{}) &\le \alpha((s_n')\hat{},\ (s_n')\hat{}) + c \\ &= \alpha \int s_n'(u)^2\ dm_\xi(u)\ + c \\ &\le 16\alpha + c. \end{aligned}$$

Write $((s_n)\hat{},\ \varphi(\xi)\ (s_n)\hat{}) = \int s_n(u)^2\ u\ dm_\xi(u)$, and observe that $s_n(u)\cdot u$ is increasing to $s(u)\cdot u$ as n tends to infinity. Thus the monotone convergence theorem implies that $(\hat{s},\ \varphi(\xi)\ \hat{s}) = \int s(u)^2\ u\ dm_\xi(u)$ exists and is bounded by $16\alpha + c$. In particular we have shown that $\hat{s}$ belongs to $\mathfrak{D}(|\varphi(\xi)|^{1/2})$. Similarly, one shows the corresponding statements for $s(-\varphi(\xi))\Omega$. Let $t \in C_0^\infty(\mathbb{R})$ be such that $s(u) + s(-u) + t(u) = 1$, for all $u \in \mathbb{R}$. Then it is clear that the same statements hold also for the vector $t(\varphi(\xi))\Omega$, and therefore they hold for Ω. In particular, we have the bound

$$|(\Omega,\ \varphi(\xi)\ \Omega)| \le K,$$

where K depends only on α, β, γ.

Finally, let ξ be a general element in $\mathcal{S}(\mathbb{R})$. Then the statement of the lemma follows by linearity of φ. □

Lemma 11.11. *Assume that* (φ, H, Ω) *admit Araki's relation and a* φ*–bound. Then there exists a constant* K *which depends only on* α, β, γ, *so that for all* $\xi \in \mathcal{S}(\mathbb{R}^s)$, *and all* $n \in \mathbb{N}$,

$$|(\Omega, \varphi(\xi)^n \Omega)| \le K^n \sqrt{n!}\, |\xi|_{2,p}^n. \tag{11.28}$$

Proof. Let $\xi \in \mathcal{S}(\mathbb{R}^s)$, $\xi \neq 0$, and let $\lambda = |\xi|_2$, $e = \lambda^{-1}\xi$. Then the φ–bound (11.23) takes the following form.

$$\pm\lambda\varphi(e) \le \alpha H + \beta\lambda^2 |e|_{2,p}^2 + \gamma, \tag{11.29}$$

on $\mathfrak{D}_e = \mathfrak{D}_\xi$. It is clear that inequality (11.29) holds for any $\lambda \in \mathbb{R}$, $e \in \mathcal{S}(\mathbb{R}^s)$ with $|e|_2 = 1$. Let $F \in \mathcal{S}(\mathbb{R})$, then we obtain from (11.29) and Lemma 11.6 the following inequality

$$0 \le \alpha((F')\hat{}, (F')\hat{}) + \lambda^2\beta |e|_{2,p}(\hat{F}, \hat{F}) \pm \lambda(\hat{F}, \varphi(e)\hat{F}) + \gamma(\hat{F}, \hat{F}).$$

Optimizing with respect to λ we get

$$(\hat{F}, \varphi(e)\hat{F})^2 \le 4\beta |e|_{2,p}^2 (\hat{F}, \hat{F})\, (\alpha((F')\hat{}, (F')\hat{}) + \gamma(\hat{F}, \hat{F}))$$

$$\le C\, |e|_{2,p}^2\, (\hat{F}, \hat{F})\, (((F')\hat{}, (F')\hat{}) + (\hat{F}, \hat{F})),$$

where $C > 0$ is a constant depending only on α, β, γ. In the following, C might change from step to step.

Let $s_a(u) = e^{-au}s(u)$, $0 < a \le 1$, $u \in \mathbb{R}$, where $s \in C^\infty(\mathbb{R})$ is as in the proof of Lemma 11.10. Obviously, $s_a \in \mathcal{S}(\mathbb{R})$, and so is $F(u) = s_a(u)\, u^k$ for every $k \in \mathbb{R}_+$. Choosing this F in the above estimate and using the notation $(\Omega, \cdot\, \Omega) \equiv \langle \cdot \rangle$, we find the following:

$$\langle s_a(\varphi(e))^2\, \varphi(e)^{2k+1} \rangle \le$$

$$\leq C \mid e \mid_{2,p}^{2} <s_a(\varphi(e))^2 \, \varphi(e)^{2k}> \times$$

$$\times \, [<(-as_a(\varphi(e)) \, \varphi(e)^k + e^{-a\varphi(e)} s_a{}'(\varphi(e)) \, \varphi(e)^k +$$

$$+ ks_a(\varphi(e)) \, \varphi(e)^{k-1})^2> + <s_a(\varphi(e))^2 \, \varphi(e)^{2k}>]$$

$$\leq C \mid e \mid_{2,p}^{2} <s_a(\varphi(e))^2 \, \varphi(e)^{2k}> \times$$

$$\times \, [<s_a(\varphi(e))^2 \, \varphi(e)^{2k}> + k^2 <s_a(\varphi(e))^2 \, \varphi(e)^{2k-2}> + 1],$$

where we used $(a+b+c) \leq 4(a^2+b^2+c^2)$, the fact that supp $s_a{}' \subset [\frac{1}{2}, 1]$, and the spectral theorem in the last step. For $n \in \mathbb{N}_0$, define $a_n = (n!)^{-1/2} \mid e \mid_{2,p}^{-n}$ $<s_a(\varphi(e))^2 \, \varphi(e)^n>$. Then for half integer k the preceding inequality reads as follows:

$$(2k+1)! \mid e \mid_{2,p}^{4k+2} a_{2k+1}^2 \leq C \mid e \mid_{2,p}^{2k+2} ((2k)!)^{1/2} \, a_{2k} \times$$

$$\times \, [((2k)!)^{1/2} \mid e \mid_{2,p}^{2k} a_{2k} + k^2((2k-2)!)^{1/2} \mid e \mid_{2,p}^{2k-2} a_{2k-2} + 1].$$

Now we make use of the facts that $k^2 \, ((2k)! \, (2k\text{-}2)!)^{1/2} \leq (2k+1)!$, and $1 = \mid e \mid_2$ $\leq \mid e \mid_{2,p}$, to estimate in the following way:

$$a_{2k+1}^2 \leq C \, a_{2k}(a_{2k} + a_{2k-2} + 1)$$

$$\leq C \, (a_{2k}^2 + a_{2k-2}^2 + 1).$$

Note that $a_0 = (\Omega, s_a(\varphi(e))^2 \, \Omega) \leq 1$, and that we have shown in Lemma 11.10 that $a_1 = \mid e \mid_{2,p}^{-1} (\Omega, s_a(\varphi(e))^2 \, \varphi(e) \, \Omega) \leq \mid e \mid_{2,p}^{-1} (\Omega, \varphi(e) \, \Omega) \leq C^{1/2}$, if we choose C large enough. Observe that C is independent of the cut-off parameter $a \in (0,1]$. Furthermore, it follows from the preceding inequalities that $a_2 \leq C^{1/2}$, for large enough C. Now it follows by a straightforward induction that for $n \in \mathbb{N}$, we have $a_n \le K^{n-1} \le K^n$, where $K^2 = \max(1, 3C)$, and K is independent of a. Thus we have proved that for every $n \in \mathbb{N}$, $a \in (0,1]$,

$$|(\Omega, s_a(\varphi(e))\, \varphi(e)^n\, s_a(\varphi(e))\, \Omega)| \le K^n \sqrt{n!}\, |e|_{2,p}^n.$$

The monotone convergence theorem implies as in the proof of Lemma 11.10 that the same bound holds in the limit $a \to 0$, i.e., with s_a replaced by s. As there, we now consider a partition of unity: $s(-u) + t(u) + s(u) = 1$, $u \in \mathbb{R}$. The above bound is derived for the vector $s(-\varphi(e))\, \Omega$ in the same way as before, and for $t(\varphi(e))\, \Omega$ it is trivial. Thus we obtain for every $n \in \mathbb{N}$,

$$|(\Omega, \varphi(e)^n\, \Omega)| \le K^n \sqrt{n!}\, |e|_{2,p}^n,$$

and finally homogeneity of both sides in e give inequality (11.28). □

Now we come to the proof of Theorem 11.8.

Proof of Theorem 11.8. Let $\eta, \xi \in \mathcal{S}(\mathbb{R}^s)$, $\lambda \in \mathbb{R}$. Use the spectral theorem to write ($\varphi(\xi)$ and $\varphi(\eta)$ commute, so that their spectral projections commute)

$$(\Omega, \exp(i\varphi(\eta+\lambda\xi))\, \Omega) = \int \exp(i(v+\lambda u))\, dm_{\eta,\xi}(u,v),$$

where $dm_{\eta,\xi}$ is a probability measure on $\mathbb{R}^2$. Expand the exponential function under the integral. Fubini's theorem and the estimate of Lemma 11.11 show that the expansion and the integral can be interchanged, because

$$\sum_{n=0}^{\infty} \frac{i^n}{n!} \int (v+\lambda u)^n\, dm_{\eta,\xi}(v,u) = \sum_{n=0}^{\infty} \frac{i^n}{n!} (\Omega, (\varphi(\eta)+\lambda\varphi(\xi))^n\, \Omega),$$

and the right hand side is absolutely convergent for every $\lambda \in \mathbb{R}$. In particular, this shows that for all $\eta, \xi \in \mathcal{S}(\mathbb{R}^s)$ the function $\lambda \mapsto (\Omega, \exp(i\varphi(\eta+\lambda\xi))\, \Omega)$, $\lambda \in \mathbb{R}$, has an entire analytic extension to $\lambda \in \mathbb{C}$ (denoted by the same symbol). Moreover, we get the following bound for all $z \in \mathbb{C}$, $\xi \in \mathcal{S}(\mathbb{R}^s)$:

$$| (\Omega, \exp(z\varphi(\xi))\, \Omega) | \leq \sum_{n=0}^{\infty} \frac{| z |^n}{n!} (\Omega, \varphi(\xi)^n\, \Omega)$$

$$\leq \sum_{n=0}^{\infty} \frac{| z |^n}{\sqrt{n!}} K^n\, | \xi |_{2,p}^n$$

$$\leq \sum_{n=0}^{\infty} \frac{| z |^n}{\sqrt{n!}} K^n\, 2^{-n}\, | \xi |_{2,p+1}^n$$

$$\leq \text{const.} \exp(\tfrac{1}{2} K^2\, | z |^2\, | \xi |_{2,p+1}^2),$$

where we made use of the fact that the operator H defining the norm $| \cdot |_{2,p}$ has an inverse on $L^2(\mathbb{R}^s)$ which is bounded by $\frac{1}{2}$, and we used Schwarz inequality in the last step. We have proved that the mapping $\xi \mapsto (\Omega, \exp(i\varphi(\xi))\, \Omega)$ from $\mathcal{S}(\mathbb{R}^s)$ into $\mathbb{C}$ is a U–functional in the sense of Definition 4.37. But then the statement of Theorem 11.8 follows from Theorem 4.38. □

Let us now investigate a number of examples. In the first one, we reconsider the free field, which was constructed in Section 11.B. Although this case is somewhat trivial, we establish some facts which will be useful later.

Example 11.12. (The Free Field) As Hilbert space we choose (L^2), the vacuum is $\Omega = 1$. Consider the free field $\varphi(\xi) = 2^{-1/2}(D^*_{\omega^{-1/2}\xi} + D_{\omega^{-1/2}\xi})$, $\xi \in \mathcal{S}(\mathbb{R}^s)$, and the Hamiltonian of the free field $H_0 = \sum_k D^*_{\omega^{1/2}e_k} D_{\omega^{1/2}e_k}$ from Section 11.B, (11.14), (11.16), where ω is the pseudo–differential operator given by $(-\Delta+m^2)^{1/2}$, and $\{e_k, k \in \mathbb{N}\}$ is a CONS of $L^2(\mathbb{R}^s, dx)$ in $\mathcal{S}(\mathbb{R}^s)$. The momentum field $\pi(\xi)$ is given by $i2^{-1/2}(D^*_{\omega^{1/2}\xi} - D_{\omega^{1/2}\xi})$. Let us first show that $\exp(i\varphi(\xi))\Omega \in \mathcal{D}(H_0)$ (which implies that $\exp(i\varphi(\xi))\Omega \in \mathcal{D}(H_0^{1/2})$). Actually, because $\varphi(\xi)$ is given by multiplication with $2^{-1/2}<\cdot, \omega^{-1/2}\xi>$, we have $\exp(i\varphi(\xi))\Omega = \exp(i2^{-1/2}<\cdot, \omega^{-1/2}\xi>)$. Using the fact that for $f \in L^2(\mathbb{R}^s)$, $D_f^* = <\cdot, f> - D_f$, it is straightforward to compute by the chain rule, that

$$H_0 \exp(i\varphi(\xi))\, \Omega = [\tfrac{i}{\sqrt{2}} <\cdot, \omega^{1/2}\xi> - \tfrac{1}{2} |\, \xi \,|_2^2] \exp(i\varphi(\xi))\, \Omega.$$

Clearly, the last expression belongs to all (L^p), $p \geq 1$, and therefore $\exp(i\varphi(\xi))\Omega$ is in $\mathfrak{D}(H_0)$ for all $\xi \in \mathcal{S}(\mathbb{R}^s)$. Next we verify the Araki relation. Let $\eta, \xi \in \mathcal{S}(\mathbb{R}^s)$, then by the chain rule we find

$$(\exp(i\varphi(\eta))\Omega, H_0 \exp(i\varphi(\xi))\Omega) =$$

$$= \sum_{k=1}^{\infty} (D_{\omega^{1/2} e_k} \exp(i\varphi(\eta))\Omega, D_{\omega^{1/2} e_k} \exp(i\varphi(\xi))\Omega)$$

$$= \sum_{k=1}^{\infty} \tfrac{1}{2}(\omega^{1/2} e_k, \eta)(\omega^{1/2} e_k, \xi)\, (\exp(i\varphi(\eta))\Omega, \exp(i\varphi(\xi))\Omega)$$

$$= \tfrac{1}{2}(\eta, \xi)_{L^2(\mathbb{R}^s)} (\exp(i\varphi(\eta))\Omega, \exp(i\varphi(\xi))\Omega).$$

Thus Araki's relation holds. Consider the following operators $H_{\xi,\pm}$:

$$H_{\xi,\pm} = \sum_{k=1}^{\infty} [D_{\omega^{1/2} e_k} \pm 2^{-1/2}(\omega^{-1}\xi, e_k)]^* [D_{\omega^{1/2} e_k} \pm 2^{-1/2}(\omega^{-1}\xi, e_k)].$$

It is clear that $H_{\xi,\pm}$ is defined on $\mathcal{P}$ (or $(\mathcal{S})$), and it is not hard to see that one can repeat the argument in the proof of Theorem 11.1 to prove that $H_{\xi,\pm}$ is essentially self-adjoint on (L^2). Its self-adjoint extension is denoted by the same symbol. Obviously, $H_{\xi,\pm}$ is positive. On the other hand, it is straightforward to show that

$$0 \leq H_{\xi,\pm} = H_0 \pm \varphi(\xi) + \tfrac{1}{2} |\, \omega^{-1}\xi \,|_2^2,$$

and therefore a φ–bound holds. Therefore Theorem 11.8 implies that the vacuum measure of the free field is represented by an element in $(\mathcal{S})^*$. Of course, this has already been established in Chapter 4 by the explicit form of its characteristic functional.

From now on we shall assume that $s = 1$, i.e., we are considering models of quantum fields in two space–time dimensions. We shall discuss various models which arise from special choices of the function F which defines the (informal) potential $V = \int_{\mathbb{R}} F(\varphi(x))\, dx$ in the Feynman–Kac formula (11.10), (11.11), or the above described ansatz for H as $H_0 + V$.

Example 11.13. (The Høegh–Krohn Model) The Høegh–Krohn model is defined by the following choice for the function F in the potential V: $F(u) = \int_{\mathbb{R}} \exp(\alpha u)\, d\rho(\alpha)$, where ρ is a Borel measure on $\mathbb{R}$ with support in $(-\sqrt{4\pi}, \sqrt{4\pi})$. For more details, in particular for the facts that we are using in the following arguments we refer to Høegh–Krohn (1971), Albeverio and Høegh–Krohn (1974), Simon (1974) and the literature quoted there. For simplicity, we shall work with the concrete choice that ρ is the Dirac measure concentrated at $\lambda \in (0, \sqrt{4\pi})$. Thus $F(u) = \exp(\lambda u)$. We take the same realization of $\varphi(\xi)$ on $\mathcal{D}^{2,1}$ as before: $\varphi(\xi) = 2^{-1/2} \langle \cdot, \omega^{-1/2}\xi \rangle \cdot$. Then it is not hard to see that the Wick ordering of $F(\varphi(\xi)) = \exp(\lambda\varphi(\xi))$ (see above), coincides with multiplication by

$$:\exp(2^{-1/2}\lambda \langle \cdot, \omega^{-1/2}\xi \rangle): = \exp[2^{-1/2}\lambda \langle \cdot, \omega^{-1/2}\xi \rangle - \tfrac{1}{4}\lambda^2(\xi, \omega^{-1}\xi)].$$

Note that this function is positive for all $\xi \in \mathcal{S}(\mathbb{R}^s)$. Therefore, if we form the corresponding cut–off interaction $V_{n,\Lambda}$ as indicated at the beginning of this section, then $V_{n,\Lambda} \geq 0$ for all $n \in \mathbb{N}$, and all Λ, being a bounded Borel set. Consequently, if we put $H_0 + V_{n,\Lambda}$, then from Example 11.12 we find that we have a φ–bound for every cut–off theory, with constants α, β, γ independent of the cut–offs. Araki's relation holds by the same argument as before, since $V_{n,\Lambda}$ commutes with $\varphi(\xi)$. It has been shown in Høegh–Krohn (1971), that $H_0 + V_{n,\Lambda}$

is essentially self-adjoint on $\mathcal{P}$ (and again we denote the closure by the same symbol), and that it has a unique ground state $\Omega_{n,\Lambda}$. By Corollary 11.9, we obtain a bound on the moments of the cut-off vacua $\Omega_{n,\Omega}$, which is uniform in n and Λ. On the other hand, it has been shown in the above quoted articles that the moments of the vacua converge, as $n \to +\infty$, and $\Lambda \to \mathbb{R}$. (Actually, it has been proved in Albeverio and Høegh-Krohn (1974) that the associated Schwinger functions converge, which - by restriction to time zero - implies the convergence of the moments of the vacua.) Thus the moments of the limiting vacuum Ω, which belongs to some Hilbert space (see Albeverio and Høegh-Krohn (1974)) satisfy the same bound 11.26. This bound in turn implies (as in the proof of Theorem 11.8) that $(\Omega, \exp(i\varphi(\xi))\ \Omega)_{\mathcal{H}}$ is a U-functional, with the result that there exists a $\Xi \in (\mathcal{S})^*_+$, so that (11.24) and (11.25) hold true. Thus, the vacuum expectation values of the Høegh-Krohn model are representable by expectations with respect to a generalized white noise functional. This result has been established in Albeverio, Hida et al. (1990) using correlation inequalities from Albeverio and Høegh-Krohn (1974) instead.

Example 11.14. ($P(\varphi)_2$ models) We make the same choices as in Example 11.12, and in particular we identify $\varphi(\xi)$ with multiplication by $2^{-1/2}\langle\cdot, \omega^{-1/2}\xi\rangle$. Moreover, we choose the function F above as a real polynomial so that either of the following assumptions holds:

(a) F is of the form $F(u) = \lambda \sum_{k=0}^{2n} a_k u^k$, $n \in \mathbb{N}$, with $\lambda, a_{2n} > 0$, and λ/m^2 small enough so that the cluster expansion in Glimm et al. (1973) converges;

(b) F is of the form $F(u) = \sum_{k=0}^{n} a_k u^{2k} + bu$, $n \in \mathbb{N}$, $a_n > 0$.

As before, let $V_{n,\Lambda}$ denote the corresponding cut-off interaction term. It has been proved in Guerra et al. (1975), Nelson (1973), Simon (1974) that the limit of $V_{n,\Lambda}$ as n tends to infinity converges in every (L^p), $p \geq 1$, and therefore the limit, denoted by V_Λ, defines a multiplication operator which is essentially self-adjoint, say on $\mathcal{P}$. We define the cut–off Hamiltonian $\hat{H}_\Lambda$ by $\hat{H}_\Lambda := H_0 + V_\Lambda$. In the before mentioned papers it has been shown that $\hat{H}_\Lambda$ is essentially self–adjoint on $\mathcal{P}$, and its spectrum is bounded from below. Set $H_\Lambda = \hat{H}_\Lambda - \inf \operatorname{spec}(\hat{H}_\Lambda)$ and denote the closure of H_Λ by the same symbol. H_Λ has a unique, strictly positive ground state $\Omega_\Lambda \in (L^2)$ (*loc. cit.*). It follows from the computations in Example 11.12 for H_0, and from the facts that V_Λ and φ commute, and $V_\Lambda \in (L^p)$, $p \geq 1$, that $\exp i(\varphi(\xi))\Omega_\Lambda \in \mathcal{D}(H_0)$. Moreover, one shows that Araki's relation holds by using the same facts and calculations (i.e., basically by applications of the chain and product rules). Fröhlich (1974) has proved the following φ–bound:

$$\pm\varphi(\xi) \leq \alpha H_\Lambda + \beta \|\!|\, \xi \,|\!\|^2 + \gamma,$$

where α, β, γ do not depend on Λ, and the norm $\|\!|\, \xi \,|\!\|$ is determined by

$$\|\!|\, \xi \,|\!\|^2 = \int_{\mathbb{R}} (2+x^2) \,|\, \xi(x) \,|^2 \, dx.$$

Obviously we have the estimation $\|\!|\, \xi \,|\!\| \leq \sqrt{2} \,|\, \xi \,|_{2,p}$, and so a φ–bound holds in the sense of Definition 11.7. Thus we can apply Theorem 11.8 and Corollary 11.9 with the result that there exists $\Xi_\Lambda \in (\mathcal{S})^*_+$, so that for all $\xi \in \mathcal{S}(\mathbb{R})$:

$$(\Omega_\Lambda, \exp(i\varphi(\xi))\, \Omega_\Lambda) = \langle\!\langle \Xi_\Lambda, \exp(i2^{-1/2} \langle \cdot, \xi \rangle) \rangle\!\rangle, \tag{11.30}$$

and the moments of both sides are bounded as in (11.26) with K independent of Λ. It follows immediately (cf. also Theorem 4.38 and its proof) that for some K_1, $K_2 > 0$, $p \in \mathbb{N}$, independent of Λ, and all $z \in \mathbb{C}$, $\xi \in \mathcal{S}(\mathbb{R})$, we have the uniform bound

$$|\, \langle\!\langle \Xi_\Lambda, \exp(iz \langle \cdot, \xi \rangle) \rangle\!\rangle \,| \leq K_1 \exp(K_2 \,|\, z \,|^2 \,|\, \xi \,|^2_{2,p}). \tag{11.31}$$

Fröhlich has proved the convergence of the right hand side of (11.30) as Λ tends to $\mathbb{R}$, where the limit Ω belongs to some Hilbert space on which $\varphi(\xi)$ acts as a self–adjoint operator. Now we apply Theorem 4.43, and it follows that as Λ tends to $\mathbb{R}$ through a sequence, Ξ_Λ converges strongly to some element $\Xi \in (\mathcal{S})^*_+$, so that (11.30) and (11.31) hold true in the limit. Therefore we have proved that also for all $P(\varphi)_2$ theories satisfying the above assumptions, the vacuum measure is represented by a generalized white noise functional.

We conclude this chapter with the remark that it has been proved in Albeverio et al. (1990a) that also the Sine–Gordon model has vacuum expectation values with can be expressed as pairings with an element in $(\mathcal{S})^*_+$. Moreover, it has been proved there on the basis of moment inequalities that also the Euclidean measures of all the above mentioned models can be represented through elements in $(\mathcal{S})^*_+$. In particular, this makes it possible to write the Schwinger functions of these models as the moments of generalized white noise functionals in $(\mathcal{S})^*$.

12. Feynman Integrals

As a practical tool, the Feynman integral has had an enormous range of applications in virtually all branches of quantum physics. At the same time, on the mathematical level it is a rather elusive object.

There have been innumerous attempts to give a mathematical meaning to formal expressions ('path integrals', 'sums over histories') such as

$$\int \mathfrak{D}^{\infty} y(\tau) \exp\Big(\frac{i}{\hbar} \int_0^t L(\dot{y},y,\tau)\, d\tau \Big) \tag{12.1}$$

where L is the Lagrangian of classical mechanics, where suitable initial and final conditions are imposed on the 'paths' $y(\,\cdot\,)$, and where $\hbar$ is Planck's constant. The goal is to 'quantize' the dynamics given by L, i.e., to construct solutions of the corresponding Schrödinger equation. To handle the informal infinite dimensional integral finite dimensional approximations have been proposed, while analytic continuation techniques have been useful to cope with the oscillatory nature of the integral. A comprehensive account is impossible in the present framework; we limit ourselves to mentioning Albeverio and Høegh-Krohn (1976b) as an illustration and for its extensive list of further references.

In white noise analysis there has been a first and to some extent heuristic discussion of Feynman integrals in Hida and Streit (1983). The guiding principle was — and is — to follow closely Feynman's original construction of averages over paths. With the help of the characterization results in Section 4.C we shall show that our construction is meaningful both mathematically and physically: (1) the resulting objects are well defined generalized functionals of white noise, and (2) they produce the correct quantum mechanical propagators as well as more general transition matrix elements.

For transparency of the argument, we shall first do so for the simple and physically trivial case of a particle without interaction. As a subsequent step we shall take a look at the scope of the method; first by discussing classes of interactions that are tractable within the present framework, and then by exploiting white noise calculus to manipulate quantum mechanical quantities.

For notational convenience we should like to observe that, through $\mathbb{E}(\Phi\cdot\varphi) \equiv <\overline{\Phi}, \varphi>$ we may extend the concept of expectation to arbitrary products of a generalized and a test function, with the convenient property of bilinearity. In particular $\mathbb{E}(\Phi) \equiv <\overline{\Phi}, 1>$. With respect to this bilinear pairing we denote dual operators by a dagger: $\mathbb{E}(\Phi\cdot M\varphi) = \mathbb{E}(M^{\dagger}\Phi\cdot\varphi)$.

12.A Construction of the Feynman Integrand: Free Particles

To begin with we shall construct the Feynman integral for the quantum mechanical propagator $K_0(y, t| y_0, t_0)$ of a free particle. K_0 obeys the Schrödinger equation

$$i\hbar\, \partial_t\, K_0(y, t| y_0, t_0) = H_0\; K_0(y, t| y_0, t_0), \tag{12.2}$$

with the initial condition

$$K_0(y, t_0| y_0, t_0) = \delta(y - y_0).$$

For the simple model at hand the Hamiltonian operator H is nothing other than the differential operator $H_0 = -\hbar^2/2m \cdot \partial^2/\partial y^2$, the corresponding Lagrangian of the classical free motion y=y(t) being $L = L_0 = \frac{1}{2m}\,(\frac{dy}{dt})^2$.

Let us now go ahead with Feynman's program, informally at first. As a first step we shall model the trajectories by Brownian paths starting at y_0

$$y(t) \equiv y_0 + \sqrt{\frac{\hbar}{m}} \int_{t_0}^{t} x(\tau)\, d\tau, \tag{12.3}$$

a slight generalization of (1.14). Substituting such paths y into the kinetic energy, Feynman's exponential $\exp(\, i/\hbar \cdot S\,)$ of the (free) action $S = \int L\, d\tau$ would take the form

$$\exp\left(\tfrac{i}{2}\int_{t_0}^{t} x^2(\tau)\,d\tau\right)$$

as in Example 3.2. Before we formalize this ansatz we proceed to mimic Feynman's non-existing 'flat' measure $\mathfrak{D}^\infty y(\tau)$ through the introduction of a factor $\exp\left(\tfrac{1}{2}\int x^2(\tau)\,d\tau\right)$, designed to compensate the Gaussian fall-off of the white noise measure $d\mu$. Clearly the resulting expression

$$\exp\left(\tfrac{1+i}{2}\int_{t_0}^{t} x^2(\tau)\,d\tau\right)$$

requires an (infinite) renormalization, as indicated in Example 3.2, to be a well-defined generalized functional. For later convenience we here describe the normalized exponential $\mathcal{N}\exp(\cdot)$ from Example 3.2 by its $\mathcal{T}$-transform

$$\mathcal{T}\colon \mathcal{N}\exp\left(\tfrac{1+i}{2}\int_\Delta x^2(\tau)\,d\tau\right) \to \exp\left(-\tfrac{i}{2}\,\|\,\xi_\Delta\|^2\right)\exp\left(-\tfrac{1}{2}\,\|\,\xi_{\Delta^c}\|^2\right) \qquad (12.4)$$

where $\Delta = [t_0,t] \subset \mathbb{R}$, Δ^c its complement, and ξ_B, for $B \subset \mathbb{R}$, means $\xi\cdot 1_B$.

To fix the end point of the trajectories we shall use a Donsker δ-function

$$\delta(y(t) - y) = \delta(y_0 + \int_\Delta x(\tau)d\tau - y);$$

we have set $\hbar$=m=1 for simplicity of notation.

But if we want to consider

$$\mathcal{N}\exp\left(\tfrac{1+i}{2}\int_\Delta x^2(\tau)\,d\tau\right)\cdot\delta(y_0 + \int_\Delta x(\tau)\,d\tau - y)\,.$$

we need to justify multiplication of these two generalized functions. This can be done in various ways, see, e.g., de Faria et al. (1991), probably most simply through the Fourier representation of the δ-function given in Example 4.52; an alternate and quite general approach for this multiplication problem will be discussed at the end of Section 12 B. In any case, the result can be conveniently described through the $\mathcal{T}$-transform of the above expression

$$\mathcal{T}: \mathcal{N}\exp\left(\tfrac{1+i}{2}\int_\Delta x^2(\tau)\,d\tau\right)\delta\left(y_0+\int_\Delta x(\tau)\,d\tau - y\right) \to \tag{12.5}$$

$$\frac{1}{\sqrt{2\pi i\,|\Delta|}}\exp\left(-\tfrac{1}{2}\,|\,\xi_{\Delta^c}\,|^2-\tfrac{i}{2}|\,\xi_\Delta\,|^2+\tfrac{i}{2|\Delta|}\left(\int_\Delta \xi(t)\,dt + y - y_0\right)^2\right).$$

Evidently the right hand side is a U-functional, hence we have arrived at a definition of the Feynman integrand

$$I_0(y_0,y;t_0,t) \equiv \mathcal{N}\exp\left(\tfrac{1+i}{2}\int_{t_0}^{t} x^2(\tau)\,d\tau\right)\delta(y(t) - y) \tag{12.6}$$

as a well-defined element in $(\mathcal{S})^*$. So much for the mathematical requirements. From the physics point of view we would want that the expectation of I_0 solves the Schrödinger equation. But this is an immediate consequence of (12.5); for $\xi \equiv 0$ we obtain

$$\mathbb{E}(I_0) = (\mathcal{T}I_0)(0) = \frac{1}{\sqrt{2\pi i\,|t-t_0|}}\exp\left(\frac{i}{2|t-t_0|}(y - y_0)^2\right), \tag{12.7}$$

and the right hand side is the well-known free propagator, solution of equation (12.2) and its initial condition.

In conclusion, we arrive at the following results which show, respectively, that our ansatz is reasonable, both mathematically and physically.

Theorem 12.1. *The 'Feynman integrand'*

$$I_0(y_2,t_2|\,y_1,t_1) = \mathcal{N}\exp\left(\tfrac{i+1}{2}\int x^2(t)\,dt\right)\delta(y(t) - y_2)$$

is an element in $(\mathcal{S})^*$.

Theorem 12.2. *The Feynman integral* $\mathbb{E}(\, I_0\,)$ *is the free particle propagator.*

We mention in passing that a larger class of matrix elements can be discussed in this fashion, – by substituting other, suitably localized wave functions a(y(t)) instead of the Donsker δ-function, see, e.g., de Faria et al. (1991) for more details.

To conclude this section we note that, even on a heuristic level, the conventional notation $\int \prod_{\tau} dy(\tau)$ is something of a misnomer: it is not the particle positions at each time $y(\tau)$ that are the independent variables, but rather the velocities $x=\dot{y}$ at each instant of time. This fact is of course quite obvious for the heat equation, where Brownian motion has independent increments, but its derivative, white noise, is a process with independent values at each time (which is what distinguishes it as a coordinate system for our infinite dimensional calculus).

12.B Construction of the Feynman Integrand: Interacting Particles

It is worthwhile to observe that in fact the $\mathcal{T}$-transform of the Feynman integrand has an interesting interpretation even for nonzero ξ. To this end we note that, by an informal integration by parts, we would obtain

$$(\mathcal{T}\, I_0)(\xi) = \mathbb{E}(\, I_0\, e^{-i\int y(\tau)\dot{\xi}(\tau)d\tau}) \cdot e^{iy_2\xi(t_2)-iy_1\xi(t_1)}, \qquad (12.8)$$

(for simplicity of notation we consider ξ that vanish identically outside Δ).

Equation (12.8) suggests that, up to boundary terms, evaluating the $\mathcal{T}$-transform should be the same as evaluating the Feynman integral with an additional time dependent term in the Lagrangian, $L_0(y) \to L_{0,\xi}(y,t) \equiv L_0(y) - \dot{\xi}(t)\, y$, so that we should expect

$$(\mathcal{T}\, I_0)(\xi) = K_0^{(\dot{\xi})}(y,t|y_0,t_0) \cdot e^{iy\xi(t)-iy_0\xi(t_0)}, \qquad (12.9)$$

where K_0 should be the propagator for a quantum mechanical particle with the time dependent potential $W(y;t) = \dot{\xi}(t)y$. Indeed, substituting $\mathcal{T}I_0$ from the previous section it is not hard to verify that K_0 given by (12.9) indeed obeys the appropriate Schrödinger equation and initial condition:

$$i\,\partial_t\, K_0{}^{(\dot{\xi})}(y,t|\,y_0,t_0) = (H_0 + \dot{\xi}(t)y)\, K_0{}^{(\dot{\xi})}(y,t|\,y_0,t_0) \qquad (12.10)$$

and

$$K_0{}^{(\dot{\xi})}(y,t_0|\,y_0,t_0) = \delta(y - y_0).$$

Let us now consider more general Feynman integrals, in the sense that $L = L_0 - V$, and correspondingly $H = H_0 + V$. In keeping with (12.1) we would want to consider an ansatz

$$I = I_0 \exp\Big(-\frac{i}{\hbar}\int_{t_0}^{t} V(y(\tau))\,d\tau\Big). \qquad (12.11)$$

We must recall however that I_0 is a generalized functional, hence such multiplication requires justification and will in consequence impose restrictions on the admissible V. We shall attack this problem by various methods, but before doing so the following observation may serve to set our sights.

Let us assume for a moment that we have constructed I as above. Then we should be able to repeat the arguments leading to equations (12.9) and (12.10), substituting systematically I_0, K_0, H_0, by I, K, and H. This suggests that alternatively we might use

$$(\mathcal{T}\,I)(\xi) = K^{(\dot{\xi})}(y,t|\,y_0,t_0)\cdot e^{iy_2\xi(t_2) - iy_1\xi(t_1)}\cdot e^{-\frac{1}{2}|\,\xi_{\Delta^c}|^2} \qquad (12.12a)$$

where K is given as a solution of

$$i\,\partial_t\, K^{(\dot{\xi})}(y,t|\,y_0,t_0) = (H + \dot{\xi}(t)\,y)\, K^{(\dot{\xi})}(y,t|\,y_0,t_0) \qquad (12.12b)$$

and

$$K^{(\dot{\xi})}(y,t_0 \mid y_0,t_0) = \delta(y-y_0).$$

as a definition of the Feynman integrand $I \in (\mathcal{S})^*$. By Theorem 4.38 the necessary and sufficient condition for this to be consistent is that the propagator $K^{(\dot{\xi})}$ is a U-functional as a function of ξ.

An interesting feature of this criterion is that the analyticity of the U-functional implies the existence of a power series expansion in the linear perturbation W(y;t), but not necessarily in V! Indeed one expects that V which grow rapidly for large argument will lead to transition matrix elements which are U-functionals, for a related class of matrix elements and for an extremely general class of interactions V this property has been demonstrated by de Faria et al. (1991).

Thus encouraged we shall now return to the constructive approach. A first observation is that for $V(\cdot)$ a polynomial of at most second order, by the methods of Gaussian integration, and suitable limit taking, we can calculate the action of I on various test functions explicitly, in particular its $\mathcal{S}$- and $\mathcal{T}$-transform.

For the harmonic oscillator with $V(y) = \frac{1}{2}\kappa^2 y^2$, $\kappa \geq 0$, the calculation through the explicit evaluation of oscillatory integrals is lengthy but straightforward and produces the $\mathcal{T}$-transform given in the following theorem.

Theorem 12.3.

(1) *The expression*

$$(\mathcal{T}I)(\xi) = (2\pi i\kappa^{-1}\sin(\kappa|\Delta|))^{-1/2}\exp\left(-\tfrac{i}{2} \mid \xi_{\Delta}|^2 - \tfrac{1}{2} \mid \xi_{\Delta^c}|^2 - \tfrac{i}{2}\kappa \operatorname{tg}(\kappa|\Delta|) y_1{}^2\right)\times$$

$$\times \exp\left(\frac{i\kappa}{\sin(2\kappa|\Delta|)} \left(y_1 - y_2 \cos(\kappa|\Delta|) - \int_{t_0}^{t} dt'\, \xi(t') \cos(\kappa(t'-t_0))\,\right)^2\right)\times \tag{12.13}$$

$$\times \exp\left(\frac{-i\kappa y_1}{\cos(\kappa|\Delta|)}\int_{t_0}^{t} dt'\ \xi(t')\ \sin(\kappa(t-t'))\right)\cdot \exp(-\tfrac{i}{2}(\xi,B\xi)),$$

where B *is the integral operator with kernel* $B(s_1,s_2)$, *symmetric in* s_1,s_2, *and given by*

$$B(s_1,s_2) = -\frac{\kappa}{\cos(\kappa|\Delta|)}\sin(\kappa(s_1-t))\cdot\cos(\kappa(s_2-t_0)) \text{ if } s_1>s_2, \qquad (12.14)$$

and $0 < \kappa\ |\Delta| < \pi/2$, *is a U-functional.*

(2) *It obeys equations* (12.12) *with* $H = H_0 + V$ *with* $V(y) = \frac{1}{2}\kappa^2 y^2$.

Proof. To show (1) it is sufficient to note that all the factors composing the right hand side of (12.12) are U-functionals. (2) amounts to a verification that the propagator $K^{(\xi)}$ resulting from (12.13) by (12.12a) indeed solves the Schrödinger equation (12.12b). Alternatively one may check that it agrees with the explicit expression given by Khandekar and Lawande (1986). □

Of course the expression for $\mathcal{T}I_0$ results in the limit of vanishing κ; and similar explicit expressions for (the $\mathcal{T}$-transform of) the Feynman integrand can be derived for more general linear and/or quadratic potentials V. Instead of doing this here let us go on as in Khandekar and Streit (1992) — to study a class of potentials for which the Feynman integrand may be constructed perturbatively as a generalized white noise functional, $I \in (\mathcal{S})^*$. To this end we shall use a power series expansion for the exponential in (12.11) and an ansatz

$$V(y(\tau)) = \int dy\ V(y)\ \delta(y(\tau) - y) \qquad (12.15)$$

for the potential. For $n \in \mathbb{N}$, let $\Delta_n = \{(t_1,\ldots, t_n): t_0<t_1<\cdots<t_n<t\}$. Then we obtain

$$e^{-i\int_{t_0}^{t} V(y(\tau))\,d\tau} = \sum_{n=0}^{\infty} (-i)^n \int_{\Delta_n} d^n t \prod_{i=1}^{n} V(y(t_i))$$
$$= \sum_{n=0}^{\infty} (-i)^n \int_{\Delta_n} d^n t \prod_{i=1}^{n} \int dy_i\, V(y_i)\ \delta(\, y(t_i) - y_i\,) \tag{12.16}$$

so that the full Feynman integrand for the propagator would read

$$\mathcal{N}\exp\left(\tfrac{1+i}{2}\int_{t_0}^{t} x^2(\tau)\,d\tau\right)\delta(y(t)-y)\ e^{-i\int_{t_0}^{t} V(y(t))\,dt} =$$
$$= \sum_{n=0}^{\infty}(-i)^n \int_{\Delta_n} d^n t \prod_{i=1}^{n} \int dy_i\, V(y_i)\ \mathcal{N}\exp\left(\tfrac{1+i}{2}\int_{t_0}^{t} x^2(\tau)\,d\tau\right) \times$$
$$\times\ \delta(y(t)-y) \prod_{i=1}^{n} \delta(\, y(t_i) - y_i\,). \tag{12.17}$$

Our goal will be to show that — for a suitable class of potentials — the right hand side is indeed a well-defined generalized functional of white noise. The second part $\mathcal{N}\exp(\ )\prod\delta(\ )$ is a slight generalization of I_0, with more than just one δ-function. It can be constructed by any number of equivalent regularizations and we simply characterize the result by its $\mathcal{T}$-transform:

$$\mathcal{T}\colon\ \mathcal{N}\exp\left(\tfrac{1+i}{2}\int_{\Delta} x^2(\tau)\,d\tau\right)\prod_{i=1}^{n+1}\delta(y(t_i)-y_i)\ \rightarrow$$
$$e^{-\frac{1}{2}\|\xi_{\Delta^c}\|^2}\cdot e^{i(y_{n+1}\xi(t_{n+1})-y_0\xi(t_0))}\cdot\prod_{i=1}^{n+1} K_0^{(\dot\xi)}(y_i,t_i|y_{i-1},t_{i-1}). \tag{12.18}$$

Here $t_0 < t_1 < t_2 < \ldots < t_{n+1}$ and $\Delta \wedge [t_0, t_{n+1}]$. This $\mathcal{T}$-transform clearly is a measurable function of y_i, t_i, and bounded by

$$| \mathcal{T}\Big(\mathcal{N}\exp\big(\tfrac{1+i}{2} \int_\Delta x^2(\tau)\, d\tau\big) \prod_{i=1}^{n+1} \delta(y(t_i) - y_i)\Big)(\xi) | \leq$$

$$\leq \prod_{l=1}^{n+1} (2\pi|t_i - t_{i-1}|)^{-\frac{1}{2}} \times$$

$$\times \exp\Big(\tfrac{1}{2}\| \xi \|^2 + 2 \sup_{0\leq j\leq n+1} |y_j| \cdot \sup_{t\epsilon\Delta} |\xi(t)| + \tfrac{1}{2} |\Delta| (\sup_{t\epsilon\Delta} |\xi(t)|)^2\Big)$$

(we have denoted $t = t_{n+1}$ and $y(t) = y_{n+1}$ to compactify our notation). Now we choose p such that $\sup_{t\epsilon\Delta} |\xi(t)| < \| A^p \xi \|$ and substitute this bound. Then the calculation of

$$\prod_{i=1}^{n} \int dy_i\, V(y_i)\ \mathcal{N}\exp\big(\tfrac{1+i}{2} \int_{t_0}^{t} x^2(\tau)\, d\tau\big)\, \delta(y(t) - y) \prod_{i=1}^{n} \delta(\, y(t_i) - y_i\,)$$

is just a matter of applying Theorem 4.51. Its conditions are fulfilled whenever $\int dy\ V(y)$ is integration with respect to a finite (signed) Borel measure with support in some finite interval $[-a,a] \subseteq \mathbb{R}$, and we obtain the bound

$$| \mathcal{T}\Big(\prod_{i=1}^{n} \int dy_i\, V(y_i)\ \mathcal{N}\exp\big(\tfrac{1+i}{2} \int_\Delta x^2(\tau)\, d\tau\big) \prod_{i=1}^{n+1} \delta(y(t_i) - y_i)\Big)(\xi) | \leq$$

$$\leq \prod_{l=1}^{n+1} (2\pi|t_i - t_{i-1}|)^{-\frac{1}{2}} \cdot \Big(\int dy\ V(y)\Big)^n \exp\Big(\tfrac{1}{2}(1+|\Delta|)\, \| A^p\xi \|^2 + 2a\, \| A^p \xi \|\Big)$$

In order to perform the t_i-integrations, according to Theorem 4.51 we only need to verify that

$$\int_{\Delta_n} d^n t \prod_{l=1}^{n+1} (2\pi|t_i - t_{i-1}|)^{-\frac{1}{2}} = M_n < \infty.$$

In fact the M_n are rapidly decreasing; as a consequence, Theorem 4.51 ensures the convergence of the series in (12.17). In conclusion we have the following

Theorem 12.4. *Whenever the potential V is such that $dv(y) \equiv V(y)dy$ is a signed Borel measure of bounded support, the Feynman integrand* (12.17) *is an element in* $(\mathcal{S})^*$.

Remarks.

(1) The series (12.17) is a (strong) Dyson series, not just for the propagator but for the Feynman integrand $I \in (\mathcal{S})^*$ itself, hence it entails a (weak) Dyson expansion for any expectation of the form $\mathbb{E}(I \cdot F)$ where $F \in (\mathcal{S})$. The usual Dyson series for the Feynman propagator results for $F = 1$, i.e., $\xi = 0$ in the $\mathcal{T}$-transform.

(2) The class of admissible interactions includes potentials with local singularities (as long as they are integrable), but also, e.g., $V(y) = \sum_i g_i \, \delta(y - a_i)$ (as long as the $|g_i|$ are summable and the a_i bounded).

As a final observation in this section we should like to address once more, from a different point of view, the problem of defining the product of the generalized functional $\mathcal{N}\exp(\,\cdot\,)$ with other (generalized) functionals. To focus on the essential point here, we shall consider the interval $\Delta = (-\infty, \infty)$, i.e., the functional

$$\Phi^z := \mathcal{N}\exp\Big(\tfrac{z^2}{2} \int_{-\infty}^{\infty} x^2(\tau) d\tau \Big).$$

Its $\mathcal{T}$-transform is given by

$$(\mathcal{T}\Phi^z)(\xi) = \exp(-\frac{(z')^2}{2}\int_{-\infty}^{\infty}\xi^2(\tau)d\tau),$$

with $(z')^2 = (1-z^2)^{-1}$; for the Feynman integrand we should put $z^2=1+i$. Recalling that

$$(\mathcal{T}\Phi^z)(\xi) = \mathbb{E}(\Phi^z \cdot e^{i<x,\xi>}),$$

and the mapping σ_z: $\varphi \to \varphi_z$ from $(\mathcal{S})$ to $(\mathcal{S})$ given by $\varphi_z(x) = \varphi(z\cdot x)$, we have immediately

Proposition 12.5. *Let* $z'=(1-z^2)^{-\frac{1}{2}}$. *Then* $\Phi^z = (\sigma_{z'})^\dagger 1\!\!1$. (12.19)

Of course we are free to multiply Φ^z, or for that matter any generalized functional, by arbitrary test functionals φ (see Corollary 4.11), but this is not good enough for interesting applications in quantum physics. Hence we are led to consider limits of such products $\Phi^z\varphi_n$. The following criterion serves to control their convergence.

Theorem 12.6. *Let* φ_n *be a sequence of test functionals. Then the following are equivalent:*

(1) *The sequence* $\Phi^z\varphi_n \to \Psi$ *converges in* $(\mathcal{S})^*$.

(2) *The sequence* $\sigma_{z'}\varphi_n$ *converges in* $(\mathcal{S})^*$ *as* $n \to \infty$.

(3) *The sequence* $\mathbb{E}(\psi\, \sigma_{z'}\varphi_n)$ *converges for all* $\psi \in (\mathcal{S})$.

The action of Ψ *is given by*

$$\mathbb{E}(\Psi \cdot \psi) = \lim_n \mathbb{E}(\sigma_{z'}(\varphi_n\psi)) \tag{12.20}$$

if one of the conditions (1–3) *holds.*

Proof. (2) is equivalent to (3) since $(\mathcal{S})^*$ is weakly sequentially complete. The rest of the theorem is an immediate consequence of the following identities:

$$\begin{aligned} \mathbb{E}(\Phi^z\varphi_n \cdot \psi) &= \mathbb{E}(\Phi^z \cdot \varphi_n\psi) \\ &= \mathbb{E}(\sigma_{z'}^{\dagger}\mathbb{1} \cdot \varphi_n\psi) \\ &= \mathbb{E}(\mathbb{1} \cdot \sigma_{z'}\varphi_n\psi) \\ &= \mathbb{E}(\sigma_{z'}\varphi_n \cdot \sigma_{z'}\psi). \end{aligned}$$ □

For the Feynman integrand, with $z'=\sqrt{i}$, Doss (1980) studies analytic potentials V for which $\exp\Big(-i\int_{t_0}^{t} V(y+z'B(\tau))\, d\tau\Big)$ is a random variable, and analytic functions f such that

$$\Psi(y,t) = \mathbb{E}\Big(f(y+z'B(t)) \exp(-i\int_{t_0}^{t} V(y+z'B(\tau))\, d\tau)\Big) \tag{12.21}$$

solves the Schrödinger equation with initial value $\Psi(y, 0) = f$.

White noise analysis allows us to go beyond this framework by considering *generalized* functionals. We shall illustrate this point by the following simple examples.

Example 12.7. Let ξ_n be a δ-sequence of test functions. Then $\varphi_n = \langle x,\xi_n\rangle \to x(t)$ is an admissible multiplier; likewise for arbitrary derivatives of x.

Example 12.8. Donsker's δ-function, and more generally $\delta(\lambda\langle x,\eta\rangle - a)$ are admissible multipliers. In fact we can extend the scaling directly to Donsker's δ-function $\varphi^z(x; a) = \delta(z\,B(t) - a)$. For real z we can use, e.g., the integral representation of Example 4.52 to calculate

$$(\mathcal{T}\varphi^z)(\xi) = (z\sqrt{2\pi t})^{-1}\, C(\xi)\, e^{-\frac{1}{2}(a/z - i\int_0^t \xi d\tau)^2}$$

From the right hand side we find that $\varphi^z(x;a) = z^{-1}\varphi^1(x;a/z) = z^{-1}\delta(B(t) - a/z)$ is in $(\mathcal{S})^*$ – and in fact analytic – for all z in $\mathbb{C}\backslash\{0\}$. In keeping with equation (12.20), we obtain the $\mathcal{T}$-transform of the Feynman integrand for the free propagator by calculating $(\mathcal{T}\varphi^{z'})(z'\xi)$, for $z'=\sqrt{i}$; in particular

$$K_0(y, t|\, y_0, t_0) = \mathbb{E}(\varphi^{z'}(x; y-y_0)).$$

In like fashion, we can obtain the more general expression (12.18) through analytic continuation (of the $\mathcal{T}$-transform) of $\varphi = \prod_{i=1}^{n+1} \delta(y(t_i) - y_i)$.

12.C Manipulating the Feynman Integrand: Schwinger's Variational Principle

In their monograph Feynman and Hibbs (1965) emphasize the fundamental role of a variational relation which – in their notation – reads

$$\left\langle \delta F \right\rangle = -\tfrac{i}{\hbar} \left\langle F\, \delta S \right\rangle . \tag{12.22}$$

Here S of course is the action functional $S = S_0 - \int V(y)\, dt$. By carefully chosen 'rules' they extract various useful quantum mechanical results such as, e.g., 'Ehrenfest's theorem'

$$\left\langle m\ddot{y} \right\rangle = - \left\langle V' \right\rangle .$$

from this symbolic equation which however has largely resisted attempts at a definition in mathematical terms and hence has not allowed for mathematically sound conclusions.

Within the white noise analysis framework this problem is one that has a very simple solution, first indicated by de Faria (1991).

We begin its discussion by reformulating equation (12.22) in our notation where it would read

$$\mathbb{E}(I \cdot D_\eta F) \;=\; -\,i\;\mathbb{E}(I\, D_\eta S \cdot F).$$

We can use the dual operator $D_\eta{}^* = D_\eta{}^\dagger$ (for real η) to formulate a strong version of equation (12.22), the following differential equation for I:

$$D_\eta{}^\dagger\, I = i\, D_\eta S \cdot I. \tag{12.23}$$

In the simplest version of a Feynman integrand — without interaction and without a δ-function to pin down the final position $y(t)$ — we have

$$I = I_0 = \mathcal{N} \exp\Big(\frac{i+1}{2} \int_\Delta x^2(t)\, dt\Big)$$

which obeys the equation

$$D_\eta\, I_0 = (i+1)\, X_\eta\, I_0$$

for any η with support in Δ so that, using $D_\eta{}^\dagger = X_\eta - D_\eta$, we obtain

$$D_\eta{}^\dagger\, I_0 = (X_\eta - D_\eta)\, I_0 = -\, i X_\eta \cdot I_0.$$

Now we need only observe that in the case at hand the action is simply quadratic

$$S = S_0 = \tfrac{1}{2} \int_\Delta :x^2:(t)\, dt, \quad \text{i.e., } D_\eta S_0 = X_\eta$$

which we can substitute into the right hand side of the previous equation to conclude the validity of the variational equation in its strong form (12.23).

Modifications of I to cover more general cases will be of the form $I = I_0\, \varphi$ so that

$$D_\eta^{\dagger}\, I = (X_\eta - D_\eta)\, I = -\, iD_\eta S_0 \cdot I - D_\eta \varphi \cdot I_0.$$

To obtain solutions of the Schrödinger equation corresponding to a nontrivial action $S = S_0 + S_1$ and with boundary value α, we should choose φ of the form

$$\varphi = e^{iS_1}\, \alpha(y(t)).$$

Substituting this into the previous expression we find again the variational equation (12.23),

$$D_\eta^{\dagger}\, I = -\, iD_\eta(S_0 + S_1) \cdot I$$

if η is chosen such that α is not varied, i.e., with $\int_0^t \eta(\tau)d\tau{=}0$ so that $D_\eta\alpha = \alpha' D_\eta y(t){=}0$.

Otherwise – i.e., if we 'vary the final position y(t)' so that $D_\eta y(t) \neq 0$ – there will be an extra term on the right hand side containing the derivative of α;

$$D_\eta^{\dagger}\, I = -\, iD_\eta S \cdot I - I_0\, e^{iS_1}\, D_\eta \alpha(y(t))$$

absent in the heuristic formula (12.22), it nevertheless produces useful quantum mechanical relationships, see, e.g., de Faria and Streit (1992).

Appendix A.1 Hermite Polynomials and Wick Powers

In this appendix we summarize some useful formulae about Hermite polynomials and Hermite functions.

A.1.A Hermite Polynomials

Definition:

$$H_n(u) = (-1)^n \, e^{u^2} \frac{d^n}{du^n} \, e^{-u^2} \; ; \; n \in \mathbb{N}_0, \; u \in \mathbb{R} \tag{A.1.1}$$

Generating function:

$$\sum_{n=0}^{\infty} \frac{t^n}{n!} H_n(u) = e^{-t^2+2tu} \tag{A.1.2}$$

Explicit formula:

$$\begin{aligned} H_n(u) &= \sum_{k=0}^{[\frac{n}{2}]} (-1)^k \frac{n!}{k! \, (n-2k)!} (2u)^{n-2k} \\ &= 2^n \sum_{k=0}^{[\frac{n}{2}]} \binom{n}{2k} (2k-1)!! \, (-\tfrac{1}{2})^k \, u^{n-2k} \end{aligned} \tag{A.1.3}$$

Examples:

$H_0(u) = 1$ $\qquad$ $H_1(u) = 2u$ $\qquad$ $H_2(u) = 4u^2-2$

$H_3(u) = 8u^3-12u$ $\qquad$ $H_4(u) = 16u^4-48u^2+12$

Differential equation:

$$H_n''(u) - 2u\, H_n'(u) + 2n\, H_n(u) = 0 \tag{A.1.4}$$

Recursion relation:

$$H_{n+1}(u) - 2u\, H_n(u) + 2n\, H_{n-1}(u) = 0 \tag{A.1.5}$$

Derivative:

$$H_n'(u) = 2n\, H_{n-1}(u) \tag{A.1.6}$$

Multiplication formula:

$$H_n(u)\, H_m(u) = \sum_{k=0}^{n \wedge m} 2^k\, k! \binom{n}{k} \binom{m}{k} H_{n+m-2k}(u) \tag{A.1.7}$$

Orthogonality relation:

$$\int_{\mathbb{R}} H_n(u)\, H_m(u)\, e^{-u^2} du = \delta_{n,m}\, \pi^{1/2}\, 2^n\, n! \tag{A.1.8}$$

Hermite functions:

$$\xi_n(u) = (\pi^{1/2}\, 2^n\, n!)^{-1/2}\, H_n(u)\, e^{-\frac{1}{2}u^2} \tag{A.1.9}$$

The collection $\{\xi_n, n \in \mathbb{N}_0\}$ forms a CONS of $L^2(\mathbb{R}, du)$. Moreover

$$-\xi_n''(u) + u^2\, \xi_n(u) = (2n+1)\, \xi_n(u) \tag{A.1.10}$$

Estimates (cf. Hille and Phillips (1957), (21.3.2), (21.3.3)):

$$\xi_n(u) = O(n^{-1/4}) \tag{A.1.11}$$

$$\sup_{u\in\mathbb{R}} | \xi_n(u) | = O(n^{-1/12}), \; \| \xi_n \|_1 = O(n^{1/4}) \tag{A.1.12}$$

A.1.B Hermite Polynomials with Parameter (Wick Powers)

Let $\sigma > 0$.

Definition:

$$:u^n:_{\sigma^2} = (-\sigma^2)^n \exp(\frac{u^2}{2\sigma^2}) \frac{d^n}{du^n} \exp(-\frac{u^2}{2\sigma^2}) \tag{A.1.13}$$

Relation with Hermite polynomials H_n:

$$:u^n:_{\sigma^2} = 2^{-n/2} \sigma^n H_n\Big(\frac{u}{2^{1/2}\sigma}\Big) \;,\; H_n(u) = 2^n :u^n:_{2^{-1/2}} \tag{A.1.14}$$

Generating function:

$$\sum_{n=0}^{\infty} \frac{t^n}{n!} :u^n:_{\sigma^2} = \exp(-\sigma^2\frac{t^2}{2} + tu) \tag{A.1.15}$$

Explicit formulae:

$$:u^n:_{\sigma^2} = \sum_{k=0}^{[\frac{n}{2}]} (-\tfrac{1}{2}\sigma^2)^k \frac{n!}{k!\,(n-2k)!} u^{n-2k}$$

(A.1.16)

$$= \sum_{k=0}^{[\frac{n}{2}]} (-\sigma^2)^k \binom{n}{2k} (2k-1)!! \, u^{n-2k}$$

$$u^n = \sum_{k=0}^{[\frac{n}{2}]} \sigma^{2k} \binom{n}{2k} (2k-1)!! : u^{n-2k} :_{\sigma^2} \tag{A.1.17}$$

Examples:

$$:u^0:_{\sigma^2} = 1 \qquad :u^1:_{\sigma^2} = u \qquad :u^2:_{\sigma^2} = u^2 - \sigma^2$$

$$:u^3:_{\sigma^2} = u^3 - 3\sigma^2 u \qquad :u^4:_{\sigma^2} = u^4 - 6\sigma^2 u^2 + 3\sigma^4$$

Differential equation:

$$\left(\sigma^2 \frac{d^2}{du^2} - u \frac{d}{du} + n\right) : u^n :_{\sigma^2} = 0 \tag{A.1.18}$$

Recursion relation:

$$:u^{n+1}:_{\sigma^2} - u : u^n :_{\sigma^2} + \sigma^2 n : u^{n-1} :_{\sigma^2} = 0 \tag{A.1.19}$$

Derivative:

$$\frac{d}{du} : u^n :_{\sigma^2} = n : u^{n-1} :_{\sigma^2} \tag{A.1.20}$$

$$\left(u - \sigma^2 \frac{d}{du}\right) : u^n :_{\sigma^2} = : u^{n+1} :_{\sigma^2} \tag{A.1.21}$$

Multiplication formula:

$$:u^n:_{\sigma^2} : u^m :_{\sigma^2} = \sum_{k=0}^{n \wedge m} k! \binom{n}{k} \binom{m}{k} \sigma^{2k} : u^{n+m-2k} :_{\sigma^2} \tag{A.1.22}$$

Orthogonality relation:

$$\int_{\mathbb{R}} :u^n:_{\sigma^2} :u^m:_{\sigma^2} \exp(-\frac{u^2}{2\sigma^2})\, du = \delta_{n,m}\, (2\pi)^{1/2}\, n!\, \sigma^{2n} \qquad (A.1.23)$$

Fourier-Gauss transformation:

$$\int_{\mathbb{R}} :u^n:_{\sigma^2} e^{iyu} \exp(-\frac{u^2}{2\sigma^2})\, du = (2\pi\sigma^2)^{1/2}\, i^n\, \sigma^{2n} y^n \exp(-\frac{\sigma^2 y^2}{2}) \quad (A.1.24)$$

Integral representation:

$$:u^n:_{\sigma^2} = (-i)^n\, \sigma^{2n+1} \exp(\frac{u^2}{2\sigma^2}) \int_{\mathbb{R}} y^n\, e^{iyu} \exp(-\frac{\sigma^2 y^2}{2})\, \frac{dy}{(2\pi)^{1/2}} \quad (A.1.25)$$

Note: in case that X is a Gaussian random variable with variance σ^2, we set $:X^n: \equiv\, :X^n:_{\sigma^2}$.

Appendix A.2. Fock Space

In this appendix we collect some more or less well–known facts about tensor products of Hilbert spaces and Fock spaces over Hilbert spaces. For more details about these structures we refer the interested reader to the works by Cook (1953), Murray and von Neumann (1936), Nelson (1973c), Reed and Simon (1972), Segal (1966) and Simon (1974). Sometimes our conventions differ slightly from those of the mentioned authors. (Mainly, this is due to the fact that in this book we often do not identify a Hilbert space with its dual.) First we recall the construction of tensor products of Hilbert spaces, and of Fock spaces, and construct bases for these. Then we sketch the theory of creation and annihilation operators on Fock space. Finally we provide some facts (some with rather explicit proofs) concerning second quantized operators.

Let $\mathcal{H}_1$, $\mathcal{H}_2$ be two complex Hilbert spaces with scalar products $(\cdot,\cdot)_1$, $(\cdot,\cdot)_2$, respectively[1]. By $\mathcal{H}_i^*$, i=1,2, we denote their duals, and by $<\cdot, \cdot>_i$, i=1,2, the dual pairings. The *tensor product* $\mathcal{H}_1 \otimes \mathcal{H}_2$ of $\mathcal{H}_1$ and $\mathcal{H}_2$ is constructed as follows.

Φ is the space of bilinear forms F on $\mathcal{H}_1^* \times \mathcal{H}_2^*$, i.e. for f, $h \in \mathcal{H}_1^*$, g, $k \in \mathcal{H}_2^*$, $\lambda \in \mathbb{C}$, we have

$$F(\lambda f, g) = F(f, \lambda g) = \lambda F(f, g) \tag{A.2.1.a}$$

$$F(f+h, g) = F(f, g) + F(f, h) \tag{A.2.1.b}$$

$$F(f, g+k) = F(f, g) + F(f, k). \tag{A.2.1.c}$$

Let $\mathcal{K}_i$, be a linear subspace of $\mathcal{H}_i$, i=1,2. By $\mathcal{K}_1 \otimes' \mathcal{K}_2$ we denote the subspace of Φ which consists of finite linear combinations of elements of the form $\varphi \otimes \psi$,

[1]Throughout we make the convention that scalar products are antilinear in the first factor.

where $\varphi \in \mathcal{H}_1$ and $\psi \in \mathcal{H}_2$, and with

$$\varphi \otimes \psi(f, g) = \langle f, \varphi\rangle_1 \langle g, \psi\rangle_2, \quad (f, g) \in \mathcal{H}_1^* \times \mathcal{H}_2^*. \tag{A.2.2}$$

$\varphi \otimes \psi$ is called the *tensor product of* φ *and* ψ.

Now consider the subspace $\mathcal{H}_1 \otimes' \mathcal{H}_2$ of Φ. It becomes a pre-Hilbert space when equipped with the following scalar product. For $\varphi \otimes \psi$, $\xi \otimes \eta \in \mathcal{H}_1 \otimes' \mathcal{H}_2$ of the form $\varphi \otimes \psi = \sum_{k=1}^{m} \varphi_k \otimes \psi_k$, $\xi \otimes \eta = \sum_{l=1}^{n} \xi_l \otimes \eta_l$, m, $n \in \mathbb{N}$, φ_k, $\xi_l \in \mathcal{H}_1$, ψ_k, $\eta_l \in \mathcal{H}_2$, we put

$$(\varphi \otimes \psi, \xi \otimes \eta) := \sum_{k=1}^{m} \sum_{l=1}^{n} (\varphi_k, \xi_l)_1 \, (\psi_k, \eta_l)_2. \tag{A.2.3}$$

$\mathcal{H}_1 \otimes \mathcal{H}_2$ is by definition the completion of $\mathcal{H}_1 \otimes' \mathcal{H}_2$ with respect to the norm induced by the scalar product (A.2.3), and it is called the *tensor product of* $\mathcal{H}_1$ *and* $\mathcal{H}_2$.

Given a complex Hilbert space $\mathcal{H}$, we can now define by induction for $n \in \mathbb{N}$

$$\mathcal{F}^{(n)}(\mathcal{H}) := \mathcal{H} \otimes \cdots \otimes \mathcal{H} \qquad \text{(n factors)}. \tag{A.2.4}$$

We also set $\mathcal{F}^{(0)} := \mathbb{C}$. By $\| \cdot \|_{\mathcal{F}^{(n)}(\mathcal{H})}$ we denote the natural norm on $\mathcal{F}^{(n)}(\mathcal{H})$.

If $\mathcal{H}$ is separable, $\mathcal{F}^{(n)}(\mathcal{H})$ can equivalently be defined as the space of n-multilinear forms F on $\mathcal{H}$ with finite Hilbert-Schmidt norm $\|F\|_{H.S.}$. Here the Hilbert-Schmidt norm of $F \in \mathcal{F}^{(n)}(\mathcal{H})$ is given by

$$\|F\|_{H.S.}^2 := \sum_{\mathbf{k}} | F(e_{k_1}, \dots, e_{k_n}) |^2, \tag{A.2.5}$$

where $(e_k, k \in \mathbb{N})$ is any complete orthonormal system in $\mathcal{H}$ and the sum in (A.2.5) is over all $\mathbf{k} = (k_1, \dots, k_n) \in \mathbb{N}^n$. It is easy to check that in this case

$\| F \|_{H.S.} = \| F \|_{\mathcal{F}^{(n)}(\mathcal{H})}$.

For $f_1, \ldots, f_n \in \mathcal{H}$, let $\hat{\otimes}$ denote the symmetrization of the tensor product $\otimes$:

$$f_1 \hat{\otimes} \ldots \hat{\otimes} f_n := (n!)^{-1} \sum_{\pi \in \mathfrak{S}(n)} f_{\pi(1)} \otimes \ldots \otimes f_{\pi(n)}, \tag{A.2.6}$$

where $\mathfrak{S}(n)$ is the permutation group of n letters. The left hand side of (A.2.6) will also be denoted by $\hat{\otimes}_{i=1}^{n} f_i$. For two symmetric tensors $\hat{\otimes}_{i=1}^{n} f_i$, $\hat{\otimes}_{i=1}^{n} g_i$, f_i, $g_i \in \mathcal{H}$, $i=1,\ldots,n$, of degree n we define a scalar product by

$$(\hat{\otimes}_{i=1}^{n} f_i \, , \, \hat{\otimes}_{i=1}^{n} g_i)_{\Gamma^{(n)}(\mathcal{H})} := \sum_{\pi \in \mathfrak{S}(n)} (f_1, g_{\pi(1)}) \cdots (f_n, g_{\pi(n)}), \tag{A.2.7}$$

and extend it by (anti-) linearity to linear combinations of such tensors. The completion of the resulting pre–Hilbert space under the corresponding norm $\| \cdot \|_{\Gamma^{(n)}(\mathcal{H})}$ is denoted by $\Gamma^{(n)}(\mathcal{H})$. $\Gamma^{(n)}(\mathcal{H})$ is a Hilbert subspace of $\mathcal{F}^{(n)}(\mathcal{H})$, and for $F \in \Gamma^{(n)}(\mathcal{H})$ it is easily checked that

$$\|F\|^2_{\Gamma^{(n)}(\mathcal{H})} = n! \, \|F\|^2_{\mathcal{F}^{(n)}(\mathcal{H})}. \tag{A.2.8}$$

The *Fock space* $\mathcal{F}(\mathcal{H})$ *over* $\mathcal{H}$ is the direct sum of the spaces $\mathcal{F}^{(n)}(\mathcal{H})$, n = 0, 1, ... The direct sum of the spaces $\Gamma^{(n)}(\mathcal{H})$, n=0,1, ... ($\Gamma^{(0)} = \mathbb{C}$) is called the *symmetric Fock space over* $\mathcal{H}$ and it is denoted by $\Gamma(\mathcal{H})$.

From now we assume that $\mathcal{H}$ is separable and that $(e_k, k \in \mathbb{N})$ is a complete orthonormal system (CONS) in $\mathcal{H}$. Then we can construct orthonormal bases for $\mathcal{F}(\mathcal{H})$ and $\Gamma(\mathcal{H})$ in the following way.

a) CONS for $\mathcal{F}(\mathcal{H})$:

Let $n \in \mathbb{N}$, let and $\mathbf{k}=(k_1, \ldots, k_n)$ be an element in $\mathbb{N}^n$. Set

$$e_{\mathbf{k}} := \otimes_{i=1}^{n} e_{k_i}. \tag{A.2.9}$$

Then

$$(e_{\mathbf{k}}\ ,\ e_{\mathbf{m}})_{\mathcal{F}^{(n)}(\mathcal{H})} = \delta_{\mathbf{k},\mathbf{m}}. \tag{A.2.10}$$

By construction of $\mathcal{F}^{(n)}(\mathcal{H})$, ($e_{\mathbf{k}}$, $\mathbf{k}\in\mathbb{N}^n$) forms a CONS of $\mathcal{F}^{(n)}(\mathcal{H})$. Thus the set of all sequences of the form

$$(0,\ 0,\ \dots,\ e_{\mathbf{k}},\ 0,\ \dots), \tag{A.2.11}$$

with $n\in\mathbb{N}$, $\mathbf{k}\in\mathbb{N}^n$ and $e_{\mathbf{k}}$ in the n+1-st entry, together with (1, 0, 0, ...) forms a CONS of $\mathcal{F}(\mathcal{H})$.

b) CONS for $\Gamma(\mathcal{H})$:

For $n\in\mathbb{N}$, let $I^n\subset\mathbb{N}^n$ denote the set of all naturally ordered elements $\alpha = (\alpha_1, \dots, \alpha_n)$ of $\mathbb{N}^n$. If $\alpha\in I^n$ and $i\in\mathbb{N}$, let $n_i(\alpha)$ be the number of entries of α equal to i. Set

$$n(\alpha)! := \prod_{i=1}^{\infty} n_i(\alpha)!, \tag{A.2.12}$$

and

$$e_\alpha := \hat{\otimes}_{i=1}^{n}\, e_{\alpha_i}. \tag{A.2.13}$$

Then

$$(e_\alpha,\ e_\beta) = \delta_{\alpha,\beta}\ n(\alpha)!. \tag{A.2.14}$$

Thus an orthonormal basis of $\Gamma(\mathcal{H})$ is given by the collection of all sequences of the form

$$(n(\alpha)!)^{-\frac{1}{2}} (0, 0, \dots, e_\alpha, 0, \dots), \tag{A.2.15}$$

with $\alpha \in I^n$, $n \in \mathbb{N}$, e_α in the n+1-st entry, together with the sequence (1, 0, 0, ...).

Let us remark that in the case $\mathcal{H}=L^2(\mathbb{R}^d, du)$ the Hilbert spaces $\Gamma^{(n)}(\mathcal{H})$ and $L^2((\mathbb{R}^d)^{\hat{}n}, d^n u)$ are isomorphic (and isometric up to a factor $\sqrt{n!}$, cf. (A.2.8)), where the latter stands for the space of L^2-functions f on $\mathbb{R}^{d \cdot n}$ with the property that for all $\pi \in \mathfrak{S}(n)$

$$f(u_1, \dots, u_n) = f(u_{\pi(1)}, \dots, u_{\pi(n)}), \quad \text{(Lebesgue–a.e.)} \tag{A.2.16}$$

$u_i \in \mathbb{R}^d$, i=1, 2, ... , n.

Next we introduce creation and annihilation operators on $\Gamma(\mathcal{H})$.

Consider $F^{(n)} \in \Gamma^{(n)}(\mathcal{H})$ of the form

$$F^{(n)} = \hat{\otimes}_{i=1}^n f_i, \tag{A.2.17}$$

where $f_i \in \mathcal{H}$, i=1,2, ... ,n. Let $f \in \mathcal{H}$. Then the *annihilation operator* $\partial(f)$ *of* f acts on $F^{(n)}$ according to

$$\partial(f)\, F^{(n)} := \sum_{j=1}^{n} (f, f_j)\, \hat{\otimes}_{\substack{i=1 \\ i \neq j}}^{n} f_i. \tag{A.2.18}$$

Note that $\partial(f)\, F^{(n)} \in \Gamma^{(n-1)}(\mathcal{H})$. It is almost obvious, that (A.2.18) does not depend on the particular representation (A.2.17) of the $F^{(n)}$. Thus (A.2.18) is well-defined. Now we extend (A.2.18) linearly to the dense subspace in $\Gamma^{(n)}(\mathcal{H})$ consisting of finite linear combinations of elements of the form (A.2.17). One easily finds the following inequality for such elements

$$\left\| \partial(f) F^{(n)} \right\|_{\Gamma^{(n-1)}(\mathcal{H})} \leq n^{1/2}\, \|f\|_{\mathcal{H}}\, \left\| F^{(n)} \right\|_{\Gamma^{(n)}(\mathcal{H})}. \tag{A.2.19}$$

(An easy way to prove (A.2.19) is to check it first on the e_α's (A.2.14) and then on their linear combinations.) The estimate (A.2.19) shows that $\partial(f)$ extends to all of $\Gamma^{(n)}(\mathcal{H})$ as a bounded operator of norm $n^{1/2}\,\|f\|_{\mathcal{H}}$. (In fact, from (A.2.19) we conclude that the norm of $\partial(f)$ is bounded by $n^{1/2}\,\|f\|_{\mathcal{H}}$, however choosing $F^{(n)} = f\otimes\cdots\otimes f$ the bound is attained.)

Now let $\Gamma'(\mathcal{H})$ denote the dense subspace of $\Gamma(\mathcal{H})$ consisting of those sequences $(F^{(n)},\, n\in\mathbb{N}_0)$ having only a finite number of non-vanishing entries. Clearly, by (A.2.19), $\partial(f)$ is defined on $\Gamma'(\mathcal{H})$ for each $f\in\mathcal{H}$, and so there exists an adjoint operator $\partial^*(f)$ of $\partial(f)$ on $\Gamma(\mathcal{H})$. It is straightforward to compute its action on elements $F^{(n)}\in\Gamma^{(n)}(\mathcal{H})$:

$$\partial^*(f)\,F^{(n)} = f\,\hat{\otimes}\,F^{(n)}. \tag{A.2.20}$$

One has the estimate

$$\left\|\partial^*(f)\,F^{(n)}\right\|_{\Gamma^{(n+1)}(\mathcal{H})} \le (n+1)^{1/2}\,\|f\|_{\mathcal{H}}\,\left\|F^{(n)}\right\|_{\Gamma^{(n)}(\mathcal{H})}. \tag{A.2.21}$$

Since both, $\partial(f)$ and $\partial^*(f)$, are densely defined on $\Gamma(\mathcal{H})$, they are closable and their closures are denoted by the same symbols. $\partial^*(f)$ is called the *creation operator of* f. The maximal complete domain for both is the subspace $\Gamma_1(\mathcal{H})$ of $\Gamma(\mathcal{H})$ consisting of those sequences $(F^{(n)},\, n\in\mathbb{N}_0)$ for which

$$\sum_{n=1}^{\infty} n\,\left\|F^{(n)}\right\|^2_{\Gamma^{(n)}(\mathcal{H})} < \infty. \tag{A.2.22}$$

It is important to note the following so-called *canonical commutation relations*:

$$[\partial^*(f)\,,\,\partial^*(g)] = [\partial(f)\,,\,\partial(g)] = 0 \tag{A.2.23}$$

$$[\partial(f), \partial^*(g)] = (f, g). \tag{A.2.24}$$

In order to describe another important property of the annihilation operators, we have to introduce the symmetric tensor product of two elements in $\Gamma'(\mathcal{H})$. First let $F^{(n)} \in \Gamma^{(n)}(\mathcal{H})$, $G^{(m)} \in \Gamma^{(m)}(\mathcal{H})$. They can be written as

$$F^{(n)} = \sum_{\alpha \in I^n} F_\alpha^{(n)} e_\alpha, \tag{A.2.25}$$

$$G^{(m)} = \sum_{\beta \in I^m} G_\beta^{(m)} e_\beta, \tag{A.2.26}$$

with certain complex numbers $F_\alpha^{(n)}$, $G_\beta^{(m)}$. The meaning of $e_\alpha \hat{\otimes} e_\beta$ is as before (cf. (A.2.7) and (A.2.13)), and we set

$$F^{(n)} \hat{\otimes} G^{(m)} := \sum_{\alpha \in I^n, \beta \in I^m} F_\alpha^{(n)} G_\beta^{(m)} e_\alpha \hat{\otimes} e_\beta. \tag{A.2.27}$$

It is an easy exercise to prove that the sum in (A.2.27) converges in $\Gamma^{(n+m)}(\mathcal{H})$. Note that

$$F^{(n)} \hat{\otimes} G^{(m)} = \sum_{\gamma \in I^{n+m}} H_\gamma^{(n+m)} e_\gamma, \tag{A.2.28}$$

with coefficients given by

$$H_\gamma^{(n+m)} = \frac{n!\, m!}{(n+m)!} \sum_{(\alpha,\beta)} F_\alpha^{(n)} G_\beta^{(m)}. \tag{A.2.29}$$

The last sum extends over all partitions (α, β) of $\gamma \in I^{n+m}$ into a subset α containing n and a subset β containing m elements.

Given now F, $G \in \Gamma'(\mathcal{H})$ we set

$$(F \hat{\otimes} G)^{(n)} := \sum_{m=0}^{n} F^{(n-m)} \hat{\otimes} G^{(m)}, \tag{A.2.30}$$

and

$$F \hat{\otimes} G := ((F \hat{\otimes} G)^{(n)}, n=0,1, \dots). \tag{A.2.31}$$

It is plain to show that $F \hat{\otimes} G = G \hat{\otimes} F$, so that $(\Gamma'(\mathcal{H}), \hat{\otimes})$ forms a (graded) commutative algebra.

Now we can compute the action of $\partial(f)$, $f \in \mathcal{H}$, on the symmetric tensor product $F \hat{\otimes} G$, F, $G \in \Gamma'(\mathcal{H})$, and the result is

$$\partial(f)\, F \hat{\otimes} G = (\partial(f)\, F) \hat{\otimes} G + F \hat{\otimes} \partial(f)\, G. \tag{A.2.32}$$

In other words: $\partial(f)$ is a derivation on $\Gamma'(\mathcal{H})$ with respect to the symmetric tensor product.

Next we are going to describe "second quantized" operators, i.e., we shall construct operators on $\mathcal{F}(\mathcal{H})$ and on $\Gamma(\mathcal{H})$ from operators on $\mathcal{H}$. We shall follow the strategy of Cook (1953), but refer also to Reed and Simon (1972). Moreover we provide some details which are useful in the text of this book.

Lemma A.2.1. *Assume that* $\mathcal{H}_i$, i=1,2, *are complex, separable Hilbert spaces with inner products* $(\cdot\,,\,\cdot)_i$, *and that* $\mathcal{D}_i$, i=1,2, *are dense linear subspaces. Then* $\mathcal{D}_1 \otimes' \mathcal{D}_2$ *is dense in* $\mathcal{H}_1 \otimes \mathcal{H}_2$.

Proof. For simplicity, we shall denote the inner product of $\mathcal{H}_1 \otimes \mathcal{H}_2$ by $(\cdot\,,\,\cdot)$, its norm by $|\cdot|$.

$\mathcal{H}_1 \otimes' \mathcal{H}_2$ is by construction dense in $\mathcal{H}_1 \otimes \mathcal{H}_2$, and hence it is sufficient to show that $\mathfrak{D}_1 \otimes' \mathfrak{D}_2$ is dense in $\mathcal{H}_1 \otimes' \mathcal{H}_2$. Since $\mathcal{H}_1 \otimes' \mathcal{H}_2$ consists of finite linear combinations of elements of the form $h_1 \otimes h_2$, $h_i \in \mathcal{H}_i$, it suffices to prove that for all $h_1 \in \mathcal{H}_1$, $h_2 \in \mathcal{H}_2$, $h_1 \otimes h_2$ can be approximated in $|\cdot|$ by elements in $\mathfrak{D}_1 \otimes \mathfrak{D}_2$. Choose sequences $\{h_{i,n}, n \in \mathbb{N}\}$ in $\mathfrak{D}_i$, i=1,2, which converge in $\mathcal{H}_i$ to h_i:

$$|h_i - h_{i,n}|_i \to 0, \text{ as } n \to +\infty,$$

where $|\cdot|_i$ is the norm of $\mathcal{H}_i$. Note that for every $n \in \mathbb{N}$, $h_{1,n} \otimes h_{2,n} \in \mathfrak{D}_1 \otimes' \mathfrak{D}_2$. An immediate calculation gives that

$$|h_1 \otimes h_2 - h_{1,n} \otimes h_{2,n}|^2 =$$

$$= |h_1|_1^2 \, |h_2|_2^2 + |h_{1,n}|_1^2 \, |h_{2,n}|_2^2 - 2 \operatorname{Re} (h_1, h_{1,n})_1 \, (h_2, h_{2,n})_2,$$

which clearly converges to zero as n tends to infinity. □

It is clear that a similar result holds also for symmetric tensor products, and by induction also for n–fold tensor products. Consider a complex separable Hilbert space $\mathcal{H}$ with dense linear subspace $\mathfrak{D}$. For $n \in \mathbb{N}$ let $\mathfrak{D}^{\otimes' n}$ denote the space of all finite linear combinations of elements $h_1 \otimes h_2 \otimes \cdots \otimes h_n$, with $h_j \in \mathfrak{D}$, j=1,2, … ,n. Similarly, $\mathfrak{D}^{\hat{\otimes}' n}$ denotes the space of all finite linear combinations of symmetric tensors $h_1 \hat{\otimes} h_2 \hat{\otimes} \cdots \hat{\otimes} h_n$. Then we have the following result:

Corollary A.2.2. *For every* $n \in \mathbb{N}$, $\mathfrak{D}^{\otimes' n}$ *is dense in* $\mathcal{H}^{\otimes n}$, *and* $\mathfrak{D}^{\hat{\otimes}' n}$ *is dense in* $\mathcal{H}^{\hat{\otimes} n}$.

It is clear that we can now pass to Fock spaces: let $\mathcal{F}'(\mathfrak{D})$ denote the space of all sequences (indexed by $\mathbb{N}_0$) having in the n+1–st entry an element from $\mathfrak{D}^{\otimes' n}$, $n \in \mathbb{N}_0$, but only a finite number of non–zero entries. Analogously $\Gamma'(\mathfrak{D})$ is defined using $\mathfrak{D}^{\hat{\otimes}' n}$ instead.

Corollary A.2.3. *$\mathcal{F}'(\mathcal{D})$ is dense in $\mathcal{F}(\mathcal{H})$ and $\Gamma'(\mathcal{D})$ is dense in $\Gamma(\mathcal{H})$.*

Consider again two complex, separable Hilbert space $\mathcal{H}_1$, $\mathcal{H}_2$. Assume that $(A_i, \mathcal{D}(A_i))$, is a densely defined, closed operator on $\mathcal{H}_i$, $i=1,2$. It follows that A_i has a unique densely defined adjoint A_i^* with domain $\mathcal{D}(A_i^*) \subset \mathcal{H}_i^*$. Note that by Corollary A.2.2, $\mathcal{D}(A_1^*) \otimes' \mathcal{D}(A_2^*)$ is dense in $\mathcal{H}_1^* \otimes \mathcal{H}_2^* = (\mathcal{H}_1 \otimes \mathcal{H}_2)^*$. Let $h_i \in \mathcal{D}(A_i^*)$, $i=1,2$, and set

$$(A_1^* \otimes' A_2^*)\ h_1 \otimes h_2 := A_1^* h_1 \otimes A_2^* h_2,$$

and extend this linearly to all of $\mathcal{D}(A_1^*) \otimes' \mathcal{D}(A_2^*)$ (for a proof that this definition leads to a well-defined operator, see, e.g., Reed and Simon (1972, Chap. VIII)). Thus $A_1^* \otimes' A_2^*$ is densely defined on $(\mathcal{H}_1 \otimes \mathcal{H})^*$, and it has a unique, closed adjoint $A_1 \otimes A_2$ with domain denoted by $\mathcal{D}(A_1 \otimes A_2)$. It is not hard to see that $\mathcal{D}(A_1) \otimes' \mathcal{D}(A_2)$ is a subspace of $\mathcal{D}(A_1 \otimes A_2)$, and therefore $A_1 \otimes A_2$ is densely defined.

By induction, we can now construct for $n \in \mathbb{N}$ a closed, densely defined operator $A^{\otimes n} = A \otimes A \otimes \cdots \otimes A$ (n factors) on $\mathcal{H}^{\otimes n}$, from a closed operator A on $\mathcal{H}$. By restriction, $A^{\otimes n}$ can also be considered as a closed, densely defined operator on $\mathcal{H}^{\hat{\otimes} n}$. Also we want to mention here that in case that we identify $\mathcal{H}$ with its adjoint $\mathcal{H}^*$, self-adjointness of A on $\mathcal{H}$ implies that $A^{\otimes n}$ is self-adjoint on $\mathcal{H}^{\otimes n}$ and on $\mathcal{H}^{\hat{\otimes} n}$ (cf. Reed and Simon (1972).

Next we extend these constructions to the Fock spaces over $\mathcal{H}$. We confine ourselves to the symmetric Fock space $\Gamma(\mathcal{H})$, since this space plays a more important role in this book. The case $\mathcal{F}(\mathcal{H})$ is done in a similar way. Consider the the dense subspace $\Gamma'(\mathcal{D}(A^*))$ (see above) of $\Gamma(\mathcal{H}^*) = \Gamma(\mathcal{H})^*$. Then we can define on $\Gamma'(\mathcal{D}(A^*))$ an operator $\Gamma'(A^*)$ by linear extension of

$$\Gamma'(A^*) \big|_{\mathcal{D}(A^*)^{\hat{\otimes}' n}} = A^* \otimes' A^* \otimes' \cdots \otimes' A^*, \text{ (n factors)}.$$

We define a closed, densely defined operator $\Gamma(A)$ with domain $\mathfrak{D}(\Gamma(A))$ on $\Gamma(\mathcal{H})$ as the unique adjoint of $(\Gamma'(A^*), \Gamma'(\mathfrak{D}(A^*)))$. Note that

$$\Gamma(A)\Big|_{\mathfrak{D}(A)\hat{\otimes}'n} = A\otimes A\otimes \cdots \otimes A, \quad \text{(n factors)}. \tag{A.2.33}$$

Again it is a fact that if we identify $\mathcal{H}$ with its dual, and if A is self-adjoint on $\mathcal{H}$, then $\Gamma(A)$ is self-adjoint on $\Gamma(\mathcal{H})$ and on $\mathfrak{F}(\mathcal{H})$, and $\Gamma'(\mathfrak{D}(A))$ and $\mathfrak{F}'(\mathfrak{D}(A))$ are cores of essential self-adjointness for $\Gamma(A)$.

In a similar manner, one can construct from a densely defined, closed operator A from a Hilbert space $\mathcal{H}$ into a Hilbert space $\mathcal{K}$, a densely defined, closed operator $\Gamma(A)$ from $\Gamma(\mathcal{H})$ into $\Gamma(\mathcal{K})$. The details are left to the interested reader.

The following result can be found in Nelson (1973c).

Proposition A.2.4. *Assume that* A *is a contraction from* $\mathcal{H}$ *into* $\mathcal{K}$. *Then* $\Gamma(A)$ *is a contraction from* $\Gamma(\mathcal{H})$ *into* $\Gamma(\mathcal{K})$.

Sometimes another type of second quantized operators on $\mathfrak{F}(\mathcal{H})$ or $\Gamma(\mathcal{H})$ is important. Again, we shall only consider its construction on $\Gamma(\mathcal{H})$. Let again A be a densely defined, closable operator on $\mathcal{H}$ with domain $\mathfrak{D}(A)$. Consider the operator $d\Gamma'(A^*)$ defined on $\Gamma'(\mathfrak{D}(A^*))$ by linear extension of

$$d\Gamma'(A^*)\Big|_{\mathfrak{D}(A^*)\hat{\otimes}'n} f_1\hat{\otimes}f_2\hat{\otimes} \cdots \hat{\otimes}f_n =$$

$$= A^*f_1\hat{\otimes}f_2\hat{\otimes} \cdots \hat{\otimes}f_n + f_1\hat{\otimes}A^*f_2\hat{\otimes}f_3\hat{\otimes} \cdots \hat{\otimes}f_n + \ldots$$

$$+ f_1\hat{\otimes} \cdots \hat{\otimes}f_{n-1}\hat{\otimes}A^*f_n,$$

where $f_j \in \mathfrak{D}(A^*)$, $j=1,2, \ldots ,n$. Then $d\Gamma'(A^*)$ is densely defined on $\Gamma(\mathcal{H})^*$, and its unique closed adjoint is denoted by $d\Gamma(A)$ with domain $\mathcal{D}(d\Gamma(A))$. It is easy to see that $\Gamma'(\mathcal{D}(A)) \subset \mathcal{D}(d\Gamma(A))$, and hence $d\Gamma(A)$ is densely defined. If $\mathcal{H} \equiv \mathcal{H}^*$, and A is self–adjoint then so is $d\Gamma(A)$, and $\Gamma'(\mathcal{D}(A))$ is a core (e.g., Cook (1953), Reed and Simon (1972, Chap. VIII)). Note that

$$d\Gamma(A)\big|_{\mathcal{D}(A)^{\hat{\otimes}' n}} = A\otimes id\otimes \cdots \otimes id + id\otimes A\otimes id\otimes \cdots \otimes id + \ldots \\ \ldots + id\otimes \cdots \otimes id\otimes A, \tag{A.2.34}$$

where each term in the last sum has n factors.

The following useful facts are easy to derive.

The number operator N on $\Gamma(\mathcal{H})$, defined by (the linear extension of)

$$N\big|_{\Gamma^{(n)}(\mathcal{H})} = \text{multiplication by } n \tag{A.2.35}$$

is equal to $d\Gamma(id)$. If B is the generator of a strongly continuous contraction semigroup $(\exp(-tB),\ t\in\mathbb{R}_+)$ on $\mathcal{H}$, then $\Gamma(\exp(-tB))$, $t\in\mathbb{R}_+$, defines a strongly continuous contraction semigroup on $\Gamma(\mathcal{H})$ with generator $d\Gamma(B)$:

$$\Gamma(\exp(-tB)) = \exp(-t\,d\Gamma(B)). \tag{A.2.36}$$

(For strongly continuous unitary groups on $\mathcal{H}$ an analogous statement holds.)

Moreover, if A is selfadjoint with purely discrete spectrum $(\lambda_k,\ k\in\mathbb{N})$ and a complete orthonormal set of eigenvectors $(e_k,\ k\in\mathbb{N})$, $Ae_k=\lambda_k e_k$, in $\mathcal{H}$, then

$$d\Gamma(A) = \sum_{k=1}^{\infty} \lambda_k\ \partial^*(e_k)\ \partial(e_k). \tag{A.2.37}$$

Appendix A.3 Reproducing Kernel Hilbert Spaces

In this appendix we provide some basic notions about reproducing kernel Hilbert spaces, necessary for some arguments in Chapter2. For more details, we refer to the article by Hida and Ikeda (1967), and the references quoted there. The contents of this appendix is based on Hida (1975).

Let T be a set, and K a skew symmetric, positive definite, complex–valued function on T×T, i.e., for all u, v∈T, $K(u,v) = \overline{K(v,u)}$, and for all n∈ℕ, $a_1, a_2, \ldots, a_n \in \mathbb{C}$, $u_1, u_2, \ldots, u_n \in T$, we have

$$\sum_{k,l=1}^{n} \overline{a_k}\, a_l\, K(u_k, u_l) \geq 0. \tag{A.3.1}$$

Define a linear space $\mathcal{F}'$ of complex valued functions on T as follows. $F \in \mathcal{F}'$ if and only if there exist n∈ℕ, $a_1, a_2, \ldots, a_n \in \mathbb{C}$, $u_1, u_2, \ldots, u_n \in T$, so that F is of the form

$$F(v) = \sum_{k=1}^{n} a_k\, K(v, u_k), \quad v \in T. \tag{A.3.2}$$

Define a Hermitean sesquilinear form $(\cdot\,,\,\cdot)$ on $\mathcal{F}'$ by

$$(F, G) := \sum_{k=1}^{n} \sum_{l=1}^{m} \overline{a_k}\, b_l\, K(u_k, v_l),$$

where we assume that F is as in (A.3.2), and G is of the form

$$G = \sum_{l=1}^{m} b_l\, K(\cdot, v_l).$$

Note that $(\cdot\,,\,\cdot)$ is non–negative:

$$(F, F) = \sum_{k,l=1}^{n} \overline{a_k}\, a_l\, K(u_k, u_l) \geq 0.$$

Consider the linear subspace $\mathcal{F}^0$ of functions in $\mathcal{F}'$, consisting of those functions for which $(F, F) = 0$. Form $\mathcal{F}'' = \mathcal{F}'/\mathcal{F}^0$. Then $(\mathcal{F}'', (\cdot\ ,\ \cdot))$ is a pre–Hilbert space, and we denote its completion by $\mathcal{F}$. The Hilbert space $\mathcal{F}$ constructed above is called the *reproducing kernel Hilbert space* (RKHS) *of* K.

Consider $F \in \mathcal{F}$. F is of the form

$$F = \sum_{k=1}^{\infty} a_k\ K(\cdot\ ,\ u_k),$$

where $\{a_k,\ k \in \mathbb{N}\}$ is a sequence of complex numbers and $\{u_k,\ k \in \mathbb{N}\}$ is a sequence in T so that

$$\sum_{k,l=1}^{\infty} \overline{a_k}\ a_l\ K(u_k,\ u_l) < +\infty.$$

Let us compute $(K(\cdot\ ,\ u),\ F)$ for $u \in T$, and F as above. We get

$$\begin{aligned}(K(\cdot\ ,\ u),\ F) &= \sum_{k=1}^{\infty} a_k\ K(u,\ u_k) \\ &= F(u).\end{aligned}$$

Reproducing kernel Hilbert spaces arise naturally in probability theory as the characteristic functionals of probability measures. Let us work out the example of a general Gaussian measure on the dual $\mathcal{N}^*$ of a separable, real, nuclear pre–Hilbert space $\mathcal{N}$, with compatible inner product $(\cdot\ ,\ \cdot)$. Let $\mathcal{H}$ denote the Hilbert space which is the completion of $\mathcal{N}$ with respect to the norm $|\cdot|$ defined by $(\cdot\ ,\ \cdot)$. For $\xi \in \mathcal{N}$ define

$$C(\xi) := \exp(-\tfrac{1}{2}|\ \xi\ |^2).$$

Then it follows from Minlos' theorem (e.g., Hida (1980)) that C is the characteristic functional of a (unique) Gaussian measure μ on $(\mathcal{N}^*, \mathcal{B})$, where $\mathcal{B}$ is the Borel σ-algebra of $\mathcal{N}^*$ as equipped with the weak (or strong) topology. In particular, because C is the Fourier transform of a positive measure, C is positive definite:

$$\sum_{k,l=1}^{n} \overline{a_k}\, a_l\, C(\xi_1-\xi_n) \geq 0,$$

for all choices of $n\in\mathbb{N}$, $a_1, a_2, \dots, a_n\in\mathbb{C}$, $\xi_1, \xi_2, \dots, \xi_n\in\mathcal{N}$. Therefore we may choose the set T above as $\mathcal{N}$, and K as given by $C(\xi-\eta)$, for $(\xi,\eta)\in\mathcal{N}\times\mathcal{N}$. Let us denote the corresponding RKHS by $\mathcal{R}(\mathcal{N},C)$.

Let $\varphi\in L^2(\mathcal{N}^*, \mathcal{B}, \mu) \equiv (L^2)$ and define for $\xi\in\mathcal{N}$

$$\mathcal{T}\varphi(\xi) := \int_{\mathcal{N}^*} \exp(i\langle x, \xi\rangle)\, \varphi(x)\, d\mu(x),$$

where $\langle\cdot\,,\cdot\rangle$ denotes the dual pairing between $\mathcal{N}^*$ and $\mathcal{N}$. Now let $\varphi\in(L^2)$ be of the following form

$$\varphi(x) = \sum_{k=1}^{n} a_k \exp(-i\langle x, \xi\rangle), \quad x\in\mathcal{N}^*$$

where $a_1, a_2, \dots, a_n\in\mathbb{C}$, $\xi_1, \xi_2, \dots, \xi_n\in\mathcal{N}$. Observe that the collection of all such φ is dense in (L^2) (cf. Theorem 1.9). It is straightforward to obtain the following result:

$$\mathcal{T}\varphi(\xi) = \sum_{k=1}^{n} a_k\, C(\xi-\xi_k),$$

that is $\mathcal{T}\varphi$ belongs to $\mathcal{R}(\mathcal{N},C)$. Moreover, if $\psi\in(L^2)$ is of the form $\psi(x) = \sum_{l=1}^{m} b_l \exp(-i\langle x, \eta_l\rangle)$, then another simple computation shows that

$$(\psi, \varphi)_{(L^2)} = (\mathcal{T}\psi, \mathcal{T}\varphi)_{\mathcal{R}(\mathcal{N},C)}.$$

Therefore we see that $\mathcal{T}$ is an isometry from a dense set of (L^2) onto a dense set of $\mathcal{R}(\mathcal{N},C)$. Thus $\mathcal{T}$ extends to a unitary mapping from (L^2) onto $\mathcal{R}(\mathcal{N},C)$.

Appendix A.4. Fréchet and Gâteaux Derivatives on $\mathcal{S}(\mathbb{R})$

Differentiation and differential operators play an important role in this book. For the convenience of the reader we present in this appendix some (more or less well-known) facts about Fréchet and Gâteaux derivatives. After the definitions (which we borrow from the book of Schwartz (1969)), we specialize to the case where we consider derivatives of mappings on a countably Banach topological vector space. Finally, we discuss the Fréchet functional derivative for functionals on $\mathcal{S}(\mathbb{R})$.

Assume that X is a topological vector vector space over $\mathbb{K} = \mathbb{R}$ or $\mathbb{C}$. Let f be a mapping from an open subset $U \subset X$ into a Banach space B over $\mathbb{K}$, with norm $\| \ \|$[1].

Definition A.4.1. *We say that* f *is* horizontal at zero, *if and only if for every* $t_0>0$, *there exists a neighborhood* U *of zero in* X *and a positive function* o, *defined on* $[0,t_0]$, *with* $t^{-1}\cdot o(t) \to 0$ as $t \to 0$, *so that for every* $x \in U$ *and for all* $t \in [0,t_0]$

$$\| f(tx) \| \leq o(t). \tag{A.4.1}$$

Definition A.4.2. f *is said to be* Fréchet differentiable at $x \in X$, *if and only if there exists a continuous linear mapping*

$$\begin{aligned} <df(x), \cdot> : X &\to B \\ y &\mapsto <df(x), y>, \end{aligned}$$

and a function o_x *which is horizontal at zero, so that for all* $y \in X$,

---

[1]Without much more effort one can discuss general topological vector spaces as the range of f, cf. the book of Schwartz (1969). However, since we shall not need this generality we avoid this here.

$$f(x+y) - f(x) = \langle df(x), y \rangle + o_x(y). \qquad (A.4.2)$$

We call df(x) *the* Fréchet derivative of f at x.

Proposition A.4.3. (*Chain Rule.*) *Assume that* f: U → V, g: V → B′, V *open in* B, *and* B′ *another Banach space. Moreover suppose that* f *is Fréchet differentiable at* x∈U, *and* g *is Fréchet differentiable at* f(x). *Then* g∘f *is Fréchet differentiable at* x, *and for all* y∈B′, $\langle dg\circ f(x), y\rangle = \langle dg(f(x)), \langle df(x), y\rangle\rangle$.

For a proof of the last proposition see e.g. the book of Schwartz (1969).

Definition A.4.4. *Let* y∈X. f *is called* Gâteaux differentiable at x∈X in direction y, *if and only if the function* s ↦ f(x+sy), s∈ℝ, *is differentiable at* s=0, *and we call*

$$D_y f(x) := \frac{d}{ds} f(x+sy) \Big|_{s=0} \qquad (A.4.3)$$

the Gâteaux derivative of F at x in direction y.

It is clear, that the chain rule holds also for Gâteaux derivatives.

From now we assume that X is a complete countably Banach space (cf. e.g. Gel'fand and Shilov (1968), vol. II). I.e., X is topologized by the projective limit of a consistent sequence $\{\|\cdot\|_p, p\in\mathbb{N}\}$ of norms $\|\cdot\|_p$ on X. Without loss of generality we may assume that the system of norms is ordered: $\|\cdot\|_p \leq \|\cdot\|_q$, if $p \leq q$. We have by definition that $X = \bigcap_p X_p$, where X_p is the completion of X with respect to the norm $\|\cdot\|_p$. Note that $X^* = \bigcup_p X_{-p}$, $X_{-p} = (X_p)^*$. The Schwartz space $\mathcal{S}(\mathbb{R}^d)$ of test functions on $\mathbb{R}^d$, $d\in\mathbb{N}$, is a complete countably Banach space (even countably Hilbert, cf. Appendix A.8).

In the case of a countably Banach space X, we have the following criterium for a function to be horizontal at zero which is slightly more convenient to apply than Definition A.4.1.

Proposition A.4.5. *Let* f *be a mapping from a countably Banach space* X *into a Banach space* B. f *is horizontal at zero in* X *if and only if there exist* $p \in \mathbb{N}$, $\delta > 0$, *such that for all* $x \in X$ *with* $\| x \|_p < \delta$

$$\| f(x) \| \leq o(\| x \|_p) \tag{A.4.4}$$

where $o(\| x \|_p)$ is such that $\| x \|_p^{-1} o(\| x \|_p) \to 0$ *as* $\| x \|_p \to 0$.

Proof. Assume that f is horizontal at zero in X. Choose $t_0 = 1$. Then there exist a function o', $p \in \mathbb{N}$ and $\eta > 0$, such that for all $y \in X$ with $\| y \|_p < \eta$ and all $t \in [0,1]$

$$\| f(ty) \| \leq o'(t)$$

Set $U_{p,\delta} := \{x = ty;\ y \in S_{p,\delta},\ t \in [0,1)\}$, with $\delta = \eta/2$ and $S_{p,\delta} := \{y \in X;\ \| y \|_p = \delta\}$. Then $U_{p,\delta}$ is an open neighborhood of zero in X, described by $\| x \|_p < \delta$, $x \in X$. Moreover, we have

$$\| f(x) \| \leq o'(2\,\eta^{-1} \| x \|_p) \equiv o(\| x \|_p)$$

for all $x \in U_{p,\delta}$. The converse implication is trivial. □

Proposition A.4.6. *Assume that* f: $U \to B$, $U \subset X$, U *open*, X *countably Banach, and* f *is Fréchet differentiable at* $x \in U$. *Then* f *is Gâteaux differentiable at* x, *and for all* $y \in X$, $D_y f(x) = \langle df(x), y \rangle$.

Proof. This is a direct consequence of the definitions and Proposition A.4.4. □

The converse statement can be proved under some additional assumptions. For convenience, we only formulate and prove it for the case $B = \mathbb{K}$.

Proposition A.4.7. *Assume that* $f: U \to \mathbb{K}$, $U \subset X$, U *open*, X *countably Banach, and that* f *is Gâteaux differentiable in* U. *Suppose that the mapping* $y \mapsto D_y f(x)$, $x \in U$, $y \in X$, *is continuous, so that* $D_y f(x) = \langle \nabla f(x), y \rangle$, *for some* $\nabla f(x) \in X^*$. *If the mapping* $x \mapsto \nabla f(x)$ *from* U *into* X^* *is continuous with respect to the inductive limit topology* τ_{ind} *on* X^*, *then* f *is Fréchet differentiable at* x, $x \in U$, *and* $D_y f(x) = \langle df(x), y \rangle$, *for all* $x \in U$, $y \in X$.

Proof. Since

$$\frac{d}{dt} f(x+ty) = D_y f(x+ty) = \langle \nabla f(x+ty), y \rangle,$$

we have

$$f(x+y) - f(x) = \int_0^1 \langle \nabla f(x+ty), y \rangle \, dt$$

$$= \langle \nabla f(x), y \rangle + \int_0^1 \langle \nabla f(x+ty) - \nabla f(x), y \rangle \, dt.$$

It remains to show that the last integral is horizontal at zero in $y \in X$. Due to Proposition A.4.5 it is sufficient to show that

$$\left| \int_0^1 \langle \nabla f(x+ty) - \nabla f(x), y \rangle \, dt \right| \leq o(\| y \|_p)$$

for some $p \in \mathbb{N}$. Since we assume that $x \mapsto \nabla f(x)$ is continuous from U into (X^*, τ_{ind}), there exists $r \in \mathbb{N}$, $\delta > 0$, so that $\| y \|_r < \delta$ implies that $t \mapsto \nabla f(x+ty)$ is continuous from $[0,1]$ into (X^*, τ_{ind}). Let $q \in \mathbb{N}$ be such that $\nabla f(x) \in X_{-q}$. Then it follows that $\nabla f(x+ty) \in X_{-q}$ for all y with $\| y \|_r < \delta$ and all $t \in [0,1]$. Let $p = q \vee r$. Then we have that for all y with $\| y \|_p < \delta$, $\nabla f(x+ty) \in X_{-p}$ for all $t \in [0,1]$. Now we can estimate as follows

$$| \int < \nabla f(x+ty) - \nabla f(x), y > dt \, | \leq \| y \|_p \int_0^1 \| \nabla f(x+ty) - \nabla f(x) \|_{-p} \, dt.$$

Therefore it suffices to prove that the last integral vanishes as y tends to zero in X_p (cf. Proposition A.4.5). That this is true can for example be seen as follows. Set $y_0 = \| y \|_p^{-1} y$. Then the last integral is equal to

$$\| y \|_p^{-1} \int_0^{\| y \|_p} \| \nabla f(x+ty_0) - \nabla f(x) \|_{-p} \, dt$$

Now the continuity of $t \mapsto \nabla f(x+ty_0)$ and the mean value theorem show that the last expression vanishes as $\| y \|_p$ tends to zero. □

Finally, let us consider the choice $X = \mathcal{S}(\mathbb{R}^d)$ (cf. also Appendix A.8). For convenience, we consider a function f on $\mathcal{S}(\mathbb{R}^d)$ with values in $\mathbb{R}$. Assume that f is (everywhere) Fréchet differentiable. By definition, this means that for every $\xi \in \mathcal{S}(\mathbb{R}^d)$, the Fréchet derivative $df(\xi)$ at ξ is an element in $\mathcal{S}'(\mathbb{R})$. If $df(\xi)$ is a regular distribution, i.e. if $df(\xi)$ can be represented by an element in $L^p(\mathbb{R}^d)$, $1 \leq p \leq +\infty$, we write

$$< df(\xi), \eta > = \int_{\mathbb{R}^d} \frac{\delta}{\delta \xi(t)} f(\xi) \, \eta(t) \, dt, \quad \eta \in \mathcal{S}(\mathbb{R}^d). \tag{A.4.5}$$

The element $\frac{\delta}{\delta \xi(t)} f(\xi)$ in $L^p(\mathbb{R}^d)$ is called the *Fréchet functional derivative of* f *at* $\xi \in \mathcal{S}(\mathbb{R}^d)$.

Appendix 5. Topologies on $\mathcal{S}(\mathbb{R}^d)$ and $\mathcal{S}'(\mathbb{R}^d)$

The topological vector spaces $\mathcal{S}(\mathbb{R}^d)$, $\mathcal{S}'(\mathbb{R}^d)$, $(\mathcal{S})$ and $(\mathcal{S})^*$ play a central role in this book. $\mathcal{S}(\mathbb{R}^d)$ is the Schwartz space of rapidly decreasing C^∞-functions, $\mathcal{S}'(\mathbb{R}^d)$ its dual, the Schwartz space of tempered distributions. $(\mathcal{S})$ and $(\mathcal{S})^*$ have been introduced in Chapter 3, and have been investigated systematically in Chapter 4. Considered as topological vector spaces, $\mathcal{S}(\mathbb{R}^d)$ and $(\mathcal{S})$ have the same structure (and consequently so have their respective dual spaces): both are nuclear Fréchet spaces, whose topology is the projective limit topology induced by a decreasing sequence of (compatible) Hilbert spaces, i.e., both are nuclear countably Hilbert spaces in the sense of Gel'fand and Vilenkin (1968). For such spaces and their duals a very rich amount results is known. In this appendix we shall collect the most important properties, particularly those which are used repeatedly in this book. We begin with the description of locally convex inductive and projective limit topologies (in the very general form given in Robertson and Robertson (1973)), and we shall specialize step by step until we reach a discussion of $\mathcal{S}(\mathbb{R}^d)$ and $\mathcal{S}'(\mathbb{R}^d)$. Moreover, we include a sketch of polar and dual topologies. For general background, details and proofs, we refer the interested reader for example to the following books: Bourbaki (1987), Gel'fand and Shilov (1968), Gel'fand and Vilenkin (1968), Köthe (1971), Robertson and Robertson (1973), and Trèves (1967).

Suppose that $\{F_\alpha, \alpha \in I\}$, I some index set, is a family of locally convex topological vector spaces[1]. Suppose furthermore that we are given another vector space F, and a family $\{g_\alpha, \alpha \in I\}$ of linear mappings, so that g_α maps F_α into F, and that the union of the spaces $g_\alpha(F_\alpha)$ spans F. There exists a finest locally convex topology τ_{ind} on F, called the *inductive limit topology*, so that for every

[1] Throughout this book we consider only vector spaces over the fields $\mathbb{R}$ of real numbers, or over the field $\mathbb{C}$ of complex numbers. All the statements in this appendix are independent of the choice of one of these two fields. It goes without saying that within context of one statement, the same field has to be used everywhere.

$\alpha \in I$ the mapping g_α is continuous. We also simply say that F is the inductive limit of the spaces F_α. A base of neighborhoods of zero for τ_{ind} is given by the family of all absolutely convex[2] subsets U of F, such that for each $\alpha \in I$, $g_\alpha^{-1}(U)$ is a neighborhood of zero in F_α. In particular, a base of neighborhoods of zero for τ_{ind} is given by the family of absolutely convex envelopes[3] of sets of the form $\bigcup_\alpha V_\alpha$, where the V_α range over a base of neighborhoods of zero in F_α. A linear map t from F into a locally convex vector space G is continuous, if and only if for every $\alpha \in I$, the mapping $t \circ g_\alpha$ is continuous from F_α into G.

Consider a family $\{E_\alpha, \alpha \in I\}$ of locally convex vector spaces. Assume that E is a vector space, and that we are given a family $\{h_\alpha, a \in I\}$ of linear mappings h_α from E into E_α, such that $\bigcap_\alpha h_\alpha^{-1}(0) = 0$. Then there exists a coarsest locally convex topology τ_{pro} on E, called *projective limit topology*, so that for every $\alpha \in I$, the mapping h_α is continuous. We also call E the projective limit of the spaces E_α. A base of neighborhoods of zero is given by the family of sets which are finite intersections of sets $h_\alpha^{-1}(V_\alpha)$, $\alpha \in I$, where for every $\alpha \in I$ the sets V_α range over a base of absolutely convex neighborhoods of zero in E_α. If the spaces E_α are Hausdorff then (E, τ_{pro}) is Hausdorff.

In this book we only need the following special cases: the vector spaces $\{F_\alpha, \alpha \in I\}$ form an increasing sequence $F_0 \subset F_1 \subset \ldots \subset F_p \subset \ldots$ of subspaces of F, F is their union, and the corresponding mappings g_p ($p \in \mathbb{N}_0 = I$) are given by the canonical injections of the spaces F_p into F. Moreover, we assume from now on that for every $p \in \mathbb{N}_0$ the topology induced by F_{p+1} on F_p is coarser than the topology on F_p. In other words, we assume that the injection of F_p into F_{p+1} is continuous. The vector spaces $\{E_\alpha, \alpha \in I\}$ form a decreasing sequence $E_0 \supset E_1 \supset \ldots \supset E_p \supset \ldots$ of vector spaces, all containing E, E being their intersection, and the

---

2 A set A is called *convex* if with any two points in A their line segment belongs to A. A set A is called *balanced* if $x \in A$ and $|\lambda| \leq 1$ implies $\lambda x \in A$. A set which is convex and balanced is called *absolutely convex*.

3 The *(absolutely) convex envelope* of a set A is the smallest (absolutely) convex set containing A.

mappings h_p ($p \in \mathbb{N}_0 = I$) are the canonical injections of E into the E_p. In addition we suppose from now on that for every $p \in \mathbb{N}_0$, the topology induced by E_p on E_{p+1} is coarser than the topology on E_{p+1}. I.e., the injections of E_{p+1} into E_p, $p \in \mathbb{N}_0$, are continuous. In this case, it is easy to see that a linear mapping t from E into a locally convex vector space G is continuous, if and only if it is continuous on some E_p, $p \in \mathbb{N}_0$. Moreover, a *sequence* converges in F, if and only if it converges in some F_p, $p \in \mathbb{N}_0$.

In our applications in this book, spaces of "smooth" functions are typically given by a projective limit space like E above, and its dual E′ is a space of "generalized" functions. For example, the four spaces mentioned at the beginning arise this way. It is important to consider various topologies that can be defined on E′. Let us quickly sketch a standard way of constructing a whole family of topologies on the dual of a locally convex topological vector space. To this end we need to recall the notion of bounded sets.

Consider a locally convex topological vector space E. A set $A \subset E$ is said to *absorb* the set $B \subset E$, if and only if there exists $\lambda_0 > 0$, so that for all $|\lambda| \geq \lambda_0$ we have $B \subset \lambda A$. A set is called *bounded* if it is absorbed by every neighborhood of zero. If the topology of E is given by a set of seminorms, then a set is bounded if and only if it is bounded in every seminorm of that set. The family of topologies on the dual E′ mentioned above arises by various choices of systems of bounded sets in E. Consider a family $\mathcal{A}$ of bounded subsets of E with the properties: i) if A, $B \in \mathcal{A}$ then there is $C \in \mathcal{A}$, so that $A \cup B \subset C$; ii) if $A \in \mathcal{A}$ and λ is a scalar, then $\lambda A \in \mathcal{A}$. Let $A \in \mathcal{A}$, and define the *polar* $A^0 \subset E'$ of A by setting

$$A^0 := \{x \in E';\ \sup_{\xi \in A} |<x, \xi>| \leq 1\},$$

where $<\cdot, \cdot>$ denotes the canonical dual pairing between E′ and E. By $\mathcal{A}^0$ denote the family of subsets of E′ consisting of all polars of sets in $\mathcal{A}$. Then $\mathcal{A}^0$ has the properties of a base of neighborhoods of zero in E′, and the topology on E′ generated by this base is called the *topology of uniform convergence on* $\mathcal{A}$. Indeed, it turns out that in the respective topologies convergence is uniform on the sets in the family $\mathcal{A}$. Let us consider the two most important cases for this book:

1. *Weak Topology.* Choose $\mathcal{A}$ as the family of finite subsets $\{\xi_1, \ldots, \xi_n\}$, $n \in \mathbb{N}$, in E. This topology on E′ is also called the *topology of pointwise convergence*, and is commonly denoted by $\sigma(E', E)$;

2. *Strong Topology.* Choose $\mathcal{A}$ as the family of all bounded subsets of E. This topology on E′ is also called the *topology of bounded convergence*, and it is usually denoted by $\beta(E', E)$.

Obviously, the weak topology is the coarsest and the strong topology the finest of all such *polar topologies.*

Conversely, one can introduce other locally convex topologies than the initial one on a locally convex vector space E which is Hausdorff. An important class consists of the *dual topologies*, which are such that the dual E′ of E (as equipped with its initial topology) is also the dual of E under a topology in this class. The weakest such topology on E is the *weak topology*, denoted by $\sigma(E, E')$, which is the coarsest topology on E which makes all elements in E′ (under the canonical pairing) continuous. More systematically (and more in the above spirit), we consider E_σ', the dual of E equipped with the weak topology $\sigma(E', E)$ from above, and families $\mathcal{A}'$ of bounded sets in E_σ' with the properties mentioned before when we constructed the polar topologies. As before, the polars in E of such a family of bounded subsets in E_σ' generate a topology on E. If one requires in addition that $\mathcal{A}'$ is a covering of E′ consisting of absolutely convex weakly compact sets, then (and only then) this new topology on E is such that the dual of E (in this topology) is equal to E′. The finest such topology on E, i.e., the case when $\mathcal{A}'$ consists of all absolutely convex weakly compact sets in E′ is called the *Mackey topology*, and it is denoted by $\tau(E, E')$. (Note that necessarily the initial topology of E is finer than the weak topology $\sigma(E, E')$ and coarser than the Mackey topology $\tau(E, E')$.)

E' equipped with the weak topology $\sigma(E', E)$ has E as its dual: this allows to introduce dual topologies also on the space E', and we shall denote the Mackey topology on E' by $\tau(E', E)$ accordingly. Essentially as a consequence of the Banach–Steinhaus theorem, it turns out that the boundedness of a set in E (or in E') is the same in all dual topologies. This implies the following well-known fact which will be of use later: let E be a Hausdorff locally convex topological vector space with dual E', then the Mackey topology $\tau(E', E)$ on E' is coarser than the strong topology $\beta(E', E)$, since the weakly compact sets in E are weakly bounded and hence bounded. Below we shall also use the fact that the inductive limit F of a family of metrizable locally convex vector spaces has the Mackey topology, i.e., we have $\tau_{\text{ind}} = \tau(F, F')$, where F' is the dual of F.

Let us now return to projective and inductive limit spaces. We make our next step of specialization by assuming from now on that the spaces E_p, $p \in \mathbb{N}_0$, of which E is the projective limit, are separable Hilbert spaces with scalar products $(\cdot\,,\,\cdot)_p$, and induced norms $|\cdot|_p$. The (topological) inclusion relations $H_p \supset H_{p+1}$, $p \in \mathbb{N}_0$, are implemented by requiring that $|\cdot|_p \leq |\cdot|_{p+1}$ for all $p \in \mathbb{N}_0$. Furthermore, we ask that the norms $|\cdot|_p$ be *compatible*, in the sense that if $\{u_n, n \in \mathbb{N}\}$ is a sequence in E, which converges to zero with respect to $|\cdot|_p$, $p \in \mathbb{N}_0$, and if it is Cauchy with respect to $|\cdot|_q$, $q \in \mathbb{N}_0$, then it also converges to zero with respect to q. (Compatibility has also been discussed at the beginning of Chapter 1, and in Chapter 3.) Therefore E becomes a space which is topologized by a countable family of norms, and so it is metrizable. Moreover, the condition of compatibility ensures (indeed, is equivalent with the fact) that E is complete in the projective limit topology. Hence E is a Fréchet space. A vector space E with the just described topological structure is called a *countably Hilbert space.* As mentioned above, the bounded sets of E are precisely the sets which are bounded in every norm $|\cdot|_p$, $p \in \mathbb{N}_0$.

Let us denote the dual of E_p, $p \in \mathbb{N}_0$, by E_{-p}, and let us identify $E_0 \equiv E$ with its dual. Hence we obtain for every $p \in \mathbb{N}$ the continuous inclusions

$$E_{-p} \supset E \supset E_p.$$

It can be shown that the dual E' of E is the union $\bigcup_p E_{-p}$ of the Hilbert spaces E_{-p}. Note that we have for every $p \in \mathbb{N}_0$, $E_{-(p+1)} \supset E_{-p}$. Thus E' carries in natural way the inductive limit topology. Since the spaces E_{-p}, $p \in \mathbb{N}$, are Hilbert spaces, and therefore metrizable, it follows from the remark made above that the inductive limit topology on E' is the same as the Mackey topology. Moreover, the strong topology $\beta(E', E)$ on E' is finer than the Mackey topology, i.e., finer than the inductive limit topology. Thus we have the following inclusions $\sigma(E', E) \subset \tau_{ind} = \tau(E', E) \subset \beta(E', E)$. We remark that as the dual of a Fréchet space, E' is strongly complete.

One of the important properties of countably Hilbert spaces is the fact that they are *reflexive*, i.e., that the dual of $(E', \beta(E', E))$ is equal to E. Thus the strong topology is a dual topology for the dual pair (E', E), and hence must be coarser than the Mackey topology. Consequently, the Mackey topology, the inductive limit topology, and the strong topology coincide all in this case:

$$\tau_{ind} = \tau(E', E) = \beta(E', E).$$

(Another way to see this is the following observation: in the case of a countably Hilbert space E, it is straightforward to check that for every $p \in \mathbb{N}_0$, the injection of E_{-p} into $(E', \beta(E', E))$ is continuous, because the bounded sets of E are bounded in all norms $|\cdot|_p$, $p \in \mathbb{N}_0$. Since τ_{ind} is the finest topology making all these injections continuous, it follows that $\tau_{ind} \equiv \tau(E', E) \supset \beta(E', E)$.)

Finally we specialize to *nuclear* countably Hilbert spaces. These are countably Hilbert spaces E as described above with the additional property that for every $p \in \mathbb{N}_0$ there exists $q \in \mathbb{N}$, $q \geq p$, so that the injection ι_p^q of E_q into E_p is a trace class operator. I.e., there exist orthonormal bases $\{e_n, n \in \mathbb{N}\}$ and $\{f_m, m \in \mathbb{N}\}$ for E_p and E_q, respectively, so that ι_p^q can be expressed as $\sum_n \lambda_n (f_n, \cdot)_q \, e_n$, with

$\lambda_n>0$, $n\in\mathbb{N}$, and $\sum_n \lambda_n<+\infty$. There are two important properties of nuclear (countably Hilbert) spaces which make them particularly attractive. The first is that they are *Montel* spaces: they are barreled (cf., e.g., the above quoted literature) and every closed, bounded set is compact. (And this implies that in the strong dual E′ every closed, bounded set is compact.) This entails the equivalence of weak and strong convergence of *sequences* in the dual E′ of E. The second property (actually a whole set of properties which run under the name of the "nuclear theorem" or "kernels theorem") is that nuclear spaces behave "nice" under tensor products. For example, the projective and the equicontinuous topological tensor products coincide, and the completions of the spaces of tensor products of nuclear spaces in these topologies are again nuclear. We refer the interested reader to the works mentioned at the beginning of this appendix for more details. Examples of nuclear countably Hilbert spaces are the Schwartz spaces $\mathcal{S}(\mathbb{R}^d)$ of rapidly decreasing C^∞–functions on $\mathbb{R}^d$, $d\in\mathbb{N}$, and the space $(\mathcal{S})$ of smooth white noise random variables which is studied in Chapters 3 and 4. As a consequence all the above statements about the locally convex topological vector space E and its dual E′, respectively, hold true for $\mathcal{S}(\mathbb{R}^d)$, $(\mathcal{S})$, and $\mathcal{S}'(\mathbb{R}^d)$, $(\mathcal{S})^*$, respectively.

Now we shall discuss the Schwartz spaces $\mathcal{S}(\mathbb{R}^d)$ and $\mathcal{S}'(\mathbb{R}^d)$ in somewhat more detail. In the following let $\alpha=(\alpha_1,\dots,\alpha_d)$, $\beta=(\beta_1,\dots,\beta_d)$, ... denote multiindices in $\mathbb{N}_0^d$, and set

$$D^\alpha = \partial_1^{\alpha_1}\cdots\partial_d^{\alpha_d}, \tag{A.5.1}$$

where ∂_j, $j=1,2,\dots,d$, is the partial derivative on $\mathbb{R}^d$ with respect to the j-th coordinate. Similarly, if $u\in\mathbb{R}^d$, $u=(u_1,\dots u_d)$, then let

$$u^\alpha = u_1^{\alpha_1}\cdots u_d^{\alpha_d}. \tag{A.5.2}$$

For any two multiindices α, β define the seminorm

$$\| \varphi \|_{\alpha,\beta} := \sup_{u \in \mathbb{R}^d} | u^\alpha D^\beta \varphi(u) |. \tag{A.5.3}$$

The *Schwartz space* $\mathcal{S}(\mathbb{R}^d)$ *of smooth functions of rapid decrease* is the set of all (real– or complex–valued) functions in $C^\infty(\mathbb{R}^d)$, having finite seminorm (A.5.3) for every α, $\beta \in \mathbb{N}_0^d$, and it is toplogized by the system of seminorms (A.5.3). Thus $\mathcal{S}(\mathbb{R}^d)$ is a Fréchet space, and the chain rule and the triangle inequality imply that it is an algebra under pointwise mulitplication (the multiplication being obviously continuous).

The *Schwartz space* $\mathcal{S}'(\mathbb{R}^d)$ *of tempered distributions* is the dual of $\mathcal{S}(\mathbb{R}^d)$.

The above defined topology on $\mathcal{S}(\mathbb{R}^d)$ is known to be equivalent to the one described in the following (cf., e.g., Simon (1971), Reed and Simon (1972) or Hida (1980)). Let H be the so-called Hamiltonian of the harmonic oscillator. It is the (unique) self-adjoint operator on $L^2(\mathbb{R}^d)$, which on $\mathcal{S}(\mathbb{R}^d)$ is given by

$$H\varphi(u) = (-\Delta + (| u |^2 + 1))\, \varphi(u), \quad \varphi \in \mathcal{S}(\mathbb{R}^d), \tag{A.5.4}$$

where Δ is the Laplacean on $\mathbb{R}^d$. (We have included the slightly unconventional 1 in the definition of H in order to be consistsent with the definition in Chapter 3, which is justified there.) Define a system of norms $\{| \cdot |_{2,p}, \; p \in \mathbb{N}_0\}$ by setting $| \varphi |_{2,p} = | H^p \varphi |_2$ on $\mathcal{S}(\mathbb{R}^d)$, where the last norm is the one of $L^2(\mathbb{R}^d)$. Then this system is equivalent to the initial system of seminorms on $\mathcal{S}(\mathbb{R}^d)$ and therefore it defines the same topoloy on $\mathcal{S}(\mathbb{R}^d)$.

Given $p \in \mathbb{N}_0$, we denote by $\mathcal{S}_p(\mathbb{R}^d)$ the Hilbert space which is the completion of $\mathcal{S}(\mathbb{R}^d)$ (more precisely of the equivalence class of $\mathcal{S}(\mathbb{R}^d)$ in $L^2(\mathbb{R}^d)$) under the norm $| \cdot |_{2,p}$. Denote by $\check{\mathcal{S}}(\mathbb{R}^d)$ the projective limit of the chain

$$L^2(\mathbb{R}^d) \supset \mathcal{S}_1(\mathbb{R}^d) \supset \cdots \; \mathcal{S}_p(\mathbb{R}^d) \supset \cdots \tag{A.5.5}$$

I.e., we put $\check{\mathcal{S}}(\mathbb{R}^d) = \bigcap_p \mathcal{S}_p(\mathbb{R}^d)$ as a set, and equip this space with the topology generated by the neighborhoods of zero which are determined by the choice of $p \in \mathbb{N}_0$, $\epsilon > 0$, and are given by $U_{p,\epsilon} = \{\check{\varphi} \in \check{\mathcal{S}}(\mathbb{R}^d), \ |\check{\varphi}|_{2,p} < \epsilon\}$. (Note that the system of seminorms $\{|\cdot|_{2,p}, p \in \mathbb{N}_0\}$ is ordered}).

Then $\check{\mathcal{S}}(\mathbb{R}^d)$ and $\mathcal{S}(\mathbb{R}^d)$ are in one-to-one correspondence: every equivalence class $\check{\varphi}$ in $\check{\mathcal{S}}(\mathbb{R}^d)$ contains exactly one representative φ belonging to $\mathcal{S}(\mathbb{R}^d)$, and to each $\varphi \in \mathcal{S}(\mathbb{R}^d)$ corresponds one class in $\check{\mathcal{S}}(\mathbb{R}^d)$. Moreover, since on $\mathcal{S}(\mathbb{R}^d)$ the topologies defined by both systems of seminorms above are equivalent, we see that this correspondence is a topological isomorphism. Unless there is danger of confusion, we shall identify these two spaces.

It is well-known that the spectrum of H is given by the set

$$\{2(n_1 + \ldots + n_d) + d + 1; n_1, \ldots, n_d \in \mathbb{N}\}.$$

Thus $H^{-(d+1)}$ is a trace class operator on $L^2(\mathbb{R}^d)$, and it follows that indeed $\mathcal{S}(\mathbb{R}^d)$ is a nuclear countably Hilbert space.

We may now identify the dual space $\mathcal{S}'(\mathbb{R}^d)$ of $\mathcal{S}(\mathbb{R}^d)$ with the union of the spaces $\mathcal{S}_{-p}(\mathbb{R}^d)$, $p \in \mathbb{N}_0$, where $\mathcal{S}_{-p}(\mathbb{R}^d)$ is the dual of the Hilbert space $\mathcal{S}_p(\mathbb{R}^d)$. Clearly $\mathcal{S}_{-p}(\mathbb{R}^d)$ is itself a Hilbert space with Hilbertian norm given by

$$|T|_{2,-p} := \sup_{\varphi \in \mathcal{S}(\mathbb{R}^d), \ |\varphi|_{2,p} \leq 1} |<T, \varphi>|. \tag{A.5.6}$$

As always, we have denoted here dual pairing by $<\cdot, \cdot>$.

As argued above, the strong and the inductive limit topology coincide on $\mathcal{S}'(\mathbb{R}^d)$, and strong and weak convergence of sequences are equivalent.

Tensor products of elements in $\mathcal{S}(\mathbb{R}^d)$ and $\mathcal{S}(\mathbb{R}^k)$ ($\mathcal{S}'(\mathbb{R}^d)$ and $\mathcal{S}'(\mathbb{R}^k)$, respectively) belong to $\mathcal{S}(\mathbb{R}^{d+k})$ ($\mathcal{S}'(\mathbb{R}^{d+k})$, respectively), and such tensor products are actually dense in $\mathcal{S}(\mathbb{R}^{d+k})$ (strongly and weakly sequentially dense in $\mathcal{S}'(\mathbb{R}^{d+k})$, respectively), cf., e.g., Trèves (1967).

Bibliography

Adams, R.A. (1975), *Sobolev Spaces*, Academic Press, New York

Albeverio, S., Fukushima, M., Karwowski, W. and Streit, L. (1981), Capacity and quantum mechanical tunneling. *Commun. Math. Phys.* 81, 501–513

Albeverio, S., Hida, T., Potthoff, J., Röckner, M. and Streit, L. (1990a), Dirichlet forms in terms of white noise analysis I – Construction and QFT examples. *Rev. Math. Phys.* 1, 291–312

Albeverio, S., Hida, T., Potthoff, J., Röckner, M. and Streit, L. (1990b), Dirichlet forms in terms of white noise analysis II – Closability and diffusion processes. *Rev. Math. Phys.* 1, 313–323

Albeverio, S., Hida, T., Potthoff, J., and Streit, L. (1989), The vacuum of the Høegh-Krohn model as a generalized white noise functional, *Phys. Lett. B* 217, 511–514

Albeverio, S. and Høegh-Krohn, R. (1973), Uniqueness of the physical vacuum and the Wightman functions in the infinite volume limit for some non–polynomial interactions. *Commun. Math. Phys.* 30, 171–200

Albeverio, S. and Høegh-Krohn, R. (1974), The Wightman axioms and the mass gap for strong interactions of exponential type in two-dimensional space-time. *J. Funct. Anal.* 16, 39–82

Albeverio, S. and Høegh-Krohn, R. (1976a), Quasi-invariant measures, symmetric diffusion processes and quantum fields. In: *Les Méthodes Mathématiques de la Théorie Quantique des Champs, Colloques Internationaux du C.N.R.S.* 248

Albeverio, S. and Høegh-Krohn, R. (1976b), *Mathematical Theory of Feynman Path Integrals*, Lecture Notes in Mathematics no.523, Springer, Berlin, Heidelberg, New York

Albeverio, S. and Høegh-Krohn, R. (1977a), Dirichlet forms and diffusion processes on rigged Hilbert spaces. *Z. Wahrscheinlichkeitstheorie verw. Gebiete* 40, 1–57

Albeverio, S. and Høegh-Krohn, R. (1977b), Hunt processes and analytic potential theory on rigged Hilbert spaces. *Ann. l'Inst. Henri Poincaré Sect. B* 13, 269–291

Albeverio, S. and Høegh-Krohn, R. (1984), Diffusion fields, quantum fields and fields with values in Lie groups. In: *Stochastic Analysis and Applications*, ed. by M. Pinski, Marcel Dekker, New York and Basel

Albeverio, S., Høegh-Krohn, R. and Streit, L. (1977), Energy forms, Hamiltonians and distorted Brownian paths. *J. Math. Phys.* **18**, 907–917

Albeverio, S., Høegh-Krohn, R. and Streit, L. (1980), Regularization of Hamiltonians and processes. *J. Math. Phys.* **21**, 1636–1642

Albeverio, S., and Röckner, M. (1989), Classical Dirichlet forms on topological vector spaces – Construction of an associated diffusion process. *Probab. Th. Rel. Fields* **83**, 405–434

Albeverio, S., and Röckner, M. (1990a), Classical Dirichlet forms on topological vector spaces – Closability and a Cameron–Martin formula. *J. Funct. Anal.* **88**, 395–436

Albeverio, S., and Röckner, M. (1990b), New developments in the theory and application of Dirichlet forms. In: *Stochastic Processes, Physics and Geometry*, S. Albeverio et al. (eds.), World Scientific, Singapore

Araki, H. (1960), Hamiltonian formalism and thecanonical commutation relations in quantum field theory. *J. Math. Phys.* **1**, 492-504

Araki, H. (1971), On the representation of the canonical commutation relations. *Commun. Math. Phys.* **20**, 9–25

Asch, J. and Potthoff, J. (1987), A generalization of Itô's lemma. *Proc. Japan Acad.* **63 A**, 289–291

Asch, J. and Potthoff, J. (1990), White noise and stochastic analysis. In: *Stochastics, Algebra and Analysis in Classical and Quantum Dynamics*, S. Albeverio et al. (eds.), Kluwer Academic Publishers, 1–14

Asch, J. and Potthoff, J. (1991), Itô's lemma without non-anticipatory conditions. *Probab. Th. Rel. Fields* **88**, 17–46

Averbuh, V.I., Smoljanov, O.G. and Fomin, S.V. (1971), Generalized functions and differential equations in linear spaces, I. Differentiable measures. *Trans. Moscow Math. Soc.* **24**, 140–184

Averbuh, V.I., Smoljanov, O.G. and Fomin, S.V. (1972), Generalized functions and differential equations in linear spaces, II. Differential operators and their Fourier transforms. *Trans. Moscow Math. Soc.* **27**, 255–270

Badrikian, A. (1970), *Séminaire sur les Fonctions Aléatoires Linéaires et les Mesures Cylindriques*, Springer Verlag, Berlin, Heidelberg, New York

Bakry, D. (1985a), Transformation de Riesz pour les semi-groupes symétriques. Première partie: Etude de la dimension 1. In: *Séminaire de Probabilités* **XIX**, ed.

by J. Azéma and M. Yor. Springer, Berlin, Heidelberg, New York

Bakry, D. (1985b), Transformation de Riesz pour les semi-groupes symétriques. Seconde partie: Etude sous la condition $\Gamma_2 \geq 0$. In: *Séminaire de Probabilités* XIX, ed. by J. Azéma and M. Yor. Springer, Berlin, Heidelberg, New York

Bakry, D. (1985c), Une remarque sur les inégalités de Littlewood–Paley sous l'hypothèse $\Gamma_2 \geq 0$. In: *Séminaire de Probabilités* XIX, ed. by J. Azéma and M. Yor. Springer, Berlin, Heidelberg, New York

Berezansky, Yu. M. (1986), *Selfadjoint Operators in Spaces of Functions of Infinitely Many Variables*, Amer. Math. Soc., Providence

Berezansky, Yu. M. and Kondratiev, Yu. G. (1988), *Spectral Methods in Infinite-Dimensional Analysis*, (in Russian), Naukova Dumka, Kiev

Berezin, F.A. (1966), *The Method of Second Quantization*, Academic Press, New York and London

Berger, M.A. and Mizel, V.J. (1980), An extension of the stochastic integral. *Ann. Probability* **10**, 435–450

Bismut, J.-M. (1981a), Martingales, the Malliavin calculus and Hörmander's theorem. In: *Stochastic Integrals*, ed. by D. Williams. Springer, Berlin, Heidelberg, New York

Bismut, J.-M. (1981b), Martingales, the Malliavin calculus and hypoellipticity under general Hörmander's conditions, *Z. Wahrscheinlichkeitsth. Verw. Geb.* **56**, 469–505

Bjørken, J. and Drell, S. (1965), *Relativistic Quantum Fields*, McGraw - Hill, New York

Bogoliubov, N.N., Logunov, A.A. and Todorov, I.T. (1975), *Introduction to Axiomatic Quantum Field Theory*, Benjamin, Reading

Bogoliubov, N.N., Shirkov, P.V. (1959), *Introduction to the Theory of Quantized Fields*, Interscience, New York

Bourbaki, N. (1987), *Topological Vector Spaces*. Springer, Berlin, Heidelberg, New York

Buckdahn, R. (1987), Skorohod's integral and linear stochastic differential equations. *Preprint*

Buckdahn, R. (1989), Anticipating linear stochastic differential equations. In: *Lecture Notes in Control and Information Sciences* **136**, 18–23

Cameron, R.H. and Martin, W.T. (1945), Transformation of Wiener integrals under a general class of linear transformations. *Trans. Amer. Math. Soc.* **58**, 184–219

Chow, P.L. (1972), Application of function space integrals to problems in wave propagation in random media. *J. Math Phys.* **13**, 1224–1236

Chow, P.L. (1978), Stochastic partial differential equations in turbulence related problems. In. *Probabilistic Analysis and Related Topics*, Vol.1, ed. by A. T. Bharucha-Reid. Academic Press, New York

Chow, P.L. (1989a), Stochastic differential equations in infinite dimensions. In: *Mathematics + Physics*, Vol. 3, ed. by L. Streit, World Scientific, Singapore

Chow, P.L. (1989b), Generalized solution of some parabolic equations with a random drift. *J. Appl. Math. Optimization* **20**, 81–96

Coester, F. and Haag, R. (1960), Representations of states in a field theory with canonical variables. *Phys. Rev.* **117**, 1137–1145

Cook, J. (1953), The mathematics of second quantization. *Trans. Amer. Math. Soc.* **74**, 222–245

Dellacherie, C. and Meyer, P.A. (1978), *Probabilities and Potential*, North–Holland: Amsterdam, New York, Oxford

Doss, H. (1980), Sur une résolution stochastique de l'équation de Schrödinger à coefficients analytiques. *Comm. math. Phys.* **73**, 247-264

Elson, C.M. (1974), An extension of Weyl's lemma to infinite dimensions. *Trans. Amer. Math. Soc.* **194**, 301–324

de Falco, D. (1987), Lévy's stochastic area in stochastic mechanics. *Lett. Math. Phys.* **14**, 41–45

de Falco, D. and Khandekar, D.C. (1988), Applications of white noise calculus to the computation of Feynman integrals. *Stoch. Proc. Appl.* **29**, 257–266

de Faria, M. and Kuo, H.–H. (1989), A delta white noise functional. *Acta Appl. Math.* **17**, 287-298

de Faria, M., Potthoff, J. and Streit, L. (1991), The Feynman integrand as a Hida distribution. *J. Math. Phys.* **32**, 2123-2127

Feyel, D. and de la Pradelle, A. (1989), Espaces de Sobolev Gaussiens, *Ann. Inst. Fourier* **39**, 875-908

Feyel, D. and de la Pradelle, A. (1991), Capacités Gaussiennes, *Ann. Inst. Fourier* 41, 49-76

Fleming, W.H. and Rishel, R.W. (1975), *Deterministic and Stochastic Optimal Control*, Springer-Verlag

Folland, G. (1976), *Introduction to Partial Differential Equations*, Princeton University Press, Princeton

Fomin, S.V. (1968a), Generalized functions of infinitely many variables and their Fourier transform. *Uspehi Mat. Nauk* 23, 215–216

Fomin, S.V. (1968b), A Fourier transform method for functional differential equations (engl. translation). *Soviet Math. Dokl.* 9, 951–953

Fomin, S.V. (1968c), Differentiable measures in linear spaces. *Uspehi Mat. Nauk* 23, 221–222

Fröhlich, J. (1974), Schwinger functions and their generating functionals I. *Helvetica Phys. Acta* 47, 265–306

Fröhlich, J. (1977), Schwinger functions and their generating functionals II. *Adv. Math.* 23, 119–180

Fukushima, M. (1979), A decomposition of additive functionals of finite energy. *Nagoya Math. J.* 74, 137–168

Fukushima, M. (1980), *Dirichlet Forms and Markov Processes*, Kodansha and North- Holland, Amsterdam and Tokyo

Fukushima, M. (1985), Energy forms and diffusion processes. In: *Mathematics + Physics*, Vol. 1, ed. by L. Streit. World Scientific, Singapore

Gaveau, B. and Trauber, P. (1982), L'intégrale stochastique comme opérateur de divergence dans l'espace fonctionnel. *J. Funct. Anal.* 46, 230–238

Gel'fand, I.M. and Shilov, G.E.. (1968), *Generalized Functions*, Vol. I - III, Academic Press, New York and London

Gel'fand, I.M. and Vilenkin, N.Ya. (1968), *Generalized Functions*, Vol. IV, Academic Press, New York and London

Gel'fand, I.M. and Yaglom, A.M. (1960), Integration in functional spaces and its applications in quantum physics. *J. Math. Phys.* 1, 48–69

Glimm, J. and Jaffe A. (1970), The $\lambda(\phi^4)_2$ quantum field theory without cutoffs, II. *Ann. Math.* 91, 362–401

Glimm, J. and Jaffe, A. (1981), *Quantum Physics, A Functional Integral Point of View*, Springer, Berlin, Heidelberg, New York

Glimm, J., Jaffe, A. and Spencer, T. (1973), The particle structure of the weakly coupled $P(\phi)_2$ model and other applications of high temperature expansions. In: *Constructive Quantum Field Theory*, G. Velo and A. Wigthman (eds.), Springer, Berlin, Heidelberg, New York

Goodman, V. (1972), A divergence theorem for Hilbert space. *Trans. Amer. Math. Soc.* **164**, 411–426

Gross, L. (1963), Harmonic analysis on Hilbert space. *Mem. Amer. Math. Soc.* **46**

Gross, L. (1967a), Potential theory on Hilbert space. *J. Funct. Anal.* **1**, 123–181

Gross, L. (1967b), Abstract Wiener spaces. *In: Proc. 5th Berkeley Symp. on Math. Statist. and Probability*, Univ. California Press, Berkeley

Gross, L. (1975), Logarithmic Sobolev inequalities. *Amer. Math. J.* **97**, 1061–1083

Guerra, F., Rosen, L. and Simon, B. (1975), The $P(\phi)_2$ Euclidean quantum field theory as classical statistical mechanics. *Ann. Math.* **101**, 111–259

Gundy, R.F. (1989), *Some Topics in Probability and Analysis*, Providence: American Mathematical Society

Hamza, M.M. (1975), *Détermination des Formes de Dirichlet sur* $\mathbb{R}^n$, Thèse de 3ème Cycle, Orsay

Hida, T. (1960), Canonical representations of Gaussian processes and their applications. *Mem. Coll. Sci. Univ. Kyoto, Ser. A* **33**, 109–155

Hida, T. (1967), Finite dimensional approximations to white noise and Brownian motion. *J. Math. Mech.* **16**, 859–866

Hida, T. (1970a), Harmonic analysis on the space of generalized functions. *Theor. Probab. Appl.* **15**, 117–122

Hida, T. (1970b), *Stationary Stochastic Processes*, Princeton University Press, Princeton

Hida, T. (1971), Quadratic functionals of Brownian motion. *J. Multivar. Anal.* **1**, 58–69

Hida, T. (1973), A role of Fourier transform in the theory of infinite dimensional unitary group. *J. Math. Kyoto Univ.* **13**, 203–212

Hida, T. (1975a), *Analysis of Brownian Functionals*, Carleton Math. Lecture Notes No. 13, Carleton

Hida, T. (1975b), White noise analysis and nonlinear filtering problems. *Appl. Math. Optimization* 2, 82–89

Hida, T. (1978), Generalized multiple Wiener integrals. *Proc. Japan Acad.* 54, 55–58

Hida, T. (1979), Nonlinear Brownian functionals. In: *Proc. 18th IEEE Conf. on Decision and Control* 1979, Vol 1. 326–328

Hida, T. (1980a), *Brownian Motion*, Springer, Berlin, Heidelberg, New York

Hida, T. (1980b), Causal analysis in terms of Brownian motion. In: *Multivariate Analysis*, ed. by P. R. Krishnaia. North-Holland, Amsterdam, 111–118

Hida, T. (1980c), Causal analysis in terms of white noise. In: *Quantum Fields - Algebras, Processes*, ed. by L. Streit. Springer, Berlin, Heidelberg, New York, 1–19

Hida, T. (1981), Causal calculus of Brownian functionals, and its applications. In: *Proc. Int. Symp. on Statistics and Related Topics*, ed. by M. Cörgö et al. North-Holland, Amsterdam, 353–363

Hida, T. (1982a), White noise analysis and its applications. In: *Proc. Int. Mathematical Conf.*, ed. by L.H.Y. Chen et al. North-Holland, Amsterdam, 43–48

Hida, T. (1982b), The role of exponential functions in the analysis of generalized Brownian functionals. *Teor. Veroyatnost. i Primenen.* 27, 569–573

Hida, T. (1983a), Generalized Brownian functionals. *In: Theory and Application of Random Fields*, ed. G. Kallianpur. Springer, Berlin, Heidelberg, New York, 89–95

Hida, T. (1983b), Causal calculus and an application to prediction theory. In: *Prediction Theory and Harmonic Analysis*, ed. by V. Mandrekar and H. Salehi. 123–130

Hida, T. (1984), White noise analysis and its application to quantum dynamics. *Physica* 124 A, 399–412

Hida, T. (1985), Brownian motion and its functionals. *Ricerche di Matematica* 34, 183–222

Hida, T. (1989), Infinite-dimensional rotation group and unitary group. *Lecture Notes in Math* 1379, 125–134

Hida, T. and Ikeda, N. (1967), Analysis on Hilbert space with reproducing kernel arising from multiple Wiener integral. In: *Proc. 5th Berkeley Symp. on Math. Statist. and Probability*, Vol 2, Part 1, 117–143

Hida, T. and Kallianpur, G. (1975), The square of a Gaussian Markov process and nonlinear prediction. *J. Multivar. Anal.* 5, 451–461

Hida, T., Kubo, I., Nomoto, H. and Yoshizawa, H. (1968), On projective invariance of Brownian motion. *Publ. Res. Inst. Math. Sci. Ser A* 4, 595–609

Hida, T., Kuo, H.-H. and Obata, N. (1991), Transformations for white noise functionals. *Preprint*, to appear in *J. Func. Anal*

Hida, T., Kuo, H.–H., Potthoff, J. and Streit, L. (eds.) (1990), *White Noise – Mathematics and Applications.* World Scientific, Singapore

Hida, T. and Nomoto, H. (1964), Gaussian measure on the projective limit space of spheres. *Proc. Japan Acad.* **40**, 301–304

Hida, T., Obata, N. and Saitô K. (1991), Infinite dimensional rotations and Laplacians in terms of white noise calculus. *Preprint*

Hida, T. and Potthoff, J. (1990), White noise analysis – an overview. In: *White Noise Analysis – Mathematics and Applications*, T. Hida, H.–H. Kuo, J. Potthoff and L. Streit (eds.). World Scientific, Singapore

Hida, T. Potthoff, J. and Streit, L. (1988), Dirichlet forms and white noise analysis. *Commun. Math. Phys.* **116**, 235–245

Hida, T., Potthoff, J. and Streit, L. (1989), White noise analysis and applications. In: *Mathematics + Physics, Vol.* 3, L. Streit (ed.). World Scientific, Singapore

Hida, T. and Saito, K. (1988), White noise analysis and the Lévy Laplacian. In: *Stochastic Processes in Physics and Engineering*, S. Albeverio et al. (eds.). 177–184

Hida, T. and Streit, L. (1977), On quantum theory in terms of white noise. *Nagoya Math. J.* **68**, 21–34

Hida, T. and Streit, L. (1981), Generalized Brownian functionals. In: *Lecture Notes in Physics* **153**, 285–287, Springer, Berlin, Heidelberg, New York.

Hida, T. and Streit, L. (1983), Generalized Brownian functionals and the Feynman integral. *Stoch. Proc. Appl.* **16**, 55–69

Hille, E. and Phillips, R.S. (1957), *Functional Analysis and Semigroups*, Amer. Math. Soc. Colloq. Publ. vol. **XXXI**

Hitsuda, M. (1972), Formula for Brownian partial derivatives. In: *Proc. 2nd Japan–USSR Symp. Probab. Th.* 2, Academy of Science USSR, 111–114

Høegh-Krohn, R. (1971), A general class of quantum fields without cutoffs in two space-time dimensions, *Commun. math. Phys.* 21, 256-260

Hu, Y.Z. and Meyer, P.A. (1988), Chaos de Wiener et intégrale de Feynman. In: *Séminaire de Probabilités* XXII, J. Azéma, P.A. Meyer and M. Yor (eds.). Berlin, Heidelberg, New York: Springer Verlag

Ikeda, N. and Watanabe, S. (1981), *Stochastic Differential Equations and Diffusion Processes*, Kodansha and North-Holland, Tokyo and Amsterdam

Ikeda, N. and Watanabe, S. (1983), An introduction to Malliavin's calculus. In: *Taniguchi Symp. SA Katata* 1983, 1–52

Isobe, E. and Sato, S. (1983), Wiener-Hermite expansion of a process generated by an Itô stochastic differential equation. *J. Appl. Probab.* **20**, 754–765

Itô, K. (1951), Multiple Wiener integral. *J. Math. Soc. Japan* **3**, 157–169

Itô, K. (1978a), Stochastic analysis in infinite dimensions. In: *Stochastic Analysis*, ed. by A. Friedman and M. Pinsky. Academic Press, New York and London

Itô, K. (1978b), Extension of stochastic integrals, In: *Proc. Intern. Symp. Stochastic Differential Equations.* K. Itô (ed.). Tokyo: Kinokuniya, New York: Wiley

Itô, K. (1990), Positive generalized functions on ($\mathbb{R}^\infty$, $\mathfrak{B}^\infty$,N^∞). In: *White Noise Analysis – Mathematics and Applications*, T. Hida, H.–H. Kuo, J. Potthoff and L. Streit (eds.). World Scientific, Singapore

Itô, K. and McKean Jr., H. P. (1965), *Diffusion Processes and Their Sample Paths*, Springer, Berlin, Heidelberg, New York.

Itô, Y. (1988), Generalized Poisson functionals. *Probab. Theory and Related Fields* 77, 1–28

Itô, Y., Kubo, I. and Takenaka, S. (1990), Calculus on Gaussian white noise and Kuo's Fourier transformation. In: *White Noise Analysis – Mathematics and Applications*, T. Hida, H.–H. Kuo, J. Potthoff and L. Streit (eds.). World Scientific, Singapore

Itzykson, C. and Zuber, J.–B. (1980), *Quantum Field Theory*, McGraw – Hill, New York

Jeulin, T. (1981), *Semi–Martingales et Grossissement d'une Filtration.* Springer,

Berlin, Heidelberg, New York

Jondral, F. (1981), Some remarks about generalized functionals of complex white noise. *Nagoya Math. J.* 81, 113–122

Jost, R. (1965), *The General Theory of Quantized Fields*, Amer. Math. Soc., Providence

Kallianpur, G. and Kuo, H.-H. (1984), Regularity properties of Donsker's delta function. *Appl. Math. Optim.* 12, 89–95

Kato, T. (1976), *Perturbation Theory for Linear Operators*, Springer, Berlin, Heidelberg, New York

Khandekar, D. C. and Lawande, S. V., (1986), Feynman path integrals: Some exact results and applications. *Phys. Rep* 137, no.2,3, 115-229

Khandekar, D. C. and Streit, L. (1992), Constructing the Feynman integrand. Universidade da Madeira preprint, to appear in *Ann. Physik*

King, R. (1975), *Stochastic integrals and metadistributions. Applications to stochastic partial differential equations and quantum field theory.* Ph. D. dissertation, Cornell University

Kondratiev, Yu.G. (1980), Nuclear spaces of entire functions in problems of infinite – dimensional analysis. *Soviet Math. Dokl.* 22, 588–592

Kondratiev, Yu.G. and Samoilenko, Yu.S. (1976), Integral representation of generalized positive definite kernels of an infinite number of variables. *Soviet Math. Dokl.* 17, 517–521

Kondratiev, Yu.G. and Streit, L. (1991a), A remark about a norm estimate for White Noise distributions. Universidade da Madeira preprint, to appear in *Ukrainian Math. J.*

Kondratiev, Yu.G. and Streit, L. (1991b), Positive definite White Noise distributions. Universidade da Madeira preprint UMa-Mat 5/91

Kondratiev, Yu.G. and Streit, L. (1992), Spaces of White Noise distributions: Constructions, Descriptions, Applications. I. BiBoS preprint no. 510, to appear in *Rep. Math. Phys.*

Korezlioglu, H. and Ustunel, A.S. (1990), Distributions, Girsanov and degree theorems on Wiener space. In: *White Noise Analysis – Mathematics and Applications*, T. Hida, H.–H. Kuo, J. Potthoff and L. Streit (eds.). World Scientific, Singapore

Köthe, G. (1971), *Topological Vector Spaces.* Springer, Berlin, Heidelberg, New York

Krée, M. (1977), Propriété de trace en dimension infinie d'espaces du type Sobolev, *Bull. Soc. Math. France* **105**, 141–163

Krée, M. and Krée, P. (1983), Continuité de la divergence dans les espaces de Sobolev relatifs à l'espace de Wiener, *Comptes Rendues Acad. Sc. Paris* **296**, 833–836

Krée, P. (1974), Application des méthodes variationnelles aux équations aux dérivée partielles sur un espace de Hilbert. *Comptes Rendues Acad. Sci. Paris* **278**, 753–755

Krée, P. (1988), La théorie des distributions en dimension quelconque et l'intégration stochastique. *Lecture Notes in Math.* **1316**, 170–233, Springer, Berlin, Heidelberg, New York

Krylov, N.V. and Veretennikov, A.Yu. (1976), On explicit formulas for solutions of stochastic equations. *Mat. Sbornik* **29**, 239–256

Kubo, I. (1983), Itô formula for generalized Brownian functionals. In: *Theory and Application of Random Fields*, ed. G. Kallianpur. Springer, Berlin, Heidelberg, New York

Kubo, I. and Kuo, H.-H. (1991), Fourier transform and cylindrical Hida distributions. *Preprint*

Kubo, I. and Takenaka, S. (1980a), Calculus on Gaussian white noise I. *Proc. Japan Acad.* **56**, 376–380

Kubo, I. and Takenaka, S. (1980b), Calculus on Gaussian white noise II. *Proc. Japan Acad.* **56**, 411–416

Kubo, I. and Takenaka, S. (1981), Calculus on Gaussian white noise III. *Proc. Japan Acad.* **57**, 433–437

Kubo, I. and Takenaka, S. (1982), Calculus on Gaussian white noise IV. *Proc. Japan Acad.* **58**, 186–189

Kubo, I. and Yokoi, Y. (1989), A remark on the space of testing random variables in the white noise calculus. *Nagoya Math. J.* **115**, 139–149

Kuo, H.-H. (1974), Integration by parts for abstract Wiener measures. *Duke Math. J.* **41**, 373–379

Kuo, H.-H. (1975), *Gaussian Measures in Banach Spaces*, Springer, Berlin,

Heidelberg, New York

Kuo, H.-H. (1976), Distribution theory on Banach space. In: *Proc. 1st Int. Conf. on Prob. in Banach Spaces*, Springer, Berlin, Heidelberg, New York

Kuo, H.-H. (1978a), Differential calculus for measures on Banach spaces. In: *Proc. Vector Space Measure Conf.*, Springer, Berlin, Heidelberg, New York

Kuo, H.-H. (1978b), The chain rule for differentiable measures. *Studia Math. T.* **LXIII**, 145–155

Kuo, H.-H. (1980), Integration in Banach spaces. In: *Notes in Banach Spaces*, ed. by H. E. Lacey, Univ. of Texas Press, Austin, 1–38

Kuo, H.-H. (1982), On Fourier transform of generalized Brownian functionals. *J. Multivar. Anal.* **12**, 415–431

Kuo, H.-H. (1983a), Donsker's delta function as a generalized Brownian functional and its application. *Lecture Notes in Control and Information Sciences* **49**, 167–178, Springer-Verlag

Kuo, H.-H. (1983b), Fourier-Mehler transforms of generalized Brownian functionals. *Proc. Japan Acad.* **59A**, 312–314

Kuo, H.-H. (1983c), A Fourier transform characterization of Gaussian Brownian functionals. *Bull. Inst. Math. Academia Sinica* **11**, 407–413

Kuo, H.-H. (1983d), Brownian functionals and applications. *Acta Appl. Math.* **1**, 175–188

Kuo, H.-H. (1986), On Laplacian operators of generalized Brownian functionals. *Lecture Notes in Math.* **1203**, 119–128, Springer-Verlag

Kuo, H.-H. (1988a), Brownian motion, diffusions and infinite dimensional calculus. *Lecture Notes in Math.* **1316**, 130–169, Springer-Verlag

Kuo, H.-H. (1988b), White noise calculus. in: *Algebra, Analysis and Geometry*, M.C. Kang and K.W. Lih (eds.) 77-119, World Scientific

Kuo, H.-H. (1989), The Fourier transform in white noise calculus. *J. Multivariate Analysis* **31**, 311-327

Kuo, H.-H. (1990), Fourier-Mehler transforms in white noise analysis. in: *Gaussian Random Fields*, K. Itô and T. Hida (eds.) 257–271, World Scientific

Kuo, H.-H. (1991a), Convolution and Fourier transform of Hida distributions. *Preprint*

Kuo, H.-H. (1991b), An infinite dimensional Fourier transform. *Preprint*

Kuo, H.-H. (1992), Lectures on white noise analysis. *Soochow J. Math.* 18, 229–300

Kuo, H.-H., Obata, N. and Saitô, K. (1990), Lévy Laplacian of generalized functionals on a nuclear space. *J. Func. Anal.* **94**, 74–92

Kuo, H.–H. and Piech, M.A. (1973), Stochastic integral and parabolic equation in abstract Wiener space. *Bull. American Math. Soc.*79, 478–482

Kuo, H.–H. and Potthoff, J. (1989), Anticipating stochastic differential equations. *Preprint.* To appear In: *Proc. 5th Int. Vilnius Conf. on Probab. Th. and Math. Stat.*

Kuo, H.–H. and Potthoff, J. (1990), Anticipating stochastic integrals and stochastic differential equations. In: *White Noise Analysis – Mathematics and Applications*, T. Hida, H.–H. Kuo, J. Potthoff and L. Streit (eds.). World Scientific, Singapore

Kuo, H.–H., Potthoff, J. and Streit, L. (1991), A characterization of white noise test functionals. *Nagoya Math. J.* 121, 185–194

Kuo, H.–H., Potthoff, J. and Yan, J.-A. (1992), Continuity of affine transformations on white noise test functionals and applications. *Stoch. Proc. Appl.* **43**, 85–98

Kuo, H.-H. and Russek, A. (1988), White noise approach to stochastic integration. *J. Multivar. Anal.* 24, 218–236

Kuo, H.-H. and Shieh, N.-R. (1987), a generalized Itô's formula for multidimensional Brownian motion and its application. *Chinese J. Math.* 15, 163–174

Kuo, H.-H. and Smolenski, W. (1982), On admissible shifts of generalized white noise. *J. Multivar. Anal.* 12, 80–88

Kusuoka, S. (1982), Dirichlet forms and diffusion processes on Banach spaces. *J. Fac. Sci. Univ. Tokyo Sect. IA* **29**, 79–95

Kusuoka, S. and Stroock, D. (1982), Applications of the Malliavin calculus, part I. *Taniguchi Symp.*, Katata, 271–306

Kusuoka, S. and Stroock, D. (1985), Applications of the Malliavin calculus, part II. *J. Fac. Sci. Univ. Tokyo, Sec.* IA, **32**, 1–76

Kusuoka, S. and Stroock, D. (1984), The partial Malliavin calculus and its application to non-linear filtering. *Stochastics* 12, 83–142

Lee, J.–S. (1992), Fourier transform and heat equation in white noise analysis.

Ph.D. dissertation, Louisiana State University

Lee, Y.-J. (1989), Generalized functions on infinite dimensional spaces and its application to white noise calculus. *J. Func. Anal.* **82**, 429–464

Lee, Y.-J. (1990), A reformulation of white noise calculus. In: *White Noise Analysis – Mathematics and Applications*, T. Hida et al. (eds.), World Scientific, Singapore (1990)

Lee, Y.-J. (1991), Analytic version of test functionals, Fourier transform and a characterization of measures in white noise calculus. *J. Funct. Anal.* **100**, 359–380

Lévy, P. (1922), *Leçons d'Analyse Fonctionelle*, Gauthier-Villars, Paris

Lévy, P. (1925), Analyse fonctionelle. *Mem. Sci. Math. Acad. Sci. Paris*, fasc. 5

Lévy, P. (1937), *Théory de L' Addition des Variables Aléatoires*, Gauthier-Villars, Paris

Lévy, P. (1948), *Processus Stochastiques et Mouvement Brownien*, Gauthier-Villars, Paris (2ème éd. revue et augmentée (1965))

Lévy, P. (1951), *Problèmes Concrets d' Analyse Fonctionelle*, Gauthier-Villars, Paris

Lévy, P. (1954), Le mouvement Brownien. In: *Mém. des Sci. Math.* CXXXVI, Gauthier-Villars, Paris

Lévy, P. (1956), Random functions: a Laplacian random function depending on a point of Hilbert space. *Univ. Calif. Publ. Statistics* 2, No. 10, 195–206

Lindstrøm, T., Øksendal, B. and Ubøe, J. (1991), Stochastic differential equations involving positive noise. In: *Stochastic Analysis*, M. T. Barlow and N. H. Bingham (eds.). Cambridge University Press, Cambridge

Lindstrøm, T., Øksendal, B. and Ubøe, J. (1992), Wick multiplication and Ito-Skorohod stochastic differential equations; in: *Ideas and Methods in Math. Anal.*, S. Albeverio et al. (eds.), Cambridge University Press

Lions, J.L. and Magenès, E. (1972), *Non-Homogeneous Boundary Value Problems and Applications*, Vol. 1, Springer-Verlag, Berlin, Heidelberg, New York

Lukacs, E. (1970), *Characteristic Functions*, 2nd edition, Griffin, London

Malliavin, P. (1978), Stochastic calculus of variation and hypoelliptic operators. In: *Proc. Int. Symp. on Stochastic Differential Equations*, ed. by K. Itô. Kinokuniya, Tokyo, and Wiley, New York

Malliavin, P. (1984), Calcul des variations, intégrales stochastiques et complexes de Rham sur l'espace de Wiener. *C. R. Acad. Sci. Paris* **299**, 347–350

McKean Jr., H. P. (1969), *Stochastic Integrals*, Academic Press, New York

Meyer, P.A. (1976), Démonstration probabiliste de certaines inégalités de Littlewood–Paley. In: *Séminaire de Probabilités* X, ed. by J. Azéma and M. Yor. Springer, Berlin, Heidelberg, New York

Meyer, P.A. (1982), Note sur les processus d'Ornstein-Uhlenbeck. In: *Séminaire de Probabilités* XVI, ed. by J. Azéma and M. Yor. Springer, Berlin, Heidelberg, New York

Meyer, P.A. (1983), Quelques résultats analytiques sur le semigroupe d'Ornstein-Uhlenbeck en dimension infinie. In: *Theory and Application of Random Fields*, ed. G. Kallianpur. Springer, Berlin, Heidelberg, New York

Meyer, P.A. (1988), Distributions, noyaux, symboles d'après Krée. *Lecture Notes in Math.* 1321, 467–476, Springer-Verlag

Meyer, P.A. and Yan, J.-A. (1987), À propos des distributions sur l'espace de Wiener. In: *Séminaire de Probabilités* XXI, ed. by J. Azéma and M. Yor. Springer, Berlin, Heidelberg, New York

Meyer, P.A. and Yan, J.–A. (1989), Distributions sur l'espace de Wiener (suite). *Séminaire de Probabilités* XXIII, J. Azéma, P.A. Meyer and M. Yor (eds.). Springer, Berlin, Heidelberg, New York

Meyer, P.A. and Yan, J.–A. (1990), Les "fonctions caractéristiques" des distributions sur l'espace de Wiener. *Preprint*

Minlos, R.A. (1962), Generalized random processes and their extension to a measure. *Sel. Transl. in Math. Statist. and Probability*, Vol. 3, 291–313

Nelson, E. (1967), *Dynamical Theories of Brownian Motion*, Princeton Univ. Press, Princeton

Nelson, E. (1972), Time–ordered operator products of sharp time quadratic forms. *J. Funct. Anal.* 11, 211–219

Nelson, E. (1973a), Construction of quantum fields from Markoff fields. *J. funct. Anal.* 12, 97–112

Nelson, E. (1973b), The free Markov field. *J. Funct. Anal.* 12, 211–227

Nelson, E. (1973c), Probability theory and Euclidean quantum field theory. In: *Constructive Quantum Field Theory*, ed. by G. Velo and A. Wightman, Springer,

Berlin, Heidelberg, New York

Neveu, J. (1964), *Bases Mathématiques du Calcul des Probabilités*, Masson et Cie, Paris

Nualart, D. and Pardoux, E. (1988), Stochastic calculus with anticipating integrands. *Probab. Theory and Rel. Fields* **78**, 535–581

Nualart, D. and Zakai, M. (1986), Generalized stochastic integrals and the Malliavin calculus. *Probab. Th. Rel. Fields* **73**, 255–280

Nualart, D. and Zakai, M. (1988), Generalized multiple stochastic integrals and the representation of Wiener functionals. *Stochastics* **23**, 311–330

Nualart, D. and Zakai, M. (1989), Generalized Brownian functionals and the solution to a stochastic partial differential equation. *J. Funct. Anal.* 84, 279–296

Obata, N. (1988), A note on certain permutation groups in the infinite dimensional rotation group. *Nagoya Math. J.* **109**, 91–107

Obata, N. (1988), Analysis of the Lévy Laplacian. *Soochow J. Math.* 14, 105–109

Obata, N. (1988), Density of natural numbers and the Lévy group. *J. Number Th.* **30**

Obata, N. (1989), The Lévy Laplacian and mean value theorem. *Lecture Notes in Math.* **1379**, 242–253, Springer, Berlin, Heidelberg, New York

Obata, N. (1990), A characterization of the Lévy Laplacian in terms of infinite dimensional rotation groups. *Nagoya Math. J.* 118, 111–132

Obata, N. (1991), Rotation-invariant operators on white noise functionals. *Preprint*, to appear in *Math. Z.*

Obata, N. (1992), Elements of white noise analysis. *Manuscript*, submitted

Ocone, D. (1984), Malliavin's calculus and stochastic integral representation of functionals of diffusion processes. *Stochastics* 12, 161–185

Ocone, D. and Pardoux, E. (1988), Non–adapted solutions of stochastic differential equations. *Preprint*

Ocone, D. and Pardoux, E. (1988), Linear stochastic differential equations with boundary conditions. *Preprint*

Ogawa, S. (1984), Quelques propriétés de l'integrale stochastique du type noncausal. *Japan J. Appl. Math.* 1, 405–416

Osterwalder, K. and Schrader, R. (1973), Axioms for Euclidean Green's functions. *Commun. Math. Phys.* 31, 83–112

Pardoux, E. and Protter, P. (1987), A two-sided stochastic integral and its calculus. *Probab. Theory and Rel. Fields* **76**, 15–49

Pardoux, E. and Protter, P. (1990), Stochastic Volterra equations with anticipating coefficients. *Annals of Probability* 18, 1635–1655

Piech, M.A. (1975), The Ornstein-Uhlenbeck semigroup in an infinite dimensional L^2-setting. *J. Funct. Anal.* 18, 271–285

Piech, M.A. (1979), Differentiability of measures associated with parabolic equations on infinite dimensional spaces. *Trans. Amer. Math. Soc.* **253**, 191–209

Potthoff, J. (1986), On the connection of the white-noise and Malliavin calculi. *Proc. Japan Acad.* **62**, 43–45

Potthoff, J. (1987a), White noise approach to Malliavin calculus. *J. Funct. Anal.* 71, 207–217

Potthoff, J. (1987b), On positive generalized functionals. *J. Funct. Anal.* 74, 81–95

Potthoff, J. (1988a), Littlewood-Paley theory on Gaussian spaces. *Nagoya Math. J.* **109**, 47–61

Potthoff, J. (1988b), On Meyer's equivalence. *Nagoya Math. J.* 111, 99–109

Potthoff, J. (1992), White noise methods for stochastic partial differential equations. In: *Stochastic Partial Differential Equations and Their Applications*, B.L. Rozovskii and R.B. Sowers (eds.), Springer, Berlin, Heidelberg, New York

Potthoff, J. and Röckner, M. (1990), On the contraction property of energy forms on infinite–dimensional space. *J. Funct. Anal.* **92**, 155–165

Potthoff, J. and Streit, L. (1990), Invariant states on random and quantum fields: ϕ–bounds and white noise analysis. *BiBoS Preprint* no. 422, to appear in *J. Funct. Anal.*

Potthoff, J. and Streit, L. (1991a), A characterization of Hida distributions. *J. Funct. Anal.* **101**, 212–229

Potthoff, J. and Streit, L. (1991b), Generalized Radon – Nikodym derivatives and Cameron – Martin theory. In: *Gaussian Random Fields*, K. Itô and T. Hida (eds.), World Scientific, Singapore

Potthoff, J. and Yan, J.-A., (1992), Some results about test and generalized functionals of white noise. In: *Probability Theory*, L.H.Y. Chen et al. (eds.), de Gruyter, Berlin, New York

Razafimanantena, E. (1991), Constructions of a general class of Dirichlet forms in terms of white noise analysis. *Stoch. Proc. Appl.* 39, 263–276

Redfern, M. (1991), White noise approach to multiparameter stochastic integration. *J. Multivariate Analysis* 37, 1-23

Reed, M. and Simon, B. (1972), *Methods in Mathematical Physics*, Vol. I, Academic Press, New York and London

Reed, M. and Simon, B. (1975), *Methods in Mathematical Physics*, Vol. II, Academic Press, New York and London

Reed, M. and Simon, B. (1978), *Methods in Mathematical Physics*, Vol. IV, Academic Press, New York and London

Röckner, M. (1990), Lectures on "Dirichlet forms on infinite dimensional state space and applications", to appear in *Proc. Silivri*

Robertson, A. and Robertson, W. (1973), *Topological Vector Spaces.* Cambridge University Press, Cambridge

Rullkötter, K. and Spönemann, U. (1983), *Dirichletformen und Diffusionsprozesse*, Diplomarbeit, Bielefeld

Saitô, K. (1987), Itô's formula and Lévy Laplacian. *Nagoya Math. J.* 108, 67–76

Saitô, K. (1988), Lévy's Laplacian in the infinitesimal generator. *Research Report, Meijo Univ.* 28, 1–5

Saitô, K. (1991), Itô's formula and Lévy Laplacian II. *Nagoya Math. J.* 123, 153–169

Schmuland, B. (1990), *Energy forms with polynomial domain*, *Preprint* Univ. Alberta

Schwartz, J. (1969), *Nonlinear Functional Analysis*, Gordon and Breach, New York

Schwartz, L. (1950), *Théorie des Distributions*, Hermann Cie, Paris (2ème éd. 1966)

Schwartz, L. (1973), *Radon Measures on Arbitrary Topological Spaces and Cylindrical Measures*, Oxford University Press, London

Schweber, S.S. (1966), *An Introduction to Relativistic Quantum Field Theory*, 2nd ed., Harper and Row, New York

Segal, I. (1956), Tensor algebras over Hilbert spaces. *Trans. Amer. Math. Soc.* 81, 106–134

Segal, I. (1958), Distributions in Hilbert space and canonical systems of operators. *Trans. Amer. Math. Soc.* 88, 12–41

Shieh, N. R. and Yokoi, Y. (1990), Positivity of Donsker's delta function. In: *White Noise Analysis - Mathematics and Applications*, T. Hida et al. (eds.). World Scientific, Singapore

Shigekawa, I. (1980), Derivatives of Wiener functionals and absolute continuity of induced measures. *J. Math. Kyoto Univ.* 20, 263–289

Shigekawa, I. (1990), Sobolev spaces over the Wiener space based on an Ornstein-Uhlenbeck operator, *Preprint*

Shigekawa, I. and Yoshida, N. (1990), Littlewood–Paley–Stein inequality for a symmetric diffusion, *Preprint*

Shiota, Y. (1986), A linear stochastic integral equation containing the extended Itô integral. *Math. Rep. Toyama Univ.* 9, 43–65

Silverstein, M. L. (1975), *Symmetric Markov Processes*, Lecture Notes in Mathematics no.426, Springer, Berlin, Heidelberg, New York

Simon, B. (1971), Distributions and their Hermite expansions. *J. Math. Physics* 12, 140–148

Simon, B. (1974), *The $P(\phi)_2$ Euclidean (Quantum) Field Theory*, Princeton Univ. Press, Princeton

Simon, B. (1979), *Functional Integration and Quantum Physics*, Academic Press, New York

Simon, B. and Høegh-Krohn, R. (1972), Hypercontractive semigroups and two–dimensional self–coupled Bose fields. *J. Funct. Anal.* 9, 121–180

Skorohod, A.V. (1975), On a generalization of a stochastic integral. *Theor. Prob. Appl.* 20, 219–233

Stein, E.M. (1970a), *Singular Integrals and Differentiability Properties of Functions*, Princeton Univ. Press, Princeton

Stein, E.M. (1970b), *Topics in Harmonic Analysis*, Princeton Univ. Press,

Princeton

Streater, R. and Wightman, A. (1964), *PCT, Spin , Statistics and All That*, Benjamin, New York

Streit, L. (1981), Energy forms: Schrödinger theory, processes. In: *New Stochastic Methods in Physics*, ed. by C. DeWitt-Morette and K. D. Elworthy. *Phys. Rep.* 77

Streit, L., (1991), The characterization theorem for Hida distributions. Generalizations and Applications. Universidade da Madeira preprint UMa-Mat 3/91

Streit, L. and Westerkamp, W. (1991), A generalization of the characterization theorem for generalized functionals of White Noise. *BiBoS preprint no.* 480

Stroock, D. (1981a), The Malliavin calculus and its applications. In: *Stochastic Integrals*, ed. by D. Williams. Springer, Berlin, Heidelberg, New York

Stroock, D. (1981b), The Malliavin calculus, a functional analytic approach. *J. Funct. Anal.* **44**, 212–257

Sugita, H. (1985a), Sobolev spaces of Wiener functionals and Malliavin's calculus. *J. Math. Kyoto Univ.* **25**, 31–48

Sugita, H. (1985b), On a characterization of the Sobolev spaces over an abstract Wiener space. *J. Math. Kyoto Univ.* **25**, 717–725

Sugita, H. (1988), Positive generalized Wiener functions and potential theory over abstract Wiener spaces, *Osaka J. Math.* **25**, 665–696

Symanzik, K. (1969), Euclidian Quantum Field Theory, in *Local Quantum Theory*, (R. Jost. ed.), Academic Press, New York

Takeda, M. (1984), (r,p)-capacity on the Wiener space and properties of Brownian motion. *Z. Wahrscheinlichkeitstheorie verw. Gebiete* **68**, 149–162

Takeda, M. (1985), On the uniqueness of Markovian sef-adjoint extensions of diffusion operators on infinite dimensional spaces. *Osaka J. Math.* **22**, 733–742

Titchmarsh, E.C. (1939), *The Theory of Functions*, 2nd edition, Oxford Univ. Press, Oxford

Trèves, F. (1967), *Topological Vector Spaces, Distributions and Kernels*, Academic Press, New York and London

Ustunel, A. S. (1986), La formule de changement de variables pour l'integrale anticipante de Skorokhod. *C. R. Acad. Sci. Paris* **303**, 329-331

Velo, G. and Wightman, A. (ed.s) (1973), *Constructive Quantum Field Theory*, Springer, Berlin, Heidelberg, New York

Watanabe, S. (1983), Malliavin's calculus in terms of generalized Wiener functionals. In: *Theory and Application of Random Fields*, ed. G. Kallianpur. Springer, Berlin, Heidelberg, New York

Watanabe, S. (1984), *Stochastic Differential Equations and Malliavin Calculus*, Bombay: Tata Inst. of Fundamental Research

Watanabe, S. (1990), Donsker's δ-function in the Malliavin calculus. To appear In: *Proc. Conf. in Honor of M. Zakai*

Wiener, N. (1923), Differential space. *J. Math. and Phys.* **2**, 131–174

Wiener, N. (1938), The homogeneous chaos. *Amer. J. Math.* **60**, 897–936

Wiener, N. (1948), *Cybernetics, or Control and Communication in the Animal and the Machine*, Wiley, New York (2nd ed. 1962)

Yan, J.-A. (1989), Sur la transformée de Fourier de H. H. Kuo. In: *Séminaire de Probabilités* XXIII, J. Azéma et al. (eds.), Springer, Heidelberg, Berlin, New York

Yan, J.-A. (1990a), An elementary proof of a theorem of Lee. *Preprint*

Yan, J.-A. (1990b), A characterization of white noise functionals. *Preprint*

Yan, J.-A. (1990c), Products and transforms of white noise functionals. *Preprint*

Yan, J.-A. (1990d), Notes on the Wiener semigroup and renormalizations. *Preprint*

Yokoi, Y. (1990a), Positive generalized white noise functionals. *Hiroshima Math. J.* **20**, 137–157

Yokoi, Y. (1990b), Positive generalized Brownian functionals. In: *White Noise Analysis – Mathematics and Applications*, T. Hida, H.-H. Kuo, J. Potthoff and L. Streit (eds.). World Scientific, Singapore

Yoshizawa, H. (1970), Rotation group of Hilbert space and its application to Brownian motion. In: *Proc. Int. Conf. Functional Analysis and Rel. Topics*, 414–423

Yosida, K. (1951), *Functional Analysis*, Springer, Berlin Heidelberg, New York

Zakai, M. (1985), The Malliavin calculus. *Acta Appl. Math.* **3**, 175–207

Notations and Conventions

$\mathcal{A}$	algebra of complex functions over $\mathcal{N}^*$, $\mathcal{S}'(\mathbb{R})$,... generated by $\mathcal{P}$, $\mathcal{E}$ and $\mathcal{E}_r$
$\mathcal{B}$	σ-algebra over $\mathcal{N}$, generated by the weak topology, i.e., by the cylinder sets
$B(t)$	Brownian motion at time $t \geq 0$, Chapt. 1
C	characteristic functional, Chapt. 1
$\mathcal{C}$	algebra of complex functions on $\mathcal{N}^*$, $\mathcal{S}'(\mathbb{R})$, ... generated by $\sin \circ X_\xi$, $\cos \circ X_\xi$, $\xi \in \mathcal{N}$, $\mathcal{S}(\mathbb{R})$, ...
δ_t	Dirac distribution concentrated at $t \in \mathbb{R}$, $\mathbb{R}^d$
D_y	Gâteaux derivative (and its extensions) in direction $y \in \mathcal{S}'(\mathbb{R})$, Chapt. 5
D_y^*	adjoint of D_y, Chapt. 5
$dB(\pm)$	integrator of the forward and backward integrals, Chapt. 8.C
$d\Gamma(A)$	second quantization of an operator A acting on $\mathcal{H}$, $L^2(\mathbb{R})$, ... , Chapt. 3, App. 2
$\mathcal{D}^{p,k}_{(A)}$	spaces of smooth and generalized functionals on $\mathcal{N}^*$, Chapt.'s 3, 7
$\mathcal{D}_{(A)}$	space of smooth functionals on $\mathcal{N}^*$, Chapt.'s 3, 7
$\mathcal{D}^*_A$	space of generalized functionals on $\mathcal{N}^*$, dual of $\mathcal{D}_{(A)}$, Chapt.'s 3, 7
δ_x	evaluation mapping on $(\mathcal{S})$: $<\delta_x, \varphi> = \tilde{\varphi}(x)$, $\varphi \in (\mathcal{S})$, $x \in \mathcal{S}'(\mathbb{R})$, Chapt. 4
$\delta_{t,y}$	Donsker's delta function: $\delta_y \circ B(t)$, $y \in \mathbb{R}$, $t > 0$, Chapt.'s 3, 4
$\mathcal{E}$	algebra of complex functions over $\mathcal{N}^*$, $\mathcal{S}'(\mathbb{R})$,... generated by $\exp(i<\cdot\, , \xi>)$, $\xi \in \mathcal{N}$, $\mathcal{S}(\mathbb{R})$, ...

$\mathcal{E}_r$	algebra of complex functions over $\mathcal{N}^*$, $\mathcal{S}'(\mathbb{R})$,... generated by $\exp(<\cdot\,,\xi>)$, $\xi\in\mathcal{N}$, $\mathcal{S}(\mathbb{R})$, ...
$\mathcal{E}(\varphi,\psi)$	energy (Dirichlet) form over $\mathcal{S}'(\mathbb{R})$, Chapt. 11
$f^{(n)}$	typical (complex) element in $\Gamma^{(n)}(\mathcal{H}_c)$, $L^2(\hat{\mathbb{R}}^n)$, $\mathcal{S}(\hat{\mathbb{R}}^n)$, ... , integral kernel of $\varphi^{(n)}$, Chapt.'s 2, 3, 4, App. 2
$F^{(n)}$	typical (complex) element in $\mathcal{S}'(\hat{\mathbb{R}}^n)$, "integral" kernel of $\Phi^{(n)}$, Chapt. 4
$\mathcal{F}C_b^\infty$	finitely based smooth bounded functions, Chapt. 11
φ, ψ, ...	typical (*complex*) elements in (L^2), $(\mathcal{S})_p$, $p\geq 0$, $(\mathcal{S})$, $\mathcal{D}$
Φ, Ψ, ...	typical (*complex*) elements in $(\mathcal{S})_p$, $p\leq 0$, $(\mathcal{S})^*$, $\mathcal{D}^*$
$\tilde{\varphi}$	pointwise defined version of $\varphi\in(\mathcal{S})$, Chapt. 4
$\varphi^{(n)}$	component of $\varphi\in(L^2)$ in the n-th homogeneous chaos $\mathcal{H}^{(n)}$, $n\in\mathbb{N}_0$
$\Phi^{(n)}$	n–th homogeneous component of $\Phi\in(L^2)_{-\infty}$, $(\mathcal{S})^*$, ...
$\Gamma(A)$	second quantization of an operator A on $\mathcal{H}$, $L^2(\mathbb{R})$, ... , Chapt. 3, App. 2
$\Gamma(\mathcal{H})$	symmetric Fock space over $\mathcal{H}_c$, App. 2
$\Gamma^{(n)}(\mathcal{H})$	n–th symmetric tensor product of $\mathcal{H}_c$ with itself, App. 2
H	self–adjoint extension of $H\xi(t) = -\xi''(t)+(t^2+1)\xi(t)$, $t\in\mathbb{R}$, $\xi\in\mathcal{S}(\mathbb{R})$, on $L^2(\mathbb{R})$, Chapt. 3
$\mathcal{H}$	real, separable Hilbert space with dense subspace $\mathcal{N}$; sometimes the complexification $\mathcal{H}_c$ of $\mathcal{H}$ is also denoted by $\mathcal{H}$
$\mathcal{H}^{(n)}$	n–th homogeneous chaos in (L^2), $n\in\mathbb{N}_0$

$\mathcal{H}_\alpha(\mathbb{R}^d)$	Sobolev space of order α over $\mathbb{R}^n$, $\alpha\in\mathbb{R}$, $n\in\mathbb{N}$
$L^2(\mathbb{R}^d)$	real space of (equivalence classes of) functions over $\mathbb{R}^d$ which are square integrable with respect to Lebesgue measure, sometimes its complexification is denoted also by $L^2(\mathbb{R}^d)$
(L^p)	$L^p(\mathcal{N}^*,\mathcal{B},d\mu)$, $p\geq 1$
$(L^2)_{\pm\infty}$	spaces of smooth and generalized functionals over $\mathcal{N}^*$, Chapter 3
$(L^2)^{\pm}_{\mathbf{a,c}}$	spaces of smooth and generalized white noise functionals, Chapt. 3.A
μ	centered Gaussian measure on $(\mathcal{N}, \mathcal{B})$ determined by the inner product of $\mathcal{H}$, Chapt. 1, typically the white noise measure
N	number operator: multiplication by n on $\mathcal{H}^{(n)}$, $n\in\mathbb{N}_0$, Chapt.'s 3, 4, 5, 6
$\mathcal{N}$	real, separable, nuclear topological vector space, Chapt.'s 1, 3; typically $\mathcal{S}(\mathbb{R}^d)$, $d\in\mathbb{N}$
$\mathcal{N}^*$	(topological) dual of $\mathcal{N}$
$\mathcal{P}$	algebra of complex polynomials over $\mathcal{N}^*$, $\mathcal{S}'(\mathbb{R})$, ... , generated by the smooth coordinate process X_ξ, $\xi\in\mathcal{N}$, $\mathcal{S}(\mathbb{R})$, ... , Chapt. 2
P_t	semigroup generated by N, Chapt. 7
Q_t	semigroup generated by $\sqrt{N}$, Chap. 7
$\mathcal{R}(\mathcal{N},C)$	reproducing kernel Hilbert space over $\mathcal{N}$ with kernel C, Chapt. 2, App. 3
$\mathcal{S}$	$\mathcal{S}$–transform, Chapt.'s 2, 4
$(\mathcal{S})$	space of white noise test functionals, Chapt.'s 3, 4
$(\mathcal{S})^*$	space of generalized white noise functionals, dual of $(\mathcal{S})$, Chapt.'s 3, 4

$(\mathcal{S})^*_+$	positive elements in $(\mathcal{S})^*$, Chapt. 4.B
$\mathcal{S}(\mathbb{R}^d)$	Schwartz space of rapidly decreasing C^∞ functions, App. 5
$\mathcal{S}'(\mathbb{R}^d)$	Schwartz space of tempered distributions, dual of $\mathcal{S}(\mathbb{R}^d)$, App. 5
$\mathcal{S}(\hat{\mathbb{R}}^n)$	symmetric Schwartz space of (complex) test functions over $\mathbb{R}^n$, $n\in\mathbb{N}_0$, usually appearing in the chaos decomposition of $(\mathcal{S})$, Chapt.'s 3,4
$\mathcal{S}'(\hat{\mathbb{R}}^n)$	dual of $\mathcal{S}(\hat{\mathbb{R}}^n)$
σ_λ	scaling by $\lambda\in\mathbb{R}$, Chapt. 4
$\mathcal{T}$	$\mathcal{T}$-transform, Chapt.'s 2, 4
Tr	the distribution in $\mathcal{S}'(\mathbb{R}^2)$ which maps $f\in\mathcal{S}(\mathbb{R}^2)$ into $\int f(t,t)\,dt$
τ_y	translation by $y\in\mathcal{S}'(\mathbb{R})$, Chapt. 4
$\xi, \eta, \ldots$	typical (*real*) elements in $\mathcal{N}$, $\mathcal{S}(\mathbb{R})$, ...
$x, y, \ldots$	typical (*real*) elements in $\mathcal{N}^*$, $\mathcal{S}'(\mathbb{R})$, ...
X_ξ, X_f	canonical coordinate process, $\xi\in\mathcal{N}$, $f\in\mathcal{H}$, Chapt. 1
$\langle\cdot\,,\cdot\rangle$	pairing between dual spaces, such as between $\mathcal{N}^*$ and $\mathcal{N}$, $\mathcal{S}'(\mathbb{R}^d)$ and $\mathcal{S}(\mathbb{R}^d)$, $(\mathcal{S})^*$ and $(\mathcal{S})$; by convention, this dual pairing is antilinear in the first factor, linear in the second factor
$(\cdot\,,\cdot)$	scalar (or inner) product on a Hilbert space; antilinear in the first, linear in the second factor
$(2n-1)!!$	$\frac{(2n!)}{2^n n!}$
$\|\cdot\|_p$	norm of (L^p), $p\geq 1$
$\|\cdot\|_{2,p}$	norm of $(\mathcal{S})_p$, $p\in\mathbb{Z}$, given by $\|\Gamma(H)^p\cdot\|_2$, Chapt.'s 3, 4
$\|\cdot\|_{H.S.}$	Hilbert–Schmidt norm

$\vert \cdot \vert$	absolute value, norm of $\mathcal{H}$, Chapt. 1
$\vert \cdot \vert_2$	norm of $L^2(\mathbb{R}^d)$
$\vert \cdot \vert_{2,p}$	norm of $\mathcal{S}_p(\mathbb{R})$, $p \in \mathbb{Z}$, given by $\vert H^p \cdot \vert_2$, Chapt.'s 3, 4, App. 5
$\vert \cdot \vert_{H.S.}$	Hilbert–Schmidt norm
$\vert\vert\vert \cdot \vert\vert\vert_{p,k}$	norm of $\mathcal{D}^{p,k}$, $p \geq 1$, $k \in \mathbb{N}_0$, given by $\Vert (1+N)^{k/2} \cdot \Vert_p$
$: u^n :_{\sigma^2}$	Hermite polynomial in u of order $n \in \mathbb{N}_0$ with parameter σ^2, App. 1
$:<\cdot\, , f>^n:$	Hermite polynomial in $<\cdot\, , f>$, $f \in \mathcal{H}$, of order $n \in \mathbb{N}_0$ with parameter $\vert f \vert_2^2$, Chapt. 2, App. 1
$:<\cdot, f_1> <\cdot, f_2> \cdots <\cdot, f_n>:$	projection of $<\cdot, f_1> <\cdot, f_2> \cdots <\cdot, f_n>$ onto $\mathcal{H}^{(n)}$, $f_j \in \mathcal{H}$, $j = 1, 2, \ldots, n$, Chapt. 2
$: \exp(\lambda <\cdot, f>) :$	$\exp(\lambda <\cdot\, , f> - \frac{1}{2}\lambda^2 \vert f \vert_2^2)$, $\lambda \in \mathbb{C}$, $f \in \mathcal{H}_{real}$, Chapt. 2
$n(\boldsymbol{\alpha})!$	if $\boldsymbol{\alpha} = (\alpha_1, \alpha_2, \ldots\ \alpha_n)$ is a multi-index in $\mathbb{N}^n$, $n \in \mathbb{N}$, and $n_i(\boldsymbol{\alpha})$ is the number of entries in $\boldsymbol{\alpha}$ equal to $i \in \mathbb{N}$, then $n(\boldsymbol{\alpha})! = \prod_{i \in \mathbb{N}} n_i(\boldsymbol{\alpha})!$
$\otimes$	tensor product, App. 2
$\hat{\otimes}$	symmetric tensor product, App. 2
$\hat{\cdot}$	symmetrization
$: x^{\otimes n} :$	n–th Wick power of $x \in \mathcal{S}'(\mathbb{R})$, Chapt. 2
$\circ$	composition
$\Phi : \Psi$	Wick product of Φ, $\Psi \in (\mathcal{S})^*$, Chapt. 4
$\mathcal{N}\exp(-\frac{1}{2}<x, Kx>)$	renormalized Gaussian functional, Chapt. 4
$s \wedge t$	$\min(s,t)$, $s,t \in \mathbb{R}$
∂_t	Gâteaux derivative in direction of δ_t, $\partial_t = D_{\delta_t}$, Chapt. 5

∂_t^*	adjoint of ∂_t, Chapt. 5
∇	gradient defined via ∂_t, Fréchet derivative on $(\mathcal{S})$, Chapt. 5
∇^*	adjoint of ∇ on $(\mathcal{S})^*$, Chapt. 5
$[\cdot\,,\,\cdot]$	commutator

Index

A

abstract Wiener space 138, 199
adjoint (of derivative) 163, 167, 170–172
admissible 374, 382, 384, 392–393
affine transformation 130–131
analytic (version) 136–137
annihilation operator 172, 460–463
anticipating (integrands) 289, 292
Araki relation 401, 420, 423

B

backward integral 295, 299–300
Bochner integral 133–136, 282
Bochner–Schwartz theorem 109
Brownian bridge 5
Brownian motion 8, 9, 115, 27

C

canonical commutation relations (CCR) 172, 399–401, 461
chain rule 158
chaos decomposition, expansion 17, 32, 57, 75, 98–99
characteristic function 2, 127
characterization theorem 119, 123, 131
closable (form) 371–373, 381, 382
closed (operator, form) 53, 55, 371–373
commutation relations 228
compatible 1, 52–53, 55, 72, 481
contraction property (Markovian) 367, 373
convolution 127, 347
coordinate process (canonical) 6, 21
countably Hilbert space 1, 2, 54, 481
creation operator 172, 277, 460–463

D

delta white noise functional 321
differential operators 146, 181–183
diffusion process (associated with a Dirichlet form) 382, 397
Dirichlet form (pre–) 366, 374, 405
Dirichlet strategy 403, 416–417, 419
Donsker's δ function 47–49, 113, 129, 437
Dyson series 445

E

energy form 367, 374
equally dense 215
evaluation map 86

F

Fernique (theorem) 139, 143
Feynman integral 42, 46, 435
Feynman integrand 438
Fock space (symmetric) 18, 54, 56, 456–460
forward integral 295, 299–300, 312
Fourier transform 317, 321
Fourier–Mehler transform 317, 357
Fréchet (functional) derivative 151, 154, 160, 472–476
Fréchet space 54, 75

G

Gâteaux derivative 146–152, 154, 156, 368, 472–476
Gauß kernel 103, 107–108, 112, 128–129, 386, 393
Gaussian measure 1, 3
Gaussian space 1, 3
Gaussian white noise functional 322
Gel'fand triple 37, 52
generalized (white noise) functional 35–52, 57, 62–65, 73, 97
generalized multiple Wiener integral 41
generalized Radon–Nikodym derivative 101, 116–118, 127 130–131
gradient 154, 160, 368
Gross Laplacian 199
ground state measure 404–405
growth 122

H

Hamiltonian 400, 404, 408, 418, 436
harmonic oscillator 65
Hermite polynomials 451–455
Hermite function 452
Hitsuda–Skorokhod integral 281–285, 287, 289–293, 296, 300
Høegh–Krohn model 431
homogeneous chaos 10, 17, 27
hypercontractive 235

I

inductive limit topology 61, 83, 477
integral kernel 27, 66, 75, 78
integral representation 27, 36
Itô's formula 287
Itô integral 278, 283–285, 296
Itô's Lemma 312, 316

L

L–functional 212
Lagrangian 435–436
Laplacian (operator) 184
left derivative 295
Lévy Laplacian 211
Lévy part 214
Lévy–Khintchin process 128
Littlewood–Paley functions 249
Littlewood–Paley–Stein (LPS) inequalities 232, 249, 258, 261–263

M

Mackey topology 480
Malliavin calculus 49, 73
Malliavin covariance matrix 275
Markov property (of Brownian motion) 115
martingale property (of Brownian motion) 115
Mehler's formula 237–238
Meyer's inequalities (equivalence of norms) 263, 269
Minlos' theorem 2

N

non-anticipating 283, 285, 292, 295–296
normal polynomial 215
nuclear 2, 58–59, 66, 75, 482
number operator 73, 77, 196

O

order 58, 232–233

Ornstein–Uhlenbeck operator 73, 77, 196, 232–233
Ornstein–Uhlenbeck process 131
Ornstein–Uhlenbeck semigroup 233–235
Osterwalder–Schrader axioms 402

P

$P(\varphi)_2$ model 432
Pettis integral 281–282
Poisson process 128
polarization identity 23, 124
positive definite 2, 111
positive generalized functional 109, 111
product rule 158
projective limit (topology) 2, 68 53–56, 72, 74–75, 478
propagator 435, 439

Q

quantum field (free, Euclidian, time zero) 4, 26, 389, 399, 402, 405–407, 415–416

R

rapid decrease 70
ray entire 121–122
reflection map 329
renormalization 40–43, 105
reproducing kernel Hilbert space 21, 468–471
right derivative 294, 299

S

$\mathcal{S}$–transform 13, 27, 98–99, 101, 123, 127, 176, 282
scaling (complex, operator) 92–97, 117
Schrödinger equation 436, 438, 440
Schrödinger representation 404
Schwartz space 1, 477, 483–485
Schwinger functions 402, 412–414
Schwinger's variational principle 448
second quantized operators 61–63, 465–467
semigroup (Markovian) 378
sequentially dense 64
smooth (test, white noise) functional 37, 57, 62–65, 72, 75
start afresh (Brownian motion) 115
stochastic integral 277
Stratonovich integral (generalized) 295–296, 299
strictly positive (generalized functional) 368
strong topology, convergence 60, 66, 83, 480

T

$\mathcal{T}$–transform 10, 99, 123, 127,178
Taylor expansion, 162
tensor product 456–457
trace 30, 78, 200
translation (opererator) 92–97, 116, 325

U

U–funtional 122, 127, 442
uniformly bounded (orthonormal system) 215

V

V–functional 207
vacuum 400
Volterra Laplacian 207–208
Volterra part 208

W

weak topology, convergence 60, 66, 480
Weyl relations 400
white noise 3, 7, 36
white noise kernel operator 184, 188
Wick power 30, 32, 179–181, 453–455
Wick product 101, 127
Wiener integral 27
Wiener–Itô decomposition theorem 10, 20
Wightman axioms 402
Wightman function 401